Microbial Biotechnology in Agriculture and Aquaculture

(Volume II)

Editor
Ramesh C Ray
Central Tuber Crops Research Institute (Regional Center)
Bhubaneswar, Orissa
E-mail: rc_rayctcri@yahoo.co.in

Science Publishers
Enfield (NH) Jersey Plymouth

CIP data will be provided on request

SCIENCE PUBLISHERS
An Imprint of Edenbridge Ltd., British Isles.
Post Office Box 699
Enfield, New Hampshire 03784
United States of America

Website: *http://www.scipub.net*

sales@scipub.net (marketing department)
editor@scipub.net (editorial department)
info@scipub.net (for all other enquiries)

ISBN 1-57808-443-1 [10 digits]
978-1-57808-443-2 [13 digits]

Published by Science Publishers, Enfield, NH, USA
An Imprint of Edenbridge Ltd.
Printed in India

PREFACE

Plant genetic engineering has revolutionized our ability to produce genetically improved plant varieties. A large portion of our major crops has undergone genetic improvement through the use of recombinant DNA techniques in which microorganisms play a vital role. The cross-kingdom transfer of genes to incorporate novel phenotypes into plants has utilized microbes at every step – from cloning and characterization of a gene, to the production of a genetically engineered plant. Subhash Minocha and Andrew Page in Chapter 1 have described the several roles of microbes as a source of tools (plasmids and enzymes) to identify and clone desirable genes, as hosts for gene amplification, as a source of DNA cutting and slicing tools (restriction enzyme and ligases), and mainly as a vehicle for gene transfer into plant cells.

In Chapter 2, Ioannis Arvanitoyannis, presents an overview of the types of GM (genetically modified) plants that are being introduced into cultivation and takes a realistic outlook of the consumers' belief towards GM plants, consumers' safety, post marketing monitoring of GM plants and differences that exist among various countries. In addition, other issues like unintended effects of GM plant, legislation for GM plant, labeling and bioethics have been elaborately discussed.

The importance of symbiotic nitrogen (N_2)-fixation for developing a sustainable agricultural system is enormous. This volume has two chapters covering rhizobial N_2-fixation. In Chapter 3, Neung Teamroong and Nantakorn Boonkerd have covered the various aspects of rhizobial biotechnology: genetic diversity of *Rhizobium*, strain selection, strain improvement, environment (soil pH and soil water) and biotic (secretion of exopolysaccharide, gene influencing nodulation efficiency and antirhizobial compounds) factors. The authors have outlined new developments in rhizobial inoculant technology, mass culture media, carrier preparations and various factors affecting the scale of inoculant production. Marta Laranjo and Solange Oliveira, in Chapter 7, have explained the genomic sequence of *Rhizobium* species and molecular methods for strain identification.

Phosphorous is one of the principal macrobiogenic elements necessary for plant growth and metabolism. A better use of soil microorganisms can improve the plant nutrition with phosphorous and simultaneously reduce the input of mineral fertilizers. Olga Mikanová and Jaromir Kubát in Chapter 4 have emphasized that efforts in research and technology should focus on the quality of inoculants, improvement in the inoculation technology and effective inoculant delivery system. Molecular biology and genetic manipulation of phosphate solubilizing bacteria may offer a great potential to improve phosphate solubilizing capabilities of microorganisms.

Rice is the staple food of the major population of the world. For sustaining optimum grain yield, the excessive reliance on chemical fertilizers is not a viable strategy in the long run because of the prohibitory cost and environmental hazards. In contrast, biofertilizers/organic fertilizers are cost-effective, eco-friendly and a renewable source of plant nutrients to fully or partially replace chemical fertilizers in rice cropping system. In Chapter 5, S. Kannaiyan and K. Kumar, have outlined the various biofertilizers commonly applied to the rice agro-system and the underlying biochemical principles.

The root region of plants is a special microhabitat supporting bacterial species that favorably stimulate plant growth. Known as plant growth promoting rhizobacteria (PGPR), these organisms include *Azotobacter*, *Azospirillum*, *Bacillus*, *Klebsiella*, *Rhizobium*, *Pseudomonas*, etc. The highest concentrations of these microorganisms normally exist in the rhizosphere. Direct promotion of plant growth may involve N_2-fixation, phosphorous solublization and enzymatic lowering of ethylene level. In Chapter 6, Bernard Glick and his colleagues have focused on lowering the concentration of ethylene in plants, via the action of the bacterial enzyme, 1-aminocylopropane-1-carboxylate (ACC) deaminase.

Two chapters in this volume have been devoted to the interaction between microorganisms and aquaculture. Microorganisms conduct many important functions in aquaculture ponds, which include photosynthesis, decomposition of organic matter, nutrient cycling, transformations of substances between oxidized and reduced states, and detoxification of waste products and pollutants. Claude Boyd and Orawan Silapajarn, in Chapter 8, have discussed the influence of microorganisms/ microbial products in water, and sediment quality in aquaculture ponds. Microbial products consist of living bacteria, often *Bacillus* species or products prepared from cultures of microorganisms that contain organic acid, micronutrients and enzymes. J. Oláh, Mike Poliouakis and Claude Boyd, in Chapter 9, have defined the linkage among ecotechnology, aquaculture, bioconversion and modified reconversion of wetlands by using nitrogen pathways, fluxes and nutrient cycles to illustrate the ideas.

Marine microorganisms possess unique structures, metabolic pathways, reproductive systems and sensory and defense mechanisms because they have adapted to extreme environments ranging from the cold polar seas at –20°C to the great pressures of the ocean floor, where hydrothermal fluids spew forth. P. Pandey and C. Purushothaman have provided an overview of the recent applications of microbial biotechnology in marine science and aquaculture in Chapter 10.

Pesticides, despite their benefits, have a wide range of toxic side effects with potential hazards to the environment. Most pesticides are toxic to human beings, and the toxicity can vary widely within each group of pesticides. Environmental risks of a pesticide comprises the following risk factors: (1) persistence (indicating how long the pesticide remains active in the environment), (2) mobility (indicating how easily the pesticide can move from where it is applied), and (3) non-target toxicity (indicating how toxic the pesticide is to organisms other than the pests). Michal Green and his colleagues have portrayed these aspects in Chapter 11 taking Atrazines as a case study.

Retting is defined as the process of separating the embedded fiber from the stem through partial rotting by immersion in water; the rotting is brought about by a complex enzyme action of microorganisms naturally present in retting water. M. Basak has elaborated in Chapter 12 the microbiological technology followed for extraction of jute and allied fibers i.e. Mesta, sun hemp, ramie and pineapple.

Like food, the alcoholic beverages indigenous to any part of the world depend on the crops which can grow in the climate pertinent to that region. Alcoholic beverages, when produced from cereals, are known as beers; while those produced from other raw materials are known as wine. Nduka Okafor has reviewed the various alcoholic beverages produced by fermentation of agri-horticultural substrates in Chapter 13. Besides the commonly known alcoholic beverages like wine, brandy, whiskey, etc., there are many ethnic beverages such as 'bouza', 'talla', 'busaa', etc. especially in the African continent.

With regard to pisciculture, the application of genetic engineering techniques pinpoint the need of Broodstock management aiming at the production of sufficient egg number and good quality larvae that will contribute to the genetic improvement and preservation of fish production characteristics, effective control of fish feed and stock farming management, and above all, excellent fish quality. A. Exadactylos and Ioannis Arvanitoyannis in Chapter 14 have dealt with the various avenues in the application of biotechnology and genetic engineering in aquaculture for enhancing fish production. This chapter also highlights the risk assessment approaches of Food and Agricultural Organization (FAO), Federal Development Agency (USA) and European Union (EU) towards transgenic fish and fish products.

After the processing of agricultural/horticultural produces, million of tons of low-value by-products are generated, which in general are called agricultural residues. These residues represent enormous and underutilized renewable resources which can create an adverse environmental impact unless they undergo biological decomposition. Alternately, these residues can serve as industrial feedstocks for bioprocessing into value-added products such as enzymes, organic chemicals, biofuels, etc. In Chapter 15, Ramesh Ray and others have clarified some aspect of bioprocessing of agricultural residues such as mushroom cultivation, single cell protein, microbial enzymes and food-additives (organic acid, microbial gum and other gelling agents, microbial flavor and sweeteners, etc).

An endeavor has been made to incorporate all the major aspects of microbial biotechnology relating to agriculture and aquaculture in the 15 chapters of this Volume II. Yet, some other frontier subjects like microbial pesticides i.e., microbial herbicides, *Bacillus thuringiensis* and baculoviruses based insecticides (in agriculture), and lignocellulose degradation and biological control by probiotic bacteria (in aquaculture) could not be included. I am thankful to the contributors of all the chapters for clearly depicting the recent findings on the subjects and adhering to the time schedule. I wish to place on record my heartfelt gratitude to Dr. N. Sethunathan, my Ph. D. supervisor, who was instrumental in creating my initial intrest in the field of microbial biotechnology and to my wife, Saraswati Ray, for her constant support.

Ramesh C. Ray

CONTENTS

LIST OF CONTRIBUTORS

A. Exadactylos

Department of Agriculture, Animal Production and Aquatic Environment, School of Agricultural Sciences, University of Thessaly, Volos 38446, Hellas, Greece

E-mail: exadact@apae.uth.gr

Andrew F. Page

Department of Plant Biology, University of New Hampshire, Durham, NH 03824, USA

Anup K. Sahoo

Regional Centre of Central Tuber Crops Research Institute, Bhubaneswar 751 019, India

Bernard R. Glick

Department of Biology, University of Waterloo, Waterloo, ON, Canada N2L 3G1, Tel: + 519 888-4567 ext. 5208, Fax: + 519 746-0614

E-mail: glick@sciborg.uwaterloo.ca

C. S. Purushothaman

Central Institute of Fisheries Education, Versova, Mumbai – 400061, India,

E- mail: cspuru@indiatimes.com

Carlos G. Dosoretz

Faculty of Civil and Environmental Engineering, Technion – Israel Institute of Technology, Haifa 32000, Israel

Claude E. Boyd

Department of Fisheries and Allied Aquacultures, Auburn University, Alabama 36849, USA, Fax: +-334-844-5933

E-mail: ceboyd@acesag.auburn.edu

Fusao Tomita

Laboratory of Applied Microbiology, Hokkaido University, Sapporo, Japan

Ioannis S. Arvanitoyannis

Department of Agriculture, Animal Production and Aquatic Environment, Agricultural Sciences, University of Thessaly, Fytokou Street, Nea Ionia Magnesias, 38446 Volos, Hellas, Greece, Fax: +30 24210 93144

E-mail: parmenion@uth.gr

J. Oláh

Department of Environmental Management, Tessedik High School, Szarvas, Hungary

E-mail: saker@szarvasnet.hv

Jaromir Kubát

Research Institute of Crop Production, Drnovska 507, 161 06 Prague 6- Ruzyne, Czech Republic

E-Mail: kubat@vurv.cz

K. Kumar

Department of Agricultural Microbiology, Tamil Nadu Agricultural University, Coimbatore – 641 003, India

Kozo Asano

Laboratory of Applied Microbiology, Hokkaido University, Sapporo, Japan

M. K. Basak

National Institute of Research on Jute and Allied Fibre Technology, 12, Regent Park, Kolkata – 700040, India

E- Mail: mkbasak2002@yahoo.com

Marta Laranjo

Department of Biology, University of Évora, Apartado 94, 7002-554 Évora, Portugal

Michal Green

Faculty of Civil and Environmental Engineering, Technion – Israel Institute of Technology, Haifa 32000, Israel

Mike Poliouakis

Department of Fisheries and Allied Aquacultures, Auburn University, Alabama 36849 USA

Moshe Herzberg

Faculty of Civil and Environmental Engineering, Technion – Israel Institute of Technology, Haifa 32000, Israel; Present address: Environmental Engineering Program, Department of Chemical

Engineering, Yale University, P.O. Box 208282, 9 Hillhouse Avenue, New Haven, CT 06520-8286, USA

E-mail: moshe.herzberg@yale.edu

Nantakorn Boonkerd

School of Biotechnology, Institute of Agricultural Technology, Suranaree University of Technology, Nakhon Ratchasima 30000, Thailand

E-mail: nantakon@ccs.sut.ac.th

Nduka Okafor

Department of Applied Microbiology and Brewing, Nnamdi Azikiwe University, Awka, Nigeria

E-mail: ndukaokafor@cs.com; or ndukaokafor1@yahoo.com

Neung Teamroong

School of Biotechnology, Institute of Agricultural Technology, Suranaree University of Technology, Nakhon Ratchasima 30000, Thailand

Nikos Hontzeas

Department of Biology, University of Waterloo, Waterloo, ON, Canada N2L 3G1

Olga Mikanová

Research Institute of Crop Production, Drnovska 507, 161 06 Prague 6-Ruzyne, Czech Republic

Orawan Silapajarn

Department of Fisheries and Allied Aquacultures, Auburn University, Alabama 36849 USA

P. K. Pandey

Central Institute of Fisheries Education, Versova, Mumbai – 400061, India

E-mail: pkpandey_in@yahoo.co.uk

Ramesh C. Ray

Regional Centre of Central Tuber Crops Research Institute, Bhubaneswar 751 019, India

E-mail: rc_ray @rediffmail.com

S. Kannaiyan

Department of Agricultural Microbiology, Tamil Nadu Agricultural University, Coimbatore – 641 003, India

E-Mail: skannaiyan@hotmail.com

Saleema Saleh-Lakha

Department of Biology, University of Waterloo, Waterloo, ON, Canada N2L 3G1

Solange Oliveira

Department of Biology, University of Évora, Apartado 94, 7002-554 Évora, Portugal, Tel: + 351 266 760878, Fax: + 351 266 760914

E-mail: ismo@uevora.pt

Subhash C. Minocha

Department of Plant Biology, University of New Hampshire, Durham, NH 03824, USA

E-mail: sminocha@cisunix.unh.edu

1

Microbes and Their Contributions to Plant Biotechnology[1]

Subhash C. Minocha and Andrew F. Page*

INTRODUCTION

Plant genetic engineering is perhaps the most obvious example of plant biotechnology that has revolutionized our ability to produce genetically improved plant varieties. On the other hand, it has created some of the biggest controversies in recent times. A large portion of our major crops has undergone genetic improvement through the use of recombinant DNA techniques to which the contribution of microbes has been indispensable. The cross-kingdom transfer of genes to incorporate novel phenotypes into plants has utilized microbes at every step–from the cloning and characterization of a gene, to the production of a genetically engineered plant. This chapter highlights some of the major contributions of microbes to this process. The role of microbes as a source of tools (plasmids and enzymes) to identify and clone desirable genes, as hosts for gene amplification, as a source of DNA cutting and splicing tools (restriction enzymes and ligases), as a direct source of commercially valuable genes, and finally, and as a vehicle for gene transfer into plant cells, are discussed briefly.

PLANT GENETIC ENGINEERING

Plant genetic engineering can be defined as "the process of chronologically or spatially altering the pattern of proteins in an organism by adding new

Department of Plant Biology, University of New Hampshire, Durham, NH 03824, USA
**Corresponding author: E-mail: sminocha@cisunix.unh.edu*
[1] *Scientific Contribution No. 2257 from the NH Agricultural Experiment Station*

genes or regulating the expression of pre-existing genes," and it has undoubtedly revolutionized the field of plant improvement. Although not a new concept (older forms of genetic alteration, e.g., mutations or hybridization, also constitute genetic manipulation), the availability of a handful of new techniques and unique approaches has contributed to the development of rapid, reliable and targeted changes in the genetic make up a plant (Miflin, 2000; James, 2002; Mohan Babu et al., 2003). The most profound aspect of this technology is that there is no limit to the source of genes that can be incorporated into a plant's genome to impart characteristics that would be nearly impossible to transfer by any other means. Examples of such inter-kingdom gene transfer include the production of insecticidal plants using *Bacillus thuringiensis* (*Bt*) genes from bacteria, virus resistant plants using viral coat proteins, cold-tolerant plants using antifreeze genes from fish, and production of animal and human proteins of pharmaceutical importance including antibodies and vaccines. Within a short period of less than two decades, more than 65% of all our major crops have become the product of this new technology (Miflin, 2000; James, 2002). This volume and many other books written before it highlight the techniques, approaches, products and controversies associated with the production and use of genetically engineered [commonly referred to as genetically modified (GM)] plants and microbes (Ioannis Arvanitoyannis, Chapter 2 in this volume). This chapter focuses specifically on the contributions of microbes to this revolutionizing technology in order to illustrate the point that without the direct use of microbes, genetic engineering, as we know it, would be impossible. The following steps have been identified in the production of a genetically engineered plant, and it will become obvious that microbes have directly contributed to all these steps in major ways:

- Gene isolation, cloning and reconstruction
- Gene transfer to plant cells
- Selection of transformed cells
- Source of novel genes

BACTERIA AS A SOURCE OF TOOLS FOR GENE ISOLATION, CLONING AND RECONSTRUCTION

The role of microbes in facilitating the isolation and cloning of DNA and the creation of recombinant DNA molecules is obvious. Their unique properties of possessing restriction enzymes and plasmids, their ability to readily take up and exchange plasmids and their rapid growth, have simplified all steps of working with isolated genes. In addition, specialized microbes (e.g. thermophiles) have provided a unique set of enzymes (e.g.

Taq polymerase) for specific laboratory procedures. Some of these contributions are summarized below:

Restriction Endonucleases

One of the greatest contributions of bacteria to plant genetic engineering is the restriction endonucleases they provide. These are enzymes that cleave DNA at specific sites, leaving the DNA otherwise intact. This restrictive digestion of DNA is one of the most common techniques of molecular biology that is often used for analysis of DNA sequences. In combination with DNA ligase, the technique is used to create new sequences (a molecular "cut and paste"; e.g. to recombine genes of interest with specific promoters or to create fusion proteins) in order to regulate their expression, or to alter the position of a gene in a (transgenic) construct. Restriction endonucleases are found almost exclusively in prokaryotes, and were discovered during investigation of the fact that some bacteria were resistant to certain viruses. This resistance was found to be a result of the bacteria's ability to digest ('restrict') the viral DNA using enzymes that recognized and then cleaved specific sequences. Genes encoding these enzymes were quickly cloned from sources such as *Escherichia coli* and *Haemophilus influenzae* and over-expressed in laboratory strains for commercial production; their products (e.g. EcoRI and HindIII) are among the most commonly used restriction enzymes. More than 800 such enzymes are currently available, each having specificity for recognition of a specific nucleotide sequence in a DNA molecule (Pingoud and Jeltsch, 2001).

During the course of a typical plant genetic manipulation protocol, restriction endonucleases are used several times. First, for cloning of a gene into a vector, the latter must be digested with an appropriate restriction enzyme to open it up. Following transformation, the ligation in the reconstituted plasmid is confirmed by digestion (and sequencing) of the resultant plasmid. The gene is then typically cut out (using one or more restriction enzymes) and re-ligated into a cassette (which also is restricted) suitable for plant genetic manipulation, i.e. adjacent to a suitable promoter, selectable marker and/or reporter gene(s). Following transfer to the plant and selection of putative transformants, the presence of T-DNA in the genome of transgenic plants must be analyzed for transgene integration and copy number by Southern hybridization, in which DNA digestion by restriction endonucleases is a key component. Restriction endonucleases, therefore, constitute possibly the most important and indispensable contribution of microbes to the success of plant genetic engineering.

Cloning Vehicles

As important as restriction enzymes are to cutting of the DNA molecules, so also is the rapid and reliable multiplication (i.e. amplification) of genes often using bacteria as the host. Typically the in vitro amplified DNA (e.g. by Polymerase Chain Reaction) or reverse-transcribed cDNA (from mRNA) is ligated into bacterial plasmids and transferred into bacteria (mostly *E. coli*) for natural amplification. The presence of antibiotic resistance genes (mostly derived from bacteria–another very significant contribution) on the plasmid facilitates the selection of transformed bacteria on selective growth medium, thus eliminating the need for lengthy selection mechanisms. Later, some of the same antibiotic resistance genes are often incorporated into the DNA molecules along with the gene of interest for transfer to the plant cell to aid in the selection of transformed cells.

The Polymerase Chain Reaction (PCR), which has become the cornerstone of numerous techniques of molecular biology, owes its existence to a special group of microbes called thermophiles, which live mostly in high temperature environment. The contribution of Taq polymerase (in all its commercial forms) to PCR amplification of DNA is as important as the restriction enzymes and the ligase to cut and recombine nucleic acid molecules.

GENE TRANSFER TO PLANT CELLS – THE ROLE OF *AGROBACTERIUM*

There are several methods of gene transfer into plant cells, including direct uptake of DNA by protoplasts, microinjection, electroporation, liposome fusion, viral vectors and bacterial vectors. By far the most important means of gene transfer involves a soil-borne microbe called *Agrobacterium tumefaciens*. This bacterium has not only made plant genetic engineering a routine process that can be accomplished in a high school classroom, but it has also contributed to the enhancement of our knowledge of gene structure and function in ways that were impossible to envision only a few years ago.

It all started as an oddity of what was termed 'neoplastic growth' or 'cancer of plants' (Braun 1969, 1978), and ended up revolutionizing the entire field of genetic manipulation of plants with all its commercial applications and implications (Miflin, 2000; James, 2002; Valentine, 2003). The story of *Agrobacterium* is a story of science filled with surprises, complexities and rewards. The contribution of *Agrobacterium* to the field of plant biotechnology and genetic engineering cannot be overemphasized; it has provided not only an excellent means of transferring genes into plants, but also has revealed some of the most complex interactions between microbes and plants (Tzfira and Citovsky, 2003; Valentine, 2003). Even today, it remains as the only example among eukaryotes of what Stachel and Zambryski (1989) call "trans-kingdom sex", in which a segment of prokaryotic DNA armed with the expression machinery (promoters) for use in a eukaryotic cellular environment, is transferred from a microbe to a eukaryotic cell at a high efficiency. Tzfira and Citovsky (2003) cite *Agrobacterium*-mediated gene transfer as an example of modern-day nanomachines at work in a precise manner that defies any mechanical creations that we have devised.

So, what has *Agrobacterium* done for us? In addition to providing a reliable and simple means to transfer genes into plant cells, *Agrobacterium* has revealed many deep secrets of host-pathogen interactions, hormonal regulation of plant development, and offered a powerful tool for cloning, characterization and understanding the functions of thousands of genes in plants; the latter through the T-DNA tagging technology. A bacterium that was once considered simply a pathogenic nuisance for farmers has become one of the most studied bacteria of recent times. In this chapter we focus primarily on the contributions of *Agrobacterium* to the field of plant genetic manipulation with a brief description of its role in revealing gene structure and function.

Agrobacterium tumefaciens was reported to be the causative agent of a widespread disease of plants called 'crown gall' tumors in early years of the 20th century (Smith and Townsend, 1907; Braun, 1978). The commercial impact of the disease was not considered significant with the exception of a few crops. However, the neoplastic growth caused by this bacterium and its close relative *A. rhizogenes* (causatively related to a disease called 'hairy roots' – Elliot, 1951) attracted the attention of plant physiologists, and later biochemists and molecular biologists alike (Drummond, 1979; Hooykaas and Schilperoort, 1992; Krysan et al., 1999; Valentine, 2003). The curiosity to explain the biological basis of this disease increased with the discovery of plant hormones and their role in regulating plant morphogenesis in vitro (Braun, 1978; Weiler and Schröder, 1987). Soon it was demonstrated that the tumor tissues maintained their ability to grow almost indefinitely without an exogenous supply of hormones in spite of the bacteria being eliminated from them (Braun, 1978). In other words, the mere association of *Agrobacterium* with the plant cells had caused them to be 'transformed' in a manner analogous to the animal cancers (Braun, 1969). In addition to the identification of certain unique metabolites, called opines (e.g. octopine and nopaline), the tumors were deemed to produce their wound hormones (both auxins and cytokinins) in quantities sufficient to sustain continued growth.

Two observations provided powerful clues regarding the role of bacteria in defining some of the characteristic changes in the tumors: (a) the nature of the opine produced by the tumor was dependent upon the *Agrobacterium* strain used (Bomhoff et al., 1976); and (b) certain mutations in the bacterium caused a change in the morphology of tumors indicating that the nature of the hormone produced was also dependent upon the bacterial genetics (Garfinkel et al., 1981; Ooms et al., 1981). The key question in the late 1960s and early 1970s was: "If this indeed was the case, would it be far-fetched to think that the bacteria could transfer their genes into the plant cells?" This speculation lasted for years, not the least due to the lack of efforts to study it, but more because of the lack of tools and techniques to investigate and confirm such a phenomenon. Thanks to the invention of techniques such as DNA labeling and Southern hybridization, it soon became possible to decipher the mystery, or at least its nature (Chilton et al., 1980).

It also became apparent that through an intricate mechanism involving crosstalk between the plant cell and the agrobacteria, a part of the bacterial plasmid was delivered to the plant cell as a package that was targeted directly at the nuclear DNA for integration. There are not very many phenomena comparable to this in the vast span of biological systems. The three key steps involved in T-DNA transfer are: (1) chemotactic attraction of *Agrobacterium* towards a wounded plant cell; (2) molecular response of the bacterium in terms of induction of more than a dozen chromosomal and plasmid genes collectively called *vir* genes which help replicate (excise) a single-stranded copy of the T-DNA within the left and right borders of its plasmid; and (3) packaging and transport of ssT-DNA out of the bacterial cell and into the plant cell. Once in the plant cell, the T-DNA is somehow transferred to the nucleus, released from the packaging proteins, and, by an unknown mechanism, is integrated into the genomic DNA of the plant cell. The process of integration is believed to be random, although there are data that support a preference for certain regions of chromatin (Tinland, 1996; Brunaud et al., 2002; Tzfira et al., 2003). Each of these steps has been reviewed extensively and is described here only in brief.

Recognition and Chemotactic Binding of Bacteria to Plant Cells

It was known early on that the bacterial interaction with plant cells is facilitated by wounding of the cells (Lippincott and Lippincott, 1969). In response to certain signal molecules (phenolic compounds, sugars and amino acids) produced by wounded plant cells, the bacteria bind to cell walls. Apparently, cell wall constituents of the target cell also play a critical role in the binding process (Gelvin, 2003). A number of genes in the plant, some involved in the production of signal molecules (e.g.

compounds like acetosyringone and β-galactans) and others that control a host range, and a number of chromosomal genes in *Agrobacterium*, control the interaction between bacteria and the plant cell. Based on the knowledge of the host range of *Agrobacterium* in nature, for quite some time it was believed that monocots either lacked the signal molecules or did not allow the transfer and/or integration of T-DNA into their genomes. Later studies, however, have shown that this is not the case.

Bacterial Response and T-DNA Packaging

Once the bacteria have been exposed to the signal molecules, there is a coordinated induction of several genes on the large Ti plasmid of *Agrobacterium*, called the *vir* genes, whose products perform a series of steps that lead to the production of a single stranded T-DNA-protein complex which can be translocated out of *Agrobacterium* into the plant cell. The first set of genes to be induced is the *vir*B and *vir*C gene complex. The polypeptides produced from these genes interact with RNA polymerase and turn on another set of *vir* genes (*vir*D and *vir*E) whose products are directly involved in the copying and excision of the T-DNA. The T-DNA borders (23 base pair (bp) inverted repeats) act as the boundaries for copying and excision of T-DNA. The ssT-DNA (up to 15-50 kilo base [kb]) is then complexed with virD2 and virE2 proteins (as many as several hundred virE2 and a few virD2) and the entire complex is transported out of the bacterial cell wall in a manner analogous to the phenomenon of bacterial conjugation (Gelvin 2000, 2003; Tzfira et al., 2000, 2003; Dumas et al., 2001). The products of *vir*B operon and *vir*D4, which are membrane-intercalating proteins, constitute a functional channel for export of the T-DNA-VIR protein complex through the membrane (Christie, 1997; Tzfira and Citovsky, 2003). Several dozen VIR proteins and several bacterial chromosome-encoded proteins are involved in this process.

The coordination between export of the T-DNA-protein complex from *Agrobacterium*, its entry into the plant cell, transportation into the nucleus, and integration of T-DNA into the genomic DNA are still matters of speculation and intense debate (Tzfira et al., 2003, 2004; Jakubowski et al., 2004; Somers and Makarevitch, 2004). It is, however, believed that virD2 acts as a chaperon for targeting the complex to the nucleus. While it is not known whether the T-DNA-protein complex is linear or it acquires a secondary structure (e.g. folding); it is, however, apparent that the complex remains intact until it reaches the nucleus. The downstream processes of release of the ssT-DNA from the proteins, and the timing and mechanism of integration and conversion into a double stranded DNA are poorly understood at present. Apparently the integration of ssT-DNA occurs prior to its becoming double stranded by a process termed 'illegitimate recombination' (Somers and Makarevitch, 2004; Tzfira et al., 2004). Once integrated into the host cell's genomic DNA, the genes located on T-DNA behave very much like native genes and are transmitted in a Mendelian fashion as dominant genes. It is, however, indicated that their methylation and suppression occur at a higher frequency than the native genes. Thus, *Agrobacterium* has revealed to us an intricate and complex mechanism of trans-kingdom gene transfer that can occur at a relatively high frequency.

It was clear from the earlier studies that physiologically and biochemically, the most important genes were those involved in the biosynthesis of an auxin and a cytokinin, which together contributed to the crown gall syndrome. While the nature and regulation of expression of these genes has been widely investigated, much less attention has been paid to the role and significance of several other genes that are often present on the T-DNA and the Ti plasmid. The universal occurrence of opine-synthesis and opine-catabolism genes is intriguing; their role has been speculated in terms of providing a microenvironment to the bacteria in which they thrive with some advantage (Weising and Kahl, 1996). Classification of *Agrobacterium* strains is based on the type of opine they induce in the transformed plant cells, and the corresponding opine catabolic genes that they contain in the Ti plasmid; e.g. octopine type, nopaline type and agroaminopine type (Hooykaas and Schilperoort, 1992). While none of these genes are required for growth and maintenance of *Agrobacterium* or for transformation of plant cells by *Agrobacterium* in the laboratory, the conserved nature of these complementary gene sets indicates some adaptive and/or evolutionary advantage to the bacteria in a natural environment. It is known that opine-biosynthesis genes are expressed only in the transformed host cells (which not only make these opines but also secrete them), while opine catabolic genes are expressed in the bacterium. Moreover, secretion occurs preferentially near the root-shoot transition zone into the soil where the bacteria thrive (hence the location of crown galls). It is postulated that opines facilitate and/or are required for conjugation involving Ti-plasmid transfer among bacteria. This leads to the conclusion that tumor induction is an ingenious way for *Agrobacterium* to ensure the maintenance of Ti plasmid in the species. There are, however, hypotheses for alternate roles of opines in the life of *Agrobacterium* (Weising and Kahl, 1996).

Unlike *A. tumefaciens*, *A. rhizogenes* has not received the same level of attention by the biologists (Nilsson and Olsson, 1997; Meyer et al., 2000). Although the two species are unique among microbes in their ability to routinely transfer their plasmid DNA (T or rDNA) to the host cell, and utilize a similar mechanism to achieve this, genes transferred by the latter (*rolA, rolB, rolC,* and *rolD*) are not as well characterized (with respect to the proteins they code for and their functions in the host cells) as those on the T-DNA (Nilsson and Olsson, 1997). The end product of *A. rhizogenes* infection, the 'hairy-root' phenotype is equally intriguing. At the biochemical level it seems that the phenotype is due to an increased auxin sensitivity as well as increased cytokinin production resulting from expression of some of the *rol* genes carried on the Ri plasmid. Commercially also, *A. rhizogenes* has not been exploited as much for gene transfer as *A. tumefaciens*.

A lesser known relative of the two *Agrobacterium* species mentioned above, *A. radiobacter*, which is avirulent, shows a niche competition with *A. tumefaciens*. It secretes an antibiotic that is toxic to *A. tumefaciens* cells containing the Ti plasmid because the opine-catabolizing enzymes process the non-toxic antibiotic to its toxic form (Wang et al., 1994). Thus, following crown gall tumor formation, this bacterium could colonize the tumors and eliminate the competition from *A. tumefaciens* (Weising and Kahl, 1996).

A number of genes in *Agrobacterium* as well as in plants, have been identified that affect the competence of the host pathogen interaction; however the mechanism by which they affect this interaction is not known (Gelvin 2000, 2003; Mysore et al., 2000). For quite some time it was believed that the host range of *Agrobacterium* largely excluded monocots (de Cleene and de Lay, 1976). Later, however, *Agrobacterium* successfully transformed many species of monocots, including the commercially important grains, without any major modification of the protocols (Hernalsteens et al., 1984; Weising and Kahl, 1996). Nevertheless, the apparent lack of crown gall tumor production in monocots has revealed some key differences between the two major groups of angiosperms in terms of their responses to plant hormones (Weising and Kahl, 1996).

T-DNA Tagging and Gene Cloning

Agrobacterium tumefaciens T-DNA, largely if not solely, has provided us the tools for what is called 'reverse genetics', which in contrast to physically the mutant phenotype and then determining the gene responsible for it, involves identifying the gene that has been mutated and seeking the phenotypic effects of the mutation (Azpiroz-Leehan and Feldmann, 1997; Krysan et al., 1999; Sussman et al., 2000; Sessions et al., 2002; Walden, 2002). This became possible through a process called 'T-DNA tagging'. The technique is based on the ability of T-DNA transferred by *Agrobacterium* to become inserted into the genome at random sites (Galbiati et al., 2000). The premise of the technique is that if a large population of randomly transformed plants could be created, it is likely that the T-DNA will be inserted into almost all the genes, thus causing gene interruptions. Since the sequence of the inserted DNA will be known in advance, and, therefore, it could be amplified, cloned and sequenced along with the neighboring gene sequences, this should lead to identification of the interrupted gene. Plants harboring the T-DNA insert in a homozygous state could then be analyzed for the phenotypic effects of such an insertion; the effect could be at the morphological, developmental, biochemical or molecular level, and the gene could be dominant or recessive.

A direct way to obtain information on the function of a gene identified by sequencing, is to create a loss-of-function mutation and study the phenotype of the resulting mutant. In many model organisms (e.g. microbes and mice), homologous recombination can be used efficiently to target mutations into specific genes by replacing the wild-type gene with a mutated allele. In plants, control of homologous recombination has proved extremely difficult because of the prevalence of illegitimate recombination events (Puchta and Hohn, 1996). Although some success has been reported (Kempin et al., 1997; Iida and Terada, 2004), gene replacement by recombination is not considered feasible on a large scale, and other strategies have to be implemented to allow the functional analysis of large numbers of genes.

Insertion mutagenesis differs from other forms of induced mutations (deletions, duplications, etc.) i.e. a known piece of DNA (whose sequence has already been delineated) is present within the gene that has been interrupted. This feature allows the identification of the DNA sequence flanking the insert by simple means such as PCR, thus facilitating the cloning of the interrupted gene (Allen et al. 1994; McKinney et al., 1995; Krysan et al., 1996; Parinov et al., 1999; Ortega et al., 2002). Two types of insertion mutagenesis have been used in plants--transposable element insertion (Martienssen, 1998) and T-DNA insertions (Azpiroz-Leehan and Feldmann, 1997; Walden, 2002). T-DNA tagging provides distinct advantages over insertional mutagenesis by transposable elements, because once inserted the T-DNA is highly stable within the genome. In contrast, the transposable elements have a tendency to move (jump) from one location to the other during the production of large populations of homozygous mutants.

While it is more difficult to achieve saturation with T-DNA insertional mutagenesis, it does result in fewer insertions (1-2 loci per line); insertions are stable, easy to maintain, and, as far as we know, do not show strong insertional biases (Azpiroz-Leehan and Feldmann, 1997; Walden, 2002). In *Arabidopsis* highly efficient in-planta transformation techniques have been instrumental in generating large populations of T-DNA insertion lines while minimizing the effect of somaclonal variation linked to in vitro culture and regeneration (Bechtold et al., 1993; Azpiroz-Leehan and

Feldmann, 1997). It is estimated that in order to obtain 99% confidence of finding a T-DNA insertion in any *Arabidopsis* gene, ~280,000 T-DNA inserts would be required (Krysan et al., 1999).

Several collections (libraries) of T-DNA insertion mutants have been created which are available to scientists worldwide for analysis of specific genes with respect to their structure and function in the life of the plant (Alonso et al., 2003). These so-called 'knockout' mutants have also resulted in the cloning of a large number of agriculturally important genes for genetic manipulation. Not only can one study the effect(s) of gene knockouts on morphological and developmental phenotype(s) that they affect, but, combined with the high throughput techniques of gene expression analyses (e.g. microarrays and proteomics) one can study the molecular effects (i.e. effects on gene expression) of a single gene knockout in different organs of the plant at different stages of development (de Risi et al., 1997; Greller and Tobin, 1999; Galbraith, 2003; Meyers et al., 2004).

Activation tagging, which is based on the use of an insertion element carrying a strong enhancer or promoter directing transcription into the region flanking the insertion, enables the isolation of gain-of-function mutations in which ectopic activation of a flanking gene promotes a mutant phenotype (Wilson et al., 1996; Weigel et al., 2000). Such mutations are generally dominant or semidominant. One should, however, be aware of the limitation of T-DNA based insertional mutagenesis. The process of identifying a gene may be easy but to causatively relate this gene to a particular phenotype requires extensive analysis involving other techniques such as complementation and over-expression.

MICROBES AS THE DIRECT SOURCE OF USEFUL GENES

Microbes have contributed to plant genetic improvement not only as tools for gene transfer, but also as an important source of commercially important genes. Some of the major targets of genetic improvement in plants via genetic engineering are listed in Table 1.1; barring a few exceptions, all of them involve a direct use of genes obtained from microbes. Based on their commercial value added to the major crops, these genes can be divided into three groups: (1) those conferring resistance to abiotic and biotic factors (e.g. herbicides and abiotic stress, pests and pathogens), (2) those involved in metabolic pathways, and (3) those used as a means of introducing genetically modified plants into the environment more safely by preventing transgene escape. A few of the important contributions of microbes as the source of specific genes for improvements in commercially important crops are described below.

Table 1.1. Selected areas of genetic improvement in plants via genetic engineering

Abiotic stress resistance	**Plants as bioreactors**
Herbicide resistance	Biodegradable plastics
Water (Drought) stress	Non-protein pharmaceuticals (e.g. antibiotics)
Heavy metal stress	Phytoremediation of organic hazardous waste
Oxidative stress	Foreign proteins/enzymes of industrial importance
Freezing and chilling stress	Oils and biomass for alcohol/energy
Salt stress	Vaccines
Pathogen and pest resistance	**Plant nutrition and taste enhancement**
Insect resistance	Increased amino acids
Fungal and bacterial resistance	High protein content
Viral resistance	High essential mineral content
Nematode resistance	Content and type of lipids and fats
	Improved vitamin content
	Alteration in sweetness, flavoring, taste

Herbicide Resistance

Although initially the genes for herbicide resistance (e.g. Basta) were introduced into plants along with genes conferring other desirable traits for selection of transformed plants, there are many instances of herbicide resistance genes that have been introduced on their own to produce commercial varieties of herbicide tolerant plants. Resultant plants are eminently marketable, as herbicides can be used for weed control without affecting the transgenic plants, thereby greatly enhancing productivity. By far the most common of these genes is phosphoinothricin-N-acetyltransferase (*pat*), which was isolated from the bacterium *Streptomyces viridochromogenes* (Table 1.2). Expression of this gene confers resistance to glufosinate ammonium (the active ingredient in phosphoinothricin herbicides). Glufosinate ammonium inhibits glutamine synthase, leading to an accumulation of ammonia which rapidly kills the plant; the PAT enzyme catalyzes the acetylation of ammonia, thereby detoxifying it. Use of the *pat* gene is widespread; it has been incorporated in most of the major GM crops.

An equally important gene conferring herbicide resistance is 5-enolpyruvylshikimate-3-phosphate synthase (*epsps*), which confers resistance to glyphosate, the active ingredient in Roundup®; a widely used herbicide. The EPSPS enzyme is involved in producing aromatic compounds, including aromatic amino acids, and is naturally present in all bacteria, plants and fungi. Glyphosate inhibits EPSPS, except in the form of the enzyme found in the common plant pathogen *Agrobacterium tumefaciens*. Introduction of the *epsps* gene from *A. tumefaciens* therefore confers resistance to the otherwise toxic glyphosate-containing herbicides

Table 1.2. Bacterial genes licensed for production of transgenic plants

Gene	Bacteria	Phenotype
Herbicide Resistance		
pat	*Streptomyces viridochromogenes, S. hygroscopicus*	Glufosinate ammonium tolerance
epsps	*Agrobacterium tumefaciens*	Glyphosate tolerance
goxv247	*Ochrobactrum anthropi*	Glyphosate tolerance
bxn	*Klebsiella pneumoniae*	Oxynil tolerance
Insect Resistance		
cry1Ab	*Bacillus thuringiensis*	Lepidopteran insect resistance
cry1Ac	*Bacillus thuringiensis*	Lepidopteran insect resistance
cry1F	*Bacillus thuringiensis*	Lepidopteran insect resistance
cry1Fa2	*Bacillus thuringiensis*	Lepidopteran insect resistance
cry2Ab	*Bacillus thuringiensis*	Lepidopteran insect resistance
cry3A	*Bacillus thuringiensis*	Coleopteran insect resistance
cry3Bb1	*Bacillus thuringiensis*	Coleopteran insect resistance
cry9C	*Bacillus thuringiensis*	Lepidopteran insect resistance
Limitation of Transgene Escape		
barnase	*Bacillus amyloliquefaciens*	Rnase leading to male sterility
barstar	*Bacillus amyloliquefaciens*	Rnase inhibitor
dam	*Escherichia coli*	Male sterility

(Padgette et al., 1995). Like *pat*, *epsps* has also been used in a broad range of commercial crops.

Several other herbicide resistance genes have been used to a lesser extent, including *glyphosate oxidoreductase* (*govx247*) from *Ochrobactrum anthropi*, which confers resistance to glyphosate by degrading it to aminomethylphosphonic acid, and the *bromoxynil nitrilase* (*bxn*) gene for nitrilase from *Klebsiella pneumoniae*, which hydrolyzes oxynil herbicides to inactive products (McBride et al., 1986). Use of these genes is less common and has been limited to only a few species including canola, cotton and tobacco.

There is hardly any organic molecule that the microbes cannot degrade (Parales et al., 2002), which means that microbes could be the source of some very unique genes not only for imparting resistance into plants against industrial chemicals that are used as herbicides, but also to metabolize these compounds for phytoremediation purposes.

Insect Resistance

Probably the best-known example of plant genetic modification is *Bt* corn, which involves the insertion of genes for crystal protein toxins (called *cry* genes that produce a peptide called δ-endotoxin) from the bacterium *B. thuringiensis* (*Bt*) (Mohan Babu et al., 2003). These genes produce crystal proteins that bind to the midgut wall of insect larvae and eventually degrade it, thereby allowing bacteria into the body cavity, leading to septicemia and death. More than 600 strains of *Bt* have been identified, each producing a different δ-endotoxin. A number of licensed *cry* genes exist (Table 1.2), although not all are in current commercial use (Bravo, 1997; Schnepf et al., 1998; Sharma et al., 2000). The most significant corn pest is the European Corn Borer, a Lepidopteran, which is the main target of *Bt* corn, although *cry* genes have been used to control other insects including Colorado Potato Beetle and Corn Root Worm (Table 1.2) as CRY proteins bind to specific receptors in the insect gut that are not present in mammals, the use of *cry* genes is one of the most promising avenues for insect resistance in human and livestock feeds (Schuler et al., 1998). At present, *cry* genes are used mainly in corn, cotton and potato; however, other crops are being added to this category. A key advantage of using this approach is that the crop prevents environmental damage from the use of chemical insecticides (Sharma et al., 2000).

Protection Against Pathogens

The well recognized gene-for-gene hypothesis for disease resistance mechanism involves three steps: (a) the expression of avirulence (*Avr*) genes in the pathogen, (b) recognition of AVR by the host through a resistance protein (R), and (c) the hypersensitive response (HR)-based programmed cell death (PCD) that causes both the host cell and the pathogen to be killed, thus limiting pathogenesis to a localized area (Martin et al., 2003). Any interference with the processes involved in these interactions will either enhance or impede pathogenesis; thus, genetic manipulation of the genes involved in these interactions is a major goal of plant scientists. A list of genes identified to be involved in such interactions is found in Abramovitch and Martin (2004). The fact that a majority of plants are resistant to most pathogens leads us to conclude that plants, in general, possess a variety of physical as well as chemical barriers, including the constitutive production of antimicrobial compounds (Mysore and Ryu, 2004). In other cases, the production of such compounds is initiated in response to pathogen challenge (e.g. AVR [avirulence] proteins, elicitins) and may be mediated through small molecules that can be systemically disseminated (e.g. jasmonic acid, salicylic acid–Kunkel and Brooks, 2002). The latter are often a part of the slow response of the pathogen in contrast to the so-called HR, which is relatively rapid.

Molecular analysis of the host-pathogen interaction mechanisms has revealed the existence of and led to the cloning of avirulence genes in several pathogens (Kim et al., 1997; Kunkel and Brooks, 2002, Martin et al., 2003). Two types of avirulence genes were identified in *Pseudomonas syringae* and in *Xanthomonas*. In the first group are genes that are involved in the production of syringolides (phenolic glucosides having a tricyclic ring) whose products determine the host response as well as host range of the pathogen. The second important set of AVR proteins are the 'hypersensitive response protein complexes' (HRPs), which are also produced by bacterial pathogens of mammals (Dangl, 1995). Transgenic expression of *avr* genes in the plant should thus lead to altering these important pathogen-host interactions, making the latter pathogen resistant. Alternatively, since these proteins are often secreted from the pathogen and incite a response in host cells, genetic alterations of the proteins that respond to the AVR protein could also be used as a means to alter the host response. Apparently, bacteria use the type-III secretion system to deliver these proteins (also referred to as 'effector proteins') into the host cell (Cornelis and Van Gijsegem, 2000; Hotson and Mudgett, 2004). A limitation of this approach is that nuclear targeting of some of the bacterial effector proteins is required, the mechanism of which is not clearly understood. Information on the nuclear targeting of T-DNA-VIR (virulence) protein complex could provide useful understanding of such a mechanism, leading to transgenic manipulation of the plant to intervene in this nuclear targeting.

Studies of the host during initial plant-pathogen interaction have identified plant genes for resistance (*'r'* genes) which respond to the AVR proteins. Overproduction of the R proteins in healthy plants could prevent pathogenesis. Furthermore, sequence modifications of the R proteins could be used to detect multiple pathogens. Some examples of such proteins include Arah, RPS2, Pto (tomato), RPM1, and Xa-21 (rice). In all these cases, the target is intervention in the transduction of signal from the pathogen to the host that elicits the host response to allow the pathogen to invade or to prevent invasion (Herbers et al., 1992; Panapoulos et al., 1996).

A genomic analysis of promoter sequences of several plant Pathogen Response Protein genes (*prp*) has led to the identification of certain elements in their promoters (e.g. W boxes, GCC-like boxes, Box D) that regulate the plant response to several pathogens. Recombining these regulatory elements into synthetic pathogen-responsive promoters has provided new insights into signal transduction during host-pathogen interaction and has been suggested as a means to regulate the expression of resistance genes in plants (Rushton et al., 2002).

Parallel studies with plant-fungal interaction have identified similar proteins and a variety of other soluble molecules in fungi that elicit the production of host response proteins, ultimately determining compatible and incompatible interactions between the fungus and the host. Examples of such proteins are elicitins, lipid transfer proteins, glucans and glycopeptides (Shiraishi et al., 1997; Blein et al., 2002; Bouarab et al., 2002).

Some fungi and bacteria employ toxins as a part of their pathogenicity. A good example of this is the tripeptide toxin called phaseolotoxin, that is produced by *Pseudomonas syringae* pv. *phaseolicola,* and causes the halo blight disease in bean. The toxin causes extensive chlorosis by inhibiting ornithine carbamoyltransferase (OTC), which makes citrulline from carbamoyl phosphate; a reaction involved in polyamine metabolism. Whereas *P. syringae* produces two ornithine carbamoyltransferases, one of which is phaseolotoxin-resistant; the other bacteria do not possess the resistant enzyme. Cloning of the gene encoding the phaseolotoxin-resistant OTC and transfer into tobacco plants has resulted in resistance to halo blight (de la Fuente-Martinez et al., 1992; Hatziloukas and Panopuolos, 1992).

Symbiotic Interactions

As important as an understanding of the microbe-plant interaction in pathogenesis is, equally intriguing and economically important is the understanding of the host determination and symbiotic association of *Rhizobium* (a nitrogen-fixing microbe) with its host plant (Freiberg et al., 1997). While significant progress seems to have been made in the host-pathogen interaction area (both pathogenesis and T-DNA delivery mechanism of *Agrobacterium*) leading to some experimental manipulation of the host range, the *Rhizobium* interactions have not been successfully altered to broaden the range of limited number of suitable host taxa (Cullimore and Jean Dénarié, 2003). While nodule formation by *Rhizobium* may appear to be similar to *Agrobacterium*-induced tumorigenesis, it is quite different in detail. The interaction that begins with the induced expression of several genes in Rhizobia by flavonoids produced by plant roots, involves a reverse induction of several genes in the host plant cells by lipo-chitooligosaccharide signals (*a.k.a.* Nod factors) that are produced in the bacterium. As a result, specialized nodular structures are produced in the plant roots that harbor highly modified forms of Rhizobia that reduce nitrogen gas into ammonia (Limpens and Bisseling, 2003).

While the nodulation response may be the result of localized host-Rhizobial interactions leading to nodule development from specific cell types in the roots, there is no reason why such an interaction could not be extended to other organs, provided the molecular responses of the cells to the Nod factors were better understood (Boivin et al., 1997; Esseling and Emons, 2004; Riely et al., 2004). An outright genetic manipulation of the plants using nitrogenase genes to impart nitrogen-fixing ability, although a dream of the scientists in the early days of genetic engineering, will most likely remain elusive for some time to come; however, considerable enhancement of plant productivity and environmental safety (from the adverse effects associated with the use of nitrogen fertilizers) are quite feasible in the foreseeable future.

Perhaps the least recognized and studied symbiotic relationship of plants is the development of mycorrhizal associations between plants and fungi. Ecologists have always considered the arbuscular mycorrhizal symbiosis to be the most important cross-kingdom interaction in higher plants (Newman and Reddell, 1987; Jacobsen et al., 2002). This symbiosis is an integral component of most ecosystems and involves the transfer of billions of tons of photosynthetically-fixed carbohydrates and lipids from plants to the fungi (Bago et al., 2000). The fungi in turn help the plant to increase its nutrient uptake capability, particularly phosphates (Bago et al., 2003). In many cases the symbiotic interaction is so important that neither of the partners can successfully complete their life cycle independent of the other. The symbiotic interaction is limited mostly to roots, and little is known about the cross-talk between the fungus and the root cells. In a manner analogous to the Nod factors of *Rhizobium*, diffusible molecules have been postulated to be produced by the fungus partner that cause specific gene activation in the plant roots to establish this relationship (Kosuta et al., 2003). Neither the identity of these molecules and the stimulatory factors (perhaps produced by the plant) that induce them in the fungus, nor the nature of the genes induced in the root cells in response to the diffusible factors, if any, is known.

It is hoped that sequencing of the genome of *Glomus intraradices*, a representative mycorrhizal fungus, yields extremely useful information that will lead to experimental improvements in establishing mycorrhizal interaction between selected strains of the fungus and the host plant. This will be particularly valuable in cultivated forestry where the establishment of such symbiosis is a limiting factor. Improvements of the host range of selected fungi should facilitate establishment of tree plantations at desirable sites, transplantations of improved tree species among distant places, better utilization of fertilizers, and improvement of nutrient uptake in poor quality soils (Sawaki and Saito, 2001; Hijri and Sanders, 2004).

Metabolic Engineering

Metabolic engineering, involving the introduction of entire pathways that do not exist in a plant species or its close relatives (thus limiting their transfer by cross hybridization), is one of the most coveted goals of genetic engineering (Della Penna, 2001; Verpoorte and Memelink, 2002). Some examples of such targets are the production of specific secondary compounds for plant protection against abiotic or biotic stresses, enhancement of nutritional quality of food and animal feed, production of antibiotics and other non-proteinaceous pharmaceuticals in plants, ability to metabolize organic pollutants in air, soil or water, etc. Microbes may provide, and in some cases already have provided, the necessary genes (either as an easy source or a sole source) to achieve these goals. An excellent example of such an improvement with tremendous potential benefit to millions of humans is the production of 'golden rice', a variety of rice that is genetically improved for accumulation of β-carotene. This could act as a staple food source of vitamin A for people who suffer from this deficiency and are, therefore, subject to widespread blindness (Ye et al., 2000; Potrykus, 2001).

Engineering plants to over-express a feedback insensitive bacterial dihydrodipicolinate synthase, greatly increased flux through the Lys biosynthetic pathway. However, in most cases this did not result in increased steady-state Lys levels as the plants also responded by increasing flux through the Lys catabolic pathway (Falco et al., 1995; Brinch-Pedersen et al., 1996). Substantial increases in Lys only occurred in plants where flux increased to such a level that the first enzyme of the catabolic pathway became saturated.

Protection against Transgene Escape

A major argument opposing the release of genetically modified plants into the environment is the fact that most transgenic plants contain superfluous genes such as antibiotic resistance genes that were incorporated in order for selection of the transformed plants during their production. Various strategies to remove the selection marker gene from a transgenic population of plants have been suggested. The use of bacterial genes for limitation of transgene escape has facilitated the release of a range of transgenic crops. Of these, the most commonly used is the combination of *barnase* and *barstar* genes from *Bacillus amyloliquefaciens*. Inclusion of the *barnase* gene (which encodes a ribonuclease) in a GM crop under the control of a tapetum- or pollen-specific promoter, leads to male-sterile flowers, however fertility can be restored by expression of the *barstar* gene (which encodes a ribonuclease inhibitor), which may be under the control of an inducible promoter. By spraying the crop with a chemical to induce expression of *barstar*, plant reproduction can be restored, while any escapes into the wild will not be able to reproduce. In another system utilizing male sterility, the DNA adenine methylase (*dam*) gene from *E. coli* has been introduced under the control of an anther-specific promoter, leading to male-sterile plants. These plants must be grown close to plants without transgenes, in order that the transgenic plants may be fertilized; the progeny will of course be male-sterile.

CONCLUSION

It is obvious from the foregoing discussion that microbes have provided some crucial tools as well as genes, for genetic manipulation of plants. It is almost certain that without these tools progress in this regard would have been nearly impossible. It goes without saying that certain microbes stand out in terms of their unique and indispensable contributions, the most prominent among them are: *E. coli* for restriction enzymes, cloning plasmids and host for cloning vectors; *A. tumefaciens* as a vehicle for gene transfer; and *B. thuringiensis* for the *cry* genes. While the list is small, it is very impressive in terms of its commercial value. The future role of microbes as direct contributors to the genetic manipulation of plants looks even more promising for a variety of commercially important improvements, the most significant of which may be the use of plants for bioremediation of organic pollutants and metabolic engineering with new pathways. Likewise, a better understanding of the plant-microbe interaction will be instrumental in devising future strategies for pathogen resistance. One can only say, "Thanks microbes!! – We couldn't have done without you."

REFERENCES

Abramovitch, R.B. and Martin, G.B. (2004). Strategies used by bacterial pathogens to suppress plant defenses. Curr. Opin. Plant Biol. 7: 356-364.

Allen, M.J., Collick, A. and Jeffreys, A.J. (1994). Use of vectorette and subvectorette PCR to isolate transgene flanking DNA. PCR Methods Appl. 4: 71-75.

Alonso, J.M, Stepanova, A.N., Leisse, T.J., Kim, C.J, Chen, H., Shinn, P., Stevenson, D.K., Zimmerman, J., Barajas, P. and Cheuk, R. (2003). Genome-wide insertional mutagenesis of *Arabidopsis thaliana*. Science 301: 653-657.

Azpiroz-Leehan, R. and Feldmann, K.A. (1997). T-DNA insertion mutagenesis in *Arabidopsis*: going back and forth. Trends Genet. 13: 152-156.

Bago, B., Pfeffer, P.E. and Shachar-Hill, Y. (2000). Carbon metabolism and transport in arbuscular mycorrhizas. Plant Physiol. 124: 949-957.

Bago, B., Pfeffer, P.E., Abubaker, J., Jun, J., Allen, J.W., Brouillette, J., Douds, D.D., Lammers, P.J. and Shachar-Hill, Y. (2003). Carbon export from arbuscular mycorrhizal roots involves the translocation of carbohydrate as well as lipid. Plant Physiol. 131: 1496-1507.

Bechtold N., Ellis J. and Pelletier G. (1993). In planta *Agrobacterium* mediated gene transfer by infiltration of adult *Arabidopsis thaliana* plants. C. R. Acad. Sci. Ser. III 316: 1194-1199.

Blein, J.-P., Coutos-Thévenot, P., Marion, D. and Ponchet, M. (2002). From elicitins to lipid-transfer proteins: a new insight in cell signalling involved in plant defense mechanisms. Trends Plant Sci. 7: 293-296.

Boivin, C., Ndoye, I., Molouba, P., de Lajudie, P., Dupuy, N. and Dreyfus, B. (1997). Stem nodulation in legumes: diversity, mechanisms, and unusual characteristics. Crit. Rev. Plant Sci. 16: 1-30.

Bomhoff, G., Klapwijk, P.M., Kester, H.C.M., Schilperoort, R.A., Hernalsteens, J.P. and Schell, J. (1976). Octopine and nopaline synthesis and breakdown genetically controlled by a plasmid of *Agrobacterium tumefaciens*. Mol. Gen. Genet. 145: 177-181.

Bouarab, K., Melton, R., Peart, J., Baulcombe, D. and Osbourn, A. (2002). A saponin-detoxifying enzyme mediates suppression of plant defences. Nature 418: 889-892.

Braun, A.C. (1969). The Cancer Problem, A Critical Analysis and Modern Synthesis. Columbia University Press, NY, USA.

Braun, A.C. (1978). Plant tumors. Biochim. Biophys. Acta 516: 167-191.

Bravo, A. (1997). Phylogenetic relationships of *Bacillus thuringiensis* δ-endotoxin family proteins and their functional domains. J. Bacteriol. 179: 2793-2801.

Brinch-Pedersen, H., Galili, G., Knudsen, S. and Holm, P.B. (1996). Engineering of the aspartate family biosynthetic pathway in barley (*Hordeum vulgare* L.) by transformation with heterologous genes encoding feed-back-insensitive aspartate kinase and dihydrodipicolinate synthase. Plant Mol. Biol. 32: 611-620.

Brunaud, V., Balzergue, S., Dubreucq, B., Aubourg, S., Samson, F., Chauvin, S., Bechtold, N., Cruaud, C., DeRose, R. and Pelletier, G. (2002). T-DNA integration into the *Arabidopsis* genome depends on sequences of pre-insertion sites. EMBO Rep. 3: 1152-1157.

Chilton, M.-D., Saiki, R.K., Yadav, N., Gordon, M.P. and Quetier, F. (1980). T-DNA from *Agrobacterium tumefaciens* is in the nuclear DNA fraction of crown gall tumor cells. Proc. Natl. Acad. Sci., USA; 86: 4060-4066.

Christie, P.J. (1997). *Agrobacterium tumefaciens* T-complex transport apparatus: a paradigm for a new family of multifunctional transporters in eubacteria. J. Bacteriol. 179: 3085-3094.

Cornelis, G.R. and Van Gijsegem, F. (2000). Assembly and function of type III secretory systems. Annu. Rev. Microbiol. 54: 735-774.

Cullimore, J. and Dénarié, J. (2003). How legumes select their sweet talking symbionts. Science 302: 575-578.

Dangl, J.L. (1995). The enigmatic avirulence gene of phytopathogenic bacteria. In: Bacterial Pathogenesis of Plants and Animals, Molecular and Cellular Mechanisms. (ed.) J.L. Dangl. Springer-Verlag, Berlin, Germany pp. 91-118.

de Cleene, M. and de Lay, J. (1976). The host range of crown gall. Bot. Rev. 42: 389-466.

de La Fuente-Martinez, G., Mosqueda-Cano, S., Alvarez-Morales, L. and Herrerra-Estrella, L. (1992). Expression of a bacterial phaseolotoxin-resistant ornithyltranscarbamylase in transgenic tobacco confers resistance to *Pseudomonas syringae* pv. *phaseolicola*. Bio/Technology 10: 905-909.

Della Penna, D. (2001). Plant metabolic engineering. Plant Physiol. 125: 160-163.

de Risi, J.L., Iyer, V.R. and Brown, P.O. (1997). Exploring the metabolic and genetic control of gene expression on a genomic scale. Science 278: 680-686.

Drummond, M. (1979). Crown gall disease. Nature 281: 343-347.

Dumas, F., Duckely, M. and Pelczar, P. (2001). An *Agrobacterium* VirE2 channel for transferred-DNA transport into plant cells. Proc. Natl. Acad. Sci. USA; 16: 485-490.

Elliot, C. (1951). Manual of Bacterial Plant Pathogens, 2nd ed. Chronica Botan Waltham, MA, USA.

Esseling, J.J. and Emons, A.M.C. (2004). Dissection of Nod factor signalling in legumes: cell biology, mutants and pharmacological approaches. J. Microscopy 214: 104-113 Part 2.

Falco, S.C., Guida, T., Locke, M., Mauvais, J., Sanders, C., Ward, R.T. and Webber, P. (1995). Transgenic canola and soybean seeds with increased lysine. Biotechnology 13: 577-582.

Freiberg, C., Fellay, R., Bairoch, A., Broughton, W.J., Rosenthal, A. and Perret, X. (1997). Molecular basis of symbiosis between *Rhizobium* and legumes. Nature 387: 352-354.

Galbiati, M., Moreno, M.A., Nadzan, G., Zourelidou, M. and Dellaporta, S.L. (2000). Large-scale T-DNA mutagenesis in *Arabidopsis* for functional genomic analysis. Funct. Integr. Genom. 1: 25-34.

Galbraith, D.W. (2003). Global analysis of cell type-specific gene expression. Comp. Funct. Genomics 4: 208-215.

Garfinkel, D.J., Simpson, R.B., Ream, L.W., White, F.F., Gordon, M.P. and Nester, E.W. (1981). Genetic analysis of crown gall: fine structure map of the T-DNA by site-directed mutagenesis. Cell 27: 143-153.

Gelvin, S.B. (2000). *Agrobacterium* and plant genes involved in T-DNA transfer and integration. Annu. Rev. Plant Physiol. Plant. Mol. Biol. 51: 223-256.

Gelvin, S.B. (2003). *Agrobacterium*-mediated plant transformation: the biology behind the "gene-jockeying" tool. Mol. Biol. Rev. 67: 16-37.

Greller, L.D. and Tobin, F.L. (1999). Detecting selective expression of genes and proteins. Genome Res. 9: 282-296.

Hatziloukas, E. and Panopoulos, N.J. (1992). Origin, structure, and regulation of *Argk,* encoding the phaseolotoxin-resistant ornithine carbamoyltransferase in *Pseudomonas syringae* pv *phaseolicola,* and functional expression of *Argk* in transgenic tobacco. J. Bacteriol. 174: 5895-5909.

Herbers, K., Conrads-Strauch, J. and Bonas, U. (1992). Race-specificity of plant resistance to bacterial spot disease determined by repetitive motifs in a bacterial avirulence protein. Nature 356: 172-174.

Hernalsteens, J.P., Thia-Toong, L., Schell, J. and van Montagu, M. (1984). An *Agrobacterium*-transformed cell culture from the monocot *Aspergillus officinalis.* EMBO J. 3: 3039-3041.

Hijri, M. and Sanders, I.R. (2004). The arbuscular mycorrhizal fungus *Glomus intraradices* is haploid and has a small genome size in the lower limit of eukaryotes. Fungal Gen. Biol. 41: 253-261.

Hooykaas, P.J.J. and Schilperoort, R.A. (1992). *Agrobacterium* and plant genetic engineering. Plant Mol. Biol. 19: 15-38.

Hotson, A. and Mudgett, M.B. (2004). Cysteine proteases in the phytopathogenic bacteria: identification of plant targets and activation of innate immunity. Curr. Opin. Plant Biol. 7: 384-390.

Iida, S. and Terada, R. (2004). A tale of two integrations, transgene and T-DNA: gene targeting by homologous recombination in rice. Curr. Opin. Biotechnol. 15: 132-138.

Jakobsen, I., Smith, F.A. and Smith, S.E. (2002). Function and diversity in arbuscular mycorrhizas. In: Mycorrhizal Ecology (eds.) M. van der Heijden and I.R. Sanders. Springer Verlag, Berlin and Heidelberg, Germany, pp. 75-92.

James, C. (2002). Global Status of Commercialized Transgenic Crops. ISAAA, Metro Manila, Philippines.

Jakubowski, S.J., Krishnamoorthy, V., Cascales, E. and Christie, P.J. (2004). *Agrobacterium tumefaciens* VirB6 domains direct the ordered export of a DNA substrate through a Type IV secretion system. J. Mol. Biol. 341: 961-977.

Kempin, S.A., Liljegren, S.J., Block, L.M., Rounsley, S.D., Lam, E. and Yanofsky, M.F. (1997). Inactivation of the *Arabidopsis* AGL5 MADS-box gene by homologous recombination. Nature 389: 802-803.

Kim, E. Hammond-Kosack and Jones, J.D.G. (1997). Plant disease resistance genes. Annu. Rev. Plant Physiol. 48: 575-607.

Kosuta, S., Chabaud, M., Lougnon, G., Gough, C., Dénarié, J., Barker, D.G. and Bécard, G. (2003). A diffusible factor from arbuscular mycorrhizal fungi induces symbiosis-specific MtENOD11 expression in roots of *Medicago truncatula.* Plant Physiol. 131: 952-962.

Krysan, P.J., Young, J.C., Tax, F. and Sussman, M.R. (1996). Identification of transferred DNA insertions within *Arabidopsis* genes involved in signal transduction and ion transport. Proc. Natl. Acad. Sci. USA; 93: 8145-8150.

Krysan, P.J., Young, J.C. and Sussman, M.R. (1999). T-DNA as an insertional mutagen in *Arabidopsis*. Plant Cell 11: 2283-2290.

Kunkel, B.N. and Brooks, D.M. (2002). Cross talk between signaling pathways in pathogen defense. Curr. Opin. Plant Biol. 5: 325-331.

Limpens, E. and Bisseling, T. (2003). Signaling in symbiosis. Curr. Opin. Plant Biol. 6: 343-350.

Lippincott, B.B. and Lippincott, J.A. (1969). Bacterial attachment to a specific wound site as an essential stage in tumor initiation by *Agrobacterium tumefaciens*. J. Bacteriol. 97: 620-628.

Martin, G.B., Bogdanova, A.J. and Sessa, G. (2003). Understanding the function of plant disease resistance proteins. Annu. Rev. Plant Biol. 54: 23-61.

Martienssen, R.A. (1998). Functional genomics: Probing plant gene function and expression with transposons. Proc. Natl. Acad. Sci. USA; 95: 2021-2026.

McBride, K.E., Kenny, J.W. and Stalker DM. (1986). Metabolism of the herbicide bromoxynil by *Klebsiella pneumoniae* subsp. *ozaenae*. Appl. Environ. Microbiol. 52(2): 325-330.

McKinney, E.C., Ali, N., Traut, A., Feldmann, K.A., Belostotsky, D.A., McDowell, J.M. and Meagher, R.B. (1995). Sequence-based identification of T-DNA insertion mutations in *Arabidopsis*: actin mutants *act2*-1 and *act4*-1. Plant J. 8: 613-622.

Meyer, A., Tempé, J. and Costantino, P. (2000). Hairy root: a molecular overview. Functional analysis of *Agrobacterium rhizogenes* T-DNA genes. In: Plant Microbe Interactions (eds.) G Stacey and NT Keen. APS Press, St. Paul, USA, pp. 93-139.

Meyers, B.C., David, W., Galbraith, T.N. and Agrawal, V. (2004). Methods for transcriptional profiling in plants. Be fruitful and replicate. Plant Physiol. 135: 637-652.

Miflin, B.J. (2000). Crop Biotechnology. Where Now? Plant Physiol. 123: 17-28.

Mohan Babu, R., Sajeena, A., Seetharaman, K. and Reddy, M.S. (2003). Advances in genetically engineered (transgenic) plants in pest management-an overview. Crop Protec. 22: 1071-1086.

Mysore, K.S., Nam, J. and Gelvin, S.B. (2000). An *Arabidopsis* histone H2A mutant is deficient in *Agrobacterium* T-DNA integration. Proc. Natl. Acad. Sci. USA. 97: 948-953.

Mysore, K.S. and Ryu, C.M. (2004). Nonhost resistance: how much do we know? Trends Plant Sci. 9: 97-104.

Newman, E.I. and Reddell, P. (1987). The distribution of mycorrhizas among families of vascular plants. New Phytol. 106: 745-751.

Nilsson, O. and Olsson, O. (1997). Getting to the root: the role of the *Agrobacterium rhizogenes rol* genes in the formation of hairy roots. Physiol. Plant. 100: 463-473.

Ooms, G. Hooykaas, P.J.J., Moolenaar, G. and Schilperoort, R.A. (1981). Crown gall plant tumors of abnormal morphology, induced by *Agrobacterium tumefaciens* carrying mutated octopine Ti plasmids: analysis of T-DNA functions. Gene 14: 33-50.

Ortega, D., Raynal, M., Laudie, M., Llauro, C., Cooke, R., Devic, M., Genestier, S., Picard, G., Abad, P., Contard, P., Sarrobert, C., Nussaume, L., Bechtold, N., Horlow, C., Pelletier, G. and Delseny, M. (2002). Flanking sequence tags in *Arabidopsis thaliana* T-DNA insertion lines: a pilot study. Crit. Rev. Biol. 325: 773-780.

Padgette, S.R., Kolacz, K.H., Delannay, X., Re, D.B., LaVallee, B.J., Tinius, C.N., Rhodes, W.K., Otero, Y.I., Barry, G.F. and Eichholz, D.A. (1995). Development, identification, and characterization of a glyphosate-tolerant soybean line. Crop Sci. 35: 1451-1461.

Panopoulos, N.J., Hatziloukas, E. and Afendra, A.S. (1996). Transgenic crop resistance to bacteria. Field Crops Res. 45: 85-97.

Parales, R.E., Bruce, N.C., Schmid, A. and Wackett, L.P. (2002). Biodegradation, biotransformation, and biocatalysis (B). Appl. Environ. Microbiol. 68: 4699-4709.

Parinov, S., Sevugan, M., Ye, D., Yang, W.C., Kumaran, M. and Sundaresan, V. (1999). Analysis of flanking sequences from dissociation insertion lines: a database for reverse genetics in *Arabidopsis*. Plant Cell 11: 2263-2270.

Pingoud, A. and Jeltsch, A. (2001). Structure and function of type II restriction endonucleases source. Nucl. Acids Res. 29: 3705-3727.

Potrykus, I. (2001). Golden rice and beyond. Plant Physiol. 125: 1157-1161.

Puchta, H. and Hohn, B. (1996). From centimorgans to basepairs: Homologous recombination in plants. Trends Plant Sci. 1, 340–348.

Riely, B.K., Ané, J.-M., Penmetsa, R.V. and Cook, D.R. (2004). Genetic and genomic analysis in model legumes bring Nod-factor signaling to center stage. Curr. Opin. Plant Biol. 7: 408-413.

Rushton, P.J., Reinstädler, A., Lipka, V., Lippok, B. and Somssich, I.E. (2002). Synthetic plant promoters containing defined regulatory elements provide novel insights into pathogen- and wound-induced signalling. Plant Cell 14: 749-762.

Sawaki, H. and Saito, M. (2001). Expressed genes in the extraradical hyphae of an arbuscular mycorrhizal fungus, *Glomus intraradices*, in the symbiotic phase. FEMS Microbiol. Lett. 195: 109-113.

Schnepf, E., Crickmore, N., Van Rie, J., Lereclus, D., Baum, J., Feitelson, J., Zeigler, D.R. and Dean, D.H. (1998). *Bacillus thuringiensis* and its pesticidal crystal proteins. Microbiol. Mol. Biol. Rev. 62: 775-806.

Schuler, T.H., Poppy, G.M., Kerry, B.R. and Denholm, I. (1998). Insect-resistant transgenic plants. Trends Biotech. 16: 168-175.

Sessions, A., Burke, E., Presting, G., Aux, G., McElver, J., Patton, D., Dietrich, B., Ho, P., Bacwaden, J., Ko, C., Clarke, J.D., Cotton, D., Bullis, D., Snell, J., Miguel, T., Hutchison, D., Kimmerly, B., Mitzel, T., Katagiri, F., Glazebrook, J., Law, M. and Goff, S.A. (2002). A high-throughput *Arabidopsis* reverse genetics system. Plant Cell 14: 2985-2994.

Sharma, H.C., Sharma, K.K., Seetharama, N. and Ortiz, R. (2000). Prospects for using transgenic resistance to insects in crop improvement. EJB Electron. J. Biotechnol. 3: 76-95.

Shiraishi, T., Yamada, T., Ichinose, Y., Kiba, A. and Toyoda, K. (1997). The role of suppressors in determining host-parasite specificities in plant cells. Intern. Rev. Cell Biol. 172: 55-93.

Smith, E.F. and Townsend, C.O. (1907). A plant tumor of bacterial origin. Science 25: 671-673.

Somers, D.A. and Makarevitch, I. (2004). Transgene integration in plants: poking or patching holes in promiscuous genomes? Curr. Opin. Biotechnol. 15: 126-131.

Sussman, M.R., Amasino, R.M., Young, J.C., Krysan, P.J. and Austin-Phillips, S. (2000). The *Arabidopsis* knockout facility at the University of Wisconsin-Madison. Plant Physiol. 124: 1465-1467.

Stachel, S.E. and Zambryski, P.C. (1989). Generic trans-kingdom sex? Nature 340: 190-191.

Tinland, B. (1996). The integration of T-DNA into plant genomes. Trends Plant Sci 1: 178-184.

Tzfira, T., Rhee, Y., Chen, M.H., Kunik, T. and Citovsky, V. (2000). Nucleic acid transport in plant-microbe interactions: the molecules that walk through the walls. Annu. Rev. Microbiol. 54: 187-219.

Tzfira, T. and Citovsky, V. (2003). Partners-in-infection: host proteins involved in the transformation of plant cells by *Agrobacterium*. Trends Cell. Biol. 12: 121-129.

Tzfira, T., Frankman, L.R., Vaidya, M. and Citovsky, V. (2003). Site-Specific Integration of *Agrobacterium tumefaciens* T-DNA via double-stranded intermediates. Plant Physiol. 133: 1011-1023.

Tzfira, T., Li, J., Lacroix, B. and Citovsky, V. (2004). *Agrobacterium* T-DNA integration: molecules and models. Trends Genet. 20: 375-383.

Valentine, L. (2003). *Agrobacterium tumefaciens* and the Plant: The David and Goliath of Modern Genetics. Plant Physiol. 133: 948-955.

Verpoorte, R. and Memelink, J. (2002). Engineering secondary metabolite production in plants Curr. Opin. Biotechnol. 13: 181-187

Walden, R. (2002). T-DNA Tagging in a genomics era. Crit. Rev. Plant Sci. 21: 143-165.

Wang, C.-L., Farrand, S.K. and Hwang, I. (1984). Organization and expression of the genes on pAgK84 that encode production of agrocin 84. Mol. Plant-Microbe Interact. 7: 472-481.

Weigel, D., Ahn, J.H., Blázquez, M.A., Borevitz, J., Christensen, S.K., Fankhauser, C., Ferrándiz, C., Kardailsky, I., Malancharuvil, E.J., Neff, M.M., Nguyen, J.T., Sato, S., Wang, Z., Xia, Y., Dixon, R.A., Harrison, M.J., Lamb, C.J., Yanofsky, M.F. and Chory, J. (2000). Activation tagging in *Arabidopsis.* Plant Physiol. 122: 1003–1014.

Weiler, E.W. and Schröder, J. (1987). Hormone genes and crown gall disease. TIBS 12: 271-275.

Weising, K. and Kahl, G. (1996). Natural genetic engineering of plant cells: the molecular biology of crown gall and hairy root. World J. Microbiol. Biotechnol. 12: 327-351.

Wilson, K., Long, D., Swinburne, J. and Coupland, G. (1996). A Dissociation insertion causes a semidominant mutation that increases expression of *TINY*, an *Arabidopsis* gene related to *APETALA2*. Plant Cell 8: 659-671.

Ye, X., Al-Babili, S., Klöti, A., Zhang, J., Lucca, P., Beyer, P. and Potrykus, I. (2000). Engineering the provitamin a (β-carotene) biosynthetic pathway into (carotenoid-free) rice endosperm. Science 287: 303-305.

2

Genetically Modified Plants: Applications and Issues

Ioannis S. Arvanitoyannis

INTRODUCTION

Objectives

Although "biotechnology" and "genetic modification" are used interchangeably, the latter is a special set of technologies that alter the genetic structure and sequence of living organisms such as animals, plants, or bacteria. Biotechnology stands for a more general term referring to using living organisms or their components, such as enzymes, to make products that include wine, cheese, beer, and yoghurt. Combining genes from different organisms is known as recombinant (r) DNA technology, and the resulting organism is said to be "genetically modified (GM)," "genetically engineered (GE)," or "transgenic", depending mainly on whether the person who reviews them is positively or negatively inclined towards them. GM products (currently available or in the pipeline) include medicines and vaccines, foods and food ingredients, feeds, and fibres (http://www.ornl.gov/sci/techresources/Human_Genome/elsi/gmfood.shtml).

In 2003, about 167 million acres (67.7 million hectares [ha]) grown by 7 million farmers in 18 countries were planted with transgenic crops, the principal ones being herbicide- and insecticide-resistant soybeans, corn, cotton, and canola. Other crops grown commercially or field-tested are sweetpotato resistant feathery mottle virus, which can reduce yields by 20-80%, thus decimating most of the African harvest, rice with increased iron

Department of Agriculture, Animal Production and Aquatic Environment, Agricultural Sciences, University of Thessaly, Fytokou Street, Nea Ionia Magnesias, 38446 Volos, Hellas (Greece)
Fax: +30 24210 93137; E-mail: parmenion@uth.gr

and vitamins that may alleviate chronic malnutrition in Asian countries, and a variety of plants able to survive weather extremes. In 2003, countries that grew 99% of the global transgenic crops were the United States (63%), Argentina (21%), Canada (6%), Brazil (4%), China (4%), and South Africa (1%). Although growth is expected to plateau in industrialized countries, it is gradually increasing in developing countries. The next decade will see exponential progress in GM product development as researchers gain increasing and unprecedented access to genomic resources that are applicable to organisms beyond the scope of individual projects (http://www.ornl.gov/sci/techresources/Human_Genome/elsi/gmfood.shtml).

There seem to be at least four major objectives being currently pursued in crop plant genetic engineering research (http://www.yale.edu/ynhti/curriculum/units/2000/7/00.07.02.x.html):

(1) To improve biological protection of crops against insects, weeds, and fungi by inserting genes for the natural production of an insecticide (Feder, 1996), or for resistance to fungi or herbicide (Hinchee, 1988).
(2) To elevate levels of important nutrients, e.g. methionine levels in soybeans (Beardsley, 1996) so as to make crops more nutritious.
(3) To obtain better control of ripening and post-harvest storage life to assure that produces are in peak condition when taken to market (Maryanski, 1995).
(4) To specifically modify genomes to produce a specific product, e.g. a caffeine-less coffee bean, and edible vaccines in potatoes (Pollack, 2000).

CURRENT SITUATION IN AGRICULTURAL BIOTECHNOLOGY

As discussed under 'objectives', six principal countries grew 99% of the global GM crops in 2003. The USA grew 42.8 million ha (63% of global total), followed by Argentina with 13.9 million ha (21%), Canada with 4.4 million ha (6%), Brazil with 3.0 million ha (4%), China with 2.8 million ha (4%), and South Africa with 0.4 million ha (1%). Globally, the principal GM crops were soybeans (41.4 million ha; 61% of global area), maize (15.5 million ha; 23%), cotton (7.2 million ha; 11%), and canola (3.6 million ha; 5%) [International Life Sciences Institute (ILSI), 2004]. The breakup by crop and country from 1996 to 2003 is illustrated in Fig. 2.1 (A and B) [James (2003), data from ISAAA (International Service for the Acquisition of Agri-Biotech Applications) Briefs]. During the eight years since introduction of commodity GM crops (1996 to 2003), a cumulative total of over 300 million

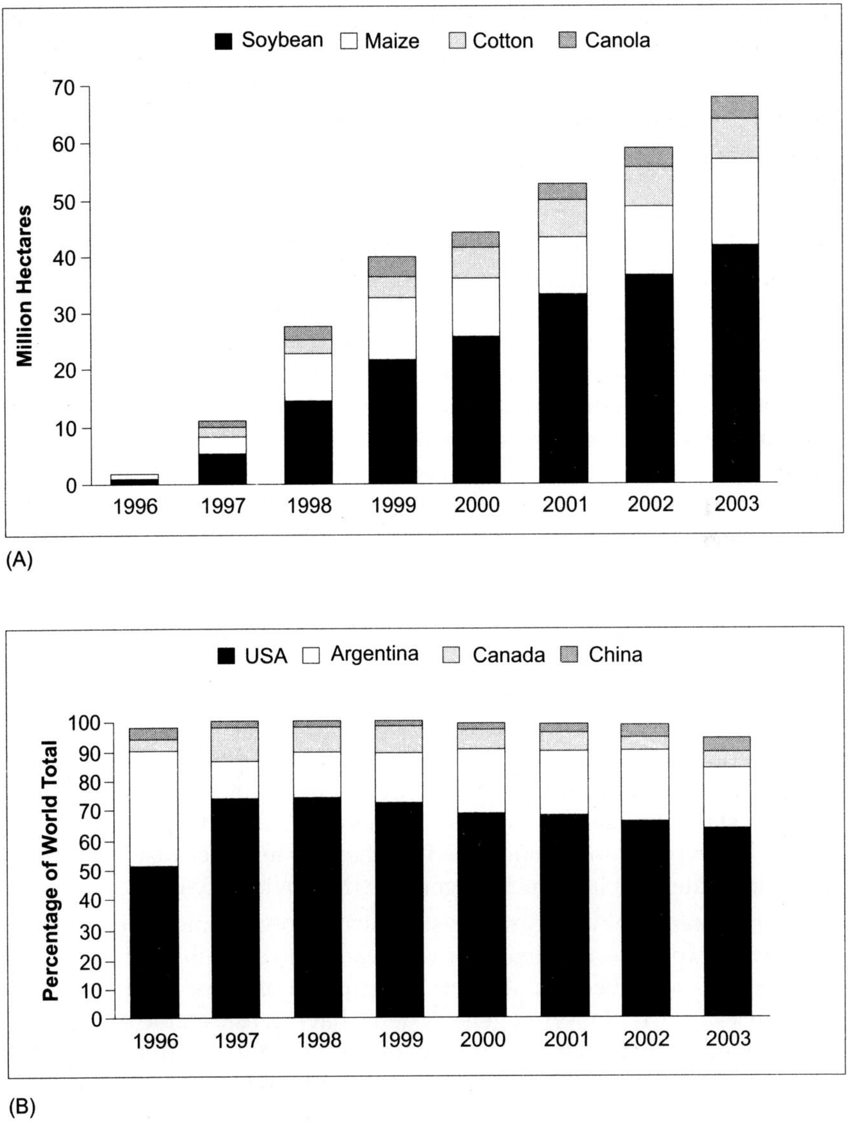

Fig. 2.1 (A) Areas planted with four primary GM crops, and (B) Areas planted with GM crops in four principal countries (Source: ISAAA Briefs)

ha (almost 750 million acres) of GM crops were planted globally by millions of large- and small-scale farmers (James, 2003). Rapid adoption and planting of GM crops by millions of farmers around the world; growing global, political, institutional, and country support for GM crops and data from independent sources confirm and support the benefits associated with GM crops (James, 2003).

PREPARATION AND SYNTHESIS OF GENETICALLY MODIFIED ORGANISMS (GMOs)

Generally, there are two main approaches for GMOs synthesis: the one utilizing bacterial (usually *Agrobacterium tumefaciens*) and viral vectors (Minocha and Page, Chapter 1 in this Volume) and the other utilizing either chemical or physical methods to introduce specific genes into the target cells. The first one, the gene transfer by the bacterium, *A. tumefaciens* has been dealt within Chapter 1; hence not discussed here. The viral gene transfer is a viral-mediated process referred to as an infection. The second, non-viral gene transfer, involves treatment of cells by chemical or physical means and the process itself is named transfection. Gene delivery by infection is more complicated than transfection. Infection requires more steps and more time than does transfection, and biosafety issues may also arise, depending on the virus used. A much safer and faster alternative to infection is transfection. The latter requires only a few reagents, including plasmid DNA containing the gene of interest under the control of a strong cell-specific promoter. However, inefficient gene delivery and poor sustained gene expression are its major drawbacks. Transfection can be categorized into two major types: transient and stable. Transient transfection is temporary, i.e. expression of foreign gene lasts for several days and is lost as DNA never integrates into the host cell DNA. In contrast, stable transfection occurs with a lower frequency (10 to 100-fold lower), but expression is maintained for the long term because the foreign DNA does integrate into the host genome (Mitrovic, 2003).

Gene transfer methods can be distinguished in viral and antiviral ones. The former exploit recombinant viruses, leading to recombinant viral short term or long term vectors. A short description of these viral vectors is given in Table 2.1. Non-viral gene systems have considerable advantages over the recombinant viruses in view of their highly immunogenic, non-infectious and non-toxic character in conjunction with accommodation of large DNA plasmid and large-scale production. However, their two main disadvantages reside in their lower gene transfer efficiency (compared to viruses) and transient gene expression. Non-viral gene transfer systems fall in the following categories: (1) chemical methods, and (2) physical methods. Chemical methods include utilization of positively charged polymers [dextran based known as Diethyl Amino Ethyl Amino (DEAE) dextran], calcium phosphate co-precipitation with DNA and cationic lipids or liposomes. Physical methods comprise electroporation, particle guns, etc. (Chang et al., 2001; Chen et al., 2003; Abead et al., 1997; Cranshaw in http://www.ext.colostate.edu/pubs/insect05556.html).

Table 2.1. Methods, advantages, disadvantages, and applications of GM production

Methods	Description of Method	Advantages	Disadvantages	Application	Reference
I. Viral gene transfer					
Ia. *Agrobacterium* mediated gene transfer	(a) Bacterial colonization (b) Induction of bacterial virulence (c) Generation of T-DNA (d) T-DNA transfer (e) Integration of T-DNA into plant genome	- Relatively large segments of DNA can be transferred with little rearrangement - Integration of low numbers of gene copies occurs in plant chromosomes	Strongly dependent on: - Stage of tissue development - Initial level of *A. tumefaciens* used - Media used to subculture the plants after transformation - Makers used to select for transformants - Type of vectors utilized, - Plant genotype	Sugarcane, corn	de la Riva et al., 1998
Ib. *Bacillus thuringiensis* (*Bt*)		- Non-toxic to people - Specific activity of *Bt* - No killing of beneficial insects - Reduced usage of conventional insecticides - Reduced requirement for energy	- Susceptible to degradation by sunlight - High specific insect activity may limit its use - *Bt* products have shorter shelf life than other insecticides - Environmental advantages	- Spray to control caterpillars on vegetable crops - Control mosquito larvae - Trees and shrubs treatment	http://www.ext.c olostate.edu/ pubs/insect 05556.html Saraswathy and Kumar, 2004

contd...

Table 2.1 contd...

II. Non-viral methods					
IIa. Physical methods					
IIa1. Electroporation	- Short electric field pulse (high voltage) - Reorganization of membrane - Diffusion of polar molecules DNA into host cells	- Versatility - Efficiency - Small-scale - In vivo	- Cell damage - Nonspecific transport	- DNA transfection - Direct plasmid transfer - Induced cell fusion - Gene therapy	http://www.bio.davidson.edu/Courses/Molbio Tsujie et al., 2001; Mitrovic, 2003
IIa2. Particle bombardment	Acceleration of DNA-coated particles (microprojectiles) directly into intact tissues or cells	- Almost any kind of cell - Easy device operation - Simplified plasmid construction - Elimination of false positive results - Small amount of plasmid DNA is required - Transient gene expression examined in few days	- Low transformation efficiency - Consumable items are expensive - For commercial use, patent royalty is required	- Most widely employed	Hagio, 1998

contd...

Table 2.1 contd...

IIa3. Micro-injection	- Glass micropipettes used for transferring DNA into protoplast - Recipient cells immobilized on a solid support	- Direct introduction - Avoidance of potentially undesirable cellular compartments (low pH)	- Extensive training - Specialized and expensive equipment - Inappropriate for large number of transfected cells	Effective with plant protoplasts and tissues	http://www.tnau.ac.in/notesbscag/notestry/abt401, http://www.dragon.zoo.utoronto.ca/~jlm gmf/T0301C/technology/microinjection, http://www.promega.com/guides/transfxn_guide/chapter_one
IIb. Chemical methods					
IIb1. DEAE-dextran based	- Positively charged polymers, as DEAE – dextran, polybrene, polyethylenimine and dendrimer, complex with negatively charged DNA molecules forming polyplex	–	- Several parameters, such as: the number of cells, polymer concentration, transfected DNA concentration and transfection duration should be optimized for a given cell line	Limited to use in transient transfections	Mitrovic, 2003

contd...

	- Enhanced ionic attraction between the net positive charge on the polycation -DNA complex and the negative charge on the cell surface enable the DNA binding and entrance in the cell by endocytosis		- Complexed DNA delivery with DEAE-dextran could be improved by osmotic shock using DMSO or glycerol or treatment with chloroquine		
IIb2. Calcium phosphate	- Mixing DNA with calcium chloride - Addition of complex to a buffered saline / phosphate - Mixture incubation at room temperature - Dispersion of precipitate	- Components are easily available - Reasonable price - Easy to use protocol - Applicable to various types of cells - Protection against intracellular and serum nucleases	–	Applicable for both transient and stable transfection of a variety of cell types	http://www.promega.com/guides/transfxn_guide/chapter_one
IIb3. Liposome mediated gene transfer	- Fusion with protoplasts using devices like PEG (polyethylene glycol)	- DNA protection from nuclease digestion - Low cell toxicity - Stability and storage of nucleic acid - High reproducibility - Applicability to all cell types	- More effective than other chemical methods (DEAE dextran and calcium phosphate) - Small amounts of DNA are required - Simple to perform to broad range of cell types	- Surface associated targeted information - Effective targeted gene delivery	http://www.tnau.ac.in/notesbscag/notestry/abt401 Mitrovic, 2003

Table 2.1 contd...

APPLICATIONS OF GENETIC ENGINEERING IN AGRICULTURE/HORTICULTURE

Plants have evolved a large variety of sophisticated and efficient defence mechanisms to prevent the colonization of their tissues by microbial pathogens and parasites. The induced defense responses can be assigned to three major categories (Table 2.2).

The incorporation of disease resistance genes into plants was first initiated with classical breeding, which yielded high crop productivity. However, currently, plant biology is oriented towards clone identification and characterization of genes involved in disease resistance. The following approaches are adopted towards developing disease resistant transgenic plants (Punja, 2004):

(1) Expression of gene products that are directly toxic to, or which reduce growth of the pathogen including pathogenesis related proteins, such as hydrolytic enzymes, antifungal proteins, antimicrobial peptides and phytoalexins.
(2) Expression of gene products that destroy or neutralize a component of the pathogen (lipase inhibition).
(3) Expression of gene products that enhance the structural defences (high levels of peroxidase and lignin).
(4) Expression of gene products that release signals regulating plant defences such as salicylic acid.
(5) Expression of resistance gene products involved in race-specific avirulence interactions and hypersensitive response.

The first commercial product of rDNA technology was the Flavr Savr tomato characterized by prolonged shelf life. This was achieved by

Table 2.2. Categorization and characteristics of plant defence responses (Kombrink and Somssich, 1995; Sticher, 1997; Stuible and Kombrink, 2004)

Category	Response Time	Expression Pattern	Result
1. Hypersensitive response (HR)	Immediate	Early defence response in the invaded plant cell	Rapid death of challenged host cells
2. De novo synthesis of proteins	Subsequent to recognition	Local activation of genes in close vicinity of infection sites	Formation of phytoalexins, structural cell wall proteins, and other protective proteins
3. Systemic acquired resistance (SAR)	Temporarily delayed	Systemic activation of genes encoding pathogenesis related proteins	Immunity to secondary infections

introducing an antisense polygalacturonase gene (Redenbaugh et al., 1995). One of the most vital contributions of genetic engineering to agriculture has been the introduction of viral genes to plants, including tomato, as a means of conferring resistance to viral diseases (Barg et al., 2001). Apart from the introduction of bacterial *Bt* (*Bacillus thuringiensis*) toxins genes in tomato, characteristics related to the quality of fruit (total soluble solids, fruit colour) were also genetically engineered.

Numerous genes of agronomic interest were introduced into *Brassica* species, generally via *A. tumefaciens* – mediated transformation of oilseed rape (Murphy, 2003). In contrast to research with oilseed and vegetable brassicas, transformation research in forage brassicas is less extensive. Genetic manipulation in brassicas has already produced plants with increased insect resistance through use of *Bt* genes. In addition, expression of a chitinase gene in oilseed rape gave field tolerance to the fungal pathogens *Leptospheria maculans, Sclerotinia sclerotiorum* and *Cylindrosporium concentricum* (Christey and Braun, 2001).

Due to its capacity for regeneration, carrot was one of the first species to be used in direct gene transfer studies and for transformation by *Agrobacterium*. As it has excellent susceptibility to infection by *Agrobacterium*, this has become the opted procedure for stable transformation in most studies. Direct gene transfer is used for the analysis of transient gene expression or to transiently produce a gene product in regeneration studies. Since carrot has proved to be quite easy to generate protoplasts from and undergoes cell wall regeneration quickly and efficiently, it has been used in several transfection studies with electroporation and polyethylene glycol (Gallie, 2001). Soybean meals are currently used for animal feeding, however, feeding experiments need to be an integrated part of a GM food/feed safety assessment. So far, only 10 scientific (peer reviewed) feeding studies on animals have been published and most of these articles have been published by Pryme (2003). Only one feeding study involving fish has been published (Sanden et al., 2004). In this study, Atlantic salmons were fed with feed containing 17.2% GM soybean for six weeks, and the purpose was to investigate the fate of GM soy DNA fragments. The authors could only detect small DNA fragments [120 and 195 bp (length unit)] by using PCR (Polymerase Chain Reaction) methodology on the content of stomach, mid-intestine and distal intestine. No GM soy DNA fragments were detected in the liver, muscle or brain tissue (Myhra and Dalmo, 2005). The recent achievements in the field of genetic engineering for potato, tomato, soybean, carrot, and brassicas are summarized in Tables 2.3-2.7. These tables provide information to the reader on the expressed gene product, and effect on disease/composition for the above mentioned vegetables.

Table 2.3. Expressed gene product, and its effect on disease/composition of potato

Expressed Gene Product	Effect on Disease/Composition	Reference
Trichoderma harzianum endochitinase	Lower lesion numbers and size due to *Altenaria solani*; reduced mortality due to *Rhizoctonia solani*	Lorito et al., 1998
Tobacco osmotin	Delayed onset and rate of disease due to *Phytophthora infestans*	Liu et al., 1994
Potato osmotin-like protein	Enhanced tolerance to infection by *Phytophthora infestans*	Zhu et al., 1996
Pea *pr1o* gene	Reduced development of *Verticillium dahliae*	Chang et al., 2001
Potato defence response gene *sth-2*	No effect against *Phytophthora infestans*	Constabel et al., 2001
Alfalfa defensin	Enhanced resistance to *Verticillium dahliae*	Gao et al., 2000
Bacillus amyloliquefaciens barnase (RNase)	Delayed sporulation and reduced sporangia production by *Phytophthora infestans*	Strittmatter et al., 1995
Synthetic cationic peptide chimera	Reduced development of *Fusarium solani* and *Phytophthora castorum*	Osusky et al., 2000
Human lactoferrin	Not tested	Chong and Langridge, 2000
Cucumber peroxidase	No effect on disease due to *Fusarium sambucinum* and *Phytophthora infestans*	Ray et al., 1998
Aspergillus niger glucose oxidase	Delayed lesion due to *Phytophthora infestans*; reduced disease development due to *Altenaria solani* and *Verticillium dahliae*	Wu et al., 1995, 1997
Tobacco catalase	Reduced lesion size	Yu et al., 1999
Bacterial salicylate hydroxylase	No effect on *Phytophthora infestans*	Yu et al., 1997

contd...

Table 2.3 contd...

ADP glucose pyrophosphorylase *E. coli*	Increased starch content	Stark et al., 1992
Inhibition of SBE A and B	Very high amylase content of starch	Schwall et al., 2000
1-SST (sucrose: sucrose 1-fructosesyltransferase and the 1-FFT (fructan: fructan 1-fructosyltransferase)	Increased inulin content	Hellwege et al., 2000
Amaranthus hypochondriacus	Enhanced sulphur rich protein content	Chakraborty et al., 2000
Antisense sterol glucotransferase (*Sgt*)-gene	Decreased solanine content	McCue et al., 2003
Rx gene	Resistance to Potato virus X (PVX)	Bendahmane et al., 1999
Harpin protein gene from *Erwinia amylovora*	Resistance to *Phytophthora infestans*	Li and Fan, 1999
Pectate lyase gene (*pl3*) of *Erwinia carotovora* ssp. *atroseptica*	Enhanced resistance to *E. carotovora* Higher systemic levels of salicylic acid	Weber et al., 1996
Halobacterium halobium	Enhanced resistance against *Phytophthora infestans*	Abead et al., 1997
Lytic peptide (chicken lysozyme)	Resistance to black leg disease caused by *Erwinia carotovora* spp. *atroseptia*	Serrano et al., 1996
Lytic peptide (tachyplesin)	Reduction in tuber rot caused by *E. carotovora atroseptica*	Allefs et al., 1996
Bacteriophage T4 lysozyme	Enhanced resistance to *E. carotovora* spp. *atroseptica*	During et al., 1993
Bacteriophage T4 lysozyme	Enhanced resistance to *E. carotovora* spp. *carotovora*	Ahrenholz et al., 2000
Potato leafroll virus (PLV)	Enhanced resistance against PLV	Prufer et al., 1999; Lawson et al., 2001;

Table 2.4. Expressed gene product, and its effect on disease/composition of tomato

Expressed Gene Product	Effect on Disease/Composition	Reference
S-adenosyl-methionine hydrolase	Reduced ethylene synthesis and prolonged shelf life	Kramer et al., 1996
S-adenosyl-methionine	Ethylene evolution reduction, extended shelf life	Good et al., 1994
Rolb- gene	Parthenocarpy	Carmi et al., 1997
Endopolygalacturonase inhibitory protein gene	Fungal resistance	Cocci et al., 1994
Coat protein gene of cucumber mosaic virus (CMV) strain WL	Enhanced CMV resistance	Gielen et al., 1996; Fuchs et al., 1996
Tomato spotted wilt virus	Enhanced resistance to spotted wilt virus	Kim et al., 1994
Gene altering starch accumulation in young fruits	Enhanced TSS (total soluble solids)	Stark et al., 1996
Silencing the gene for acid invertase	Enhanced TSS (total soluble solids)	Klann et al., 1966
Wild tomato (*Lycopersicon chilense*) chitinase	Reduced development of *Verticillium dahliae* races 1 and 2	Tabaeizadeh et al., 1999
Radish defensin	Reduced number and size of lesions due to *Altenaria solani*	Parashina et al., 2000
Grape stilbene (resveratrol) synthase	Reduced lesion development by *Phytophthora infestans*; no effect on *Altenaria solani* and *Botrytis cinerea*	Thomzik et al., 1997
Bean galactorunase inhibiting protein	No effect on disease due to *Fusarium oxysporum*, *Botrytis cinerea* and *Altenaria solani*	Desiderio et al., 1997
Pear polygalactorunase inhibiting protein	Reduced rate of development of *Botrytis cinerea*	Powell et al., 2000
Collybia velutipes oxalate decarboxylase	Enhanced resistance to *Sclerotinia sclerotiorum*	Kesarwani et al., 2000

contd...

Table 2.4 contd...

Tobacco anionic peroxidase	No effect on disease due to *Fusarium oxysporum* and *Verticillium dahliae*	Lagrimini et al., 1993
Enterobacter ACC deaminase	Reduced symptom development due to *Verticillium dahliae*	Robinson et al., 2001
Tobacco chitinase + β-1,3- glucanase	Reduced disease severity due to *Fusarium oxysporum* f. sp. *lycopersici*	Van den Elzen et al., 1993; Jongdijk et al., 1995
Antisense polygalacturonase gene	Prolonged shelf life	Redenbaugh et al., 1995
Suppression of ACC synthase + ethylene forming enzyme	Ethylene evolution reduction	Hamilton et al., 1990, 1995
Tocacco N-gene	Resistance against TMV virus	Whitham et al., 1996
Bt -toxin	Pest resistance	Kuiper and Noteborn, 1996; Fischhoff et al., 1987
Cloned *Pto* gene	High resistance to *P. syringae* pv. *tomato*	Martin et al., 1993
PR-5 gene	Resistance to *Botrytis cinerea, Oidium lycopersicum, Leveillula taurica,* and *Phytophthora infestans*	Veronese et al., 1999; Vidhyasekaran, 2004

Table 2.5. Expressed gene product, and its effect on disease/composition of *Brassica napus* L. and *B. juncea* L

Expressed Gene Product	Effect on Disease/Composition	Reference
Trichoderma endochitinase	Reduced lesion due to *Altenaria brassicae*	Mora and Earle, 2001
Pea chitinase *PR 10.1* gene	No effect on *Leptosphaeria maculans*	Wang et al., 1999
Pea defence response gene (defensin)	Reduced infection and development of *Leptosphaeria maculans*	Wang et al., 1999
Tomato chitinase	Lower number of diseased plants due to *Cylindrosporium concentricum* and *Sclerotininia sclerotiorum*	Grison et al., 1996
Hevia chitin binding lectin (hevein) (*B. juncea* L.)	Smaller lesion and reduced rate of development due to *Altenaria brassicae*	Kanrar et al., 2002
Macadamia antimicrobial peptide	Reduced lesion size due to *Leptosphaeria maculans*	Kazan et al., 2002
Bean chitinase	Reduced rate and total seedling mortality due to *Rhizoctonia solani*	Broglie et al., 1991
Barley oxidase oxalate	Not tested	Thompson et al., 1995
Tomato *Cf9* gene	Delayed disease development due to *Leptosphaeria maculans* Resistance against *Cladosporium fulvum*	Hennin et al., 2001 Punja, 2004
γ-tocopherol methyl transferase (*Arabidopsis*)	Enhanced vitamin E content	Shintani and de la Penna, 1998
Lauroyl ACP thioesterase bay tree	Enhanced lauric acid content	del Vecchio, 1996
δ-6 and δ-12 desaturases	Increased γ-linoleic acid content	Liu et al., 2002

contd...

Table 2.5 contd...

δ-6 desaturase gene (Mortierella)	Increased ω-3-Fatty acid content	Ursin, 2000; James, 2003
Phytoensynthase (Daffodil)	Increased β-carotene content	Ye et al., 2000
Ch Fat B2, a thioesterase DNA	Greater number of medium chain fatty acids	Dehesh et al., 1996
Shiva protein	Resistance to *Xanthomonas campestris* pv. *campestris*	Braun et al., 2000
Agrobacterium	Resistance to *Xanthomonas campestris* pv. *campestris*	Braun et al., 2000
cryIA gene from *Bacillus thuringiensis*	Resistance to *Helicoverpa zea* Boddie and *Spodoptera exigua* (Hubner)	Stewart et al., 1996

Table 2.6. Expressed gene product, and its effect on disease/composition of soybeans

Expressed Gene Product	Effect on Disease/ Composition	Reference
Synthetic proteins	Improved amino acid composition	Rapp, 2002
Over-expressing the maize 15kDa zein protein	Increased sulphur amino acids	Dinkins et al., 2001
Δ-12 desaturase	Higher oleic acid content	Kinney and Knowlton, 1998
Ribozyme termination of RNA transcripts down regulate seed fatty acid	Higher oleic acid content	Buhr et al., 2002
Gene silencing of cysteine protease P34 (34kDa)	Lower immunodominant allergen content	Herman, 2002
Isoflavone synthase	Greater isoflavones content	Jung et al., 2000
Wheat oxalate oxidase (germin)	Reduced lesion length and disease progression due to *Sclerotina sclerotiorum*	Donaldson et al., 2001
CP4 EPSPS gene	Occasional reactivity with sera from Brazil nut-allergic patients	Teshima et al., 2000 Tutel'yan et al., 1999

Table 2.7. Expressed gene product, and its effect on disease/composition of carrot

Expressed Gene Product	Effect on Disease/ Composition	Reference
Tobacco chitinase	Reduced rate and final disease incidence due to *Botrytis cinerea, Rhizoctonia solani* and *Sclerotina rolfsii.* No effect on *Thielaviopsis basicola* and *Altenaria radicina*	Punja and Raharjo, 1996
Rice thaumatin like protein	Reduced rate and final disease incidence due to *Botrytis cinerea* and *Sclerotina sclerotiorum*	Chen and Punja, 2002
Tobacco chitinase + β 1,3-glucanase osmotin	Enhanced resistance to *Altenaria dauci, A. radicina, Cercospora carotae* and *Erysiphe heraclei*	Melchers and Stuiver, 2000
Human lysozyme	Enhanced resistance to microorganisms	Takaichi and Oeda, 2000

DETECTION METHODS AND UNINTENDED EFFECTS OF GMOs

Detection Methods

The food industry has suffered for a long time from loss of income due to extensively occurring frauds. Since adulteration is always ahead of analysis and detection methods, there is an urgent need for inventing a faster, more accurate and lower cost technology for obtaining more reliable results. Biotechnological advances have paved the way for very effective and reliable authenticity testing. The latter can be focused either on variety and geographical origin determination, or traceability testing of a wide range of food products, produces of agricultural and animal origin and package counterfeiting (Table 2.8). Apart from the widely employed DNA methods, other instrumental methods include SNIF (Site Specific Natural Isotope Fractionation) – NMR (Nuclear Magnetic Resonance), the more updated technology micro-satellite marker, RFLP (Restricted Fragment Length Polymorphism), and SSCP (Single Strand Conformation Polymorphism). Fish and meat expensive species in conjunction with orange juice and olive oil attracted the researchers' interest because of their adulteration with other species of lower quality and price. Another important issue is the detection of geographical origin of agricultural produces of added value and special properties like rice, olive oil and wine. Post-commercialization tracking of GM crops requires three types of tests (Auer, 2003):

(1) A rapid detection assay to determine whether a GM crop is present in a sample of raw ingredients or food products.

(2) An identification assay to determine which GM crop is present.

(3) Quantitative methods to measure the amount of GM material in the sample.

The first stage can be accomplished by qualitative methods (presence or absence of transgene), whereas the third stage uses semi-quantitative (above or below a threshold level) or quantitative (weight/weight percentage or genome/genome ratio) methods. Currently, the two most important approaches are immunological assays using antibodies that bind to the novel proteins and PCR-based methods using primers that recognize DNA sequences unique to the GM crop. The two most common immunological assays are enzyme-linked immunosorbent assays (ELISA) and immunochromatographic assays (lateral flow strip tests). ELISA can produce qualitative, semi-quantitative and quantitative results in 1-4 hours of laboratory time (http://www.ers.usda.gov/publications/aib762 and http://www.anrcatalog.ucdavis.edu/pdf/8077.pdf).

Unintended Effects of GMOs

The occurrence of unintended effects (undesired effects, but it is a well accepted terminology) is not specific to genome modification through rDNA technology because unintended effects also occur frequently as a result of conventional breeding and during the use of chemical and irradiation mutagenesis to produce new phenotypes. Insertion in plant DNA must, therefore, be evaluated with an awareness of such natural DNA events, an issue that has been reviewed recently (Cellini et al., 2004).

Table 2.8. Most commonly employed methods for the testing of GMOs in foods

Requirement/ Methods	Chemical Fingerprinting			Protein Based				DNA-Based					
	SNIF-NMR	LC-NMR	NIR	Affinity gel electrophoresis	Western Blot	ELISA	Lateral flow strip	Micro-arrays	Southern Blot	Qualitative PCR	QC-PCR	Real time PCR	RAP-DNA
Ease of use	Difficult	Difficult	Easy	Medium	Difficult	Moderate	Simple	Moderate	Difficult	Difficult	Difficult	Difficult	Difficult
Requires special equipment	Yes	Yes	Yes	Yes	Yes	Yes	No	Yes	Yes	Yes	Yes	Yes	Yes
Equipment cost	High	High	Low	Low	High	Medium	Low	High	High	Medium	Medium	Medium	Medium
Sensitivity	High	High	Medium	High	Medium	High	High	High	Medium	High	High	High	High
Duration	Depends on isotope measured	Depends on isotope measured	Short	Short	Medium	Short	Short	Short	Medium	Long	Medium	Medium	Medium
Measurement cost	Medium	Medium	Low	Medium	High	Low	Low	Low	High	High	High	High	High
Quantification	Yes	Yes	No	No	No	Yes	No	Yes	No	No	Yes	Yes	Yes
Suitability for field testing	No	No	No	No	No	Yes	Yes	No	No	No	No	No	No
Employed by	Universities, Research Centres	Universities, Research Centres	Producers	Universities, Research Centres	Universities, Research Centres	Research Centres, Producers	Research Centres, Producers	Universities, Research Centres	Universities, Research Centres	Research Centres, Producers	Research Centres, Producers	Research Centres, Producers	Universities, Research Centres

Source: Ahmed 2002a, b; Noteborn et al., 1995; Andrews et al., 2000; Sachar-Hill, 2002; Wu et al., 2001; Charlton, 2001;

SNIF-NMR: Site specific natural isotope fractionation –Nuclear magnetic resonance, LC- NMR: Liquid chromatography, –Nuclear magnetic resonance

NIR: near infra red, ELISA: enzyme – Linked immunosorbent assay, PCR: Polymerase chain reaction, QC-PCR: Quantitative competitive – polymerase chain reaction, RAPD: Random amplified polymorphic DNA

Some representative examples of unintended effects that occurred either during conventional breeding or genetic engineering of food crops are summarized in Table 2.9. Parallels exist between natural recombination and DNA insertion in GM crops. The molecular changes that occur during conventional breeding may nevertheless be less easy to trace than changes caused by insertion, rearrangement, or mutagenesis of exogenous, distinguishable DNA originating from the same or other species (ILSI, 2004). Natural chromosomal recombination mechanisms play an important role in plant breeding, and mechanisms can be grouped into: (1) homologous recombination, and (2) illegitimate recombination, a non-homologous end-joining process. Both processes are characterized by double-strand break repair mechanisms. Non-homologous recombination is the predominant form in plants (Siebert and Puchta, 2002). Although recombination events could theoretically occur at random along the length of the chromosomes, the presence of preferential sites for recombination breakpoints is well known (Schnable et al., 1998).

Upon random insertion of specific DNA sequences into the plant genome (intended effect), the disruption, modification or silencing of active genes or the activation of silent genes may occur, which may result in the formation of either new metabolites or altered levels of existing metabolites. Unintended effects may be partly predictable on the basis of

Table 2.9. Unintended effects in traditional breeding and genetic engineering breeding (Kuiper et al., 2001; ILSI, 2004)

	Host Plant	Unintended Effect	Reference
Traditional Breeding	Celery	High furanocoumarins content	Beier, 1990
	Potato	Low yield, high glycoalkaloid content	Harvey et al., 1985
	Squash, zucchini	High curcubitac in content	Coulston and Kolbye, 1990
Genetic Engineering Breeding	Potato	Reduced glycoalkaloid content (–37 to +48%)	Engel et al., 1998
	Potato	Increased glycoalkaloid content (+16 to 88%)	Hashimoto et al., 1999a, b
	Potato	Adverse tuber tissue perturbations	Turk and Smeekens, 1999
		Impaired carbohydrate transport in the phloem	Dueck et al., 1998
	Soybean	Higher lignin content at normal soil temperatures (20°C)	Gertz et al., 1999
	Canola	·Multiple metabolic changes (tocopherol, chlorophyll, fatty acids, phytoene)	Shewmaker et al., 1999

ILSI: International Life Sciences Institute

knowledge of the place of the transgenic DNA insertion, the function of the inserted trait, or its involvement in metabolic pathways; while other effects are unpredictable due to the limited knowledge of gene regulation and gene-gene interactions. It should be emphasized that the occurrence of unintended effects is not specific for genome modification through rDNA technology, but it occurs frequently in conventional breeding programmes. Unintended effects may be identified by an analysis of the agronomical/morphological characteristics of the new plant and an extensive chemical analysis of key nutrients, anti-nutrients and toxicants typical for the plant. Limitations of this analytical, comparative approach are the possible occurrence of unknown toxicants and anti-nutrients, in particular in food plant species with no history of (safe) use, and the availability of adequate detection methods (Kuiper et al., 2001).

TOXICITY STUDIES PERFORMED WITH GM FOOD CROPS

The selection of animal studies is based on considerations including a molecule's structure, function and in vitro toxicity results, as well as ethical criteria. Of these three distinct approaches, evidence from animal tests is usually most indicative of potential toxic effects of a test substance in humans. However, the extrapolation of results from animal tests to humans is uncertain, unpredictable differences can include inter-species and inter-individual differences in metabolism, physiological processes, and lifestyle. These uncertainties are usually addressed through the use of uncertainty factors (Konig et al., 2004).

The difficulties encountered in assessing the safety of foods derived from GM crops in bioassays such as animal tests are well recognized [OECD (Organization for Economic Cooperation and Development)], 1993, 1998, 2000 and 2002; LSRO (Limited Standard Reactor Operator), 1998; FAO/WHO, 2000). It has been pointed out on numerous occasions that animal feeding studies with whole foods or feeds must be designed and conducted with great care to avoid problems encountered with nutritional imbalance from overfeeding a single whole food, which itself can lead to adverse effects. In undertaking such tests, a balance must be struck between feeding enough of the test material to have the possibility of detecting a true adverse effect and, on the other hand, not inducing nutritional imbalance. In any event, what one would like to extrapolate by means of animal testing, for anticipated human intake is simply not achievable for practical reasons, and enhanced margins of safety, of one to three times higher have to be accepted (WHO, 1987; Hattan, 1996; Munro et al., 1996a, b). This limits the sensitivity of animal bioassays to detect small differences in composition, which may be more readily detected with thorough analytical characterization (Novak and Halsberger, 2000).

According to a review by Munro et al., (1996a, b) based on 120 rat bioassays (each of 90 days duration) of chemicals of diverse structure including food additives, pesticides, and industrial chemicals, it was found that the lowest observed adverse effect levels (LOAEL) ranged from 0.2 to 5,000 mg kg^{-1} body weight with a median of 100 mg kg^{-1} and a 5th percentile of 2 mg kg^{-1}. To achieve the 5th percentile of exposure from a toxic constituent present in, say, a food crop in a rodent bioassay (at a food incorporation rate of 30%) the toxin would have to be present at a level of 80 μg g^{-1}. To achieve the median exposure of 100 mg kg^{-1} it would have to be present at 5,000 μg g^{-1}. These concentrations fall well within the range of existing analytical techniques for detection of inherent toxicants in food. The concentrations should also be readily detected during compositional analysis of the known toxicants in the host organism used to generate the improved nutrition crop. The broiler chicken has emerged as a useful animal model for assessing nutritional value of foods and feeds derived from GM crops. It should be noted, however, that, contrary to laboratory rodents, the rapidly growing broiler has been obtained through breeding efforts with the aim to create an efficient food-producing animal. This may, therefore, not render it optimal for toxicological testing of foods and feeds. In fact, disorders such as "sudden death syndrome" and "ascites," (an abnormal accumulation of serous fluid in the abdominal cavity) are considered related to metabolic disorders associated with its rapid growth (Olkowski and Classen, 1995). On the other hand, broiler chickens have been optimized for growth relative to highly characterized diets such that small changes in nutrients or antinutrients in the diet are readily manifested in reduced growth. In addition, one of the first indications of an ill animal is loss of appetite or reduced growth rate (ILSI, 2004). Table 2.10 summarizes some representative toxicity studies carried out on rats and mice (unless indicated) with GM vegetables.

CONSUMERS' BELIEFS TOWARDS GMO – DIFFERENCES AMONG OUNTRIES

Genetic engineering (GE) has been used for many years in the production of consumer goods, such as pharmaceuticals and detergents. While consumers do not pay particular attention to these applications, the implementation of GE in food production is met with intense consumer mistrust. Nowadays, biotechnology is developing in an environment where public concerns about food safety and environmental protection are steadily increasing. The scientific claims related to biotechnological advantages in terms of social benefits are not accepted without criticism. The ethical issues arising from the foreseeable changes due to the applications of biotechnology in daily life can result in reluctance to adopt

Table 2.10. Representative toxicity studies carried out on rats and mice (unless indicated) with GM vegetables

Vegetable	Trait	Duration	Parameters	Reference
Potato	Lectin (*Galanthus nivalis*)	10 days	Histopathology of intestines	Ewen and Pusztai, 1999
Potato	Cry1 endotoxin (*B. thuringiensis* var *kurstaki* HD1)	14 days	Histopathology of intestines	Fares and El Sayed, 1998
Potato	Glycinin (soybean)	28 days	• Feed consumption • Body weight • Blood chemistry • Blood count • Organ weights • Liver and kidney histopathology	Hashimoto et al., 1999a, b
Soybean GTS 40-3-2	CP4 EP SPS (*Agrobacterium*)	105 days	• Feed consumption • Body weight • Intestines histopathology • Serum IgE and IgE levels	Teshima et al., 2000, 2002
Soybean GTS 40-3-2	CP4 EP SPS (*Agrobacterium*)	150 days	• Blood chemistry • Urine composition • Hepatic enzyme activities	Tutel'yan et al., 1999
Soybean GTS 40-3-2	2S albumin (Brazil nut)		Reactivity with sera from Brazil nut-allergic patients	Nordlee et al., 1996
Soybean*[1] GTS 40-3-2	Herbicide resistant	29 days	• Body weight • Milk production • Milk composition • Dry matter digestibility • Ruminal fluid composition	Hammond et al., 1996
Soybean*[2] GTS 40-3-2	Herbicide resistant	42 days	• Weight increase • Feed consumption • Breast muscle weight • Fat pads weight • Mortality	Hammond et al., 1996

contd...

Table 2.10 contd...

Soybean*[3] GTS 40-3-2	Herbicide resistant	70 days	• Weight increase • Feed consumption • Filet composition	Hammond et al., 1996
Tomato	Cry1 Ab endotoxin (*B. thuringiensis* var *kurstaki*)	91 days	• Feed consumption • Body weight • Blood chemistry • Organ weights • Histopathology	Noteborn et al., 1995
Tomato	Antisense polygalacturonase (tomato)	28 days	• Feed consumption • Body weight • Blood chemistry • Organ weights • Histopathology	Hattan, 1996

*[1] animal: lactating cow, *[2] animal: broiler chicken, *[3] animal: channel catfish

the new technology. Questions may arise from hazards that these changes result in (Kerr, 1999), although these concerns may have less impact than the expected benefits.

The attempts to introduce GM food into the market must be supported by the scrupulous analysis of consumer evaluation procedures for GM products. The realization of the variations in consumer beliefs with respect to biotechnology is indispensable for decision-makers to foresee potential problems of biotechnology acceptance and to take consumer concerns into consideration in the development of novel products. Consumers will be the ultimate judges of success of the upcoming new foods produced from agricultural biotechnology (Saba et al., 1998), while they increasingly question whether any further technological treatment is beneficial or necessary, especially when it is considered to have a serious impact on food safety and environmental protection (Hoban, 1996a; Saba et al., 1998, 2000; Saba and Vassalo, 2002).

Attitudes towards GE crops were found to vary considerably among countries. American and Canadian consumers seem to be much more favourably disposed towards such products than their counterparts in Europe, and Japanese are somewhere in the middle (Gaskell et al., 1998) (Table 2.11). It is rather surprising that the already existing negative attitude of European consumers towards GM foods was further reinforced from 1996 to 1999 (Gaskell et al., 2000) as opposed to Canadian consumers who are very much interested in obtaining more information on GM foods (Magnusson and Hursti, 2002). However, even within the European Union, Scandinavian consumers and, in particular, Swedish, appear to be the most sceptical and negatively inclined towards GM foods. Between the two sexes, women were shown to be even less willing to try these new GM products foods (Magnusson and Hursti, 2002).

CONSUMER SAFETY – POST MARKETING MONITORING

Consumer Safety

There has been a very heated debate over consumer safety issues vis-a-vis the consumption of GM foods. This controversy was primarily fuelled by the differences in terms of political orientation (see, legislation differences between EU and US) in conjunction with several incidents like the potato incident where a lectin called concanavalin A (known toxin) was identified in the GM potato. Another issue stands for the presence of a strong allergen in soybean. In this case Brazil nut genes were transferred to soybeans (Leighton, 1999; Arvanitoyannis and Krystallis, 2005). Every side (EU and USA) adopted the respective pros and cons (social, environmental, or economic nature) (Table 2.12) in order to justify their

Table 2.11. Comparison of values relevant to GE crops and foods among EU, Japan, Canada, and the USA

Values	EU	Japan	Canada	USA
Importance of food safety	Highly important but occurrence of diseases and contamination undermined the public trust	Highly important and public supports regulatory agencies' actions	Highly important and public encourages regulatory agencies' actions	Highly important and public favours regulatory agencies' actions
Environmental consciousness	Very strong	Very strong	Strong	Moderate
Approach to science and technology	Cautious	Innovative	Positive	Enthusiastic
Attitude towards risk taking	Medium	Medium	Strong	Very strong
Attitude towards food supply and trade	Strong but heavily opposed by environmental awareness	Strong and linked with environmental awareness	Strong but mitigated by environmental awareness	Strong

Table 2.12. Potential economic and social pros and cons of genetic engineered (GE)/modified (GM) crops and foods (Mepham, 2000; Engel et al., 2002; The Royal Society, 2002, 2003; Weick and Walchli, 2002)

Factor	Pros	Cons
Economic	• Cost savings and higher crop yields • Reduced need for spraying with pesticides/herbicides • Drought resistance • Precision agriculture • Plants resistant to high mineral content and salinity	• Initial high cost of GE grain (e.g. Flavr Savr tomato) • HRCs may be transformed into weeds acting as a source of pests and diseases undermining the principles of crop rotation • Transfer of genes from GM to non-GM crops may have undesirable effects upon the latter (e.g. development of *Bt* resistance through use in GM organisms might undermine organic farming)
Social	• Reduced use of chemical pesticides used in grain production will reduce the environmental pollution • Enhanced food production may help avert world food shortages • Food products endowed with enhanced cooking properties and medicinal value will be available	• Long term environmental risks are unknown • GE crops are perceived as unnatural and unethical • Private organizations may place profit before safety and world needs • Risk of transferring antibiotic resistance from crop genes to humans will eventually compromise the effective treatment of patients • Consumers can hardly understand the labelling wherever the latter is available • Production cost savings are not really passed on to the consumers • Effective destruction of weeds may reduce the availability of habitats for various insects and invertebrates

HRC: Herbicide Resistant Crop

claims and adherence to that particular side. If substantial equivalence can be established except for a single or few specific traits of the genetically modified plant, further assessment focuses on the newly introduced trait itself. Demonstration of the lack of amino acid sequence homology to known protein toxins/allergens, and a rapid proteolytic degradation under simulated mammalian digestion conditions, was deemed to be sufficient to assume the safety of the new protein (FAO/WHO, 1996). However, according to Kuiper et al., (2001) there may be circumstances that require more extensive testing of the new protein, such as: (i) the specificity and biological function/mode of action of the protein is partly known or unknown; (ii) the protein is implicated in mammalian toxicity; (iii) human and animal exposure to the protein is not documented; or (iv) modification of the primary structure of naturally occurring forms. Bacterial *Bt* proteins are examples of proteins that have been introduced into crop varieties by genetic modification. *Bt* proteins (Cry proteins) from *B. thuringiensis* strains have been introduced into genetically modified crop plants for their insecticidal properties in the larvae of target herbivoral insect species (Peferoen, 1997).

Post-marketing Monitoring

Post-market monitoring systems have been established by several food companies for certain food products to act as early warning systems and to ensure effective product recall in the event where health concerns might be associated with a specific food. The organization of post-market monitoring is primarily the responsibility of the manufacturer of the food. Methods vary from establishing channels of communication in the firm to receive direct consumer feedback on the product, to the repurchase of products to determine the quality of the product on the supermarket shelf. Post-market monitoring programmes may serve to confirm the absence of specific adverse health effects of certain products after they have been marketed. The feasibility and validity of post-market monitoring depends on the health end point of interest and on the way the product is marketed (Konig et al., 2004).

The potential allergenicity of foods derived from biotechnology has been recognized as relevant for particular consideration because some of the modifications in these foods may include the expression of proteins not otherwise present in that food (Taylor and Hefle, 2002). The presence of proteins per se does not merit attention as a concern regarding the safety of foods derived from biotechnology. However, the current definition of food allergens (for example, proteins eliciting an IgE-mediated hypersensitivity response in sensitive individuals following consumption) dictates that an assessment for allergenic potential is in order when the biotechnologically derived food or food ingredient contains a novel protein that would otherwise not be present in food (ILSI, 1997, 2003a, b and 2004). A pre-market decision tree strategy (a number of questions organized in a tree diagram in order to facilitate decision taking for market issues) for assessing the allergenicity of GM foods has been recommended by FAO and WHO (FAO/WHO, 2000, 2001) and has proceeded to Step 8 of the Codex Alimentarius Commission's assessment of the safety of foods derived through biotechnology (Codex, 2002). According to ILSI (2004) the methodological considerations to be taken into account for effective post-market monitoring are as follows:

(1) Measuring population exposure to foods and food ingredients.
(2) Tracking the disappearance of biotechnology-derived foods into the food supply.
(2) Integrating disappearance data with food consumption Databases.
(3) Demonstration of causality.

LEGISLATION FOR GM – LABELLING – BIOETHICS

Legislation

The introduction of rDNA technology of new genes into major crops consumed by animals has raised serious questions about the safety of novel feeds. The European Council Directive 2001/18/EC requires an assessment of risks for human, animals and the environment before viable seeds can be imported or the plant itself can be cultivated in Europe. Furthermore, the Novel Food and the Novel Food Ingredient Regulation (EC, 2003 http://europa.eu.int/comm/food/fs/sc/ssc/out327en.pdf) covers the use of non-viable products of any GM plant intended for food purposes. In practice, compositional analysis of key nutrients and key toxicants used to compare a GM plant with its conventional counterpart is the major source of data used to establish substantial equivalence (MacMahon in http://www.yale.edu/ynhti/curriculum/units/2000/7/00.07.02.x).

The current US process for assessing feed and food safety of crops produced using technology comprises the meticulous coordination of the USDA (United States Department of Agriculture), FDA (Federal Department of Agriculture), and EPA (Environmental Protection Agency) (for crops endowed with pesticidal properties). Reviews carried out by FDA for food safety have adopted the decision tree approach (several questions to be answered with a yes/no answer in order to make up a decision) for assessing host plants, gene donors, proteins introduced by donors, new or modified fats or oils, and carbohydrates (Faust, 2002). It is encouraging that various independent studies converge to the fact that there is no detrimental effect as such for livestock fed biotechnology-derived crops and no differences of livestock fed with GM plants and conventional crops (Aumaitre et al., 2002).

The UN Cartagena Protocol on Biosafety adopted in Montreal (January 2000) has already started to affect the international trade in GMOs and their products. Article 4 makes it clear that this protocol shall apply to the trans-boundary movement, transit, handling and use of all Living Modified Organisms (LMOs) that may have an adverse effect on the conservation and sustainable use of biodiversity, taking also into account risks to human health (Gupta, 2000; Xue and Tisdell, 2000, 2002). LMOs are all living organisms that possess a novel combination of genetic material obtained through the use of modern biotechnology. According to the Protocol, trade in GMOs requires approval by a specified competent authority in the importing country. The international movement of GMOs will be reviewed more strictly not only from the angle of agricultural production, but mainly from the viewpoint of environmental protection, human health, and other aspects. Increased cost will be involved in the export and import of GMOs. The risk assessment steps mainly include identification and documentation of any novel genotypic and phenotypic characteristics associated with LMOs that may adversely affect biodiversity and human health; direct use as food or feed and introduction into the environment (Xue and Tisdell, 2002).

Recently, Zedalis (2002) published a review referring to GMO food measures as "Restrictions" under General Agreement on Tariffs and Trade (GATT) Article XI. GATT Article XI (1) condemns the use of prohibitions on imports and, thus, may attack GMO measures to flatly prevent importation. In case GMO is to be released to the environment, the following information is required (EEC, 2001; Directive EU 2001/18/EC):

(1) Purpose of the release.
(2) Foreseen dates and duration of the release.
(3) Method by which the genetically modified plants will be released.
(4) Method for preparing and managing the release site, prior to, during and post-release, including cultivation practices and harvesting methods.
(5) Approximate number of plants or plants per square metre.

Another EU regulation (COM/2002/0085 – COD 2002/0046) aims at establishing a common system of notification and information for exports to third ccountries of GMOs in order to contribute towards ensuring an adequate level of protection in the field of the safe transfer, handling and use of GMOs that may have adverse effects on the conservation and the sustainable use of biological diversity, also taking into account risks to human health. A synoptical presentation of the most important currently valid laws (in conjunction with the ones not valid any more) with regard to food safety, GMOs, and their labelling is given in Table 2.13.

Labelling

Recent times have seen the emergence of a new concept for food labelling: the consumers' right to information, allowing "informed choice" in full knowledge of the facts. This right has taken many forms: real or "perceived" safety information on ingredients and additives, philosophical or ethical concerns (mode of production, absence/presence of given ingredients, including genetically modified foods), nutrition information, and declaration of potential allergens (Hollingsworth et al., 2003; Cheftel, 2005). Further information, such as nutrition and health claims (with relevance to obesity and the risk of various diseases), is on its way. The recent occurrence of several food crises has emphasized food safety and protection of consumers' health as main objectives for the food legislation. One of the most general rules of the European (and other) legislation can be stated as "no misleading the consumer" (the protection of consumers' interests is one of the principles of food law, as reiterated in Regulation 2002/178/EC). This applies to information concerning the characteristics of foods (nature, identity, properties, composition, quantity, storage life, origin, and method of production or manufacture). The label should not attribute to the food effects or properties which it does not possess, nor suggest that the food possesses special characteristics that are possessed by other similar foodstuffs (Cheftel, 2005).

Labelling emerged as a major issue in the EU – US GMO confrontation. According to US authorities (FDA) and as per several reports published by FAO/WHO, ILSI, EU, and NZRC (Table 2.14), GM foods could be consumed provided the principle of substantial equivalence is proved. However, in EU, a recent series of directives and regulations aims at reassuring the consumer that the GMO consumption will continue to remain his choice thanks to the strict traceability and labelling system [Regulation (EC) No.641/2004, Directive EU 2004/657/EC, Directive EU 2004/204/EC].

The proposal for a Regulation COM/2002/0085 – COD 2002/0046 aimed at controlling the transboundary movement of GMOs, in particular, to third world countries. Regulation (EC) No.1829/2003 laid down the provisions for labelling food and feed. In fact this regulation applies to:

(1) GMOs for food use.

(2) Food containing or consisting of GMOs.

(3) Food produced from or containing ingredients produced from GMOs.

Table 2.13. EU Legislation for GMOs

Directive – Title	Main Points	Comments
E.U. 2001/18/EC (entry into force 17/4/2001) Deliberate release into the environment of GMOs Proposal for a regulation COM/2002/0085 – COD 2002/0046 (entry into force 27/10/2002) The transboundary movement of GMOs	• Authorize the release and disposal of GMOs in market • Application of authorizations for 10 years with potential of renewal • Institution of obligatory controls after the disposal of GMOs in the market • Common method of Risk Assessment and amendment, suspension or interval of GMOs' release in case of danger • Obligatory consultations with the public and labelling of GMOs • Publication of summary of measures every three years with regard to the directive from the Member states • Imposition of sanctions for delinquencies of national provisions • Establishment of a notifying system and exchanging information on the exports of GMO to Third Countries • Environmental and healthy inspection (negative repercussions in the biodiversity and human health) • Not application for pharmaceuticals for human use • Notification in the contributing and not contributing parts of import • Information on the contributing part of export • Notification in the Biosafety Control Organization • Measures for not transboundary movement of GMOs • Surveillance, submission of reports and imposition of sanctions on any infringement of the regulation	Repeal • Directive EU 90/220/EEC since 17/10/2002 Amendment • Regulation (EC) No.1830/2003 (entry into force 7/11/2003)
Regulation (EC) No.1829/2003 (entry into force 7/11/2003) GM food and feed	• Measures for human and animal health protection, Community procedures of approval, inspection and labelling of GM food and feed • Definitions (food, feed, GMO, GM food etc.) • Amendment, interval and repeal of the approvals	Replacement • Regulation (EC) No.1139/98 • Regulation (EC) No.49/2000 • Regulation (EC) No.50/2000

contd...

Table 2.13 contd...

	• Application for approvals for 10 years and potentiality of renewal • Measures for products' use not only as food but also as feed
Regulation (EC) No.1830/2003 (entry into force 7/11/2003). Traceability and labelling of GMOs and traceability of food and feed products produced from GMOs	• Framework of traceability of products consisting of, or containing GMOs and foodstuffs, feed produced from GMOs • Application for all stages of disposal in the market • Definitions (GMO, traceability etc.) • Demands on labelling of products constituted of, or containing GMOs • Inspection and control measures and sanctions in case of infringement
Regulation (EC) No.65/2004 (entry into force on the date of its publication in the *Official Journal of the European Union*) Establishing a system for the development and assignment of unique identifiers for GMOs	• Not application for pharmaceuticals intended for human and veterinary use • Existence of a unique identifier for each GMO which is placed in the market
E.U. 2004/204/EC (entry into force 23/3/2004). Arrangements for the operation of the registers for recording information on genetic modifications in GMOs	• Register lists of information of genetic modification in GMOs • Detailed report of documents contained in the lists • Availability of the lists to the public
Regulation (EC) No.641/2004 (entry into force 18/4/2004)	• Transformation of applications and statements in the applications • General requirements of import in the market of certain products

contd...

Table 2.13 contd...

The authorization of new GM food and feed, the notification of existing products and adventitious or technically unavoidable presence of GM material which has benefited from a favourable risk evaluation	• Transitional measures for adventitious or technically unavoidable presence of GM material which has benefited from a favourable risk evaluation	
EU 2004/657/EC Placing on the market of sweet corn from GM maize line *Bt*11 as a novel food or novel food ingredient	• Product as safe as conventional • Obligatory labelling as "GM sweet corn" • No limitations or provisions for placing in the market • No more controls after placing in the market • Specific notification in food and food ingredients derived from genetic modification. • Obligatory recording of the code SYN-BT11-1 (unique)	Replacement • Directive EU 90/219/EC and 90/220/EC.

Table 2.14. Recent milestones in the international consensus on the safety assessment of biotechnology-derived foods (EU 2004/643 EC; EU Regulation 1139/98/EC)

Year	Organization	Item Derived	Reference
2000	FAO/WHO	Expert consultation on safety assessment in general, including the principle of substantial equivalence	FAO/WHO, 2000
2001	ILSI	Europe concise monograph series of genetic modification technology and food consumer health and safety	Robinson, 2001
2001	EU	EU-sponsored Research on Safety of Genetically Modified Organisms. "GMO research in perspective." Report of a workshop held by External Advisory Groups on the "Quality of Life and Management of Living Resources Program"	EU, 2001
2001	NZRC	Report on New Zealand Royal Commission on Genetic Modification	NZRC, 2001
2000-2003	FAO/WHO	Guidelines for Codex Alimentarius Committee, developed by Task Force for Foods Derived from Biotechnology. Codex Ad Hoc Intergovernmental Task Force on Foods Derived from Biotechnology, Food and Agriculture Organization of the United Nations, Rome, Italy	FAO/WHO, 2002, 2003
2003	ILSI	Crop composition database (http://www.cropcomposition.org)	ILSI, 2003b
2003	EU	GM food and feed (monitoring traceability and authenticity)	Regulation (EC) No.1829/2003
2004	EU	Arrangements for the operation of the registers for recording information on genetic modifications in GMOs	Directive EU 2004/204/ EC
2004	EU	The authorization of new GM food and feed, the notification of existing products and adventitious or technically unavoidable presence of GM material which has benefited from a favourable risk evaluation	EU Regulation (EC) No.641/2004

Regulation (EC) No.1830/2003 is targeted at providing traceability and labelling of GMOs and traceability of food and feed products produced from GMOs. This Regulation applies to the following, at all stages of placing the product in the market:

(1) Products consisting of/or containing GMOs placed in the market in accordance with community legislation.

(2) Food produced from GMOs, placed in the market in accordance with community legislation and feed produced from GMOs placed in the market in accordance with Community legislation.

To be more specific, in terms of traceability, at the first stage of placing a product (consisting of/or containing GMOs) in the market, including bulk quantities, operators shall ensure that the following information is transmitted in writing to the operator receiving the product:

(1) That it contains or consists of GMOs.

(2) The unique identifier assigned to those GMOs is in accordance with the regulation.

As for labelling of the products consisting of/or containing GMOs, operators shall ensure that:

(1) For pre-packaged products consisting of/or containing GMOs, the words "This product contains GMOs" or "This product contains GM... (name of organism)" appear on the label.

(2) For non-pre-packaged products offered to the final consumer the words "This product contains GMOs" or "This product contains GM... (name of organism)" shall appear on, or in connection with the display of the product.

Finally, a recent EU directive No.65/2004 aimed at establishing a system for the development and assignment of unique identifiers for GMOs. According to this directive, where consent or authorization is granted for the placing a GMO product in the market:

(1) The consent or authorization shall specify the unique identifier for that GMO.

(2) The Commission, on behalf of the Community, or, where appropriate, the competent authority that has taken the final decision on the original application shall ensure that the unique identifier for that GMO is communicated as soon as possible, in writing, to the biosafety clearing house.

(3) The unique identifier for each GMO concerned shall be recorded in the relevant registers of the commission.

Directive E.U. 2004/204/EC specifies the required arrangements for the operation of the registers for recording information on genetic modifications in GMOs. Furthermore, the registers shall be available to the public. The information recorded shall be divided to a set of data accessible to the public and to a set of data comprising additional confidential data, accessible only to the Member States, the Commission and the EFSA (European Food Safety Authority). Another more recent Regulation (EC) No.641/2004 covers the topic of the authorization of new GM food and feed, the notification of existing products and adventitious or technically unavoidable presence of GM material, which has benefited from a favourable risk evaluation. The application shall include the following:

- The monitoring plan.
- A proposal for labelling, in compliance with the requirements of the regulation.
- A proposal for a unique identifier for the GMO in accordance with Commission Regulation.
- A proposal for labelling in all official Community languages.
- A description of methods of detection, sampling and specific identification.
- A proposal for post-market monitoring regarding the use of the food for human consumption or the feed for animal consumption is not necessary.

And last but not least, is the Directive EU 2004/657/EC for placing in the market of sweet corn from GM maize line *Bt*11 as a novel food or novel food ingredient. Information to be entered in the Community Register of GM food and feed should comprise: (1) authorization holder (name, address, company), (2) designation and specification of the product (sweet maize, fresh or canned, that is progeny from traditionally crosses of traditionally bred maize with genetically modified maize line *Bt*11), and (3) wording of the label ("GM sweet corn"). The milestones of the recent international consensus on the safety awareness of biotechnology derived foods are summarized in Table 2.14.

Bioethics

Ethics can be defined as the branch of philosophy concerned with how one may decide what is morally right or wrong. It is a specific discipline which attempts to analyze the concepts and principles used to justify our moral choices and actions in particular situations. Ethics is a specific discipline trying to probe the reasoning behind our moral life. The first question one is asked is whether biotechnology actually raises new ethical questions in food production. One has to take into account that ethical implications can be divided in the so-called pre-farm and post-farm gate. The former comprises environmental ethics (principle of non-interference with nature), biodiversity threat (very few plant varieties will continue to grow), sustainability (spread of weedy plants brought from overseas), animal rights (young animals suffering from viruses or malformations) and socio-economic impacts (threat to small farm survival and impact on developing countries economy) (Thompson, 1997). The post-farm gate is focused on issues like principles relevant to the responsibility of the producer (provide consumer with correct information), consumers' rights (to be informed to take the risk to consume a novel food), right to be informed (proper labelling), right to choose (related to religion, morality, taste, opinion) (Straughan, 1998). The second bioethics issue of great importance refers to the environment. In fact, application of the precautionary principle (Rio Declaration) states "in order to protect the environment, the precautionary approach shall be widely applied by States according to their capabilities". Where there are threats of serious or irreversible damage, lack of full scientific certainty shall not be used as a reason for postponing cost-effective measures to prevent environmental degradation". Environment sustainability and biodiversity are the two major issues which are at stake with the introduction of GM products (Cockburn, 2002).

CONCLUSION

The use of GMOs raises several questions on the morality of the technology, the balance of risk in society, public involvement in decision making and the appropriateness of using patents in an area linked to life processes (Atkinson, 1998). Although the usage of GM plants appears to be a promising route out of the imminent nutritional crisis in view of the occurring world population explosion, especially in the third world countries, there seems to be no accurate experimental data on the long-term effects of genetic engineering. There are still several basic questions unanswered:

(1) Will people over a period of time develop allergic reactions to the transgenic proteins produced from genetic engineering?
(2) Will the horizontal movement of genetic materials have a negative impact on ecosystems?
(3) Will some virulent new pathogen develop from the transfer and transformation of microbial DNA made available by genetic engineering?
(4) Will humans be able to make the correct ethical choices so that all of humanity may share the potential benefits of genetic engineering?

Table 2.15. Anticipated objectives, strategies, and suggested methodologies for achieving genetic and environmental traceability

Field	Objectives	Strategies	Steps/Phases	Suggested Methodology	Possible Targets
Genetic traceability	- Implement analytical methods and strategies to assist in certification of origin, authenticity, composition, processes - Provision of technical solutions in support of Common Policies and regulatory Governmental bodies	- Develop diagnostics based on DNA, proteins, metabolites	- Genomics - Proteomics - Metabolomics - Molecular markers - Genetic diversity and genotyping - Influence of processing technology on isotopical values - Statistical evaluation - Establish comprehensive databases	- Access to products and raw materials "omics" technologies panels of probes and molecular descriptors - Genetic Identification (GI) - Qualitative and quantitative PCR	- Differentiation between GM and non-GM plant products - Species determination in mushroom products - Species determination in berrys and nut products - High value EU import food products from outside the Community - Absence of allergenic species - Ethnical-religious food

contd...

Table 2.15 contd...

Environmental traceability	- Assurance that chemicals, intentionally or accidentally present, do not pose unacceptable risks to human health and to the environment - Nutritional imbalance due to primary production	- Improved understanding of the type and nature of chemicals occurring in plants and their impact on public health - Dietary consideration of macro-, micro-nutrients - Risk analysis of chemicals in plants - Develop stable isotope analysis with focus on light stable isotopes ($^{13}C/^{12}C$; $^{15}N/^{14}N$; $^{18}O/^{16}O$)	- Systematic data collection - Evaluation of nutritional and anti-nutritional components - Evaluation of health promoting elements - Evaluation of toxic compounds - Determination and assessment of source of toxic elements contamination (e.g. soil, irrigation water, atmospheric pollution and packaging materials) in the food/plant chain	- Spectroscopic methods (ICP, AAS, ICP-MS) - Radioactivity determination, and isotope analysis - Separation techniques (GC-LC, GC-MS, GC-MS, LC-MS) - Mathematical models for simulation of pollutants behaviour - Molecular monitoring on/off-line - Biosensors - Life Cycle Analysis of agroalimentary processes - Monitoring inorganic nutrients content and correlation with genotype, culture techniques Evaluation of Pb, Cr, Ni, Mn, V, Cd, As, Hg contents in food and beverages by Atomic Spectroscopy Techniques	- Natural toxins - Mycotoxins, aflatoxin, marine biotoxins - Metal contaminants - Nitrites, Nitrates - Environmental pollutants - Chemicals from production practices (plant protection) - Hormone residues and "phyto-hormones" - Added chemicals - Food packaging contaminants - Berries and nuts - Mushrooms - Vegetables and Fruits - Cereals - Herbs

It is crystal clear that such answers can be given in the near future, whereas the GM production will continue to increase exponentially fuelled by the US, Canada, Argentina and more recently China, as already stated in the relevant statistics. Although EU states seemed to be very sceptical in the beginning and a series of EU directives reflected this tendency, there has been a change in the latest Directives Regulation (EC) No.641/2004 and EU 2004/657/EC regarding the permission of placing in the market of sweet corn from GM maize line *Bt*11 as a novel food or novel food ingredient. These new directives could be regarded as an effort on the part of the EU towards reaching a consensus with the US and Canada vis-a-vis the production and consumption of GM foods. Anticipated objectives, strategies and suggested methodologies for achieving genetic and environmental traceability in the near future are summarized in Table 2.15.

REFERENCES

Abead, M.S., Hakimi, S.M. and Kaniewski, W.K. (1997). Characterization of acquired resistance in lesion-mimic transgenic potato expressing bacterio-opsin. Mol. Plant Microbe 10(5): 635-645.

Ahmed, F.E. (2002a). Testing for GMOs in food products. Lab Plus Int. 16(4): 8-17.

Ahmed, F.E. (2002b). Detection of GMOs in foods. Trends Biotechnol. 20: 215-223.

Ahrenholz, I., Harms, K., De Vries, J. and Wackernagel, W. (2000). Increased killing of *Bacillus subtilis* on the hairy roots of transgenic Ta lysozyme- producing potatoes. Appl. Environ. Microbiol. 66: 1862-1865.

Allefs, S.J.H.M., DeJong, E.R. and Florack, D.E.A. (1996). Erwinia soft rot resistance of potato cultivars expressing antimicrobial peptide tachyplesin I. Mol. Breeding 2(2): 97-105.

Andrews, A.T., Harris, D.P., Wright, G., Pyle, D.L. and Assenjo, J.A. (2000). Affinity gel electrophoresis as a predictive technique in the fractionation of transgenic sheep milk proteins by affinity aqueous two phase partitioning. Biotechnol. Lett. 22: 1349-1353.

Arvanitoyannis, I.S. and Krystallis, A. (2005). Consumers' beliefs, attitudes and intentions towards genetically modified foods, based on the perceived safety vs. benefits perspective. Int. J. Food Sc. Technol. 40: 343-360.

Atkinson, D. (1998). Genetically modified organisms. Available from http://www.sac.ac.uk/info/External?publications/GMO ASP., Scottish Agricultural College, Scottland.

Auer, C.A. (2003). Tracking genes from seed to supermarket: Techniques and trends. Trends Plant Sci. 18(12): 591-597.

Aumaitre, A., Aulrich, K., Chesson, A, Flachowsky, G. and Piva, G. (2002). New feeds from genetically modified plants: Substantial equivalence, nutritional equivalence, digestibility, and safety for animals in the food chain. Livestock. Prod. Sci. 74: 223-238.

Barg, R., Shabtai, S. and Salts, Y. (2001). Transgenic tomato. In: Biotechnology in Agriculture and Forestry, 47. Transgenic Crops II (ed.) Y.P.S. Bajaj, Springer, Berlin, Germany, pp. 212-233.

Beardsley, T. (1996). Advantage: Nature. Scientific American 274: (May) 33.

Beier, R.C. (1990). Natural pesticides and bioactive components in food. Rev. Environ. Contam. Toxicol. 113: 47-137.

Bendahmane, A., Kanyuka, K. and Baulcombe, D.C. (1999). The *Rx* gene from potato controls separate virus resistance and cell death responses. Plant Cell 11(5): 781-791.

Braun, R.H., Reader, J.K. and Christey, M.C. (2000). Evaluation of cauliflower transgenic for resistance against *Xanthomonas campestris* pv. *campestris*. Acta Hortic. 539: 137-143.

Broglie, K., Chet, I. and Holliday, M. (1991). Transgenic plants with enhanced resistance to the fungal pathogen *Rhizoctonia solani*. Science 254: 1194-1197.

Buhr, T., Sato, S., Ebrahim, F., Xing, A., Zhou, Y., Mathiesen, M., Schweiger, B., Kinney, A., Staswick, P. and Clemente, P. (2002). Ribosome termination of RNA transcripts down regulate seed fatty acid genes in transgenic soybean. Plant J. 30: 155-163.

Carmi, N., Salts, Y., Shabtai, S., Pilowsky, M., Dedicova, B. and Barg, R. (1997). Transgenic parthenocarpy due to specific oversensitization of the ovary to auxin. In: Horticuture Biotechnology: In Vitro Culture and Breeding (eds.) A. Altman and M. Ziv, Acta Hortic. 447: 579-581.

Cellini, F., Chesson, A., Colquhoun, I., Constable, A., Davies, H.V., Engel, K.H., Gatehouse, A.M.R., Kärenlampi, S., Kok, E.J., Leguay, J.J., Lehesranta, S., Noteborn, H.P.J.M., Pedersen, J. and Smith, M. (2004). Unintended effects and their detection in genetically modified crops. Food Chem. Toxicol. 42(7): 1089-1125.

Chakraborty, S., Chakraborty, N. and Datta, A. (2000). Increased nutritive value of transgenic potato by expressing a nonallergenic seed albumin gene from *Amaranthus hypochondriacus*. Proc. Natl. Acad. Sci., USA 97: 3724-3729.

Chang, M.M., Chiang, C.C., Martin, M.W. and Hadviger, L.A. (2001). Expression of a pea disease resistance response gene in the potato cultivar Shepody. Am. Potato J. 70: 635-647.

Charlton, A. (2001). NMR- novel metabolite research? Lab Plus Int. 15(6): 10-12.

Cheftel, J.C. (2005). Food and nutrition labelling in the European Union. Food Chem. 93: 531-550.

Chen, F., Duran, A., Blount, J.A., Sumner, L.W. and Dixon, R.A. (2003). Profiling phenolic metabolites in transgenic alfalfa modified in lignin biosynthesis. Phytochemistry 64: 1013-1021.

Chen, W.P. and Punja, Z.K. (2002). Transgenic herbicide and disease tolerant carrot plants obtained through *Agrobacterium* mediated transformation. Plant Cell Rep. 20: 929-935.

Chong, D.K.X. and Langridge, W.H.R. (2000). Expression of a full length bioactive antimicrobial human lactoferrin in potato plants. Transgenic Res. 9: 71-78.

Christey, M.C. and Braun, R.H. (2001). Transgenic vegetable and forage *Brassica* species. In: Biotechnology in Agriculture and Forestry 47: Transgenic Crops II. (ed.) Y.P.S. Bajaj, Springer, Berlin, Germany, pp. 87-101.

Cocci, C., Mezetti, B. and Rosati, P. (1994). Regeneration and transformation of strawberry. In: VIIIth Int. Cong. of Plant Tissue and Cell Culture, Firenze, Italy, pp. 152.

Cockburn, A. (2002). Assuring the safety of GM food. J. Biotechnol. 98: 79-106.

Codex. (2002). Codex Alimentarius Commission. Report of the third session of the Codex ad hoc intergovernmental task force on foods derived from biotechnology. Yokohama, Japan 4-8 March 2002. Alinorm 03/34. Available from: ftp://ftp.fao.org/codex/alinorm03/Al03_34e.pdf. Accessed on 24 June, 2003.

COM/2002/0085 – COD 2002/0046 (entry into force 27/10/2002). The transboundary movement of GMOs [Proposal for regulation].

Constabel, P.C., Bertrand, C. and Brisson, N. (1993). Transgenic potato plants over expressing the pathogenesis related *STH-2* gene show unaltered susceptibility to *Phytophthora infestans* and potato virus X. Plant Mol. Biol. 22: 775-782.

Coulston, F. and Kolbye, A.C. (1990). Biotechnologies and food: Assuring the safety of foods produced by genetic modification. Regul. Toxicol. Pharmacol. 12: S1-196.

Cranshaw, W.S. (2000) "Bacillus Thuringiensis" in http://www.ext.colostate.edu/pubs/insect05556.html. Accessed on 21 December 2004.

Dehesh, K., Jones, A., Knutzon, D.S. and Voelker, T.A. (1996). Production of high levels of 8:0 and 10:0 fatty acids in transgenic canola by over expression of Ch FatB2, a thioesterase cDNA from *Cuphea hookeriana*. Plant J. 9: 167-172.

De la Riva, G.A., Gonzalez-Cabrera, J., Vasquez-Padron, R. and Ayra-Pardo, C. (1998). *Agrobacterium tumefaciens*: A natural tool for plant transformation. Electronic J. Biotechnol. 1(3): 118-133.

Del Vecchio, A.J. (1996). High laurate canola. How Calgene's program began, where it's headed. [INFORM] International News on Fats, Oils and Related Materials 7: 230.

Desiderio, A., Aracri, B., Leckie, F., Mattei, B. and Cervone, B. (1997). Polygalacturonase inhibiting proteins with different specificities are expressed in *Phaseolus vulgaris*. Mol. Plant Microbe Interact. 10: 852-860.

Dinkins, R.D., Reddy, M.S.S., Meurer, C.A., Yan, B., Trick, H., Thibaud-Nissen, F., Finer, J.J., Parrott, W.A. and Collins, G.B. (2001). Increased sulphur amino acids in soybean plants over expressing the maize 15 kDa zein protein. In vitro cell development. Biol. Plant 37: 742-747.

Donaldson, P.A., Anderson, T., Lane, B.G. and Simmonds, D.H. (2001). Soybean plants expressing an active oligomeric oxalate oxidase from the wheat *gf-2.4* (germin) gene are resistant to the oxalate secreting pathogen *Sclerotininia sclerotiorum*. Physiol. Mol. Plant Pathol. 59: 297-307

Dueck, T.A., Van Der Werf, A., Lotz, L.A.P. and Jordi, W. (1998). Methodological approach to a risk analysis for polygene–genetically modified plants (GMPs): A mechanistic study (AB nota 50). Wageningen, Netherlands: Research Institute for Agrobiology and Soil Fertility (AB-DLO), p. 46.

During, K., Porsch, P., Fladung, M. and Lorz, H. (1993). Transgenic potato plants resistant to the phytopathogenic bacterium *Erwinia carotovora*. Plant J. 3: 587-598.

EC. (2003). Guidance document for the risk assessment of genetically modified plants and derived food and feed, health and consumer protection. Directorate-General, European Commission. Available from: http://europa.eu.int/comm/food/fs/sc/ssc/out327_en.pdf. Accessed on 2 April 2005.

EEC. (2001). Directive 2001/18/CE of the European Parliament and of the Council of 12 March 2001 on the deliberate release into the environment of genetically modified organisms and repealing Council Directive 90/220/EEC.

Engel, K.H., Gerstner, G. and Ross, A. (1998). Investigation of glycoalkaloids in potatoes as example for the principle of substantial equivalence. In: Novel Food Regulation in the EU - Integrity of the Process of Safety Evaluation. Federal Institute of Consumer Health Protection and Veterinary Medicine, Berlin, Germany, pp. 197-209.

Engel, K.H., Frenzel, T. and Miller, A. (2002). Current and future benefits from the use of GM technology in food production. Toxicol Lett. 127: 329-336.

EU. 90/219/EEC (entry into force 23/10/1991). Contained use of GM microorganisms.

EU. 90/220/EEC (entry into force 23/10/1991). Deliberate release into the environment of GMOs.

EU. 2001/18/EC (entry into force 17/4/2001) Deliberate release into the environment of GMOs.

EU. (2001). EC-sponsored Research on Safety of Genetically Modified Organisms; 5th Framework Program - External Advisory Groups. "GMO research in perspective." Report of a workshop held by External Advisory Groups on the "Quality of Life and Management of Living Resources" Program. Available from: http://www.europa.eu.int/comm/research/quality-of-life/gmo/index.html and http://www.europa.eu.int/comm/research/fp5/eag-gmo.html. Accessed on 22 July 2003.

EU. Regulation (EC) No.1829/2003 (entry into force 7/11/2003). GM food and feed.

EU. Regulation (EC) No.1830/2003 (entry into force 7/11/2003). Traceability and labelling of GMOs and traceability of food and feed products produced from GMOs.

EU. Regulation (EC) No.65/2004 (entry into force on the date of its publication in the Official Journal of the European Union). Establishing a system for the development and assignment of unique identifiers for GMOs.

EU. 2004/204/EC (entry into force 23/3/2004). Arrangements for the operation of the registers for recording information on genetic modifications in GMOs.

EU. Regulation (EC) No.641/2004 (entry into force 18/4/2004). The authorization of new GM food and feed, the notification of existing products and adventitious or technically unavoidable presence of GM material which has benefitted from a favourable risk evaluation.

EU. 2004/643/EC Placing on the market of a maize product (*Zea mays* L. line NK603) GM for glyphosate tolerance.

EU. 2004/657/EC Placing on the market of sweet corn from GM maize line *Bt*11 as a novel food or novel food ingredient.

EU. Regulation (EC) No.1139/98 (entry into force 1/9/1998). The compulsory indication of the labelling of certain foodstuffs produced from GMOs.

Ewen, S.W.B. and Pusztai, A. (1999). Effect of diet containing genetically modified potatoes expressing *Galanthus nivalis* lectin on rat small intestine Lancet 354: 1353-1354.

FAO/WHO. (1996). Biotechnology and food safety. Report of a joint FAO/WHO consultation; 30 Sept. - 4 Oct. 1996. FAO Food and Nutrition Paper 61. Rome, Italy: Food and Agriculture Organization of the United Nations. Available from: ftp://ftp.fao.org/es/esn/food/biotechnology.pdf. Accessed on 22 July 2003.

FAO/WHO. (2000). Safety Aspects of Genetically Modified Foods of Plant Origin. Report of a Joint FAO/WHO Expert Consultation on Foods Derived from Biotechnology, 29 May - 2 June 2000. Rome, Italy: Food and Agriculture Organization of the United Nations. Available from: ftp://ftp.fao.org/es/esn/food/gmreport.pdf. Accessed on 5 June 2005.

FAO/WHO. (2001). Safety assessment of foods derived from genetically modified microorganisms. Report of the Joint FAO/WHO Expert consultation on Foods Derived from Biotechnology, 24-28 September 2001 Geneva, Switzerland: World Heath Organization/Food and Agriculture Organization and the United Nations. Available from http://www.who.int/fsf/documents/gmm.

FAO/WHO. (2002). Report of the third session of the Codex Ad Hoc Intergovernmental Task Force on Foods Derived from Biotechnology (ALINORM 01/34). Rome, Italy: Codex Ad Hoc Intergovernmental Task Force on Foods Derived from Biotechnology, Food and Agriculture Organization of the United Nations. Available from: ftp://ftp.fao.org/codex/alinorm03/Al03_34e.pdf. Accessed on 22 March 2005.

Fares, N.H. and El Sayed, A.K. (1998). Fine structural changes in the ileum of mice fed on delta-endotoxin-treated potatoes and transgenic potatoes. Nat. Toxins 6: 219-233.

Faust, M.A. (2002). New feeds from genetically modified plants: the US approach to safety for animals and the food chain. Livestock Prod. Sci. 74: 239-254.

Feder, B.J. (1996). Geneticists Arm Corn Against Corn Borer, Pest May Still Win. The New York Times, Tuesday, July 23, 1986, p. C1.

Fischhoff, D.A., Bowdish, K.S., Perlak, F.J., Marrone, P.J., McCormick, S.H. and Fraley, R.T. (1987). Insect tolerant transgenic tomato plants. Biotechnology 5: 807-813.

Fuchs, M., Providentti, R., Slightom, J.L. and Gonsalves, D. (1996). Evaluation of transgenic tomato plants expressing the coat protein gene of cucumber mosaic virus strain WL under field conditions. Plant Dis. 80: 270-275.

Gallie, D.R. (2001). Transgenic carrot. In: Biotechnology in Agriculture and Forestry 47. Transgenic Crops II. (ed.) Y.P.S. Bajaj, Springer, Berlin, Germany, pp. 147-159.

Gao, A.G., Hakimi, S.M., Mittanck, C.A. and Rommens, C.M.T. (2000). Fungal pathogen protection in potato by expression of a plant defensin peptide. Nat. Biotechnol. 18: 1307-1310.

Gaskell, G., Bauer, M.W. and Durant, J. (1998). Public perceptions of biotechnology in 1996: Euro barometer 46.1 In: Biotechnology in the Public Sphere: A European Sourcebook, (eds.) J. Durant, M.W. Bauer and G. Gaskell, Science Museum, London, UK.

Gaskell, G., Bauer, M.W., Durant, J., Allansdotir, A., Bonfadeli, H., Boy, D., de Cheveigne, S., and Sakellaris, G. (2000). Biotechnology and the European public. Nat. Biotechnol. 18: 953-958.

Gertz, J.M., Vencill, W.K. and Hill, N.S. (1999). Tolerance of transgenic soybean (*Glycine max*) to heat stress In: Proceedings of the 1999 Brighton Crop Protection Conference: Weeds Vol.3, Farnham, UK British Crop Protection Council, pp. 835-840.

Gielen, J., Ultzen, T., Bontems, S., Loots, W., Van Schepen, A., Westerbroek, A., de Haan, P. and van Grinsven, M. (1996). Coat protein mediated protection to cucumber mosaic virus infections in cultivated tomato. Euphytica 88: 139-149.

Good, X., Kellogg, J,A. and Wagoner, W. (1994). Reduced ethylene synthesis by transgenic tomatoes expressing S-adenolyl methionine hydrolase. Plant Biol. 26 (3): 781-790.

Grison, R., Grezes-Besset, B., Scheider, M. and Toppan, A. (1996). Field tolerance to fungal pathogens of *Brassica napus* constitutively expressing a chimeric chitinase gene. Nat. Biotechnol. 14: 643-646.

Gupta, A. (2000). Governing trade in GMO: The Cartagena protocol on biosafety. Environment 40(4): 23-33.

Hagio, T. (1998). Optimizing the particle bombardment method for efficient genetic transformation. JARQ 32(4): 65-72.

Hamilton, A.J., Lycett, G.W. and Grierson, D. (1990). Antisense gene that inhibits synthesis of the hormone ethylene in transgenic plants. Nature 346: 284-287.

Hamilton, A.J., Fray, R.G. and Grierson, D. (1995). Sense and antisense inactivation of fruit ripening genes in tomato. Curr. Topics Microbiol. Immunol. 197: 77-89.

Hammond, B.G., Vicini, J.L., Hartnell, G.F., Naypor, M.W., Knight, C.D., Robinson, E.H., Fuchs, R.L. and Padgette, S.R. (1996). Th feeding value of soybeans fed to rats, chickens, catfish and dairy cattle is not altered by genetic incorporation of glyphosate tolerance. J. Nutr. 126: 717-727.

Harvey, M.H., McMillan M., Morgan, M.R. and Chan, H.W. (1985). Solanidine is present in sera of healthy individuals and in amounts dependent on their dietary potato consumption. Human Toxicol. 4: 187-194.

Hashimoto, W., Momma, K., Katsube, T., Ohkawa, Y., Ishige, T., Kito, M., Utsumi S, and Murata, K. (1999a). Safety assessment of genetically engineered potatoes with designed soybean glycinin: Compositional analyses of the potato tubers and digestibility of the newly expressed protein in transgenic potatoes. J. Sci. Food Aric. 9: 1607-1612.

Hashimoto, W., Momma, K., Yoon, H-J., Ozawa, S., Ohkawa, Y., Ishige, T., Kito, M., Utsumi, S. and Murata, K. (1999b). Safety assessment of transgenic potatoes with soybean glycinin by feeding studies in rats. Biosci. Biotechnol. Biochem. 63: 1942-1946.

Hattan, D. (1996). Evaluation of toxicological studies on Flavr Savr tomato. In: Food Safety Evaluation, Organization for Economic Co-operation and Development, Paris, France, pp. 58-60.

Hellwege, E.M., Czapla, S., Jahnke, A., Willmitzer, L. and Heyer, A.G. (2000). Transgenic potato (*Solanum tuberosum*) tubers synthesize the full spectrum of inulin molecules naturally occurring in globe artichoke (*Cynara scolymus*) roots. Proc. Natl. Acad. Sci., USA 97: 8699-8704.

Hennin, C., Hofte, M. and Diederichsen, E. (2001). Functional expression of *Cf9* and *Avr9* genes in *Brassica napus* induces enhanced resistance to *Leptosphaeria maculans*. Mol. Plant Microbe Interact. 14: 1075-1085.

Herman, E. (2002). Targeted gene silencing removes an immunodominant allergen from soybean seeds. Agricultural Research Magazine, September 2002 50(9): 25-31.

Hinchee, M.A.W. (1988). Production of transgenic soybean plants using *Agrobacterium*-mediated DNA transfer. Biotechnology 6: 915-922.

Hoban, T.J. (1996a). Anticipating public reaction to the use of genetic engineering in infant nutrition. Am. J. Clin. Nutr. 63: 657-662.

Hollingsworth, R.M., Bjeldanes, L.F., Bolger, M., Kimber, I., Meade, B.J., Taylor, S.L. and Wallace, K.B. (2003). The safety of genetically modified foods produced through biotechnology. Report of the Society of Toxicology. Toxicol Sci 71: 2-8.

ILSI (1997). Europe Novel Foods Task Force. The safety assessment of novel foods. Food Chem. Toxicol. 34: 931-940.

ILSI (2003a). Best practices for the conduct of animal studies to evaluate crops genetically modified for input traits. Washington, DC: International Life Sciences Institute.

ILSI (2003b). ILSI crop composition database. Washington, DC: International Life Sciences Institute. Available from: http://www.cropcomposition.org. Accessed on 25 June 2004.

ILSI (2004). Crop composition database. Available from: http://www.cropcomposition.org. Accessed on 13 February 2005.

James, C. (2003). Preview: Global Status of Commercialized Transgenic Crops: 2003. ISAAA Briefs Nr 30. Ithaca, NY: International Service for the Acquisition of Agribiotech Applications. Available from: http://www.isaaa.org. Accessed on 17 February 2005.

Jongedijk, E., Tigelaar, H. and Van Roekel, J.S.C. (1945). Synergistic activity of chitinases and 1,3-glucanase enhance fungal resistance in transgenic tomato plants. Euphylica 85: 173-180.

Jung, W., Yu, O., Lau, S.C. and MaGonigle, B. (2000). Identification and expression of isoflavone synthase, the key enzyme for biosynthesis of isoflavone in legumes. Nat. Biotechnol. 18: 208-212.

Kanrar, S., Venkateswari, J.C. and Kirti, P.B. (2002). Transgenic expression of hevein, the rubber tree lectin, in Indian mustard confers protection against *Alternaria brassica*e. Plant Sci. 162 (3): 441-448.

Kazan, K., Rusu, A., Marcus, J.P., Coulter, K.C. and Manners, J.M. (2002). Enhanced quantitative resistance to *Leptosphaeria maculans* conferred by expression of a novel antimicrobial peptide in canola. Molecular Breeding 10: 63-70.

Kerr, W.A. (1999). Genetically modified organisms, consumer scepticism and trade law: implications for the organisation of international supply chains. Supply Chain Management 4: 67-74.

Kesarwani, M., Azam, M., Natarajan, K., Mehta, A. and Datta, A. (2000). Oxalate decarboxylase from *Collybia velutipes*: Molecular cloning and its over expression to confer resistance to fungal infection in transgenic tobacco and tomato. J. Biol. Chem. 275: 7230-7238.

Kim, J.W., Sun, S.S.M. and German, T.L. (1994). Disease resistance in tobacco and tomato plants transformed with the tomato spotted wilt virus nucleocapsid gene. Plant Dis. 78: 615-621.

Kinney, A.J. and Knowlton, S. (1998). Designer oils: the high oleic acid soybean. In: Genetic Modification in the Food Industry (eds.) S. Rollerand and S. Harlander, Blackie, London, UK, pp. 193-213.

Klann, E.M., Hall, B. and Bennett, A.B. (1996). Antisense acid invertase (*TIV1*) gene alters soluble sugar composition and size in transgenic tomato fruit. Plant Physiol. 112: 1321-1330.

Kombrink, E. and Somssich, I.E. (1995). Defense responses of plants to pathogens. Adv. Bot. Res. 21: 1-34.

Konig, A., Cockburn, A., Crevel, R.W.R., Debruyne, E., Grafstroem, R., Hammerling, U., Kimber, I., Knudsen, I., Kuiper, H.A., Peijnenburg, A.A.C.M., Penniks, A.H., Poulsen, M., Schauzu, M and Wal, J.M. (2004). Assessment of the safety of foods derived from genetically modified (GM) crops. Food Chem. Toxicol. 42: 1047-1088.

Kramer, M.G., Kellogg, J., Wagoner, W., Matsumura, W., Good, X., Peters, S.T. and Bestwick, R.K. (1996). Reduced ethylene synthesis ripening control of tomatoes expressing S-adenosyl-methionine hydrolase. In: Biology and Biotechnology of the Plant Hormone Ethylene (eds.) A.G. Kanellis, Crete, Hellas, Greece, Kluwer, Dordrecht, The Netherlands, pp. 307-319.

Kuiper, H.A. and Noteborn, H.P.J.M. (1996). Food safety assessment of transgenic insect resistant *Bt* tomatoes. Proceedings of an OECD sponsored workshop held on 12-15 Sept. 1994, Oxford, UK, pp. 50-57.

Kuiper, H.A., Kleter, G.A., Noteborn, H.P.J.M. and Kok, E.J. (2001). Assessment of the food safety issues related to genetically modified foods. The Plant J. 27(6): 503-528.

Lagrimini, L.M., Vaughn, J., Erb, W.A. and Miller, S.A. (1993). Peroxidase overproduction in tomato: Wound induced polyphenol deposition and disease resistance. Hort Science 28: 218-221.

Lawson, E.C., Weiss, J.D., Thomas, P.E. and Karniewski, W.K. (2001). New leaf Plus russet burbank potatoes: Replicase mediated resistance to potato leaf roll virus. Molecular Breeding 7: 1-12.

Leighton, J. (1999). Science, medicine and the future: Genetically modified food. Br. Med. J. 318: 581-584.

Li, R.G. and Fan, Y.L. (1999). Reduction of lesion growth rate of late blight plant disease in transgenic potato expressing harpin protein. Sci. China Ser. C42(1): 96-101.

Liu, D., Ragothama, K.G., Hasegawa, P.M. and Bressan, R. (1994). Osmotin over expression in potato delays development of disease symptoms. Proc. Nat. Acad. Sci., USA 91: 1888-1892.

Liu, Q., Singh, S. and Green, A. (2002). High-oleic and high-stearic cottonseed oils: Nutritionally improved cooking oils developed using gene silencing. J. Am. Coll. Nutr. 21: 205S-211S.

Lorito, M., Woo, S.L., Fernandez, I.G., Colucci, G., Harman, G.E. and Scala, F. (1998). Genes from mycoparasitic fungi as a source for improving plant resistance to fungal pathogens. Proc. Nat. Acad. Sci., USA 95: 7860-7865.

LSRO. (1998). Alternative and traditional models for safety evaluation of food ingredients, (eds.) D.J. Raiten and M. Bethesda, Life Sciences Research Office, American Society for Nutritional Sciences. Prepared for Center for Food Safety and Applied Nutrition, Food and Drug Administration, Dept. of Health and Human Services, Washington, DC.

MacMahon, R.R. (2000). Genetic Engineering of Crop Plants in http://www.yale.edu/ynhti/curriculum/units/2000/7/00.07.02.x. Accessed on 21 April 2005.

Mangusson, M.K. and Hursti, U.K. (2002). Consumer attitudes towards genetically modified foods. Appetite 39: 9-24.

Martin, G.B., Brommonschenkel, S.H., Chunwongse, J., Frary, A., Ganal, M.W., Spivey, R., Wu, T., Earle, E.D. and Tanksley, S.D. (1993). Map-based cloning of a protein kinase gene conferring disease resistance in tomato. Science 262: 1432-1436.

Maryanski, J.H. (1995). FDA's policy for foods developed by biotechnology. In: Genetically Modified Foods: Safety Issues (eds.) T. Engels and K. Teranishi. Am. Chem. Soc. Symposium Series No. 605: Ch 2 pp. 12-22

McCue, K.F, Shepherd, L.V.T., Allen, P.V., Maccree, M.M., Rockhold, D.R., Davies, H.V. and Belknap, W.R. (2003). Modification of steroidal alkaloid biosynthesis in transgenic tubers containing an antisense sterol glyco transferase (*Sgt*) gene encoding a novel steroidal alkaloid diglycoside rhamnosyl transferase. Paper presented at the 87th Annual Meeting of The Potato Association of America; 10-14 August 2003; Spokane, Wash.

Melchers, L.S. and Stuiver, M.H. (2000). Novel genes for disease resistance breeding. Curr. Opin. Plant Biol. 3: 147-152.

Mepham, B. (2000). A framework for the ethical analysis of novel foods: the ethical matrix. J. Agric. Environ. Ethics 12: 165-176.

Mitrovic, T. (2003). Gene Transfer Systems. Facta Universitatis 10(3): 101-105 Accessed from facta.junis.ni.ac.yu/facta/mab/mab200303/mab200303-01.pdf.

Mora, A.A. and Earle, E.D. (2001). Resistance to *Altenaria brassicola* in transgenic broccoli expressing a *Trichoderma harzianum* endochitinase gene. Molecular Breeding 8: 1-9.

Munro, I.C., McGirr, L.G., Nestmann, E.R. and Kille, J.W. (1996a). Alternative approaches to the safety assessment of macronutrient substitutes. Regul. Toxicol. Pharmacol. 23: S6-S14.

Munro, I.C., Ford, R.A., Kennepohl, E. and Sprenger, J.G. (1996b). Correlation of structural class with no-observed-effect levels: A proposal for establishing a threshold of concern. Food Chem. Toxicol. 34: 829-867.

Murphy, D.J. (2003). Agricultural biotechnology and oil crops: current uncertainities and future potential. Appl. Biotechnol. Food Sci. Policy 1(1): 25-38.

Myhra, I. and Dalmo, R. (2005). Introduction of genetic engineering in aquaculture: Ecological and ethical implications for science and governance. Aquaculture. 250: 542-554.

Nordlee J.A., Taylor, S.L., Townsend, J.A., Thomas, L.A. and Bush, R.K. (1996). Identification of a Brazil-nut allergen in transgenic soybeans. N. Engl. J. Med. 334: 688-692.

Noteborn, H.P.J.M., Bienenmann-Ploum, M.E., van den Berg, J.H.J., Alink, G.M., Zolla L, Reynaerts A, Pensa, M. and Kuiper, H.A. (1995). Safety assessment of the *Bacillus thuringiensis* insecticidal crystal protein CRYIA(b) expressed in transgenic tomatoes. In: Genetically Modified Foods – Safety Aspects, (eds.) K.H.Engel, G. R. Takeoka and R. Teranishi. ACS Symposium Series 605. Washington, D.C: American Chemical Society, pp 134-147.

Novak, W.K. and Halsberger, A.G. (2000). Substantial equivalence of antinutrients and inherent plant toxins in genetically modified novel foods. Food Chem. Toxicol. 38: 473-483.

NZRC. (2001). Report of the Royal Commission on Genetic Modification. New Zealand Royal Commission on Genetic Modification. Available from: http://www.gmcommission.govt.nz/index.html. Accessed on 25 May 2005.

OECD. (1993). Safety evaluation of foods derived by modern biotechnology: Concepts and principles. Organization for Economic Co-operation and Development, Paris, France.

OECD. (1998). Report of the OECD workshop on the toxicological and nutritional testing of novel foods. Organization for Economic Co-operation and Development, Paris, France.

OECD. (2000). Report of the OECD task force for the safety of novel foods and feeds. Organization for Economic Co-operation and Development, Okinawa, Japan.

OECD. (2002). Report of the OECD workshop on nutritional assessment of novel foods and feeds. 5-7 February 2001. Organization for Economic Co-operation and Development, Ottawa, Canada, Paris, France.

Olkowski A.A. and Classen, H.L. (1995). Sudden death syndrome in broiler chickens: a review. Poult. Avian Biol. Rev. 6: 95-105.

Osusky, M., Zhou, G., Osuska, L., Tamot, B., Pauls, K.P., Moffatt, B.A. and Glick, B.R. (2000). Reduced symptoms of Verticillium wilt in transgenic tomato expressing a bacterial ACC deaminase. Molecular Plant Pathol. 2: 135-145.

Parashina, E.V., Serdobinskii, L.A., Kalle, E.G. and Naroditskii, P.V. (2000). Genetic engineering of oilseed rape and tomato plants expressing a radish defensin gene. Russian J. Plant Physiol. 47: 471-478.

Peferoen, M. (1997). Progress and prospects for field use of *Bt* genes in crops. Trends Biotech. 15: 173-177.

Pollack, A. (2000). New ventures aim to put farms in vanguard of drug production. New York Times, USA, 14 May 2000, p. 1.

Powell, A.L.T., van Kan, J., Have, A.T. and Labavitch, J.M. (2000). Transgenic expression of pear PGIP in tomato limits fungal colonization. Mol. Plant Microbe Interact. 13: 942-950.

Prufer, D., Kawchuk, L., Monecke, M., Nowok, S., Fischer, R. and Rhode, W. (1999). Immunological analysis of potato leaf roll luetovirus (PLRV) P1 expression identifies 25 KDA RNA-binding protein derived via *P1* processing. Nucleic Acid Res. 27(2): 421-425.

Pryme, I. (2003). In vivo studies on possible health consequences of GM food and feed. Nutr. Health 17: 1- 8.

Punja, Z.K. and Raharjo, S.H.T. (1996). Response of transgenic cucumber and carrot plants expressing different chitinase enzymes. Plant Dis. 80: 999-1005.

Punja, Z.K. (2004). Genetic engineering of plants to enhance resistance to fungal pathogens. In: Fungal Disease Resistance in Plant (ed.) Z.K. Punja, Haworth Press, Inc. New York, USA, pp. 207-258.

Rapp, W. (2002). Development of soybeans with improved amino acid composition. 93rd AOCS Annual Meeting & Expo; 5–8 May; Montréal, Québec, Canada. Champaign, Ill.: American Oil Chemists' Society Press, pp.79-86.

Ray, H., Douches, D.S. and Hammerschmidt, R. (1998). Transformation of potato with cucumber peroxidase: Expression and disease response Physiol. Mol. Plant Pathol. 53(2): 93-103.

Redenbaugh, K., Hiuatt, W., Martineau, B. and Emlay, D. (1995). Determination of the safety of genetically engineered crops. ACS Symp Ser 605. Genetically Modified foods safety issues. American Chemical Society. Washington, DC, USA, pp. 72-87.

Robinson, C. (2001). ILSI Europe concise monograph series. Genetic modification technology and food consumer health and safety. Washington DC, USA 4: 5.

Royal Society. (2002). Genetically modified plants for food use and human health—an update. London, UK: The Royal Society. Available from: http://www.royalsoc.ac.uk/policy/index.html. Accessed on 22 July, 2003.

Royal Society. (2003). GM crops, modern agriculture and the environment. Report of a Royal Society Discussion Meeting held on 11 February 2003. London, UK: The Royal Society. Available from: http://www.royalsoc.ac.uk/files/statfiles/document- 222.pdf. Accessed on 22 July 2003.

Saba, A., Moles, A. and Frewer, L.J. (1998). Public concerns about general and specific applications of genetic engineering: a comparative study between the UK and Italy. Nutr. Food Sci. 1:19-29.

Saba, A., Rosati, S. and Vassalo, M. (2000). Biotechnology in agriculture: Perceived risks, benefits and attitudes in Italy. British Food J. 102: 114-121.

Saba, A. and Vassalo, M. (2002). Consumer attitudes towards the use of gene technology in tomato production. Food Quality and Preference 13: 13-21.

Sachar–Hill, Y. (2002). Nuclear Magnetic Resonance and plant metabolic engineering. Metabol. Eng. 4: 90-97.

Sanden, M., Bruce, I.J., Rahman, M.A. and Hemre, G. (2004). The fate of transgenic sequences present in genetically modified plant products in fish feed: Investigating the survival of GM soybean DNA fragments during field trials in Atlantic salmon, *Salmo salar* L. Aquaculture 237: 391-405.

Saraswathy, N. and Kumar, P.A. (2004). Protein engineering of endotoxins of *Bacillus thuringiensis*. Electronic J. Biotechnol. 7(2): 180-190.

Schnable, P.S., Hsia, A.P. and Nikolau, B.J. (1998). Genetic recombination in plants. Curr. Opin. Plant Biol. 1: 123-129.

Schwall, G.P., Safford, R., Westcott, R.J., Jeffcoat, R., Tayal, A., Shi, Y.C., Gidley, M.J. and Jobling, S.A. (2000). Production of very-high-amylose potato starch by inhibition of SBE A and B. Nat. Biotechnol. 18: 551-554.

Serrano, C., Arce-Johnson, P. and Torres, H. (1996). Expression of the chicken lysozyme gene in potato enhances resistance to infection by *Erwinia carotovora* subsp. *atroseptica*. Am. J. Potato Res. 77(3): 191-199.

Shewmaker, C.K., Sheely, J.A., Daley, M., Colburn, S. and Ke, D.Y. (1999). Seed-specific over expression of phytoene synthase: Increase in carotenoids and other metabolic effects. Plant J. 22: 401-412.

Shintani, D. and De La Penna, D. (1998). Elevating the vitamin E content of plants through metabolic engineering. Science 282: 2098-2100.

Siebert, R. and Puchta, H. (2002). Efficient repair of genomic double-strand breaks by homologous recombination between directly repeated sequences in the plant genome. Plant Cell 14: 1121-1131.

Stark, D.M., Timmerman, K.P., Barry, G.F., Preiss, J. and Kishore, G.M. (1992). Role of the amount of starch in plant tissue by ADP glucose pyrophosphorylase. Science 258: 287-292.

Stark, D.M., Barry, G.F. and Kishore, G.M. (1996). Improvement of food quality traits through enhancement of starch biosynthesis. Ann. New York Acad. Sci. 792: 26-36.

Stewart, C.N., Adang, M.J. and All, J.N. (1996). Genetic transformation, recovery, and characterization of fertile soybean transgenic for a synthetic *Bacillus thuringiensis crylAc* gene Plant Physiol. 112 (1): 121-129.

Sticher, L. (1997). Systemic acquired resistance. Annu. Rev. Phytopathol. 35: 235-270.

Straughan, R. (1998). Moral concerns and the educational function of ethics. In: Genetic Modification in the Food Industry (eds.) S. Roller and S. Haslander, Blackie, London, UK, pp. 180-189.

Strittmatter, G., Janssens, J., Opsomer, C. and Botterman, J. (1995). Inhibition of fungal disease development in plants by engineering controlled cell death. Bio/Technology 13: 1085-1089.

Stuible, H.-P. and Kombrink, E. (2004). The hypersensitive response and its role in disease resistance. In: Fungal Disease resistance in Plants (ed.) Z.K. Punja, Haworth Press, Inc. New York, USA, pp. 57-92.

Tabaeizadeh, Z., Agharbaoui, Z. and Harrak, H. (1999). Transgenic tomato plants expressing a *Lycopersicon chilense* chitinase demonstrate improved resistance to *Verticillium dahliae* race 2. Plant Cell Reports 19: 197-202.

Takaichi, M. and Oeda, K. (2000). Transgenic carrots with enhanced resistance towards two major pathogens. Plant Sci. 153: 135-144.

Taylor, S.L. and Hefle, S.L. (2002). Genetically engineered foods: Implications for food allergy. Curr. Opin. Allergy Clin. Immunol. 2: 249-252.

Teshima, R., Akiyama, H., Okunuki, H., Sakushima, J.I., Goda, Y., Onodera, H., Sawada, J.I. and Toyoda, M. (2000). Effect of GM and Non-GM soybeans on the immune system of BN rats and B10A mice. J. Food Hyg. Soc. Jpn. 41: 188-193.

Teshima, R., Watanabe, T., Okunuki, H., Isuzugawa, K., Akiyama, H., Onodera, H., Imai, T., Toyoda, M. and Sawada, J. (2002). Effect of subchronic feeding of genetically modified corn (CBH351) on immune system in BN rats and B10A mice. J. Food Hyg. Soc. Jpn. 43: 273-279.

Thompson, C., Dunwell, J.C.E., Lay, V., Ray, J., Schmitt, M., Watson, H. and Nisbet, G. (1995). Degradation of oxalic acid by transgenic oilseed rape plants expressing oxalate oxidase. Euphytica 85: 169-172.

Thompson, P.B. (1997). Food Biotechnology in Ethical Perspective. Blackie Academic and Professional, London, UK.

Thomzik, J.E., Stenzel, K., Stoecker, R., Schreier, P.H., Hain, R. and Stahl, D.J. (1997). Synthesis of a grapevine phytoalexin in transgenic tomatoes conditions resistance against *Phytophthora infestans*. Physiol. Mol. Plant Pathol. 51: 265-278.

Turk, S.C.H.J. and Smeekens, S.C.M. (1999). Genetic modification of plant carbohydrate metabolism. In: Applied Plant Biotechnology (eds.) V.L. Chopra, V.S. Malik and S.R. Bhat, Science Publishers, Enfield, UK, pp. 71-100.

Tutel'yan, V.A., Kravchenko, L.V., Lashneva, N.V., Avren'eva, L.I., Guseva, G.V., Sorokina, Y.E. and Chernysheva, O.N. (1999). Medical-biological evaluation of the safety of protein concentrate obtained from genetically modified soybean. Biochemical Investigation. Vopr. Pitan 68(5-6): 9-12.

Tsujie, M., Isaka, Y., Nakamura, H., Imai, E. and Hori, M. (2001). Electroporation-mediated gene transfer that targets *Glomeruli*. J. Am. Soc. Nephrol. 12: 949-954.

Ursin, V. (2000). Genetic modification of oils for improved health benefits. Presentation at conference, Dietary Fatty Acids and Cardiovascular Health: Dietary Recommendations for Fatty Acids: Is There Ample Evidence? June 5-6, 2000. Reston Va.: American Heart Association.

Van den Elzen, P.J.M., Jongedijk, E., Melchers, L.S. and Cornelissen, B.J.C. (1993). Virus and fungal resistance: from laboratory to field. Phil. Trans. Royal Soc. London B 342: 271-278.

Veronese, P., Crino, P., Tucci, M., Colucci, F. and Saccardo, F. (1999). Pathogenesis related proteins for the control of fungal diseases of tomato In: Genetics and Breeding for Quality and Resistance (eds.) G.T. Scarascia, E. Mugnozza and M.A. Pagnotta, Kluwer, Dordrecht, Netherlands, pp. 15-24.

Vidhyasekaran, P. (2004). Concise Encyclopedia of Plant Pathology. Food Products The Haworth Reference Press, New York, USA, pp. 293-323.

Wang, H.Z., Zhao, P.J. and Zhou, X.Y. (2000). Regeneration of transgenic *Cucumis melo* and its resistance to virus diseases. Acta Phytophylactica Sinica 27: 126-130.

Wang, P., Nowak, G., Culley, D., Hadwiger, L.A. and Fristensky, B. (1999). Constitutive expression of a pea defense gene *DRR206* confers resistance to blackleg disease in transgenic canola (*Brassica napus*). Mol. Plant Microbe Interact. 12: 410-418.

Weber, J., Olsen, O. and Wegener, C. (1996). Digalacturonates from pectin degradation induce tissue responses against potato soft rot Physiol. Mol. Plant Pathol. 48 (6): 389-401.

Weick, C.W. and Walchli, S.B. (2002). Genetically engineered crops and foods: Back to the basics of technology diffusion. Technol. Soc. 24: 265-283.

Whitham, S., McCormick, S. and Baker, B. (1996). The N gene of tobacco confers resistance to tobacco mosaic virus in transgenic tomato. Proc. Natl. Acad. Sci., USA 93: 8776-8781.

WHO (World Health Organization). (1987). Principles of the safety assessment of food additives and contaminants in food. Environmental Health Criteria. No. 70. International Programme on Chemical Safety. WHO. Geneva, Switzerland.

Wu, G., Schortt, B.J., Lawrence, E.B., Leon, J., Fitzimmons, K.C., Levine, E.B., Raskin, I. and Shah, D.M. (1995). Disease resistance conferred by expression of a gene encoding H_2O_2 generating glucose oxidase in transgenic potato plants. Plant Cell 7: 1357-1368.

Wu, G., Schortt, B.J., Lawrence, E.B., Leon, J., Fitzimmons, K.C., Levine, E.B., Raskin, I. and Shah, D.M. (1997). Activation of host defense mechanisms by elevated production of H_2O_2 in transgenic plants. Plant Physiol. 115: 427-435.

Wu, W., Wildsmith, S.E., Winkey, A.J. and Bugelski, P.J. (2001). Chemometric strategies for normalisation of gene expression data obtained from cDNA microarrays. Anal. Chim. Acta 446: 451-466.

Xue, D. and Tisdell, C. (2000). Safety and socio-economic issues praised by modern biotechnology. Int. J. Soc. Econ. 27(7/8/9/10); 699-708.

Xue, D. and Tisdell, C. (2002). Global trade in GM food and the Cartagena protocol on biosafety : consequences for China. J. Agric. Environ. Ethics 15: 337-356.

Ye, X., Al-Babili, S., Klöti, A., Zhang, J., Lucca, P., Beyer, P. and Potrykus, I. (2000). Engineering the provitamin A (beta-carotene) biosynthetic pathway into (carotenoid-free) rice endosperm. Science 287(5451): 303-305.

Yu, D., Liu, Y., Fan, B., Klessig, D.F. and Chen, Z. (1997). Is the high basal level of salicyclic acid important for disease resistance in potato? Plant Physiol. 115: 343-349.

Yu, D., Xie, Z., Chen, C., Fan, B. and Chen, Z. (1999). Expression of tobacco class II catalase gene activates the endogenous homologous gene and is associated with disease resistance in transgenic potato plants. Plant Mol. Biol. 39: 477-488.

Zedalis, R.J. (2002). GMO food measures as Restrictions under GATT Article XI (1). Eur. Envir. Law Rev. Jan 2002: 16-28.

Zhu, B., Chen, T.H.H. and Li, P.H. (1996). Analysis of late blight disease resistance and freezing tolerance in transgenic potato plants expressing sense and antisense genes for an osmotin-like protein. Planta 198: 70-77.

http://www.shef.ac.uk/content/1/c6/01/42/04/TRANSFEC.pdf Accessed on 26 March 2005.

http://www.anrcatalog.ucdavis.edu/pdf/8077.pdf. Accessed on 25 March 2005.

http://www.dragon.zoo.utoronto.ca/~jlm-gmf/T0301C/technology/microinjection. Accessed on 25 March 2005.

http://www.ers.usda.gov/publications/aib762. Accessed on 22 March 2005.

3

Rhizobial Production Technology

*Neung Teamroong and Nantakorn Boonkerd**

INTRODUCTION

Legumes have a reputation for maintaining soil fertility because they can assimilate nitrogen (N) from the atmosphere through symbiotic biological N_2-fixation (BNF) with rhizobia. Legumes can derive most of their nitrogen requirement and supply some fixed nitrogen to the soil system through BNF. Thus, including legumes in the cropping system can substantially reduce the input of nitrogen fertilizer. To be more beneficial to legumes and cropping system, the rhizobial inoculant should contain high effective strains of rhizobia.

Nowadays, rhizobia, the N_2-fixing symbionts of legumes are a good model in which genetics, biochemistry and ecology can be brought together. In the past, rhizobia were divided into two groups depending on growth rate: *Rhizobium* and *Bradyrhizobium*. However, rhizobial diversity has notably progressed in the last decade, mainly due to generation of 16S rDNA sequencing and polyphasic taxonomic approaches. In the first volume of this series (Microbial Biotechnology in Agriculture and Aquaculture published in 2005), Zahran has discussed the significance of rhizobial N_2-fixation, interactions of rhizobia and legumes, and stress impacts of rhizobial N_2-fixation in agriculture (Zahran, 2005). In this chapter, new developments in *Rhizobium* ecology, recent progress towards controlling infection in the field and new inoculation technology have been explained.

School of Biotechnology, Institute of Agricultural Technology, Suranaree University of Technology, Nakhon Ratchasima 30000, Thailand

*Corresponding author: E-mail: nantakorn@ccs.sut.ac.th

CLASSIFICATION AND STRAINS SELECTION

Genetic Diversity of *Rhizobium*

Based on the small subunit (SSU) rRNA gene, the members of Rhizobiaceae are part of the α-subdivision of the purple bacteria or the α-proteobacteria. In addition, in the current taxonomy of rhizobia, a polyphasic approach has been emphasized which depends on the data from phenotypic, genomic and phylogenetic characterizations. To date, six genera, including more than 27 species have been described for these bacteria and they form several deep branches, represented by *Azorhizobium, Bradyrhizobium, Mesorhizobium, Allorhizobium, Sinorhizobium,* and *Rhizobium*.

1. Genus *Azorhizobium*

This genus consists of only one species, *A. caulinodans* which was proposed for the bacterial nodulation in the stems and roots of *Sesbania rostrata* growing in Africa (Drcyfus et al., 1988). However, *Azorhizobium* and *Xanthobacter* have very similar rRNA cistron. The close genetic relationship between *Azorhizobium* and *Xanthobacter* is supported by up to 98.2% sequence similarities of the SSU rRNA genes.

2. Genus *Bradyrhizobium*

Bradyrhizobia are clearly differentiated from the other legume symbionts because they grow slowly and produce an alkaline reaction and no serum zone in litmus milk. To date, three species, including *B. japonicum, B. elkanii,* and *B. liaoningense,* have been widely recognized, in addition to many unnamed lineages. Recently, another novel species of *Bradyrhizobium* was found in China called *B. yaunmingense,* based on legume species belonging to the genus *Lespedeza* nodulation and numerous methods for characterization. The type strain of novel species CBAU 10071 (T) (=CFNEB 101 CT) formed ineffective nodules on *Medicago sativa* and *Melilotus albus* but did not nodulate soybean (Yao et al., 2002).

3. Genus *Mesorhizobium*

This genus is recognized by its slow or moderately slow growth and acid production. It consists of species: *M. ciceri, M. loti, M. mediterraneum, M. tianshanines, M. huakii, M. amorphe* and *M. plurifarium*. However, Wang and Martinez-Romero (2000) suggested that the species in this genus have diverged and that 16S rRNA gene sequencing is not sensitive enough to distinguish them.

4. Genus *Allorhizobium*

Allorhizobium consists only of a single species *Al. undicola*, which efficiently nodulates *Neptunia natans*, found first in Senegal (de Lajudie et al., 1998). The closest relationship based on 16S rRNA gene sequence of *Al. undicola* is *Agrobacterium vitis* (Wang and Martinez-Romero, 2000).

5. Genus *Sinorhizobium*

This genus has no common phenotypic features which can be used to distinguish species from those in the genera *Allorhizobium* and *Rhizobium*. Currently the genus *Sinorhizobium* consists of six species, *S. meliloti* (formerly *R. fredii* the fast growing soybean isolates from China), *S. saheli, S. terangae, S. xinjiangense,* and *S. medicae* (nodulates alfalfa and several annual medics). The novel species which has just been proposed is *S. kummerowiae* with isolate CCBAU 71042 (T) (=AS 1.3045 (CT) as the type strain (Wei et al., 2002) isolated from *Kummerowia stipulaceae*.

6. Genus *Rhizobium*

The group of this genus was found to have similarity with the group of *Agrobacterium* based on 16S rRNA sequences. Several species belonging to this genus are *R. giardinii, R. galegae, R. huautlense, R. hainanense, R. galicum, R. mongolense, R. etli, R. leguminosarum* and *R. tropici* (type A and B).

Recently, some novel species of *Rhizobium* have been proposed, such as *R. indigoferae* sp. nov. (isolated from the root nodule of indigofera growing in the Loess Plateau in North-western China) (Wei et al., 2002), *R. yanglingense* sp. nov. isolated from wild legumes *Amphicarpaea trisperma, Coronilla varia* and *Gueldenstaedtia multiflora* growing in arid and semi-arid regions in North-western China (Tan et al., 2001). This species is mostly related to *R. galicum* and *R. mongolense, R. sullae* sp. nov. isolated from *Hedysarum coronarium* L. (Squartini et al., 2002) and *R. loessense* sp. nov. newly isolated from *Astragalus* and *Lespedeza* species growing in the Loess plateau, China. Also, this novel species is closely related to *R. galegae* and *R. hautlense* (Wei et al., 2003).

Moreover, the discovery of another rhizobial branch containing genus *Methylobacterium* was recently reported. Rhizobia isolated from *Crotalaria* legumes were assigned to a new species, *Methylobacterium nodulans*. These rhizobia facultatively grow on methanol. Genes encoding two key enzymes of methylotroph and nodulation were also found. Phylogenetic sequence analysis demonstrated that *M. nodulans* NodA is closely related to *Bradyrhizobium* NodA (Abdoulaye et al., 2001).

At present, molecular phylogeny has been extended to genus other than 16S rRNA genes. Recently, sequence data from 23S rRNA have been widely used to estimate relationships among rhizobial species. However, it seems that the average rate of sequence is faster in 23S rRNA genes than in 16S rRNA genes, thus, it should be more valuable for estimation of close relationships (Wang and Martinez-Romero, 2000). In addition, phylogeny divided from symbiotic genes is also constructed for comparative study. Both *nif* and *nod* genes have been analyzed and compared with 16S rRNA data. The results from several researchers revealed that symbiotic genes have different evolutionary histories from the 16S rRNA genes. These also support the transfer of symbiotic islands and the lateral transfer of other chromosomal genes. Nevertheless, some other sets of genes as gyraseB (*gyr*B) were also used in term of comparative 16S rRNA sequence of *B. japonicum* and *B. elkanii.* The results showed marginal difference in topology of cluster but *gyr*B is more similar to the phylogenetic tree constructed using ITS (Internal Transcribed Spacer) (Sameshima et al., 2003).

Prior to creating collections of rhizobial strains to be used as inoculants or to serve as gene pools for future inoculant development, several attempts to investigate rhizobial diversity in each geographic origin have been explored. For example, in case of Senegal, diversity of indigenous rhizobia associated with three cowpea cultivars (*Vigna unguiculata* [L.] Walp.) grown under limited and favorable water conditions indicated that more diversity of *Bradyrhizobium* was found in water-limited conditions (Karasova-Wade et al., 2003). In addition, diversity of rhizobia from 27 tropical leguminous species native of Senegal also showed that bradyrhizabia are the main population (Doignon-Bourcier, 1999).

In case of environmental effect on rhizobial diversity in Thailand, the investigation was conducted from various ecological types from 1997-1999. Soil samples were collected from undisturbed forests, forest clearance for crop cultivation for one, two and three years, and from the areas where intensive agricultural production was practiced using high rate of pesticides and fertilizers, such as vegetable plantation. In case of rhizobia, it was found that highest population persisted at the foothills of mountains and under the agricultural area of rice in rotation with leguminous plants. The important factors affecting the population were temperature and moisture. Moreover, the areas under legume plantation were also the appropriate habitat for rhizobia. However, in some areas where vegetable plantation was conducted, a negligible or non-detectable amount of rhizobial population was found. It was also observed that the dominant native genus was *Bradyrhizobium* sp. which mainly nodulated, was only from a group of cowpeas and not soybean. When these strains were analyzed with RAPD-PCR technique, the results suggested that these groups of bradyrhizobia contained close relations among the group.

According to the RAPD-PCR technique, some native bradyrhizobial strains established the ability to persist across all seasons throughout the year (Matpatawee, 1999). Moreover, 130 symbiotic and non-symbiotic strains of *Bradyrhizobium* were directly isolated from inoculated soybean and uninoculated legume-free virgin field soils in Thailand using a direct selection medium. About 47% and 58% of the isolates, obtained from inoculated and uninoculated fields, respectively, were characterized as being non-symbiotic bradyrhizobia. Partial and nearly full-length sequence analyses of regions encoding 16S rRNA indicated that the non-symbionts were closely related to *B. elkanii* (17-99% identity) and *B. japonicum* (96-100% identity). Southern hybridization analyses showed that DNA from the non-symbiotic bradyrhizobia failed to hybridize to *nif* and *nod* gene probed. rep-PCR DNA fingerprint analyses made using the BOXA1R primer, indicated the symbionts and non-symbionts could be separated into two distinct clusters. There was no relationship between the geographical origin of isolates and groups based on serological reaction or their ability to nodulate soybean, cowpea, mungbean, or siratro. These results indicated that a relatively large percentage of non-nodulating bradyrhizobia are present in Thai soils (Pongsilp et al., 2002).

Strain Selection

Inoculant production can be divided into three phases. The first relates to selection and testing of rhizobial strains; the second, inoculant preparation, including selection and processing of material carrier and mass culture of rhizobia; and the third, quality control procedures and distribution.

The first step in manufacturing inoculants is to obtain highly effective N_2-fixing strains of rhizobia for the legumes to be inoculated since rhizobial strains within a species vary in their effectiveness (Kucey et al., 1988). The effectiveness of a strain of rhizobia is due to the genetic interaction with the host plant, which is known as host-strain specificity. A strain of *Rhizobium*, which is effective on one legume cultivar, may not be highly effective on the others. Thus, the selection of strains of rhizobia for cultivated legumes is a critical step in the production of legume inoculant. In most countries, the government normally undertakes this activity through its research facilities. Actually, promotion of plant biomass, nodule dry weight and nodule number are the main criteria to select an appropriate strain. However, some studies suggested that nitrogen accumulation is a more effective criterion for selecting the strain. Several strain selection developments are being conducted in both developed and developing countries. As in New Zealand, selection of inoculant *Rhizobium* strain for improved establishment and growth of hexaploid Caucasian clover (*Trifolium ambigumm*) was employed (Heather et al., 1998). The selected strain of *R. leguminosarum* bv. *trifolii* strain ICC148 is more favorable than other recommended New Zealand commercial inoculants (Pryor et al., 1998). In Erzurum, cold highland areas in Turkey, the native rhizobia were selected for planting the chickpea (*Cicer arietinum* L. cv. Aziziye-94). *Rhizobium leguminosarum* spp. *ciceri* strains HF274 and HF177 showed promising results which could be used as inocula in a particular plant and area, since both the selected strains increased seed yields by approximately 23% over uninoculated strain and about 7% higher yield than using standard rhizobial culture (Kantar et al., 2003).

In the case of strain selection for pea (*Pisum sativum* L. spp. *sativum*) in Russia, the program of breeding for high efficiency in *Pisum sativum*/ *R. leguminosarum* symbiosis was also established. From 481 rhizobial isolates tested in pot trials on the basis of N-accumulation in plants and high competitiveness, the commercially valuable strains of *R. leguminosarum* bv. *viceae* were obtained. Under the field condition the highly effective isolate (52) significantly exceeds the former commercial strain 250a in its influence on seed mass (by 20.3%) and on nitrogen accumulation in seeds (by 25.7%) (Provorov et al., 2000). For the Thailand reforestation program, selection, and management of rhizobia for tree legumes was also elucidated. Root nodules from five trees from each of the following tropical leguminous species–*Acacia mangium*, *A. auriculaformis*, *Pterocarpus macrocarpus*, *Xylia kerrii*, and *Milletia leucantha*, from 25 reforested sites each in North, Northeast, Middle, and South of Thailand were collected. Six hundred rhizobial strains were isolated, purified and sub-cultured from these nodules. The efficiency of each individual strain on nodule formation, acetylene reduction assay and host plant biomass was evaluated using Leonard's jar technique under glasshouse condition. The results indicated that effective rhizobial strain increased plant biomass compared to ineffective and uninoculated treatments. Ten most efficient rhizobial strains for each tree species were then tested in pot culture on low fertile soil. Among the 10 strains for each tree species, three most effective strains were chosen based on their ability to perform in their host plant. Inoculation of *Rhizobium* alone or with phosphorus and potassium amendment resulted in higher plant productivity compared to uninoculated treatment. Serological and 16S rDNA sequence techniques indicated that the selected rhizobial strains belong to the genus *Bradyrhizobium*. These strains will be chosen for further inoculant production for tree legumes. Therefore, before producing inoculants the most suitable rhizobia for each target legume should be selected (Manassila et al., 2004).

TRENDS FOR RHIZOBIAL STRAIN IMPROVEMENT

Nodulation competitiveness of the inoculum strains often fails to infect target plants in natural ecosystems. When inoculum is applied to soils that already contain indigenous rhizobia, the inoculant strain is almost unsuccessful (Zahran, 2005, Chapter 3 in Volume I of this series). The reduction in percentage of nodules formed by the inoculant strain occurred when the indigenous strains were at a level of only 10 rhizobia per gram (Thies et al., 1991). Even though introduced strains were outnumbered by indigenous soil populations by as much as 250:1, they were not evenly distributed throughout the soil and often not well adapted to general soil conditions (McDermott and Graham, 1989; Brockwell et al., 1995). Therefore, inoculant rhizobia may remain only from 5-10% of the nodule produced in the first year in field grown legumes (Sadowsky and Graham, 1998).

Several attempts have been made to overcome this problem. The factors including environment and soil conditions are being improved to reduce the strain competitiveness on the basis of modern biotechnological approach.

1. Environmental Factors

Environmental factors are known to influence the competitiveness of specific rhizobial inoculants, which include soil water, nutrient limitation, soil acidity, soil temperature, etc. Some of these factors might directly affect competitiveness, as they perhaps act by altering the persistence and survival of inoculated strains and only indirectly influence competitive interaction.

Soil Water

Both establishment and activity of the legumes–*Rhizobium* symbiosis have been found to be extremely sensitive to drought stress. In dry condition, plant growth, root architecture and root exudates can be affected, which can also result in low population level of rhizobia in soil. Therefore, symbiotic N_2-fixation is highly sensitive to drought, which results in decreased nitrogen accumulation. Numerous studies have indicated that a nitrogenous signal(s), associated with nitrogen accumulation in the shoot and nodule, exists in legume plants so that N_2-fixation is inhibited early in soil drying (Serraj et al., 1999).

Genetic improvement of legumes for increased tolerance of N_2-fixation to water deficits have been employed. Since greater sensitivity to drought stress is related to relatively high petiole ureid levels, therefore using petiole ureid measurement to screen a large number of legumes (soybean) lines is widely investigated. Among the lines with low petiole ureid, lines were ultimately identified that have a substantial level of drought tolerance of N_2-fixation (Purcell et al., 1998). In addition, tropical legumes, which transport ureids, appear to be more sensitive to soil drying. This includes some species such as cowpea and pigeon pea, which have the reputation of being generally tolerant to drought in terms of plant survival (Sinclair and Serraj, 1995). Thus, existence of genetic variation among legume cultivars opens the possibility for enhancing N_2-fixation tolerant to drought through selection and breeding.

In case of rhizobia, results from several studies have suggested that bradyrhizobia are more resistant to water stress than rhizobia. However, some gave contradictory results. This might have been caused from intrageneric strains diversity differences (Sadowsky and Graham, 1998). As generally recognized that proline is accumulated in soybean and alfalfa plants when grown under salt and drought stress, van Dillewijn et al., (2001) has constructed a constitutive promoter which was placed upstream of the *put*A gene (encode for proline dehydrogenase enzyme, catalyzing the oxidation of proline to glutamate). The presence of multicopy plasmids containing this *put*A gene construct increased the competitiveness of *S. meliloti* in microcosm experiments in nonsterile soil planted with alfalfa plants subjected to drought stress altogether within the field level at Leon, Spain. The results showed the frequency of nodule occupancy was higher.

Nutritional Factors

Nutritional factors in rhizosphere influence the growth of rhizobia. Nitrogen limitation as the classic phenomenon is being investigated particularly with bradyrhizobia association. Recently, Lopez-Garcia et al., (2001) demonstrated that rhizobial nitrogen starvation has positive influence on the symbiosis of *B. japonicum* with soybean plants. The production of exopolysaccharide (EPS), capsular polysaccharide (CPS), and soybean lectin binding (SBL) resulted in increased nodulation efficiency and competitiveness. In case of plant root exudation, some percentage can be used as a carbon and/or nitrogen source. Myo- inositol is abundant in soil and several microbes including *R. leguminosarum* bv. *viciae* and *S. meliloti* can grow using inositol as the sole carbon source (Poole et al., 1994; Galbraith et al., 1998). Galbraith et al., (1998) recently found that myo- inositol dehydrogenase gene (*idh*A) from *S. meliloti* is essential for inositol catabolism.

It was again confirmed by Jiang et al., (2001) that a functional *idh*A gene is required for efficient N_2-fixation and for competitive nodulation of soybean by *S. fredii* USDA191. Thus, another approach to introduce *idh*A gene into a selected strain of rhizobia might enhance nodulation competitiveness. Moreover, an alternative approach to looking for catabolic genes important in the soil and rhizosphere is to look for genes induced by root exudates, or by specific compounds known to be abundant in the rhizosphere and plant root. Therefore, it is desirable to generate a bank of mutants made with different promoter probe transposons in order to screen for inducible genes.

In addition, chemotactic ability and/or mobility confer a competitive advantage on rhizobia. Interactions between chemo-attractants and their receptors appear to affect the regulation motility in response to source availability (Wei and Bauer, 1998). Since rhizobia are able to use such a variety of different carbon and nitrogen sources, the genes homologous to MCP (methyl-accepting chemotaxis protein) type chemoreceptors specific for compounds have been investigated. A family of at least 17 genes, and probably over 20, exhibiting homology to MCP genes was detected in plasmid of *R. leguminosarum* VF 39. The complete sequences of three genes, *mcp*B, *mcp*C and *mcp*D were elucidated. The *mcp*B gene was responsible for generalized chemotactic response, and both *mcp*C and *mcp*D genes were important for competitive nodulation (Hynes et al., 2002).

Soil pH

Soil acidity effects on legume nodulation include reduced survival and poor growth of rhizobia in soil and seed, reduced attachment and root-hair infection. Actually, acid pH has a detrimental effect on the fast growing rhizobia than bradyrhizobia (Sadowsky and Graham, 1998).

One of physiological basis in pH tolerance is calcium metabolism. For example, *Bradyrhizobium* sp. strain SEMIA G144 (peanut symbiont) showed a growth and viability diminution at low pH 5.0 and at a calcium concentration 0.05 mM (Maccio et al., 2002). Increasing concentrations of calcium significantly improved the rhizobial growth under acid stress conditions (Zahran, 2005). The biomolecule related to this phenomenon, such as expolysaccharide (EPS) and lipopolysaccharide (LPS), showed changes in their electrophoretic motility at different pH values and calcium concentrations. Therefore, a possible relationship exists between the presence of calcium in the growth medium and polysaccharide synthesis. Introduction of rhizobial strains with acid tolerance property has been conducted in the central region of Argentina. This area was found to become progressively more acidic over the last two decades. The inoculant was applied for alfalfa; however, the response to inoculation with specific rhizobia was strongly influenced by native soil rhizobia that competed with the introduced strains (Broughton et al., 2003).

The genes identified as essential for acid tolerance are probably only a small proportion of the total genes required (suggested to be between 20-

50 genes). However, the *act*S and *act*R genes from *S. meliloti* play important roles in acid tolerance. *Act*S sequences are most likely to be membrane-located, and a typical sensor C-terminal domain, whilst *Act*R which is suggestive of DNA binding role; has apostate and lysine residues typical of regulators. In addition, *phr*R and *lpi*A genes are also found to have ability to be expressed when introduced with low pH condition (Dilworth et al., 2001). The question regarding regulation of these genes remains unresolved.

Soil Temperature

Soil temperature is another variable that has important effects on nodulation of legumes. As an example, soybean is evolved under sub-tropical conditions and its optimal growth temperature is 25-30°C (Smith, 1995). Soil temperature below this range restricts nodulation and N_2-fixation. Low temperatures also decrease biosynthesis and secretion of genistein (a flavonoid) from plant roots (Zhang and Smith, 1996). Therefore, applying flavonoid to soybeans growing at low soil temperatures, might significantly improve nodulation and N_2-fixation. However, methods of making them resistant to degradation by hydrolytic enzymes while maintaining in soil have yet to be solved.

Another approach using co-inoculation with PGPR (Plant Growth Promoting Rhizobacteria) under sub-optimal root zone temperature has been found successful. Bai et al., (2002) coinoculated PGPR strains, *Serratia proteomaculans* 1-102 and *S. liquefaciens* 2-68 with *B. japonicum* on soybean under controlled root zone temperatures 25, 20, and 15°C. The results indicated that the effective dose for both PGPR strains is 10^8 cells per seedling. The co-inoculation of PGPR increased nodule number, plant biomass and fixed nitrogen, as well as shortened the nodule initiation time and increased the nodulation rate. The interactions of rhizobia with other beneficial soil microorganisms have been elaborately discussed by Singh (2005) in Chapter 6 in Volume I of this series–Microbial Biotechnology in Agriculture and Aquaculture, published in 2005.

2. Biotic Factors

Biotic factors influencing competition for nodulation include: exopolysacclaride, nod factors, some antibiotics production, hydrogen recycling system, and some indigenous substances, which are detailed below.

Expolysaccharide and Cell Surface

Exopolysaccharide (EPS) forms the outer layer of the cell surface, which contributes to cell protection against environmental influences, attachment to surfaces, nutrient gathering and antigenicity. Furthermore, the structural diversity enables EPS to function as informational molecules in cell-to-cell communications. The ability to establish an effective symbiosis is severely affected in mutants of many rhizobia deficient in EPS production. Bhagwatt and Keister (1991) found that *B. japonicum* strains deficient in expolysaccharide synthesis were less competitive than wild-type strains. Overproduction of EPS I oligosaccharide can also have severe effects on nodule development. However, mutants of *S. meliloti* SU47 overproducing EPS I are able to infect alfalfa nodule. In some symbiotic systems, addition of purified EPS might restore the ability of EPS mutants to infect nodules (Becker and Puhler, 1998). Another gene found in *R. etli*, which plays a critical role in both determination of cell surface characteristics and nodulation competitiveness, is *ros*R, which has sequence similarity to a family of transcriptional repressor involved in EPS production. When the *ros*R mutants were constructed, changes in colony characteristics and competitiveness in *Phaseolus vulgaris* nodulation were also observed (Bittinger and Handelsman, 2000).

Even then, the regulation of EPS production is not clearly understood. Many different regulators were identified in rhizobia. The environmental factors influencing EPS production and mechanisms of regulation still need to be investigated. However, previous work of Sikora and colleagues at University of Bonn, Germany, demonstrated that living and heat-killed cells of *R. etli* G12 induced in potato roots increased systematic resistance against plant-parasitic nematode *Globodera pallide*. They found that lipopolysaccharide (LPS) reduced nematode infection significantly in concentrations as low as 1-0.1 mg ml^{-1}. The LPS of *R. etli* G12 seems to act as the inducing agent of systematic resistance in potato roots (Reitz et al., 2000).

Genes Influencing Efficiency of Nodulation

In general, symbiotic interactions are controlled by a signal exchange between the two partners. Plants secrete flavonoid, phenolic compounds, which in conjunction with the bacterial activator protein NodD, induce the expression of rhizobial nodulation (*nod, nol,* and *noe*) genes. As a result, rhizobia produce Nod-factors. A few years ago, attempts were made to increase Alfalfa nodulation, N_2-fixation, and plant biomass by specific DNA amplification (SDA) in *S. meliloti*. Recombinant plasmid contained *nod*D1, *nod*ABC, and an essential *nif*N from the nod regulon of the symbiotic plasmid of *S. meliloti*. The results suggested that the copy number of symbiotic region was critical in determining the plant phenotype. In case of the strains with moderate increase in copy number, symbiotic properties were improved significantly. The inoculation of alfalfa with transconjugant strains resulted in an enhancement of plant growth (Castillo et al., 1999). Another gene, which was presumably involved in nodulation competitiveness in common bean with *R. etli*, is *slp* (stomatin like protein). Yon et al., (1998) found that *slp* gene may provide a growth or competition advantage to rhizobial strain in the infection thread. The mutant strain without *slp* may have a disadvantage in ion exchange and nutrient uptake, and loss of capacity for competition for nodule invasion. However, *slp* gene is not present in all rhizobia. *Slp* is present in *R. etli*, *R. leguminosarum* bv. *phaseoli* and *R. tropici* type A, but is absent in *R. tropici* type B, *Bradyrhizobium* and several other *Rhizobium* spp.

Another set of genes that have recently been found responsible for tropical legume specificity determinant is type III secretion system (TTSS). However, TTSS is found in several bacterial pathogens, and actually can be induced on contact with host cell. TTSS was first clarified in *Rhizobium* sp. NGR234. The genes that encode components of TTSS are in symbiotic plasmid. The secretion of at least two proteins (y4xL and NolX) is controlled by the TTSS. When both proteins are blocked, the nodulation ability is changed. In addition, the most important role of TTSS would occur after the bacteria have entered the plant with the secretion of both proteins being part of the second signal (Viprey et al., 1998). In case of TTSS in *S. fredii*, one of two genera of rhizobia that forms N_2-fixing nodules on soybean is also well elucidated. *S. fredii* USDA257 is known to form active N_2-fixing nodules on primitive soybean cultivars such as Peking, but fails to nodulate agronomically improved North American Cultivars (Keyser et al., 1982; Heron and Pueppke, 1984). A different strain of *S. fredii* (USDA191) forms nodules and fixes nitrogen with both primitive and advanced soybean cultivars (Heron et al., 1989). This host specificity determinant is most likely controlled by two types of proteins secreted from TTSS (Jiang and Krishnan, 2000). Moreover, soybean cultivar specificity is regulated by *nolXWBTUV* locus, which encodes part of TTSS. From the results of several experiments, it is suggested that some proteins secreted via the TTSS can have either deleterious or beneficial effects on nodulation depending on the host. Especially, NoLX appears to be beneficial to cultivar Peking nodulation rather than cultivar McCall nodulation. Since TTSS loci in bacterial pathogen are involved in preventing the host defense response, this NoLX may be perceived as pathogenic by MaCall soybean roots, triggering a defense response that leads to the aborting of infection threads (Krishnan, 2002).

Recently, quorum sensing, which plays a large role in rhizobia-host interaction, is being greatly emphasized. Further, the interaction in terms of population density-dependent prior to establishing the symbiosis can be more elucidated. The most widespread form of signal molecules found among members of Rhizobiaceae and Gram-negative bacteria are the acyl homoserine lactones (AHLs). These molecules accumulate with increasing population density and trigger gene expression when threshold levels are obtained (Loh et al., 2002a, b). The mutants deficient in AHL production and regulation are known to affect the ability of *Rhizobium* to nodulate their respective hosts. In case of *B. japonicum*, Loh and group discovered the novel signaling molecule (cell density factor) named bradyoxetin (Loh et al., 2002b). Bradyoxetin is different from other Gram-negative quorum signals. The proposed structure of bradyoxetin is 2-[4-[[4-(3-aminooxetan-2-2yl) pheny1]-(imino) methyl] phenyl] Oxetan-3-ylamine. *Bradyrhizobium japonicum* joins a growing list of bacteria that utilize non-AHL-type signaling molecules. In addition, bradyoxetin which regulates *nod* gene activity, has antibiotic activity and inhibits growth of competing bacteria in the rhizosphere. Besides these modes of actions, bradyoxetin is also similar in nature to munegeic acid, a known siderophore. Consistent with the fact that bradyoxetin regulates *nod* gene expression, a corresponding increase in *nolA* expression and a concomitant decrease in *nod* gene expression were noted under iron-starved condition (Loh and Stacy, 2003).

Antirhizobial Compounds and Related Substances

Production of antirhizobacterial compounds by rhizobial inoculant strains could be a useful strategy in limiting nodulation by indigenous strains. One of antirhizobacterial compounds that has a potential for enhancing the nodulation competitiveness is trifolitoxin (TFX). TFX is ribosomally synthesized and this post-translationally modified peptide is produced by *R. leguminosarum* bv. *trifolii* T24 (Breil et al., 1993). TFX inhibits members of a specific clade of the α- subdivision of the division Proteobacteria that includes legume symbionts, plant pathogens, and animal pathogens. Under laboratory conditions, the TFX production phenotype has been shown to significantly increase the nodulation competitiveness of *Rhizobium* and *Sinorhizobium* strains (Scupham et al., 1996; Robleto et al., 1997). The field efficacy studies of *R. etli* CE3 on *P. vulgaris* cv. *Azteca* were tested. The TFX production had no effect on grain yield over three years; the average yield reached 2,400 kg ha^{-1}. The nodule occupancy was increased under field conditions without adverse effects on grain yield (Robleto et al., 1998). Another antirhizobacterial compound known as bacteriocin, is also of significance. *Rhizobium leguminosarum* strains have been shown to produce bacteriocin, which has been well characterized. *R. leguminosarum* strain 248 contains the symbiotic plasmid which includes determinants for medium bacteriocin production. Analysis of the deduced amino acid sequences revealed similarly to hemolysin and leukotoxin. Bacteriocin bioactivity was correlated with the presence of a protein of approximately 100 kDa in the culture supernatants, and the bacteriocin bioactivity demonstrated a calcium dependence in both *R. leguminosarum* and *S. meliloti*. For the nodule competition, experiments under controlled conditions showed that the bacteriocin could play a statistically significant role in competition against certain test strains (Oresnik et al., 1999).

A phytotoxin, produced by certain strains of soybean bradyrhizobia i.e. so-called rhizobitoxine, also plays an important role in nodulation competitiveness. Rhizobitoxine (2-amino-4-(2-amino-3-hydropropoxy)-*trans*-but-3-enoic acid) increases nodulation and competitiveness via inhibition of endogenous ethylene synthesis in host plants, because ethylene acts as a plant hormone that restricts, nodulation in many legumes (Minamizawa and Mitsui, 2000). Recently, Okasaki et al., (2003) demonstrated that rhizobitoxine production by *B. elkanii* USDA94 gave this strain a nodulation competitiveness about 10 times greater than that of a non-rhizobitoxine producing mutant strain on siratro plant. Rhizobitoxine enhancement of competitive nodulation occurred at the late stage during nodulation. The cluster of genes involved in the biosynthesis of rhizobitoxine in *B. elkanii* is *rtx*A and *rtx*C. These genes were located in a DNA region that includes nodulation and symbiotic N_2-fixation genes. It is very interesting to note that *rtx* cluster and *noe*E gene upstream of the cluster were almost completely conserved in *B. japonicum* USDA110, which does not synthesize rhizobitoxine (Yasuta et al., 2001).

Some adequate nodulations are caused from toxic substances released from seed and root exudates. These can inhibit the growth of nodulating rhizobacteria. In case of mimosine, a free amino acid is produced in large quantities in the leaves, seeds and other parts of legume leucaena (*Leucaena leucocephala*). It is toxic not only to animals grazing on leucaena foliage, but also inhibits growth of rhizobia (Soedarjo and Borthakur, 1998). However, while making attempts to provide the nodulation competition advantage, some rhizobial strains which are able to degrade mimosine were isolated. *Rhizobium* sp. strain 1145 can degrade mimosine under two major steps: in the first step mimosine is degraded to 3-hydroxy-4-pyridone (HP), which is then converted to pyruvate, formate and ammonia in the subsequent step. The *midD* gene regulates an aminotransferase, which is required for degrading mimosine into HP, was found (Awaya et al., 2003). Therefore, this might be a possibility to use gene manipulation technique to provide the competition advantage for selected rhizobial strains.

Genes Influencing Efficiency on N_2-Fixation and Hydrogen (H_2) Uptake System

Since symbiosis requires a respiratory chain that has a high affinity for O_2 and is efficiently coupled to ATP production because N_2-fixation is an energy-consuming process, requiring up to 20 ATP molecules to reduce just one molecule of N_2. The three-subunit terminal oxidase (*cbb*$_3$) is responsible for bacterial terminal oxidase production. Recently, attempts have been made to overproduce *cbb*$_3$ terminal oxidase in *R. etli*, which can enhance symbiotic N_2-fixation in *P.vulgaris* cv. *negro jamapa* (Soberon et al., 1999). According to the N_2-fixation, to reduce N_2 to NH_3 by nitrogenase, H_2 is evolved during reduction reaction. Thus, H_2 production, due to nitrogenase activity is a cause of inefficiency in *Rhizobium*-legume symbioses. Certain rhizobial strains synthesize a H_2 uptake (Hup) system that recycles the evolved H_2. The ability to utilize H_2 is a desirable characterization for generating more productive and efficient rhizobial inoculants. However, this phenotype has been found only in a few *Rhizobium* and *Bradyrhizobium* strains. To extend this ability to other Hup strains, Hup system from *R. leguminosarum* bv. *viciae* UPM 791 was transferred by triparental mating. The final results showed the increasing of hydrogenase activity (Brito et al., 2000). Furthermore, *R. leguminosarum* bv. *viciae* UPM 791 induces hydrogenase activity in pea (*Pisum sativum* L.) bacteroids but not in free-living cells. The symbiotic induction of hydrogenase structural gene (*hup*SL) is mediated by *nif*A, the general regulator for N_2- fixation process. In addition, bacteroids from pea plants grown in low-nickel concentration induced higher levels of hydrogenase activity and were able to recycle all H_2 evolved by nodules (Brito et al., 2002).

NEW DEVELOPMENT IN INOCULANT TECHNOLOGY

1. Mass Culture

The next step of inoculant production is the mass culture of rhizobia. Rhizobia can be cultured in flasks using mechanical agitation, a glass bottle aerated with sterile air, or in large fermenters of various designs, which are also pressure vessel. Media and methods for culturing rhizobia are detailed by Burton (1979), Roughley (1970) and Date and Roughley (1977). Carbohydrate compounds such as sucrose, mannitol, glycerol, arabinose and lactate, are the major components in the media. Prices of carbohydrate compounds differ and can affect the cost of production. Utilization of carbon compounds by rhizobia varies with species and strains within a species (Graham, 1964; Stowers and Elkan, 1984). Sucrose is a cheap carbohydrate source and can be used for culturing fast growing rhizobia successfully (Balatti et al., 1987) but most *Bradyrhizobium*, the slow growing rhizobia, cannot use this compound (Stowers and Elkan, 1984). Therefore, mannitol, which is about 20 times more expensive than sucrose, is the most frequently used carbohydrate for the production of *Bradyrhizobium*. The *Bradyrhizobium* are specific for most legumes cultivated in the tropics. To reduce the cost of production, alternate sources of carbon compounds should be explored. Glycerol, which is cheaper than mannitol, can be used as an alternative source. It has been reported by Balatti et al., (1987) that *B. japonicum*, which is specific on soybean, could use glycerol as the carbon source for their growth.

Selection of appropriate and alternate types of carbon sources should be considered with respect to the cost of sugar and genus or strain of rhizobia, which will utilize it. In the past, rhizobia were divided into two groups on the basis of growth rate. Fast-growing rhizobia are diverse in carbon utilization, whereas slow-growing bradyrhizobia appear to be nutritionally fastidious (Stowers, 1985). Therefore, to determine whether different groups of rhizobia can use carbon sources differently, another available low cost carbon source to reduce the cost production should be considered. From our research, firstly various sources of carbon utilization by five genera of rhizobia were investigated (Table 3.1).

The results in part of carbon source utilization showed different genus of rhizobia used carbon sources differently. Genus *Rhizobium* and *Sinorhizobium*, which are fast growers, could utilize various kinds of sugar,

Table 3.1. Carbon source utilization by rhizobia

C-sources	Genera of Rhizobia				
	Rhizobium	*Sinorhizobium*	*Mesorhizobium*	*Azorhizobium*	*Bradyrhizobium*
Succinate	+	+	±	±	±
Malonate	±	±	±	±	±
Tartrate	±	+	±	±	±
Fumarate	+	+	-	±	±
Gluconate	+	+	±	+	+
Ribose	+	+	±	+	±
Xylose	+	+	+	±	+
Mannose	+	+	-	±	±
Galactose	+	+	±	±	+
Arabinose	+	+	+	±	+
Cellobiose	+	+	-	±	±
Raffinose	+	+	+	±	±
Fructose	+	+	±	±	+
Glucose	+	+	+	±	±
Maltose	+	+	+	±	±
Lactose	+	+	+	±	±
Sucrose	+	+	+	±	±
Propanediol	+	+	+	±	±
Myo-inositol	+	+	-	±	±
Sorbitol	+	+	±	±	±
Glycerol	+	+	±	+	+
Mannitol	+	+	+	+	+

++ all strains can use; - no strain can use; ± >1strain can use whereas genus *Mesorhizobium*, *Azorhizobium*, and *Bradyrhizobium* are fastidious (Tittabutr et al., 2003).

It is interesting to note that the pattern of carbon utilization in each genus is somewhat related to rhizobiaceae phylogeny. Since the phylogenetic analysis using 16S rRNA genes reveals the different origins of the rhizobial genera, the genera *Azorhizobium*, *Bradyrhizobium* and *Mesorhizobium* have a distant relationship to *Rhizobium* and *Sinorhizobium* in the Rhizobiaceae.

As the inoculants produced in Thailand have been emphasized particularly on soybean, mungbean, and peanut, and the rhizobia used are species of *Bradyrhizobium*, thus mannitol is the sole carbon source for producing *Bradyrhizobium* inoculant. However, mannitol is very expensive when it is used at industrial level. As such, alternative carbon sources from agricultural raw materials to replace mannitol has also been investigated. It was found in our laboratory that saccharification of cassava tuber by amylase-producing fungi strains SUT-1, followed by glycerol conversion from glucose under alkaline condition by yeast, *Saccharomyces cerevisiae* (isolate F) could produce a high yield of glycerol, which is comparable to mannitol for culturing bradyrhizobia. By using this process, *Bradyrhizobium* can be cultured with high number. *Bradyohizobium japonicum* USDA110 was used for cultivation and it was found that the population number of *B. japonicum* reached 10^9 cfu ml^{-1} in 5 % fermented medium, which gave the same yield as cultivated in Yeast-Mannitol medium. This strongly suggested that glycerol and other nutrients including vitamins provided from yeast in the medium are sufficient for bradyrhizobial growth.

2. Carrier Preparation

Carrier material is an important component of inoculant. The root nodule bacteria, rhizobia, must be in suitable carriers for application to legume seeds or seedbed. Carriers must have a high water-holding capacity, provide nutrients for growth of rhizobia, and favor their survival both during distribution and when inoculated onto seed (Date and Roughley, 1977). Although it has been reported that various carrier materials have been used in inoculant production (Giles and Atherly, 1981), finely ground peat is used extensively and is still the preferred carrier (Burton, 1967; Date and Roughley, 1977; Peterson and Loynachan, 1981; Boonkerd, 1991). Lack of high quality peat in some areas has forced scientists to examine alternate carriers. Coal, charcoal, bentonite and corn oil have been reported to be as good as peat when they were used as carriers (Paczkowski and Berryhill, 1979; Kremer and Peterson, 1983). However, Halliday and Graham (1978) did not recommend coal as an inoculant carrier since it tended to form hard lumps during storage and resisted wetting at the time of seed inoculation. Kremer and Peterson (1983) found that the quality of oil based inoculant (freeze dried rhizobia in a vegetable oil base) remained high even when stored at high temperature. In one test, the oil based inoculant had 10^5 cells per gram after storage for 56 days at 60°C compared to peat inoculant which contained only 10 rhizobia per gram. Survival of rhizobia at high temperature is important for retaining the quality of inoculant to be delivered to farmers in tropical countries.

Successful inoculation, as determined by legume response to inoculation or the inoculant strain forming many nodules in the crop, depends on inoculant quality. Inoculant quality is measured by the number of effective rhizobia per unit weight of carrier. Providing farmers with high quality inoculant is a worldwide problem. It is generally agreed that the highest quality inoculant is prepared by using a sterile peat carrier. Inoculant produced from presterilized peat contains more than 1×10^9 rhizobia per gram at manufacture (Roughley and Pulsford, 1982).

Rhizobial inoculant production is a time- and cost-consuming process. A large amount of medium and a long period of time are necessary for producing a high concentration of rhizobial cells before mixing with the carrier. However, it has been reported that peat is an organic matter, which provides some nutrients to promote the growth of rhizobia. Therefore, the low concentration of rhizobial cells would be possible to grow on peat. The rhizobial inoculant production by using dilution technique and solid state fermentation was one of the research priorities at our Institute. The rhizobia from all the six genera were used for inoculant production. The concentrated rhizobial cells from each genus were diluted to be 10, 100, and 1,000-folds. Both the diluted and undiluted rhizobial cells were injected into peat, which was adjusted to the moisture level at 0, 20, 25, and 30%, and finally at 40%. The results showed that the amount of cells from each treatment was not significantly different when compared to the same genus. In genus *Rhizobium* and *Mesorhizobium*, the number of cells dramatically increased after one week and then slightly increased during one month. Then the number was almost stable or slightly decreased to 10^7 cells g^{-1} peat after four months. However, the number of cells in genus *Bradyrhizobium* and *Azorhizobium* increased after one week, and were still stable until six months at 10^9 cells g^{-1} peat. Therefore, the dilution technique and solid state fermentation could be applied for low cost rhizobial inoculant production (Tittabutr et al., 2003). Carrier sterilization can be done by autoclaving or gamma irradiation. Somasegaran (1985) reported that inoculant produced in irradiated peat was superior to the autoclaved peat, while the study at the Thai Department of Agriculture found no difference (Boonkerd, 1991). The storage characteristics of inoculant produced in sterilized carriers are superior. Sterilized carrier inoculant can be stored at room temperature (25°C-35°C) for 8-12 months while the non-sterilized carrier can be maintained only for four months. Most carriers are the solid forms, especially peat. Recently a liquid form of inoculant has been produced by LIPHATECH company (Nethery, 1991).

Another important aspect for rhizobial inoculant production research is the appropriate carrier. Even though peat is widely used as an inoculum carrier, its limitation still exists due to high cost in the pretreatment process and complicated machineries. Thus, it renders the need for further development, particularly in introducing a novel type of carrier.

Liquid carrier is another option to replace peat utilization. Most of the liquid carriers used are polymers. Their properties are: good water-holding capacity, high binding capacity with the seeds, and the ability to detoxify some substances from seed exudates. In case of *B. japonicum* USDA110, it was found that by using tapioca starch solution, Polyethyleneglycol (PEG) 3000, Polyvinylpyrrolidone (PVP), Arabic gum and sodium alginate as liquid carriers, showed high potential to maintain the population number as 10^8 cells ml^{-1} for six months, at room temperature (32-35°C). When comparing the effect of both carriers *B. japonicum* inoculants produced from fermented and yeast-mannitol media in field-test examination, the results showed no difference in statistics in terms of nodule number, plant biomass and grain yield of soybean (Tittabutr et al., 2003).

In addition, the effect of type carrier for inoculant in the field was also investigated. The results demonstrated very promising data i.e. liquid carriers are able to replace peat-based carrier. Thus, the technology developed here is very useful for producing low cost inoculant. It is suitable for developing countries where glycerol and mannitol are quite expensive and sometimes not available. Liquid carriers have also proved useful as alternative good quality carriers to peat, which is difficult to find in some countries.

3. Factors Affecting the Scale of Inoculant Production

Production units for inoculant can vary from capacities of less than a ton to over 200 tons $year^{-1}$. Appropriate scale depends on the country's requirement, method of production and conditions in the distribution network. Most countries in Asia, except Thailand, produce less than 50 tons inoculant $year^{-1}$. The limitation of production is based on government policies, inadequate farmers' demand and extent of legume cultivation. Government policy promoting inoculant production is normally based on the evaluation of the need for inoculation in each country. Thus, scientists who are concerned with legume production and rhizobial technology must provide convincing evidence that the application of inoculant is cost effective and an appropriate technology for local farmers.

Very large-scale inoculant production normally employs a non-sterile carrier. This type of inoculant used to be produced in the USA because the demand was high and inexpensive cold storage conditions were available (inoculant is produced in the fall and winter for distribution at spring planting). This is a great advantage because cool climate protects the quality of inoculant. If this type of inoculant is produced in the tropical country, cold storage is required which results in greater investment and operating costs. Thus, small-scale production units using sterilized carriers are recommended to produce inoculant in the tropical countries.

The production of sterilized carrier inoculant requires some additional steps of sterilization and the addition of broth into the carrier. Sterilization of carrier can be made by autoclaving or gamma irradiation at 5 Mrad. In one experiment, survival of soybean rhizobia (USD110) in different sterilization treatments was conducted. The remaining rhizobial populations were compared when the peat carriers were treated with Gamma irradiation at 3.5 Mrad, 5 Mrad and autoclaved, and stored for one year. The results indicated that the initial population number at 10^7 cells ml^{-1} could be reduced to 10^5-10^6 cells ml^{-1} in non-sterilized peat, while carriers treated with Gamma irradiation able to maintain the same number as initially inoculated. Moreover, it was found that the autoclaved peat can also promote the growth from 10^7 cells ml^{-1} to be 10^8 cells ml^{-1} during one year storage. Under the same conditions, storage temperature also demonstrated the effect of changes in population number. The storage temperature at 10°C can maintain and promote the population number to 10^9 cells ml^{-1}, while storage temperature at 28-30°C affects only maintaining the population number.

In addition, a treatment called electron beam sterilization was also investigated. Bradyrhizobial strains for soybean (TAL 102) and mungbean (THA 305) were tested in peat treated with electron beam and autoclaved. After one-year storage, the population numbers of both bradyrhizobial strains were enumerated. The results clearly suggested that the treatment with electron beam could maintain the population number in the same amount as when initially inoculated, while the population number in autoclaved, peat was reduced in one-magnitude level.

The cost for preparing sterilized carrier is low and this cost can be taken care of by increasing the selling price or reducing the package size. Most small and medium-scale private inoculant producers in Australia, New Zealand, England and Brazil prefer producing sterilized carrier inoculant because its shelf life is longer. The cost of production of the sterilized carrier may be a little more than the non-sterilized carrier, but the size of a package for a unit area of application of sterilized inoculant can be smaller than the non-sterilized carrier, because the number of rhizobia per gram of sterilized inoculant is usually higher. Therefore, the overall cost of inoculant may not be much different.

Most of the extra cost of sterilized carrier inoculant production involves labor for injection broth into the carrier package. Thus, most developed countries will develop a complicated machinery to reduce the labor requirement. Since labor in many tropical countries is inexpensive, the high cost machinery may not be required.

4. Alternative Techniques for Producing Inoculant

Simple Technique

Production of high quality inoculant to date is based mostly on pre-sterilized peat as carrier, which allows pure culture of rhizobia to maintain at a high number (Roughley and Pulsford, 1982). Unlike non-sterilized peat, the number of rhizobia in sterilized carrier inoculant is not related to the size of inoculant. Roughley (1968) reported that inoculants of sterilized peat inoculated with broth containing 10^4-10^9 rhizobia per ml and incubated for three weeks contained comparable number of rhizobia. Somasegaran (1985) successfully developed a dilution technique to produce inoculant, but the procedure was a bit cumbersome. Based on those findings, a simple technique was developed.

The effective way of sterilizing peat by autoclaving is in moist condition, and the effective method of neutralization of peat is with calcium carbonate. The most convenient way of filling peat into plastic bags is also under moist condition. Thus the peat is neutralized with the suspension of calcium carbonate to provide 20-25% of moisture and pH 6.8-7.0 Then the neutralized moist peat is filled into a polypropylene bag with the desirable amount and heat-sealed after pressing air out of the bag. The sealed bags are put into a box, which is designed for this purpose and sterilized. After sterilization, the broth is injected into each bag at 15-20% to provide 40% moisture in inoculant. After injecting, the bag must be thoroughly kneaded to ensure spreading of rhizobia in the bag. After incubation for 10-15 days at room temperature (28-32°C), the population of rhizobia in inoculant will reach 10^9 cells per ml^{-1}.

NifTAL's Micro Production Unit (MPU)

NifTAL's research efforts in Biological N_2-Fixation (BNF) for international development were, in part, directed to the delivery of appropriate technology for small-scale production of rhizobial inoculants. The goal of this effort was to offer feasible, cost-effective strategies to inoculant producers by modifying and refining the existing technology. Successful adaptation of this technology to the economic and agricultural conditions of a region will create viable commercial enterprises, increase productivity of commercial legume crops, and benefit farmers who can reduce their use of costly nitrogen fertilizers.

The Micro Production Unit (MPU) approach to inoculant production relies on two key concepts: (1) the use of a Sterile Carrier (peat), and (2) the Dilution Technique applied to broth cultures of rhizobia. By integrating these concepts, decentralized small-scale commercial inoculant production is possible.

There are several advantages of producing rhizobial inoculants with these techniques that a small-scale production facility will acquire compared to a large-scale operation. Some of these are:

- Initial investment for start up, fixed, variable, and personnel costs are considerably reduced.
- The scaled-down requirement of the MPU reduces direct cost of fermentation equipment and simplifies design. Only moderate technical skills are needed for fabrication of the MPU, and most materials may be acquired locally.
- Fermentation vessel sizes and units required are reduced directly by the factor chosen for the dilution rate used in production. While reducing fermentation vessel requirements, the dilution technique directly increases potential reduction capacity by the same factor.
- Dilution of a rhizobial broth culture by up to 1,000-folds has been shown to be consistent to the production of high quality inoculants.
- Production schedules are more flexible because requirements for cultured broth are reduced. Time constraints are reduced, and producers can respond to demands for varying inoculant quantities and for a greater variety of specific inoculants.
- The use of a sterile carrier increases the shelf life of inoculants. Over a period of time (six to nine months), the quality of the sterile peat-based inoculants is higher than non-sterile inoculants stored under similar conditions.
- Storage of sterile peat based inoculants requires less stringent conditions. As refrigerated storage is not required, capital costs are considerably reduced.
- It is both possible and desirable to locate the small-scale production facility in rural farm areas. A rural locale will increase flexibility of production schedules, reduce product distribution costs, and allow "customizing" of inoculants with superior performing rhizobial strains proved for that area.
- Increasing the production capability of the MPU is relatively inexpensive compared to the initial capital invested for start-up operations. If increased market demands require the expansion of

present fermentation capacities, low cost input to additional small fermentors can adjust the production potential proportionally.

- If, because of market demand, a small-scale production facility must increase production to a large-scale capacity, fermentors used for small-scale operations will not become obsolete. Small volume fermentors may be retained in service to provide for low demand "custom" inoculants, late requests, and act as larger volume "starter" culture vessels for expanded capacity fermentors.
- Quality control procedures are significantly simpler in materials and method. Rapid direct plate enumeration may be used with sterile produced inoculants.

Some of the disadvantages associated with small-scale production utilizing the dilution technique are:

- The dilution technique requires a reliable source of sterile pre-packaged peat. If local sources are not available, problems associated with importation, foreign exchange, and transportation must be overcome.
- Some specialized equipments are also mandatory for this production approach. The MPU requires an injection system, which can be sterilized and which can deliver a set volume of diluted cultured broth. A means of effectively sterilizing large volumes of dilution water must also be integrated.
- Large volume batch blending of sterile peat and diluted broth is not economically feasible. Sterile peat bags must be injected individually.
- When employing dilution rates of up to 1:1000, aseptic conditions must be maintained to avoid contamination with faster growing organisms. This necessitates the use of higher skilled technical staff for key positions.

Application of *Rhizobium* Inoculant

Advancements in inoculant technology have provided inoculant types at a lower cost per area, with higher rhizobia concentrations and more options for planting systems. Several inoculant types are available in the market. These include powder, liquid, granular and sterile peat based.

Powder inoculates are the most common and most reliable. This kind of inoculant is lower in cost, contains higher amount of bacteria and longer life. Application equipment provides uniform coverage, as does dusting into the grain. Actually, powder inoculates are available for all types of legumes.

Granular inoculants actually contain the same concentration of *Rhizobium* as the powder inoculant, but may cost two to three times as much. Particle size of granular inoculant is larger than the powder form which allows the inoculant to flow through an insecticide, fertilizer or similar planter attachment. In addition, the inoculant can be distributed by a planter attachment and no seed mixing is required. Granular inoculation is particularly useful when a pesticide-coated seed is planted or when excursive seed handing can damage the seed coat, as with peanut.

Liquid inoculants are better than those available a decade ago, and they are certainly much easier to use than the powder inoculants. In addition, they are easy to apply with the least amount of equipment and best with air seeders.

In case of inoculation with soybean rhizobia, researchers from university of Guelph have demonstrated the effect of inoculant type on yield of soybean during 1998-2000. The results showed that liquid inoculants applied as in-furrow sprays or dribbled into the seed furrow at planting produced the highest yields when compared with liquid seed-applied, sterile-carrier powdered peat and granular inoculant (http://www.gov.onca/omafra/english/crops/field/news/croptalk/2003). However, the experiment using *R. leguminosarum* bv. *viciae* as inoculant in granule forms with pea (*Pisum sativum* 'Carneval') had more beneficial effects than the liquid inoculant, in an experiment conducted at Saskatchewan, Canada. Interaction effects were rare between seed-applied fungicide and seed-applied liquid or soil-applied granular rhizobium inoculant, which indicated compatibility of these products. Variability in the effects on inoculant types and fungicides suggests that responses are dependent on local soil and environmental conditions (Kutcher et al., 2002).

Apart from genetic improvement of nodulation, application of flavonoids and related compounds directly to the seeds or in-furrows in soil that contain adequate population of rhizobia has been conducted. A commercial product "Soya Signal" which consists mainly genistein and diadzein, has been marketed in North America for about five years (Smith and Zhang, 1999). Results of more than 100 field trials in Northern America since 1994-1999 showed that "Soya Signal" significantly improved nodulation and N_2-fixation, resulting in an average increase in grain yield of 7% (Broughton et al., 2003).

Another technique becoming more popular is seed coating. Seed coating has been used particularly on forage legumes for more than 20 years. The next generation seed coating is likely bring more beneficial substances. Seeds can be coated with a polymer that will prevent them from imbibing the water needed for germination until a desired temperature is reached. The set point at which the polymer becomes permeable to water can be adjusted by changing the polymer composition. More biocontrol agents, such as *Trichoderma hazianum or* PGPR such as *Pseudomonas* spp. and *Bacillus* spp. will be also applied, as well as genetically engineered rhizobia (Walsh et al., 2001)

QUALITY CONTROL

Along with the increase in commercial inoculant production, there is an increasing demand for its quality control, which involves the verification of the identity of the inoculant strain and a check of the number of living cells in the inoculant. Altogether, extensive data have been showing that rhizobial strains differ in their nodulating competitiveness, as estimated by the percentage of nodules formed when host legumes are inoculated with a mixture of strains, or when they are applied as a single inoculant in soil containing indigenous rhizobial populations. Therefore, many techniques have been developed for monitoring introduced rhizobial strains.

1. Conventional Method

Most countries have some forms of quality control, which may be supported by appropriate legislations or may be voluntary on the part of inoculant manufacturers. However, the quality control programs deal mainly with the quality of the strains in the inoculants and their numbers, as well as numbers of contaminating microorganisms. Generally, to determine the selected strains, their morphological and physiological characteristics have been employed, such as shape, color, and texture of colony, growth rate, etc. To quantify or enumerate the rhizobial number, viable cell evaluation technique is routinely used. The spread-plate and drop-plate methods are common techniques, but the only limitation is counting the viable rhizobia from pure culture. Therefore, quantifying *Rhizobium* in soil is difficult. Moreover, when it is necessary to quantify the occurrence of viable cells of a particular *Rhizobium* in non-sterile materials, a plant infection method must be employed. The plant infection count (also called the most probable number: MPN) is an indirect method commonly applied both in growth pouch and Leonard jar. Enumeration from this technique is analyzed based on nodule formation in each plant and least dilute member of a dilution series.

2. Genetic Analysis of Competitiveness and New Techniques for Monitoring Introduced Rhizobia in Soil

In order to understand the fate of introduced *Rhizobium* in soil-plants system, the methodology to rapidly detect and reliably quantify bacterial cells is necessary. Conventional methods which suffer from certain limitations such as these, are time consuming, laborious or inaccurate. More recently developed techniques based on molecular biology offer many advantages as they either enhance accuracy or facilitate the detection process.

In case of Thai soybean cultivar's interaction with different *B. japonicum* strains, all possible paired combinations of these strains (THA 5, THA 6, and USDA 110) were used in co-inoculation and they were investigated for nodule occupancy. The nodule competition assay was conducted by using a fluorescent antibody technique to detect rhizobial strain nodules. The results suggested that *B. japonicum* USDA 110 showed higher nodulation competitiveness than the other strains on soybean cultivars. The Thai strains THA 6 appeared to be more competitive than USDA 110 on cultivars SJ5. Thus, nodulation competitiveness of *B. japonicum* strains was affected by the cultivars of soybean used (Payakapong et al., 2004). However, this technique can be problematic in specificity of strains, it also involves some difficulties, and can take time to produce a new antibody. Recombinant marker or reporter genes, e.g. *gusA, lacZ, luc,* or *lux*, have been successfully used to monitor specific cultivable bacterial strains in field releases.

The project of "Exploitation of Chinese biodiversity resources in sustainable crop production, using a biotechnological approach for rhizobial diversity evaluation, strains improvement and risk assessment" under European Commission research project contract number IC18-CT96-0103 has shown the application of member genes of *Rhizobium* (Lindstrom, 1996). The marking was achieved by introducing suicide plasmids pDB30 and pCAM111, which carry *Tn5-luxAB* and mini *Tn5-gusA* genes into selected strains through the conjugative transfer. Both markers gave reliable detection of rhizobium recovered from soil, rhizosphere and nodules (http://www.biocenter.helsinki/groups/lindstrom/index.) A Finnish researcher group recently demonstrated another case. A non-indigenous wild-type strain *R. galegae* HAMBI 540, which specifically nodulates perennial goat's rue (*Galega orientalis*), and its marker gene-tagged derivatives *R. galegae* HAMBI 2363 (*luc*), *R. galegae* HAMBI 2368 (*gusA21*), and *R. galegae* HAMBI 2364 (*gusA30)* were used to evaluate the persistence, population dynamics and competitiveness for nodulation of rhizobia under field conditions in Finland. The lysimeters were filled with clean or diesel oil-polluted (3000 $\mu g\ g^{-1}$) agricultural soil. During the first two years of the field release *luc*- and *gusA21*-tagged strains could be effectively detected by cultivation, reinforced with colony polymerase chain reaction (PCR). The population densities remained relatively stable from 10^4 to 10^5 CFU g^{-1} dry soil from spring until late autumn. Replicated limiting dilution PCR analysis gave comparable results with cultivation with strain HAMBI 2363 until 49 weeks after inoculation. GUS (β-glucuronidase) activity of strain HAMBI 2368 could be stable, as detected in nodules and soil. On the other hand, *luc* activity weakened clearly in cold conditions along with decreased metabolic activity of rhizobia. The competitive ability for nodulation of the *gusA30*-tagged strain decreased slowly with time compared to the wild-type strain. Moderate soil pollution did not have significant effects on target bacteria or plant growth. Limited vertical movement of target bacteria outside the rhizosphere was detected from percolated water (Pitkajarvi et al., 2003).

In addition, investigations were recently conducted in Thailand, to observe the highly effective rhizobial strains for development of reforestation program using various tree legume plants. The tree legumes used in this program were *Acacia mangium, A. auriculfamis, Milletia leucantha, Pterocorpus indicus* and *Xylia xylocarpa.* Most of the effective strains belonged to the genus *Bradyrhizobium.* These strains were marked by *gusA* gene with triparental mating prior to analyzing competitiveness of selected and indigenous strains in forest soil. The results obtained from the nodule occupancy indicated the high effectiveness of the selected strain in the range 63-100% of nodule occupancy.

However, some disadvantages of this approach are linked to background activity or insensitivity. In addition, from the experience gained, construction of the recombinant strain with the use of triparental mating or conjugative transfer is more effective than eletroporation. Tagging the reporter gene by chromosomal integration is more favorable than in plasmid form. The specificity of nucleic acid probes proved to be another rapid approach to detect the target bacteria in a different sample. There are several examples in the literature of successful extraction, hybridization, and PCR amplification of target DNA from bacterial DNA community. One example of achievement on quality control on the basis of this method was conducted from the peat-product of a Finnish company. In the past, the Finish company, Elomestari Ltd., produces commercially available, peat-based inoculants for legumes cultivated in Finland. The quality of the inoculant is routinely tested by plate-counting and plant-infection test. In 1995, Tas and the group developed the procedures based on DNA hybridization and PCR for quality control of these *Rhizobium* inoculants. The inoculants contained *R. galegae* HAMBI 1174 and HAMBI 1207 and *R. leguminosarum* bv. *viciae* strain 16HSA, either solely or in combination of two or three strains. The results demonstrated that the methodology has sensitivity enough to be used for inoculant quality control. The target bacterial strains were identifiable even in the dense background of nontarget bacteria (Tas et al., 1995). As of date, the techniques known as DGGE (Denaturing Gradient Gel Electrophoresis) and T-RFLP (Terminal Restriction Length Polymorphisms) can enhance the resolution both in terms of quality and quantity of target DNA. Thus, if these techniques are combined with this methodology, it might be a very efficient tool to evaluate the competitiveness capability of selected rhizobia.

CONCLUSION

Since the rhizobial genera, which are currently re-classified on the basis of using 16S rRNA, the genera *Azorhizobium* and *Bradyrhizobium* have a distant relationship to *Rhizobium* and other genera in the Rhizobiaceae. This is, however, somewhat related to their carbon source utilization. *Rhizobium* and *Sinorhizobium* which are fast growers, could utilize various kinds of sugar while *Mesorhizobium, Azorhizobium* and *Bradyrhizobium* are fastidious. Therefore, prior to producing inoculants, appropriate types of media should be considered. The alternative type of carbon source particularly for *Bradyrhizobium* has been investigated. Using saccharification of cassava tuber by amylolytic fungi and followed by glycerol and other vitamins produced by yeast was proved to be another appropriate medium for culturing *Bradyrhizobium*. This can be replaced by the costly mannitol. However, a deeper understanding of microbial physiology under diverse environmental conditions would explain the competition for nodulation among rhizobia. Several attempts to overcome or improve more competitiveness in selected rhizobia have been conducted.

Improvement of the strain on the basis of genetic engineering to compete the abiotic factors, introduced *put*A gene to rhizobia under drought condition or *idh*A gene to catabolize inositol in soil rendering enhance competitive nodulation of soybean by *S. fredii* were also investigated. Several genetic analyses of genes involving in biotic factor have also been more emphasized, such as genes controlling the EPS production, nodulation, antirhizobial compound production, N_2-fixation and H_2 uptake system. Technology towards carrier selection and production in industrial level are also improved. The simple and low cost inoculant production technique has also been developed. The technique involves sterilized peat either with Gamma radiation or electron bean together with dilution techniques and seems to be the most promising form of rhizobial inoculant. Liquid carrier is a better option since it is convenient to apply. Moreover, application of molecular biology of rhizobia can be a new tool for initiating the novel quality control program facilitating higher accuracy of other conventional methods.

However, the best way to ensure that the inoculated strains get to the desired target host and enable them to compete with limitation factors, is by understanding and controlling the inoculant's metabolic state and ecological role in a given soil environment.

REFERENCES

Abdoulaye, S., Giraud, E., Jourand, P., Garcla, N., Willems, A., de Lajudie, P., Prin, Y., Neyra, M., Gillis, M., Boivin-Masson, C. and Dreyfus, B. (2001). Methylotrophic *Methylobacterium* bacteria nodulate and fix nitrogen in symbiosis with legumes. J. Bacteriol. 183: 214-220.

Awaya, J.D., Fox, P.M. and Borthakur, D. (2003). Genes encoding a fructose-1,6-bisphosphate aldolase and a fructose-1,6-bisphosphatse are present within the gene cluster for mimosine degradation in *Rhizobium* sp. Strain TAL1145. Plant Soil 257: 11-18.

Bai Y., Pan, B., Charles, T.C. and Smith, D.L. (2002). Co-inoculant dose and root zone temperature for plant growth promoting rhizobacteria on soybean [*Glycine max* (L.) *Merr*] grown in soil-less media. Soil Biol. Biochem. 34: 1953-1957.

Balatti, A.P., Mazza, L.A. and Moretti, E. (1987). Aeration requirements of *Rhizobium* cultures. Mircen J. 3 : 227-234.

Becker, A. and Puhler, A. (1998). Production of polysaccharide. In: The Rhizobeaceae, Molecular Biology of Model Plant-Associated Bacteria (ed.) H.P. Spaink, Kluwer Academic Publisher, Dordrecht, The Netherland, pp. 97-115.

Bhagwat, A.A. and Keister, D.L. (1991). Isolation and characterization of a competition-defective *Bradyrhizobium japonicum* mutant. Appl. Environ. Microbiol. 57: 3496-3501.

Bittinger, M.A. and Handelsman, J. (2000). Identification of genes in the *ros*R regulon of *Rhizobium etli*. J. Bacteriol. 182: 1706-1713.

Boonkerd, N. (1991). Inoculant quality control and standards in Thailand In: Expert Consulation on Legume Inoculant Production and Quality Control. FAO, Rome, pp. 121-129.

Breil, B.T., Ludden, P.W. and Triplett, E.W. (1993). DNA sequence and mutational analysis of genes involved in production and resistance to the antibiotic peptide trifolitoxin. J. Bacteriol. 175: 3693-3702.

Brito, B., Monza, J., Imperial, J., Ruiz-Argueso, T. and Palacios, J.M. (2000). Nickel availability and hupSL activation by heterologous regulators limit symbiotic expression of the *Rhizobium leguminosarum* bv. *viciae* hydrogenase system in Hup-rhizobia. Appl. Environ. Microbiol. 66(3): 937-942.

Brito, B., Palacios, J.M., Imperial, J. and Ruiz-Agrueso, T. (2002). Engineering the *Rhizobium leguminosarum* bv. *viciae* hydrogenase system for expression in free-living microaerobic cells and increased symbiotic hydrogenase activity. Appl. Environ. Microbiol. 68 (5): 2461-2467.

Brockwell, J., Bottomley, P.J. and Thies, J.E . (1995). Manipulation of rhizobia mucroflora for improving legume productivity and soil fertility, a critical assessment. Plant Soil 174: 143-180.

Broughton, W.J., Zhang, F., Perret, X. and Staehelin, C. (2003). Signals exchanged between legumes and *Rhizobium* : Agricultural uses and perspectives. Plant Soil 252: 129-137.

Burton, J.C. (1967). *Rhizobium* culture and use. In: Microbial Technology (ed.) H.J. Peppler, Reinhold Publishing Co., New York, USA, pp. 1-33.

Burton, J.C. (1979). *Rhizobium* species. In: Microbial Technology, 2nd edn. (ed.) H.J. Peppler and D. Periman , Academic Press, New York, USA, pp. 29-58.

Castillo, M., Flores, M., Mavingui, P., Martinez-Romero, E., Palacios, R. and Hernandez, G. (1999). Increase in alfalfa nodulation, nitrogen fixation, and plant growth by specific DNA amplification in *Sinorhizobium meliloti*. Appl. Environ. Microbiol. 65: 2716-2722.

Date, R.A. and Roughley, R.J. (1977). Preparation of legume seed inoculants. In: A Treatise on Dinitrogen Fixation, Section IV. Agronomy and Ecology (eds.) R.W.F. Hardy and A.H. Gibson, John Wiley SMS, New York, USA, pp. 243-275.

de Lajudie, P., Laurent-Fulele, E., Willems, A., Torck, V., Coopman, R., Collins, M.D., Kersters, K., Dreyfus, B. and Gills, M. (1998). *Allorhizobium undicola* gen. nov., sp. nov., nitrogen-fixing bacteria that efficiently nodulate *Neptumia nalans* in Senegal. Int. J. Syst. Bacteriol. 38: 392-397.

Dilworth, M.J., Ruve, W.G., Tiwari, R.P., Vicas-Marfisi, A.I., Fenner, B.J. and Glenn, A.R. (2001). Understanding acid tolerance in root nodule bacteria. 13th International Congress on Nitrogen Fixation, Hamilton, Ontario, Canada, July 2001. pp. 204-208.

Doignon-Bourcier, F., Sy, A., Willems, A., Trock, U., Dreyfus, B., Gills, M. and de Lajuudie, P. (1999). Diversity of bradyrhizobia from 27 tropical Leguminosae species native of Senegal. http://www.ncbi.nlm.nih.gov/entrez/query.fcgi?cmd=Retrieve&bd=Pub Med &lis...27/1/2547.

Drcyfus, B., Gercia, L.J., and Gillis, M. (1988). Characterization of *Azorhizobium caulinodans* gen. nov., sp. nov., a stem-nodulating nitrogen fixing bacterium isolated from *Sesbania rostrata*. Int. J. Syst. Bacteriol. 38, 89-98.

Galbraith, M.P., Feng, S.F., Bornemon, J., Triplett, E.W., de Bruijn, F.J. and Rossbach, S. (1998). A functional myo-inositol catabolism pathway is essential for *Rhizobium* utilization by *Sinorhizobium meliloti*. Microbiology 144: 2915-2924.

Giles, K.L. and Atherly, A.G. (1981). Biology of the Rhizobiaceae. In: International Review of Cytology (eds.) G.H. Bourne and J.F. Danielli, Academic Press, New York, USA, pp. 321-331.

Graham, P.H. (1964). Studies on utilization of carbohydrate and Kerb's cycle intermediates by rhizobia, using an agar plate method. Antonie von Leeuwenhoek. J. Microbiol. Serol. 30: 68-72.

Halliday, J. and Graham, P.H. (1978). Coal compared to peat as a carrier of rhizobia. Turrialba 28: 348-349.

Heather, N., Pryor, W.L., Lowther, K., Mcintyre, H.J. and Ronson C.W. (1998). An inoculant *Rhizobium* strain for improved establishment and growth of hexaploid Caucasian clover (*Trifolium ambiguum*). N.Z.J. Agric. Res. abstracts. http://www.rsnz.govt.nz/publish/nzjar/20.php.

Heron, D.S. and Pueppke, S.G. (1984). Mode of infection, nodulation specificity, and indi genous plasmid of 11 fast growing *Rhizobium japonicum* strains. J. Bacteriol. 160: 1061-1066.

Heron, D.S., Ersek, T., Krishnan, H.B. and Pueppke, S.G. (1989). Nodulation mutants of *Rhizobium fredii* USDA 257. Mol. Plant Microbe. Interact. 2: 4-10.

Hynes, M.F., Yost, C.K., Oresnik, I.J., Garcia de los Santos, A., Clark, S.R. D. and Macintosh, J.E. (2002). Catabolic and chemoreceptor genes and their role in rhizobial ecology. Proc. 13th International Congress on Nitrogen Fixation, Hamilton, Ontario, Canada, July 2001, pp. 155-157.

Jiang, G. and Krishnan, H.B. (2000). *Sinorhizobium fredii* USDA 257, a cultivar-specific soybean symbiont, carries two copies of y4yA and y4yB, two open reading frames that are located in a region that encodes the Type III protein secretion system. MPMO. 13: 1010-1014.

Jiang G., Krishnan, A.H., Kim, Y.W., Wacek, T.J. and Krishnan, H.B. (2001). A functional *myo*-inositol dehydrogenase gene is required for efficient nitrogen fixation and competitiveness of *Sinorhizobium fredii* USDA 191 to nodulate soybean (*Glycine max* [L.] Merr.). J. Bacteriol. 183: 2595-2604.

Kantar, F., Elkoca, E. Ogutcu, H. and Algur, O. F. (2003). Chickpea Yields in Relation to Rhizobium Inoculation from Wild Chickpea at High Altitudes. http://www. blackwell-synergy.com/links/doi/10.104 6/j.1439-037X.2003.00046.htm.

Karasova-Wade, T., Ndoye, L., Braconier, S., Sarr, B., de Lajudie, P. and Neyra, M. (2003). Diversity of indigenous bradyrhizobia associated with three cowpea cultivars (*Vigna unguiculata* (L.) Walp.) grown under limited and favourable water conditions in Senegal (West Africa). African J. Biotech. 2(1): 13-22.

Keyser, H.H., Bohlool, B.B., Hu, T.S., and Weber, D.F. (1982). Fast-growing rhizobia isolated from root nodules of soybean. Science 215: 1631-1632.

Kremer, R.J. and Peterson, H.L. (1983). Effects of carrier and temperature on survival of *Rhizobium* spp. In legume inocula : Development of an improved type of inoculant. Appl. Environ. Microbiol. 45: 1790-1794.

Krishnan, H.B. (2002). NolX of *Sinorhizobium fredii* USDA257, a type III-secreted protein involved in host range determination, is localized in the infection threads of cowpea (*Vigna unguiculata* [L.] Walp) and soybean (*Glycine max* [L.] Merr.) nodules. J. Bacteriol. 184: 831-839.

Kucey, R.M.N., Snitwonge, P., Chaiwanakupt, P., Wadisirisuk, P., Siripoibool, C., Arayangkool, T., Boonkerd, N. and Rennie, R.J. (1988). Nitrogen fixation (N^{15} dilution) with soybeans under Thai field conditions : I. Developing protocols for screening *Bradyrhizobium japonicum* strains. Plant Soil 108 : 33-41.

Kutcher, H.R., Lafond, G., Johnston, A.M. Miller, P.M., Gill, K.S. May, W.E., Hogg, T., Johnson, E., Biederbeck, V.O. and Nybo, B. (2002). *Rhizobium* inoculant and seed-applied fungicide effects on field pea production. Can. J. Plant. Sci. 82: 645-651.

Lindstrom K. (1996). Exploitation of Chinese biodiversity resources in sustainable crop production, using a biotechnological approach for rhizobial diversity evaluation, strain improvement and risk assessment. http://europa.eu.int/comm/research/quality-of-life/gmo/02-plantgrowth/02-05...27/1/2547.

Loh, J. and Stacey, G. (2003). Nodulation gene regulation in *Bradyrhizobium japonicum*: a unique integration of global regulatory circuits. Appl. Environ. Microbiol. 69(1): 10-17.

Loh, J.R., Carlson, W., York, S. and Stacey, G. (2002b). Bradyoxetin, a unique chemical signal involved in symbiotic gene regulation. PNAS 99: 14446-14451.

Loh, J., Pierson, A., Pierson III, L.S., Stacey, G. and Chatterjee, A. (2002a). Quorum sensing in plant associated bacteria. Curr. Opin. Plant. Biol. 5: 285-290.

Lopez-Garcia, S.L., Vazouez, T.E.E., Favelukes, G. and Lodeiro, A.R. (2001). Improved soybean root association of N-starved *Bradyrhizobium japonicum*. J. Bacteriol. 183: 7241-7254.

Maccio, D., Fabra, A. and Castro, S. (2002). Acidity and calcium interaction affect the growth of *Bradyrhizobium* sp. and the attachment to peanut roots. Soil Biol. Biochem. 34: 201-208.

Manassila, M., Nantajig, A., Kotepong, S., Boonkerd, N. and Teaumroong, N. (2004). Characterization and monitorning selected rhizobial strains isolated from tree legumes in Thailand. Proc. The 6th European Nitrogen Fixation Conference. Toulouse, France, pp. 70.

Matpatawee, D. (1999). Polygenetic diversity of rhizobial strains isolated from diverse ecosystems in Thailand. M.Sc. Thesis, Suranaree University of Technology, Thailand, pp. 112.

McDermott, T.R. and Graham, P.H. (1989). *Bradyrhizobium japonicum* inoculant mobility, nodule occupancy and acetylene reduction in the soybean root system. Appl. Environ. Microbiol. 55: 2493-2498.

Minamizawa, K. and Mitsui, H. (2000). Genetic ecology of soybean bradyrhizobia. In: Soil Biochemistry, Vol. 10. (eds.) J.M. Bollag and G. Stotsky. Marcel Dekker Inc., NY, USA.

Nethery, A.A. (1991). Inoculant production with nonsterile carrier. In: Expert Consultation on Legume Inoculant Production and Quality Control. FAO, Rome, pp. 43-50.

Okasaki, O., Yuhashi, K.I. and Minamizawa, K. (2003). Quantitative and time-course evaluation of nodulation competitiveness of rhizobitoxine-producing *Bradyrhizobium elkanii*. FEMS Microbiol. Ecol. 45: 155-160.

Oresnik, I.J., Twelker, S. and Hynes, M.F. (1999). Cloning and characterization of a *Rhizobium leguminosarum* gene encoding a bacteriocin with similarities to RTX toxins. Appl. Environ. Microbiol. 65: 2833-2840.

Paczkowski, M.W. and Berryhill, D.L. (1979). Survival of *Rhizobium phaseoli* in coal-based legume inoculants. Appl. Environ. Microbiol. 38: 612-615.

Payakapong, W., Tittabutr, P., Teamroong, N. and Boonkerd, N. (2004). Soybean cultivars affect nodulation competition of *Bradyrhizobium japonicum* strains. World J. Microbiol. Biotechnol. 20(3): 311-315.

Peterson, H.L. and Loynachan, T.E. (1981). The significance and application of *Rhizobium* in agriculture. Int. Rev. Cytol. Suppl. 13: 311-331.

Pitkajarvi, J., Rasanen, L.A., Langenskiod, J., Wallenius, K., Niemi, M. and Lindstrom, K. (2003). Persistence, population dynamics and competitiveness for nodulation of marker gene-tagged *Rhizobium galegae* strains in field lysimeters in the boreal climatic zone. FEMS Microbiol. Ecol. 46: 91-104.

Pongsilp, N., Teaumroong, N. Nuntagij, A., Boonkerd, N. and Sadowsky, M.J. (2002). Genetic structure of indigenous non-nodulating and nodulating populations of *Bradyrhizobium* in soils from Thailand. Symbiosis 33: 39-58.

Poole, P.S., Blyth, A., Reid, C.J. and Walters, K. (1994). Myo-inositol catabolism and catabolic regulation in *Rhizobium leguminosarum* bv. *viciae*. Microbiology 140: 2787-2795.

Provorov, N.A., Tikhonovich, I.A. and Kozhenmyakov, A.P. (2000). Breeding for high efficiency in "*Pisum sativum/Rhizobium leguminosarum*" symbiosis. '.ttp:// www.hermes.bionet.nsc.ru/pg/31/54.htm.

Pryor, H.N., Lowther, W.L., McIntyre, H.J. and Ronson, C.W. (1998). An inoculant *Rhizobium* strain for improved establishment and growth of hexaploid Caucasian clover (*Trifolium ambiguum*). New Zealand J. Agricul. Res. 41: 179-189.

Purcell, L.C., Vandez, V., Sinclair, T.R., Serraj, R. and Nelson, R. (1998). Screen of soybean germplasm for N_2-fixation drought tolerance. 16th North American Symbiotic Nitrogen Fixation Conference. Cancun, Mexico, February 1998.

Reitz, M., Rudolph, K., Schroder, I., Hoffmann-Hergarten, S., Hallmann, J. and Sikora, R.A. (2000). Lipopolysaccharides of *Rhizobium etli* strain G12 act in potato roots as an inducing agent of systemic resistance to infection by the cyst nematode *Globodera pallida*. Appl. Environ. Microbiol. 66: 3515-3518.

Robleto, E.A., Scupham, A.J., and Triplett, E.W. (1997). Trifolitoxin production in *Rhizobium etli* strain CE3 increases competitiveness for rhizosphere growth and root nodulation of *Phaseolus vulgaris* in soil. Mol. Plant Microbe Interact. 10: 228-233.

Robleto, E.A., Kmiecik, K., Oplinger, E.S., Nienhuis, J. and Triplett, E.W. (1998). Trifolitoxin production increases nodulation competitiveness of *Rhizobium etli* CE3 under agricultural conditions. Appl. Environ. Microbiol. 65: 2833-2840.

Roughley, R.J. (1968). Some factors influencing the growth and survival of the root nodule bacteria in peat culture. J. Appl. Bacteriol. 31: 259-265.

Roughley, R.J. (1970). Preparation and use of legume seed inoculants. Plant Soil 32: 675-701.

Roughley R.J. and Pulsford, D.J. (1982). Production and control of legume inoculants. In: Nitrogen Fixation in Legumes. Academic Press, Australia, pp. 193-209.

Sadowsky, M.J. and Graham, P.H. (1998). Soil Biology of the *Rhizobiaceae.* In: The Rhizobiaceae (eds.) H. P. Spaink, A. Kondorosi and P.J.J. Hooykaas, Kluwer Academic Press, Dordrecht, The Netherlands, pp. 155-172.

Sameshima, R., Isawa, T., Sadowsky, J. M., Hamada, T., Kasai, H., Shutsrirung, A., Mitsui, H. and Minamisawa, K. (2003). Phylogeny and distribution of extra-slow-growing *Bradyrhizobium japonicum* harboring high copy numbers of RSa, RSb and IS1631. FEMS Microbiol. Ecol. 44: 191-202.

Scupham, A.J., Bosworth, A.H., Ellis, W.R., Wacek, T.J. Albrecht, K.A., and Triplett, E.W. (1996). Inoculation with *Sinorhizobium meliloti* RMBPC-2 increases alfalfa yield compared with inoculation with a non-engineered wild-type strain. Appl. Environ. Microbiol. 62: 4260-4262.

Serraj, R., Sinclair, T.R. and Purcell, L.C. (1999). Symbiotic N_2 fixation response to drought. J. Exp. Bot. 50: 143-455.

Sinclair, T.R. and Serraj, R. (1995). Dinitrogen fixation sensitivity to drought among grain legume species. Nature 378: 344.

Singh, H.P. (2005). Interactions among beneficial soil microorganisms. In: Microbial Biotechnology in Agriculture and Aquaculture, Volume I (ed.) R.C. Ray, Science Publishers, Enfield, New Hampshire, USA, pp. 159-210.

Smith, C.W. (1995). Soybean (*Glyciae mex* [L.] Merrill), Crop Production, Evaluation, History and Technology, John Wiley & Sons, New York, USA. pp. 351-357.

Smith, D.L. and Zhang, F. (1999). Composition for enhancing grain yield and protein yield of legume grown under environmental conditions that inhibit or delay nodulation there of. US Patent 5922316.

Soberon, M., Lopez, O., Morera, C., de L. Girard, M., Tabche, M.L. and Miranda, J. (1999). Enhanced nitrogen fixation in a *Rhizobium etli ntr*C mutant that overproduces the *Bradyrhizobium japonicum* symbiotic terminal oxidase cbb_3. Appl. Environ. Microbiol. 65(5): 2015-2019.

Soedarjo, M. and Borthakur, D. (1998). Mimosine, a toxin produced by tree-legume *Leuceana* provides a nodulation competition advantage to mimosine-degrading *Rhizobium* strains. Soil Biol. Biochem. 30: 1605-1613.

Somasegaran, P. (1985). Inoculant production of *Rhizobium* spp. with diluted liquid cultures and autoclaved peat: Evaluation of dilutents, *Rhizobium* spp., peats, sterility requirements, storage and plant effectiveness. Appl. Environ. Microbiol. 50: 398-405.

Squartini, A., Struffi, P. Doring, H., Selenska-Pobell, S., Tola, E., Giacomini, A., Vendramin, E., Vetazquez, E., Mateos, P.E., Martinez-Molina, E., Dazzo, F.B., Casella, S. and Nuti, M.P. (2002). *Rhizobium sullae* sp. nov. (formerly *Rhizobium hedysari*), the root- nodule microsymbiont of *Hedysarum coronarium* L. Int. J. Syst. Evol. Microbiol. 52: 1267-1276.

Stowers, M.D. and Elkan, G.H. (1984). Growth and nutritional characteristics of cowpea rhizobia. Plant Soil 80: 191-200.

Stowers, M.D. (1985). Carbon metabolism in *Rhizobium* species. Annu. Rev. Microbiol. 39: 89-108.

Tan, Z.Y., Kan, L.F., Peng, X.G., Wang, T.E., Reinhold-Hurek, B., and Chen, X.W. (2001). *Rhizobium yanglingense* sp. nov. isolated from arid and semi-arid regions in China. Int. J. Syst. Evol. Microbiol. 51: 909-914.

Tas E., Saano, A., Leinonen, P. and Lindstrom, K. (1995). Identification of *Rhizobium* spp. in peat-based inoculants by DNA hybridization and PCR and its application in inoculant quality control. Appl. Environ. Microbiol. 61(5): 1822-1827.

Thies, J.E., Singleton, P.W. and Buhlool, B.B. (1991). Influence of size of indigenous rhizobial populations on establishment and symbiotic performance of introduced rhizobia on fieldgrown legumes. Appl. Environ. Microbiol. 57: 19-28.

Tittabutr, P., Payakapong, W., Teaumroong, N. and Boonkerd, N. (2003). Development of rhizobial inoculant production and formulation : Carbon source utilization and sugar producing from various raw starch materials by using amylase-producing fungi. Biotech. for Sust. Util. Bio. Res. in the Tropics. 15: 197-200.

van Dillewijn, P.M., Soto, J., Villadas, P.J. and Toro, N. (2001). Construction and environmental release of *Sinorhizobium meliloti* strain genetically modified to be more competitive for alfalfa nodulation. Appl. Environ. Microbiol. 67: 3860-3865.

Viprey, V., Greco, A.D., Golinowski, W., Broughton, W.J. and Perret, X. (1998). Symbiotic implications of type III protein secretion machinery in *Rhizobium*. Mol. Microbiol. 28 (6): 1381-1989.

Walsh, U.F., Morrissey, J.P. and O'Gara, F. (2001). *Pseudomonas* for biocontrol of phytopathogens : From functional genomics to commercial exploitation. Curr. Opin. Biotech. 12(3): 289-295.

Wang, E.T. and Martinez-Romero, E. (2000). Phylogeny of root-and stem-nodule bacteria associated with legumes. In: Prokaryotic Nitrogen Fixation : A Model System for Analysis of a Biological Process. Horizon Scientific Press, UK, pp. 177-186.

Wei, G.H., Wang, E.T., Tan, Z.Y., Zhu, M.E. and Chen, W.X. (2002). *Rhizobium indiqoferae* sp. nov. and *Sinorhizobium kummerowiae* sp. nov., respectively isolated from *Indigofera* spp. and *Kummerowia stipulaceae*. Int. J. Syst Evol. Microbiol. 52: 2231-2239.

Wei, G.H., Tan, Z.Y., Zhu, M.E., Wang, E.T., Han, S.Z. and Chen, W.X. (2003). Characterization of rhizobia isolated from legume species within the genera *Astragalus* and *Lespedeza* grown in the Loess Plateau of China and description of *Rhizobium loessense* sp. nov. Int. J. Syst. Evol. Microbiol. 53: 1575-1583.

Wei, X. and Bauer, W.D. (1998). Starvation-induced changes in motility, chemotaxis, and flagellation of *Rhizobium meliloti*. Appl. Environ. Microbiol. 64: 1708-1714.

Yao, Z.Y., Kan F.L., Wang, E.T., Wei, G.H., and Chen, W.X. (2002). Characterization of rhizobia that nodulate legume species of the genus *Lespedeza* and description of *Bradyrhizobium yuanmingense* sp. nov. Int. J. Syst. Evol. Microbiol. 52: 2219-2230.

Yasuta T., Okazaki, S., Misui, H., Yuhashi, K., Ezura, H. and Minamisawa, K. (2001). DNA sequence and mutational analysis of rhizobitoxine biosynthesis genes in *Bradyrhizobium elkanii*. Appl. Environ. Microbiol. 67(11): 4999-5009.

Yon Y., Gao, X., Ho, M.M. and Borthakur, D. (1998). A stomatin-like protein encoded by the *slp* gene of *Rhizobium etli* is required for nodulation competitiveness on the common bean. Microbiology 144: 2619-2627.

Zahran, H.H. (2005). Rhizobial nitrogen fixation in agriculture: biotechnological perspectives. In: Microbial Biotechnology in Agriculture and Aquaculture, Volume I (ed.) R.C. Ray, Science Publishers, Enfield, New Hampshire, USA, pp. 71-100.

Zhang, F. and Smith, D.L. (1996). Genistein accumulation in Soybean [*Glycine max* (L.) Merr.] root systems under suboptimal root zone temperatures. J. Exp. Bot. 47: 785-792.

http://www.biocenter.helsinki groups/lindstrom/index.

http://www.gov.on.ca/omafra/english/crops/field/news/croptalk/2003.

4

Phosphorus Solubilizing Microorganisms and Their Role in Plant Growth Promotion

*Olga Mikanová and Jaromir Kubát**

INTRODUCTION

Phosphorus (P) is one of the principal macrobiogenic elements necessary for almost all the metabolic processes of plant growth and development, and is also one of the key factors of production for higher crop yields. In soil, phosphorus is present both in inorganic bonds and organic compounds. Phosphates are the main natural source of phosphorus.

Phosphorus is supplied to soil in the form of mineral phosphorus fertilizers, manure, plant residues and/or domestic organic waste. Nevertheless, up to 90% of phosphorus supplied to soil by way of mineral fertilizers is transformed into forms hardly available to plants (Goldstein, 1986; Domey, 1987). In spite of the fact that soils usually contain a sufficient amount of total phosphorus, the amount of plant-available phosphorus is usually low. That is why in some cases phosphorus often becomes a limiting factor to plant growth.

IMPORTANCE OF PHOSPHORUS FOR PLANTS

Phosphorus is a necessary nutrient for plants. Among the more significant metabolic processes in plants on which phosphorus has an important effect are: photosynthesis, nitrogen (N_2)-fixation, crop maturation (flowering and fruiting, including seed formation), root development,

Research Institute of Crop Production, Drnovska 507, 161 06 Prague 6- Ruzyne, Czech Republic
**Corresponding author: E-Mail: kubat@vurv.cz*

strength of straw in cereal crops, improvement of crop quality, especially of forages and vegetables (Brady, 1990).

Phosphorus content in plant dry matter amounts to 0.1-0.5%. Phosphorus uptake in plants is uneven and depends on the plant age. It is mostly consumed during the period of flowering and fruiting. Up to 50% of phosphorus uptake is stored in seeds. Plants mostly uptake phosphorus as orthophosphate ions. Most of the phosphorus uptake is expeditiously transformed into organic compounds in roots. The most sensitive parameters that have impact on phosphorus uptake are growth and diameter of roots. Interestingly, it is not the length of roots that is decisive for uptake, but abundance of root hair, phosphorus concentration in the soil solution in the root zone, and activity of soil and root microflora (Romer et al., 1988). Phosphorus is rather mobile in plants. Organic forms of phosphorus in plants include: phospholipids, phosphoryl sugars, nucleoproteins and nucleic acids, purine and pirimidine nucleotides, flavine nucleotides, phytin, and adenosine triphosphate (Natr and Lastuvka, 1980). In fact, it is not possible to build up greater phosphorus stock in plants for future use (except for fruit). The overall plant metabolism requires an ongoing phosphorus uptake over entire ontogenesis.

Phosphorus also plays an important role in biological N_2-fixation. Leguminous plants have a greater need for phosphorus than plants dependent on nitrogen from mineral fertilization (Israel, 1993). In some cases phosphorus may become a limiting factor to the growth of and symbiotic N_2-fixation by the leguminous plants. In case of lack of available phosphorus, the number and weight of nodules as well as overall nitrogenase activity increases (Singh and Singh, 1993). At the same time, microorganisms are known to live in soil with an ongoing lack of nutrients. According to the research conducted by Cassman et al., (1981), some *Rhizobium* strains do not grow in soil with a low phosphorus concentration. It is difficult to separate the specific role of phosphorus in symbiotic fixation of N_2 from its effect on the host plant. How N_2-fixation is affected by phosphorus has been presented in the following two ways by Robinson et al., (1981). First, it improves plant growth and thus raises nitrogen concentration in plants; second, phosphorus promotes nodule formation. Mikanová and Kubát (1997a) also confirmed the positive effect of phosphorus on symbiotic fixation of N_2 in tests where inoculation by P-solubilizing strains of *Bradyrhizobium japonicum* increased the total nitrogenase activity in soils with a low content of plant-available phosphorus.

The effect of phosphorus on the symbiotic fixation of N_2 in pot experiment was confirmed in our study. An experiment was started in pots with soybean, the Sluna variety. Grains were inoculated with D504 and D574 strains of P-solubilizing bacteria from the collection of soil microorganisms in the Research Institute of Crop Production (RICP), Prague - Ruzyně and three strains (Br29, Br89 and Br96) acquired from CNPAB (Centro National de Pesquisa de Agrobiologia), EMBRAPA in Brazil. All the five strains were earlier tested in liquid cultures with addition of insoluble tricalcium phosphate on their P-solubilizing activity. The content of soluble phosphorus was determined after one week of incubation and simultaneously pH of the medium was determined (Table 4.1). As shown in Table 4.1, two strains have shown a high P-solubilizing activity, while the P-solubilizing activity of the three of them was close to zero.

The experiment was conducted in 12 variants with 10 replicates: (1) One set of five variants inoculated and a non-inoculated control with the addition of superphosphate and (2) The other set of five variants inoculated and a non-inoculated control without the addition of superphosphate. The superphosphate dose corresponded to 50 kg of P_2O_5 ha^{-1}. In five pots the total nitrogenase activity was measured and a dry matter yield of aboveground matter was evaluated at the initial flowering stage. The remaining five pots were left out and dry matter yield of grains was determined. For results see Table 4.2.

The results of this experiment clearly defined the difference between fertilized and unfertilized variants in the case of inoculation with strains incapable of releasing phosphorus from soil (Br29, Br85 and Br96) and strains with high P-solubilizing activity (D574 and D504). Variants inoculated with strains without P-solubilizing activity showed statistically significant differences between the fertilized and unfertilized variant with regard to yield of aboveground matter with the former having higher yields. The total nitrogenase activity showed the same trend, although the differences were not statistically significant. As for variants inoculated with strains with a high P-solubilizing activity, there was no statistically significant difference between the fertilized and unfertilized variants. Given the fact that the pots were not fertilized with nitrogenous fertilizers, the differences were attributable to the effect of phosphorus. In the case of

Table 4.1. Solubilized P and pH in liquid cultures

P-solubilizing strain	mg P litre^{-1}	pH
D504	58.3	4.6
D574	43.7	7.6
Br29	0	8.1
Br89	0	8.2
Br96	0	8.0

Table 4.2. Dry matter (d.m.) yields of aboveground matter, nitrogenase activity and grain yields of the soybean

Inoculant strain	Fertilization	Shoot d.m. g per pot	Seed d.m. g per pot	Nitrogen. activity (μmol R^{-1})
Br29	Fertilization*	16.80 e**	31.58 bc	46.89 cde
	No fertilization	9.68 ab	30.32 bc	38.97 abcd
Br85	Fertilization	14.87 cde	34.02 c	53.74 cde
	No fertilization	8.55 a	26.24 bc	19.16 abc
Br96	Fertilization	14.05 bcde	28.24 bc	23.06 abcd
	No fertilization	8.22 a	24.06 bc	9.09 ab
D574	Fertilization	15.71 bc	36.43 c	43.20 bcde
	No fertilization	10.74 abc	29.62 bc	58.87 de
D504	Fertilization	14.57 cde	33.65 c	78.15 e
	No fertilization	11.74 abcd	26.55 bc	75.24 e
No inoculant	Fertilization	14.84 cde	18.14 ab	5.31 a
	No fertilization	9.15 a	9.18 a	0.00 a

*The fertilized variants received KH_2PO_4 corresponding to 50 kg P_2O_5 ha^{-1}.

**Averages in the column designed by the same letter (a, b, c, d and e) do not differ in statistically significant way.

variants unfertilized with phosphorus, phosphorus might have been released if the measured values of total nitrogenase activity in unfertilized variants equalled the values in fertilized variants (D574 and D504). Nevertheless, it was worth noticing that as for grain yield, this difference was partially reconciled when plants were left till the grain ripened. It might be explained by the fact that in a pot experiment, the nutrient sources and the living space of plants were limited and growth slowed down at a certain stage, which enabled the delayed variants to catch up with the rest (Mikanová and Kubát, 1997a).

These findings were confirmed by the results achieved by some other authors (Dhingra et al., 1988; Adu-Gyamfi et al., 1989; Lauer and Blevins, 1989; Pongsakul and Jensen, 1991). These authors arrived at the conclusion that application of mineral phosphorus fertilizers to soils where phosphorus is a deficit nutrient increases symbiotic fixation of N_2 and dry matter yield in plants. Some studies also mention enhanced phosphorus concentration in plants and nodules (Israel, 1993).

PHOSPHORUS FORMS IN SOIL

As already mentioned, phosphorus is present in soil both in inorganic and organic compounds. Natural sources of phosphorus include mineral phosphates in particular, e.g. fluorapatite, chlorapatite, inorganic salts of phosphoric acid with calcium (calcium hydrogen phosphate, calcium phosphate), aluminium (variscite) or iron (strengite). Formation of salts of phosphoric acid depends on pH of the environment. There are large quantities of calcium salts in alkaline soils, and aluminium and ferric salts in acid soils.

Organic phosphorus accounts for some 25-85% of the total content of soil phosphorus. Organic phosphorus compounds are, to a large extent, present in grasslands, woodlands and tropical soils. The deeper the soil profile, the lower the content of organic phosphorus (Alexander, 1961). Organic soil phosphorus may be found in plants, soil animals and microorganisms. At the beginning of the pedogenic processes, phosphorus was present in soil only in the inorganic forms. Microorganisms, plants and animals that required phosphorus as a necessary nutrient started appearing gradually and phosphorus was integrated in organic parts of their cells. Both phosphorus forms, inorganic and organic, returned to soil through decomposition of dead bodies of these organisms and this is how the biological cycle of phosphorus transformation started.

During the process of soil organic matter mineralization, the organic phosphorus is mineralized as well. Plant residues and organic fertilizers enter the mineralization process as substances from which the soil microflora is able to release mineralize nutrients for its nutrition and for the plant nutrition, as well. The amount of mineralized phosphorus differs under various climatic and soil conditions and depends on the amount of soil organic matter. A majority of organic phosphorus forms present in natural conditions are orthophosphate acid esters. Some phosphate esters form an integral part of living organisms (nucleic acids, nucleotides, phospholipids, etc.) and appear in all living cells.

As far as phosphorus availability to plants is concerned, mineral soil phosphorus may be divided into three categories: phosphorus soluble in the soil solution (phosphorus available for plant uptake), labile phosphorus in the solid phase (complements of the soil solution), and stable phosphorus in the solid phase. Labile phosphorus complements the soil solution and thus serves as an important source of nutrients for plants. Upon exhaustion of labile phosphorus, even stable phosphorus may be released; however, it is a very slow process (Hartikainen, 1991). Only a small soil phosphorus stock contained in the soil solution serves as an immediate source of phosphorus for plants. When this pool is depleted, system balance is disturbed and some of the labile phosphorus is released into the soil solution. Mendoza et al., (1990) state that phosphate concentration in the soil solution may continuously change with the help of phosphate sorption and release from soil. The dominating process determines the net changes in the phosphorus concentration.

Phosphorus Mobilization

Importance

Regular and high volumes of mineral fertilizers are used in order to achieve high yields. Nevertheless, abundance of natural rock phosphates is rather limited and the process of excavation is extremely arduous and expensive. Besides that, apart from phosphorus, a number of accompanying elements is put into the soil, some of which could be very hazardous when entering the food chain. From the environmental point of view, in the long run it will be imperative to enhance efficiency of phosphorus uptake from the soil stock and phosphorus usage from mineral phosphorus fertilizers. One of the options to attain this goal is by solubilization of hardly soluble phosphates.

Soil Phosphorus Mobility

Phosphorus may move within the soil profile for three reasons: through soil organisms (plants, animals, microorganisms), soil water movement, and in the direction of concentration gradient (diffusion).

The effect of higher plants is of major importance. In fact, all labile phosphorus may be included in this process, while the extent and speed of movement depends on the amount of phosphorus taken by plant roots. Subsequently, phosphorus moves through plants to the aboveground part and returns back to the soil through waste matter. It results in specific distribution of organic phosphorus forms, with the highest concentration being usually in the soil surface layer and the root zone.

The amount of phosphorus that takes part in movement with soil water is determined by phosphorus concentration in the soil solution and intensity of the soil solution movement. This type of movement of soil phosphorus is important for ensuring phosphorus supply to roots and possible phosphorus wash-out into the understrata. In the case of low phosphorus concentration in the soil solution, phosphorus movement is negligible and loss caused by phosphorus washout is insignificant.

Activity of soil microorganisms contributes to a higher phosphorus concentration in the soil solution, as well as a greater potential movement of phosphorus. Penetration is higher in organically fertilized soils than in soils fertilized with an equivalent amount of phosphorus in the inorganic form (Anderson, 1980).

Diffusion is a process in which transport from one part of the system to another takes place by means of movement of molecules or ions. When the system is in balance, the particles stop moving. If there is a concentration gradient in the system, the system attempts a change in order to achieve a balanced status. This mechanism is partially applied in phosphorus movement in proximity of the root zone (Eghball et al., 1990).

Possibilities of Phosphorus Mobilization

One of the ways to prevent phosphorus from bonding into hardly soluble aluminium and ferric compounds is liming of acid soils. As a result of the presence of Al^{3+} and Fe^{3+}, stable crystalline minerals like variscite and strengite can be formed in acid soils. The importance of strong chemisorption in acid soils must be respected when using phosphorus fertilization. Soils with low pH must be limed before phosphorus fertilization. On the other hand, liming must be avoided on neutral and alkaline soils as hardly soluble calcium phosphate can be formed in soils high in calcium.

Appropriate crop rotation, particularly inclusion of leguminous plants in the cropping pattern, also facilitates phosphorus mobilization. As a result of more aggressive root exudates, most of these plants can make better use of phosphorus from soil compounds. Another advantage of including leguminous plants is that their root system reaches considerable depths (particularly that of clover and alfalfa). With the help of its roots, these plants are capable of taking up phosphorus from lower strata. It is also necessary to mention the symbiosis of leguminous plants with bacteria of the *Rhizobium* family, which infect roots of leguminous plants and some of which have the ability to release phosphorus from hardly soluble compounds. Also, rich root residues that remain in soil after harvest are a source of organic phosphorus and nutrients supporting development of soil microflora.

Another measure to mobilize phosphorus is to enrich soil with organic matter. Organic matter of plant and animal origin entering soil is a source of energy and allows nutrient cycling in soil. Application of organic matter decreases phosphorus sorption, thereby increasing phosphorus uptake by plants, particularly in well-irrigated soils and under aerobic conditions. Minhoni et al., (1991) found in their study that if a source of carbon is missing, solubilization of phosphates does not take place. Minhoni further claims that glucose increases the solubilization process more than any other organic material. Also, Lee et al., (1990) confirm the favourable effect of added organic matter on increased levels of plant-available phosphorus in soil. Brady (1990) states that specific organic compounds form complexes with Fe and Al ions, preventing them from reacting with phosphates. Release of phosphorus from rock phosphate by humic and fulvic acids has been reported (Kapoor, 1995). Humic substances are produced in soil as a result of organic matter decomposition and contain humic and fulvic acids as their main components. Organic matter is also a substrate containing a large quantity of soil microorganisms, particularly fungi and bacteria, out of which some show P-solubilizing activity.

Phosphorus mobilization is also affected by root exudates. As already mentioned above, particularly leguminous plants are well-known for their aggressive root exudates which facilitate better utilization of phosphorus from soil. Another plant well-known for releasing phosphorus using its exudates, is oilseed rape. It was also interesting to note that plants which were grown in soil low in phosphorus, had released a much higher amount of acid exudates. Hoffland et al., (1989a, b) quantified the amount of acid exudates released by oilseed rape to agar which either did or did not contain phosphorus. Oilseed rape plants grown with no source of phosphorus released acid exudates, while plants grown in a medium containing phosphorus did not release acid exudates. These exudates contained some organic acids, such as citric and malic acids. This was confirmed by Lipton et al., (1987) in alfalfa, Dinkelaker et al., (1989) and Minggang et al., (1997) in lupine. The relationship between rhizosphere acidification and phosphorus mineral nutrition was also published earlier by Marschner et al., (1986).

Soil macrofauna also plays an important role in releasing and moving phosphorus in soil. Soil invertebrates (e.g. earthworms) contribute to disintegration of organic residues and their mixing with other soil components. Crushing of the residues increases its surface and microorganisms can decompose organic substances more efficiently. This facilitates release of phosphorus from organic residues to soil.

Soil microflora plays a crucial role in phosphorus transformations in soil. In terms of nutrient cycling, the biomass of soil microorganisms is a small but very important part of soil organic matter. When obtaining phosphorus from soil, most microorganisms use the enzyme phosphatase, which breaks down organic phosphates. Approximately 2 to 5% of the total amount of phosphorus in soil can be temporarily bound in microbial biomass. After the microorganisms die and decompose, phosphorus is released and is made available to plants again. The uptake rate of phosphorus from a solution is twice as high in fungi and nine times as high in bacteria as the intake rate of phosphorus by plant roots. If the water soluble phosphorus is depleted from the soil solution, it is replaced, as a result of chemical balancing, by phosphorus bound to soil particles–labile phosphorus.

Mechanisms of P-solubilization

Solubilization of Mineral Phosphate

The primary solubilization mechanism is release of organic acids by P-solubilizing microorganisms, resulting in a decrease of the medium's pH, as was confirmed in several studies (Kubát and Mikanová, 1994; Mikanová and Kubát, 1994b) and elsewhere (Rodriguez and Fraga, 1999).

The efficiency of P-solubilizing microorganisms isolated from the rhizosphere of wheat was tested in liquid shaken cultures. During a week of incubation, about 10 out of 12 tested isolates transformed a portion of poorly soluble calcium phosphate into phosphate ions dissolved in a medium, while two strains did not, and despite that they survived and grew in a medium without soluble phosphates. Table 4.3 provides the results representing arithmetic means of three measurements. The largest measured phosphorus concentration in a medium was 73.5 mg litre^{-1} in No. 88, which accounts for 73.3% of supplied calcium phosphate. A portion of the supplied phosphorus was incorporated into newly formed microorganic biomass.

It is obvious that P-solubilizing activity of all tested organisms was, to a large extent, related to the medium's pH at the end of the experiment. These results confirm the assumption that one of the key mechanisms of P-solubilization is a decrease of pH, probably due to the production of organic acids, while the variable closeness of the relation between P-solubilizing activity and the medium's pH can be caused by other solubilization mechanisms (e.g. complexes formation, CO_2 formation, etc.). Production of organic acids as the primary solubilization mechanism is also confirmed in the studies by Sperber (1958) and Taha et al., (1969). The highest P-solubilization was recorded in case of citric acid (Gaur, 1990). The effect of citric acid on phosphate solubilization was also mentioned in the study by Gardner and Parbery (1983). According to their findings, when there is a lack of phosphorus in soil, roots of white lupine plants produce a large quantity of citric acid in root exudates.

The type of acids produced and their quantity depends on the micr台organisms involved in the P-solubilization. Organic acids, such as

Table 4.3. The amount of the solubilized phosphorus and pH value of P-solubilizing bacterial isolates from wheat rhizosphere

Isolates (No.)	mg P litre^{-1}	pH
88	73.5	3.8
86	63.0	3.4
85	63.0	3.5
22	43.4	4.5
3	40.5	5.6
30	37.1	5.9
11	36.9	5.4
202	6.5	6.7
18	6.4	8.6
27	3.7	7.6
8	0.0	8.3
23	0.0	8.8

lactic, citric, succinic, fumaric, malic, and oxalic acids, take part in solubilization of hardly soluble phosphates (Sperber, 1958; Taha et al., 1969). The efficiency of P-solubilization is affected by acid, which the microrganism produces.

Some P-solubilizing microorganisms produce 2-ketogluconic acid. The mechanism is accompanied by formation of calcium chelates. Organic acids also form stable complexes with iron and aluminium.

Other substances contributing to phosphate solubilization are humic and fulvic acids. Inskeep and Silvertooth (1988) proved in their study that presence of humic and fulvic acids inhibits precipitation of hydroxyapatite. This also explains the increased P-concentration in the presence of manure and sludge. The favourable effect of sludge on release of phosphorus was also confirmed by Rydin and Otabbong (1997).

Some authors describe a number of other mechanisms of phosphorus solubilization from hardly soluble compounds. However, these are mostly mechanisms resulting in decreased pH (Thomas, 1985; Asea et al., 1988). Kapoor (1995) describes CO_2 production, which results in decreased pH due to the formation of carbonic acid, H_2SO_4 production by sulphur bacteria as well as H_2S production, which is caused by the activity of some bacteria under anaerobic conditions. However, the effectiveness of these processes has been questioned and their contribution to phosphorus release in soil appears to be negligible (Rodriguez and Fraga, 1999).

Factors Affecting P-solubilization

Value of pH: Optimum pH for solubilization of phosphates depends on the type of insoluble phosphate present in soil. Various ion forms of phosphorus prevail in the soil solution depending on its pH (Mikanová, 2000; Fig. 4.1). It is $H_2PO_4^-$ which occurs in the maximum amount when pH equals to 3.5-5.5 that is best plant-available. Higher the pH, the higher the share of HPO_4^{2-} and PO_4^{3-}. Hartikainen and Simojoki (1997) state that if H_2SO_4 is added to soil, soluble phosphorus is released; i.e. the more the acid is added, the more phosphorus is released. Menon and Chien (1990) also found out that acidification of raw phosphates with H_2SO_4 has a favourable effect on maize yield in comparison with non-acidified phosphates. Nevertheless, this seemingly simple dependence is complicated by the presence or absence of other compounds or ions in relation to various pH levels. Presence of Fe^{3+} and Al^{3+} when soil pH is low, and presence of Ca^{2+} when soil pH is high are of paramount importance for phosphate solubility. When pH is low and free Al^{3+} and Fe^{3+} ions are present, insoluble aluminium and ferric compounds are formed. In the case of an alkaline reaction and presence of Ca^{2+} ions, even calcium phosphate, which is hardly available to plants, may be formed. In

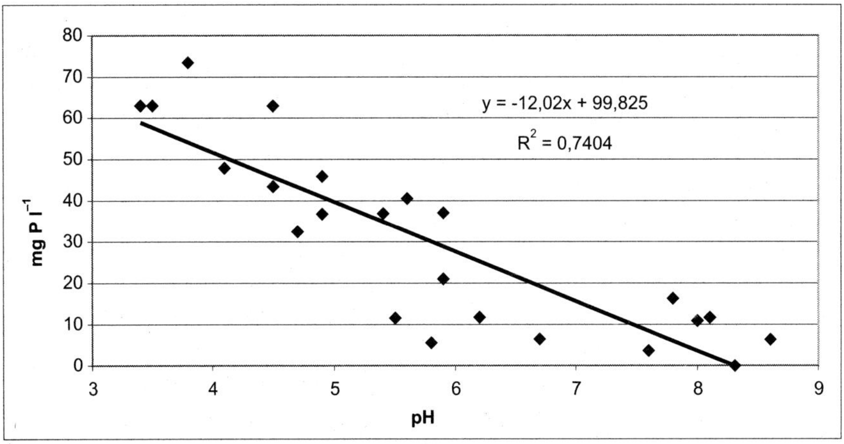

Fig. 4.1. The relation between phosphorus-concentration (mg P litre^{-1}) and pH value of the medium in P-solubilizing rhizosphere isolates (Mikánova, 2000)

spite of all these facts, Brady (1990) mentions that maximum of plant-available phosphorus in soil is present with pH being 6.0-7.0. The pH value also has an impact on microorganisms involved in solubilization. Bacteria prefer a neutral up to slightly alkaline environment for maximum solubilization, while fungi prefer a slightly acid up to neutral environment. However, bacteria vary as to their optimum pH for solubilization. Bajpai and Rao (1971) in their study compare various genera of P-solubilizing bacteria with various pH in a medium. They found that when bacteria released maximum amount of phosphorus, the pH level differed from genera to genera.

Activity of soil microorganisms: The carbon source plays a vital role in the activity of the soil microflora solubilizing phosphates. Organic carbon source is a key factor that affects the microbial growth and creation of extracellular metabolic products. Glucose, xylose and saccharose are considered the best carbon source for solubilization of phosphates in the case of fungi, whereas in the case of bacteria it is galactose, glucose, saccharose and arabinose (Kapoor, 1995). The nitrogen source used by microorganisms also affects the activity of P-solubilizing microorganisms. Goswami and Sen (1962) state that sodium nitrate was a better nitrogen source than urea and ammonium sulphate for the P-solubilizing bacteria under their study. Gaur (1990) also presents various nitrogen sources which enable *Bacillus, Pseudomonas, Aspergillus* and *Penicillium* to achieve maximum P-solubilization activity. Temperature also affects activity of microorganisms. Every microorganism taking part in solubilization of phosphates has a different optimum temperature at which its P-solubilizing activity is maximum.

Type of hardly soluble phosphate: Another factor that affects the amount of released phosphorus is the type and origin of insoluble phosphate. Kapoor (1995) mentions that calcium phosphate, variscite and strengite are better soluble than rock phosphates. Solubility of rock phosphates depends on the type and origin of a given rock phosphate.

Presence of available phosphorus: Presence of soluble phosphates in soil also greatly impacts solubilization of the hardly soluble phosphates. Goldstein (1986) found out that in the case of *Escherichia coli* bacterium, the group of genes involved in the mineral phosphate solubilization was negatively regulated by the soluble phosphate concentration in a medium. As early as in 1965, Chhonkar and Subba Rao (1967) compared P-solubilizing activity of various genera of fungi in a medium with addition of soluble KH_2PO_4. Although fungi's P-solubilizing activity was high, following the addition of KH_2PO_4 this activity ceased. Tiwari et al., (1989), who in their study point out a greater effect of inoculation with P-solubilizing bacteria in soils low in phosphorus, indirectly found out the same. Many other authors (Menge et al., 1978, Thomson et al., 1986, Amijee et al., 1989) also mention that colonization by AM fungi was inhibited by increasing phosphorus fertilization. Braunberger et al., (1991) quote a number of authors who are of the opinion that the reduction is caused by an increased phosphorus concentration in plant tissues, resulting in changed quality and quantity of root exudates, and a decreased number of fungi's entrance points in roots. Mikanová et al., (1997b) also confirm in their study that there are two types of bacteria whose P-solubilizing activity is inhibited by presence of plant-available phosphates, and other bacteria whose activity is not affected.

The effect of soluble phosphate in medium on the P-solubilizing activity of eight rhizosphere isolates from the collection of RICP was tested in liquid cultures with the addition of various concentrations of soluble KH_2PO_4 (Table 4.4) (Mikanová and Novaková, 2002). These isolates have shown a high P-solubilizing activity in previous tests. The concentration of phosphorus in uninoculated control corresponds to the amount of hardly soluble calcium phosphate added – 40 mg litre^{-1}.

It can be seen from the Table 4.4 that all of the tested isolates posses P-solubilizing activity. Six of these isolates solubilized more than half of the added tricalcium phosphate. All the tested isolates (except 39) were repressed under the presence of soluble phosphate. The isolate number 39 dissolved all the added tricalcium phosphate and its P-solubilizing activity was not inhibited by added soluble phosphate. The isolate number 24 was inhibited the most markedly. This isolate was inhibited already at the level 5 mM P litre^{-1}.

Table 4.4. The rest of insoluble phosphorus (mg P litre^{-1}) in medium with the added various concentrations of soluble KH_2PO_4 after incubation

Isolate	Concentration of soluble P in medium before incubation in mM L^{-1}					
	0 mM P L^{-1}	5 mM P L^{-1}	10 mM P L^{-1}	15 mM P L^{-1}	20 mM P L^{-1}	25 mM P L^{-1}
39	0	0	0	0	0	1.85
287	6.61	7.61	11.83	8.13	13.27	12.76
24	1.79	16.53	17.11	19.23	19.41	18.65
88	12.75	15.01	17.73	22.12	18.19	21.79
262	13.09	16.64	17.73	21.95	23.29	26.35
10	19.61	31.82	27.22	31.23	30.49	29.81
22	26.09	31.78	26.17	28.60	29.23	27.19
15	27.31	31.20	31.70	32.41	34.25	36.42
Control	40.00	40.00	40.00	40.00	40.00	40.00

The inhibition of P-solubilizing activity in the presence of soluble phosphates in medium may account for a small efficiency of the inoculation of soil with P-solubilizing microorganisms, especially soils well supplied with phosphorus.

Regulation of the P-solubilizing activity by soluble phosphate in medium was also studied by 12 strains of *Rhizobium* and *Sinorhizobium* from the Rhizobia Culture Collection of the RICP (Mikanová and Novaková, 2002). The results are shown in Table 4.5. Three strains of

Table 4.5. The rest of insoluble phosphorus (mg P litre^{-1}) in medium with the added various concentrations of soluble KH_2PO_4 after incubation by *Rhizobium leguminosarum*, *Rhizobium trifolii* and *Sinorhizobium meliloti*

Strains	Concentration of soluble P in medium before incubation in mM L^{-1}					
	0 mM L^{-1}	5 mM L^{-1}	10 mM L^{-1}	15 mM L^{-1}	20 mM L^{-1}	25 mM L^{-1}
D558 - *R. trifolii*	14.30	19.58	22.42	23.88	31.66	33.66
D659 - *R. trifolii*	14.30	17.70	20.44	23.20	28.76	38.00
D481 - *R. trifolii*	14.98	16.26	21.42	23.84	30.78	35.70
D561 - *R. legum.*	17.11	25.09	28.64	33.28	33.91	39.44
D663 - *R. trifolii*	17.98	22.50	22.66	26.38	32.04	36.96
D661 - *R. trifolii*	18.78	22.96	24.92	29.32	32.40	34.62
D600 - *R. legum.*	20.15	25.59	27.89	37.67	39.09	40.06
D662 - *R. trifolii*	22.25	26.81	28.94	31.07	36.09	40.79
D562 - *R. legum.*	23.54	30.11	30.03	32.87	36.59	42.32
D564 – *R. legum.*	24.13	30.07	35,96	37.93	40.76	40.17
D163 – *S. melilot.*	31.08	32.12	33.40	34.54	35.92	35.80
D4 - *S. melilot.*	31.10	34.80	36.30	34.30	30.90	36.45
Control	40.00	40.00	40.00	40.00	40.00	40.00

Rhizobium trifolii (D558, D659 and D481) showed high P-solubilizing activity (25-26 mg P were solubilized from 40 mg P added) without soluble phosphate in medium. Only the activity of the strain D558 was not completely inhibited even at concentration of 25 mM P litre^{-1}. The activity of other strains is already strongly negatively influenced at concentration of 25 mM P of soluble phosphorus per liter, or the solubilization is stopped completely. Five of tested strains stopped phosphorus solubilization at concentration of 20 mM P litre^{-1}. The strain D600 stopped P-solubilization at concentration of 15 mM P litre^{-1}.

In order to verify the effect of soluble phosphates on P-solubilizing activity, a pot experiment with green pea, the Sonet variety, was established. Prior to seeding, the seeds were inoculated with *Rhizobium leguminosarum* strains with a comparable nitrogenase and P-solubilizing activity which, however, differed in the degree of effect on the said activity due to presence of soluble phosphates (D561 – unaffected, and D600 – negatively affected possessing comparable P-solubilizing activity in the absence of soluble phosphates, D562 – unaffected, D564 – negatively affected with a comparable P-solubilizing activity in the absence of soluble phosphates). The available phosphorus content in soil was 91 mg P kg^{-1} (Bray and Kurtz, 1945). Each variant was in 10 replicates. For five pots in each variant the following factors were determined at the beginning of flowering: the total nitrogenase activity (TNA), the aboveground dry matter, and the root dry matter. Other five pots were left till the legume ripened and then the seed dry matter was determined (Table 4.6).

The results showed that there were statistically significant differences among the TNA values and the shoot dry matter yields of plants inoculated with different *R. leguminosarum*. The plants inoculated with strains whose P-solubilizing activity was not inhibited by soluble phosphates showed a higher nitrogenase activity. It is well-known from the literature and also from our previous studies that phosphorus is necessary for nodulation and N_2-fixation. A greater amount of fixed nitrogen ofcourse resulted in a greater amount of the aboveground dry matter. These differences were as statistically significant (at the

Table 4.6. Total nitrogenase activity (TNA), dry matter (d.m.) yield of aboveground matter, roots and seeds by peas inoculated by *Rhizobium leguminosarum*

Inoculate	TNA (μ mol C_2H_2 h^{-1})	Shoot d.m. (g)	Root d.m. (g)	Seed d.m. (g)
D561 – unaffected	33.55 c*	4.88 b	2.68	5.28
D600 – negativ.aff.	22.84 ab	4.22 a	2.44	3.61
D562 – unaffected	32.75 c	4.84 b	2.81	5.41
D564 – negativ.aff.	21.61 a	4.18 a	2.86	4.60

*Averages in the column designed by the same letter (a, b, c) do not differ significantly.

significance level of P = 0.05) for nitrogenase activity and the aboveground dry matter, whereas in the case of the root dry matter and seed dry matter, the differences were not statistically significant although the results showed a similar trend.

Solubilization of Organic Phosphorus

The decomposition of organic matter in soil is carried out by the action of saprophytes, which produce the release of radical orthophosphates e.g. nucleic acids, phospholipids, etc., from the carbon skeleton of the molecule. The mineralization of these organic phosphorus compounds is carried out by means of the action of several phosphatases (= phosphohydrolases): acid phosphatases and alkaline phosphatases. Acid phosphatases are further classified as specific or non-specific acid phosphatases, in relation to their substrate specificity. The specific acid phosphatases with different activities include: 3'- nucleotidases and 5'-nucleotidases (Va'zquez, 1996), hexose phosphatases (Pradel and Boquet, 1988) and phytase (Cosgrove et al., 1970).

ROLE OF SOIL MICROFLORA IN PHOSPHORUS TRANSFORMATIONS

Soil microflora, particularly P-solubilizing bacteria and fungi, play the most important role in releasing phosphorus from hardly soluble compounds in soil. Approximately 20-30% of soil microorganisms have the ability to transform poorly soluble phosphates into soluble forms (Domey, 1987).

P-solubilizing Bacteria

Among bacteria, these include the genera *Azotobacter* (Iswaran and Marwaha, 1981), *Erwinia* (Babenko et al., 1984), *Bacillus* (Goswami and Sen, 1962, Fernández et al., 1985, Datta et al., 1992), *Pseudomonas* (Myšków, 1960, Ralston and McBride, 1976), *Escherichia* (Sperber, 1957), *Enterobacter* (Laheurte and Berthelin, 1988), and others. These bacteria can be isolated from soil or root rhizosphere of various plants. Isolation is carried out on agar plates containing insoluble calcium phosphate. In the case of P-solubilizing microorganisms, euphotic zones are formed around colonies of microorganisms.

Methodology (Mikanová, 2000): Petri dishes with P-agar (nutrient agar containing insoluble calcium phosphate) are inoculated with the tested microorganism at several points. P-agar composition: glucose, 10.0 g; K_2SO_4 0.2 g; asparagin, 1.0 g; $MgSO_4.7H_2O$, 0.4 g; yeast autolysate, 0.2 g; agar, 17 gL^{-1} of distilled water, before distribution solutions of Na_3PO_4 7 ml and $CaCl_2$ 3 ml 200 ml^{-1} of agar are added; the solutions contain 10.9 g of $Na_3PO_4.12\ H_2O$ and 22.0 g of $CaCl_2$, respectively, and each is filled up to the level of 100 ml with distilled water.

The size of the euphotic zone indicates P-solubilizing activity of a respective strain. This method is suitable because of its low time and labour intensity, and allows for a quick and preliminary estimate of P-solubilizing activity of a large number of microorganisms.

Testing in liquid cultures with the addition of calcium phosphate and subsequent measuring of released phosphorus is used for more accurate determination of P-solubilizing activity of these microorganisms (Mikanová and Kubát, 1994b).

Testing is based on the measuring of the content of released phosphorus in liquid shake cultures inoculated with individual microorganisms. Hundred ml of the cultivation medium (the same as that for P-agar but with no agar, $CaCl_2$ and Na_3PO_4) is poured into each 250 ml Erlenmayer flask and 0.05 g of $Ca_3(PO_4)_2$ is then added to each flask. The individual flasks are inoculated with 2 ml of inoculum (suspension of microbes).

Preparation of the inoculum: Bacteria are bred on Hirte, or pea (rhizobia), agar in test tubes; two tubes are washed down by a sterile saline solution and total 10 ml of inoculum is pipetted out.

The flasks are let to incubate for seven days at a temperature of 28°C under constant shaking. Following incubation, the cultivation medium is centrifuged (for 10 minutes at 5,000 rpm), the supernatant is filtered out and the concentration of water-soluble phosphorus in the filtrate is determined.

The principle of determination: In the presence of antimonyl-potassium tartrate, a phosphatic ion reacts with ammonium molybdate in the environment of sulphuric acid, producing phosphomolybdic acid, which reduces ascorbic acid to molybdenum blue (Murphy and Riley, 1962). The intensity of colouration caused by molybdenum blue is measured colourimetrically at 710 nm. Finally supernatant pH is determined.

Abundance of P-solubilizing bacteria in soil also depends on the examined location. Yahya and Al-Azawi (1989) isolated P-solubilizing bacteria from dissimilar soil samples and found different numbers of these bacteria, in a downward trend, under the vegetation of vegetables, legumes, grass, crops and the fewest P-solubilizing bacteria were identified in orchards.

The ability to release phosphorus from sparingly soluble compounds was also identified in bacteria of the genera *Rhizobium, Bradyrhizobium* and *Synorhizobium.* P-solubilizing activity of the *Bradyrhizobium japonicum* strain, which fixes N_2 in symbiosis with soya-bean plants, was confirmed in studies by Reichlova (1972), Halder et al., (1991) and Mikanová and Kubát (1994a, 1995b, 1996, 1997c).

The experiments with leguminous plants inoculated with effective strains of the genus *Rhizobium* are based on the fact that both factors, plant-available phosphorus and inoculation, increase nodulation and nitrogenase activity and may, therefore, affect yields. Given that strains showing comparable nitrogenase activity were selected for inoculation, the increase in dry matter can at least be partially attributed to their different phosphorus solubilizing activity. The positive effect of strains of the genus *R. leguminosarum* with P-solubilizing activity on bean yield is presented by Sudhakar et al., (1989). It results from their study that the yield in the variant fertilized with 40 kg of phosphorus per hectare matches that of the unfertilized but inoculated variant.

The results corroborated with several other studies (Mikanová and Kubát, 1994a, 1995b, 1996, 1997c) according to which the savings of phosphorous mineral fertilization of leguminous plants inoculated with *R. leguminosarum* strains with P-solubilizing activity could be between 40 and 50 kg of phosphorus per hectare. It was shown in a pot experiment aimed at determining the effect of inoculation and phosphorus fertilization on the yield of pea. Green pea, the Tyrkys variety, was seeded. The pots contained 5 kg of sifted soil from the arable layer (0-25 cm) of Orthic Luvisol in the RICP Prague-Ruzyně. The content of plant-available phosphorus was 42 mg P kg^{-1} (Bray and Kurtz, 1945). Variants fertilized with superphosphate at the dose matching 45 kg of phosphorus per hectare and variants unfertilized were established. Following sowing, seeds were inoculated with a suspension of a mixture of the *R. leguminosarum* strains with high nitrogenase activity and simultaneously high phosphorus solubilizing activity (P+), a mixture of *R. leguminosarum* strains with high nitrogenase activity without phosphorus solubilizing activity (P-) and a check variant without inoculation, in six replicates, respectively. The Table 4.7 shows the results of the pot experiment.

Table 4.7. The yield of seed dry matter (d.m.) of pea

Inoculation	Fertilization	Seed d.m. (g per pot)
Mixture P+	No Fertilization*	22.5 bcd**
	Fertilization	19.0 ab
Mixture P-	No Fertilization	17.2 a
	Fertilization	23.4 d
No inoculation	No Fertilization	18.4 a
	Fertilization	22.8 d

* The fertilized variants received KH_2PO_4 corresponding to 45 kg P ha^{-1}

** Averages in the column designed by the same letter (a, b, c, d) do not differ significantly

The yield of seed dry matter was affected by both pea inoculation and the dose of the fertilizer used. There was a statistically significant increase in seed dry matter yield upon superphosphate fertilization in the variants inoculated with a mixture of ineffective strains (variant P-) and in the check variant (no inoculation). The yield was unaffected by fertilization in the variant inoculated with effective phosphorus solubilizing strains (variant P+). The yield of seed dry matter in the unfertilized variant inoculated with a mixture of phosphorus solubilizing strains is statistically significantly higher than in the other unfertilized variants and approximately the same as in the other variants fertilized with 45 kg of phosphorus per hectare (Mikanová et. al., 1995a).

P-solubilizing activity of the other strains of the genus *Rhizobium* was also identified (Kabesh et al.,1975; Sudhakar et al.,1989; Halder and Chakrabartty,1993; Abd-Alla,1994; Mikanová et al.,1995a).

P-solubilizing Fungi

The ability of fungi to release phosphorus from phosphates has been known for years (Barroso and Nahas, 2005). Venkateswarlu et al., (1984) tested different soil microorganisms for their P-solubilizing activity and found out that fungi are much more efficient in releasing phosphorus than bacteria. Increase of plant available phosphorus on calcareous soil caused by fungi of the genera *Penicillium* and *Aspergillus* is cited in the study by Salih et al., (1989). Asea et al., (1998) have described the release of phosphorus from raw phosphates in liquid cultures by isolates of the genus *Penicillium*. Higher yields of oilseed rape after inoculation by *Penicillium bilaji* are cited by Kucey and Leggett (1989). Solubilization of phosphorus from raw phosphates by fungi of the genus *Aspergillus* is also confirmed by Singal et al., (1994).

Another group of fungi in which many authors discovered the phosphorus solubilizing ability is arbuscular mycorrhiza (AM) fungi. AM fungi were found in a majority of farm crops. They rank among endomycorrhiza fungi and their positive effect on plant nutrition consists particularly in forming a network of hyphae, thereby improving uptake of nutrients from soil, including improved plant nutrition with phosphorus. These are mostly strains of the genera *Glomus* and *Gigaspora*. Khurana and Dudeja (1995) cite a number of authors who mention production of organic acids by fungal hyphae and subsequent solubilization of phosphorus as a result of decreased pH. Jayachandran et al., (1989) present another possible mechanism of phosphorus solubilization by AM fungi–production of siderophores, which are capable of forming ferric chelate compounds, thereby preventing formation of ferri-ferous phosphates. Many authors cited by Jayachandran et al., (1989) confirm that if lack of phosphorus limits plant growth, plants colonized by AM fungi are able to take out more phosphorus than plants not colonized, thus improving their growth and development. The population of AM fungi is affected by a variety of factors. There are many studies dealing with the effect of crop rotation or soil irrigation on AM population; there is also evidence of the effect of nitrogen and potassium on AM fungi in field conditions. The alterations in the colonizing activity of AM fungi isolated from soils fertilized, in the long term, by phosphorus fertilizers in relation to fungi isolated from unfertilized soils is dealt with by Gryndler and Vancura (1991). The highest intensity of infection occurred in the variants where plants suffered from only a slight lack of phosphorus. In this study, Gryndler arrived at a conclusion that the best colonizing populations are, as a rule, present in soils with low content of plant-available phosphorus. It is, therefore, possible that any intervention in soil (including fertilization) may cause shifts in the characteristics of the original populations of AM fungi and this previously functional symbiont does not find favourable conditions in rhizosphere, and its presence in soil need not have a positive impact on plants. In such a case, it is appropriate to introduce into the soil another population of AM fungi, which can replace the original population. However, it is necessary to mention the substantial disadvantage of using AM fungi as an inoculant. AM fungi cannot be bred in vitro on a nutrient medium, but always only in symbiosis with plant roots. This makes their application in large-scale agricultural production considerably more difficult (Vancura, 1980; Rea and Tullio, 2005; Chapter 4 in Volume I of this series–Microbial Biotechnology in Agriculture and Aquaculture).

PRACTICAL USE OF P-SOLUBILIZING SOIL MICROFLORA

Role of P-solubilizing Soil Microflora in Plant Nutrition

From the point of view of sustainable agriculture, at least a part of phosphorus necessary for plants needs to be supplied from other sources. Solubilization of phosphates is, besides traditional methods of phosphate mobilization, a phosphate mobilization method with very good prospects. Phosphorus supplied to soil in the form of mineral phosphorus fertilizers is transformed to such forms up to 90% that are hardly plant-available. In spite of the fact that soils usually contain a sufficient amount of total phosphorus, the amount of plant-available phosphorus is rather small and may thus become a limiting factor to plant growth. Usage of soil microflora is one of the options to provide plants with sufficient supply of phosphorus.

Data in literature make it obvious that some 20-30% of soil microorganisms are able to transform the hardly soluble phosphates to soluble forms. The ability to release phosphorus was also found with bacteria of the genera *Rhizobium, Bradyrhizobium* and *Sinorhizobium*. These bacteria, which are well-known for their ability to fix atmospheric nitrogen, are commonly used worldwide for inoculation of leguminous plants.

The procedure for verifying the size of P-solubilizing activity of soil microflora usually aims at determining P-solubilizing activity of strains in liquid shake cultures (in vitro), testing strains in association with plant roots (in vivo) in pot experiments and field experiments.

Determination of P-solubilizing Activity of Microorganisms

P-solubilizing activity on agar plates: Determination of solubilization on agar plates is used for isolation, quantification of microorganisms with this activity in soil, and for preliminary estimate of the size of this activity (for methodology, refer earlier section in this chapter: P-solubilizing Bacteria). This method is advantageous due to its low time intensity and allows for a quick and preliminary estimate of P-solubilizing activity of a large number of microorganisms.

P-solubilizing activity in liquid shake cultures: Solubilization in liquid cultures is used for testing the size of microorganisms' activity. The testing is based on measuring the content of released phosphorus in liquid shake cultures inoculated with individual microorganisms. For methodologies, again refer section: P-solubilizing Bacteria.

P-solubilizing activity in association with plant roots: In terms of practical usage of characteristics of P-solubilizing microorganisms, their activity in association with plant roots is decisive. These experiments may take place in a greenhouse, in pots or small-plot or field experiments.

The pot experiment with red clover, the Kvarta variety, was started in order to test the selected *Rhizobium trifolii* strains. First of all these strains were tested in liquid shake cultures with the addition of insoluble tricalcium phosphate. Results of determination of solubilized phosphorus and medium pH are presented in Table 4.8.

Table 4.8. The amount of solubilized P and pH value of *Rhizobium trifolii*

Strain	mg P litre^{-1}	pH
D660	76.2	4.5
D662	62.5	5.1
D661	30.7	6.7
D593	2.5	6.1

Another experiment was established to verify their efficiency in soil. Potassium phosphate (KH_2PO_4) at a dose equal to 50 kg of P_2O_5 ha^{-1} was put in one half of the pots, before sowing. The second half was not fertilized. Ten replicates were established for each variant. The weight of the dry aboveground matter was evaluated for four cuttings (Table 4.9).

Comparison of the fertilized and unfertilized variants inoculated with all the strains of *Rhizobium trifolii* produced interesting results. There were remarkable differences in the first cutting between the fertilized and non-fertilized variants because the inoculation was not active yet. The differences between the fertilized and unfertilized variants were nearly triple in favour of the former, in the first cutting. Such difference in yield was caused not only by a lack of phosphorus, but also by a lack of nitrogen. Nevertheless, in the fourth cutting, the above ground matter yields were similar in all variants. Obviously, certain start-up amount of nutrients could be beneficial, in spite of that the yields in the fourth cutting were similar in all variants. The differences between the fertilized and unfertilized variants in the aboveground matter yields outlasted in the total yield of all cuttings.

Inoculation

The objective of inoculation is to increase soil fertility by introducing specific microorganisms. However, application of inoculants to soil is not without problems. The main issue regarding introduction of microorganisms to soil is related to their survival. The introduced microorganisms survive in soil much worse than microorganisms living in soil in the long-term. Furthermore, not even the surviving microorganism may be able to compete with native soil microflora and perform those activities for which it was selected. Many studies show that specific microorganisms are reliable so as to fulfil certain functions in laboratory

Table 4.9. Yield in four cuttings of the dry aboveground matter of red clover

Inoculation	Fertilization ($kg\ ha^{-1}$)	1st Cut	2nd Cut	3rd Cut	4th Cut	Total
		Shoot dry weight (g)				
D660	50	14.42	5.75	7.72	10.82	38.71
	0	5.41	4.43	8.17	10.20	28.21
D662	50	12.92	6.08	8.08	11.03	38.12
	0	5.08	4.38	7.75	10.40	27.61
D661	50	13.04	6.13	8.25	10.14	37.56
	0	4.38	4.67	5.84	9.32	24.21
D593	50	13.60	7.47	7.97	10.89	39.93
	0	4.02	5.74	6.12	10.07	25.95

tests. Nevertheless, if further tests are conducted in non-sterile soil or in field conditions, the results are usually less convincing. It is because soil is a very complex system subject to a number of influences. It is also true that commonly used amounts of biopreparations represent only a negligible fraction of biomass of microorganisms already present in soil. Besides that, composition of microbial population in soil is rather varied. These microorganisms undergo an enormous long-term selection pressure. It is because in soil there are many types of surviving microorganisms that almost constantly suffer profound lack of nutrients. On the one hand, it is the cause of remarkable performance and adaptability of soil microflora. On the other hand, it greatly limits the introduced microorganisms' ability to survive. Another issue arises because the introduced microorganisms may face such impacts in soil that are not taken into consideration in simplified laboratory tests. Nevertheless, in spite of all these problems, greater exploitation of the biological potential of soil microorganisms has good prospects especially due to environmental reasons.

Inoculating preparations are indispensable where, for some reason, there are no original natural populations of microorganisms (e.g. in substrates of coal mine dumps or in soils damaged by other anthropogenic activities). Growing of leguminous plants at places where they were not grown before, in which there are either no or rather low native rhizobia necessary for symbiosis, may serve as another example. It must be also emphasized that inoculated preparations always contain selected effective strains that are constantly tested with regard to a given activity for which they were selected and that is why they are more active than natural populations.

Combined Inoculation

Much attention is paid to combined inoculation. The effort to combine several positive effects in one inoculant is understandable because costs remain basically the same, but the effect may be manifold. Therefore, many studies address combined inoculation of P-solubilizing bacteria with AM fungi and the genera *Azotobacter* or *Rhizobium.* Probably most studies deal with combination of P-solubilizing bacteria and mycorrhiza fungi. H.P. Singh has elaborately discussed these aspects in Chapter 6 in Volume I of this series–Microbial Biotechnology in Agriculture and Aquaculture. Nevertheless, some examples are cited here to demonstrate the beneficial role of P-solubilizing microorganisms and other microbes. Piccini and Azcon (1987) found out that the combined inoculation with P-solubilizing bacteria and AM fungi increases the dry aboveground matter of alfalfa more than by inoculation of each one separately. They assume that metabolites released by P-solubilizing bacteria transformed rock phosphate to the plant-available form of phosphorus and mycorrhiza then allowed for a fast uptake of the released phosphorus. Raj et al., (1981) also noticed improvement in yield due to inoculation with P-solubilizing bacteria of the genus *Bacillus* and mycorrhiza fungi of the genus *Glomus*. The studies often mention combined inoculation with the strains *Rhizobium* and AM fungi or other fungi. Not all the authors, however, agree with regard to the positive effect of such combined inoculation. While Kaur and Singh (1988), Morton and Yarger (1990), and Vejsadová et al., (1987) present a positive effect on nodulation improvement, rise in nitrogenase activity of nodules and higher yields, Downey and Kessel (1990), on the contrary, reported a lower amount of fixed N_2 in the case of combined inoculation with *R. leguminosarum* and *Penicillium bilaji*. Kaur and Singh (1988) explain the positive impact of AM fungi on N_2-fixation as indirect – in spite of an increase of plant-available phosphorus and also a favourable effect on N_2-fixation. Downey and Kessel (1990), however, explain lower N_2-fixation by production of organic acids with subsequent pH reduction, which inhibits nodulation.

The authors examining double inoculation – with P-solubilizing bacteria and symbiotic N_2-fixers, also failed to arrive at unambiguous conclusions. Alagawadi and Gaur (1988) observed that while the inoculation with *Rhizobium* as such increases nodulation and the inoculation with P-solubilizing bacteria as such raises the content of plant-available phosphorus in soil, the combined inoculation not only boosts all these parameters, but also increases the yield of the dry aboveground matter and seeds. Dubey (1996) also confirmed the favourable effect of double inoculation with *Bradyrhizobium* and P-solubilizing *Pseudomonas*. On the other hand, Badr El-Din et al., (1986) reported no beneficial effect of combined inoculation with *Rhizobium japonicum* and P-solubilizing bacteria.

The authors examining the effect of combined inoculation with P-solubilizing bacteria and wild N_2-fixing bacteria (*Azotobacter*) also did not arrive at unanimous conclusions. While Varma and Mathur (1990) found the repressive effect of this double inoculation, Kundu and Gaur (1980) present positive results in their study.

Composting of Raw Phosphates

Another option, which has recently been given a lot of attention, is composting of raw phosphates. Variation of phosphorus content in rock phosphates is dependent on the location. Some rock phosphates contain a small quantity of phosphorus and, therefore, efforts are being made to find a cheap way of releasing phosphorus from these phosphates. The possibility of composting rock phosphates of low quality together with agricultural waste matter was dealt with by several researchers (Singh et al.,1983; Kubăt et al., 1985; Singh, 1985; Zaharah and Bah, 1997). The authors agree that composting offers a possibility of utilizing rock phosphates of low quality, which contain only a small quantity of plant-available phosphorus. The effect of solubilization depends on the kind of waste matter used for composting, as well as the quality of the rock phosphate. Kapoor (1995) described composting of a rock phosphate with plant residues, tree leaves, cow manure and soil. The composting period was 30 days. He claimed that during this period up to 50% of plant-unavailable phosphorus contained in the rock phosphates were released. He attributes the release to formation of organic acids, humic acids and fulvic acids, during decomposition of organic matter upon composting.

Commercial Inoculant Preparations

Practical inoculation requires preparation of an appropriate inoculant. The basis of the inoculants is a high number of vital and efficient microorganisms on an appropriate carrier. For laboratory, pot or small-plot experiments, there are no requirements set for inoculants with regard to a carrier. These inoculants are prepared right before use and, therefore, need not have long-term survival rate. However, they must contain a sufficient number of efficient microorganisms. For industrially manufactured preparations, however, the appropriate carrier is such that will guarantee, as a prerequisite, longer-term survival of microorganisms.

The use of biological preparations brings along the following advantages: low price, easy application, preservation of soil and the environment, wide range of applications, utilization in places where fertilizers or chemical preparations are not allowed (water protection zones or natural preserves) and increased yields. Based on these advantages, manufacturing of biological preparations is spread all over the world. Biological preparations may also include preparations containing P-solubilizing microorganisms. As early as 1947, the former Soviet Union manufactured the preparation called 'Phosphobacterin', which contained P-solubilizing bacteria. Determination of 'Phosphobacterin' efficiency was dealt with by Smith et al., (1961). However, they found that 'Phosphobacterin' inoculation was inefficient for tomato crops. At the same time, Smith cites a number of authors who also dealt with the examination of the effect of this preparation and their results show the inoculation effect ranging from 0 to 70% yield increase. Based on current knowledge of P-solubilization, these varied results can be partly attributed to repression of P-solubilizing activity by the presence of soluble mineral phosphates, because testing was carried out on diverse soils with dissimilar content of plant-available phosphorus.

Exploitation of P-solubilizing activity of microorganisms, together with additional activity of microorganisms, seems to be a very prospective concept. Marinová (1970) dealt with the combined effect of Nitragin and Phosphobacterin. In her study, she claims that combined inoculation with these two preparations resulted in a 34% increase in soybean yields. With regard to the complexity of AM fungi cultivation (as mentioned above), this preparation is designed particularly for land reclamation of mine dumps, revitalization of biologically decontaminated soils and container plant growing. Preparations that signify improved plant nutrition with phosphorus, as well as increase the content of plant-available phosphorus in soil, as one of multiple effects, are also manufactured abroad. However, their microbial composition is usually a business secret.

In the Czech Republic application of biological preparations has had a long tradition. The preparation named Rizobin (former Nitrazon), containing nodular bacteria capable of fixing atmospheric nitrogen (rhizobia) in symbiosis with plants, has been manufactured in the Czech Republic since 1949. Traditional selection and testing of appropriate strains for the manufacturing of Rizobin is based on the measuring of nitrogenase activity. Recently, the selection and testing of the rhizobia has focused on another, already mentioned ability of these bacteria, which is the capability to dissolve hardly soluble mineral phosphates. For the innovated Rizobin preparation, the strains are selected on both symbiotic N_2-fixation and also on the ability to release soluble phosphates from hardly soluble compounds. The testing of the innovated Rizobin has brought promising results (Mikanová and Kubát, 1997a).

Efficiency of the innovated preparation was verified in a soybean pot experiment. Soybean, the Marple Arrow variety, was seeded in pots containing 5 kg of sifted soil. Before seeding, the pots were fertilized with hardly soluble Gafsa phosphate and superphosphate at doses equal to 45 kg of phosphorus per ha. After seeding, the seeds were either inoculated with the innovated Rizobin or a lab-prepared preparation from a mixture of rhizobia strains without P-solubilizing activity, while a check variant was left uninoculated. At the end of the experiment, the value of seed dry matter was determined. The results of the experiment are presented in Table 4.10. (Mikanová and Kubát, 1999)

The results showed that the yield of dry matter was affected by both soybean inoculation and the fertilizer used. The highest yield of seed dry matter was found in the variant inoculated with the innovated Rizobin. When comparing the variants fertilized with hardly soluble Gafsa phosphate on different types of inoculation, the highest yields were reported in the variant inoculated with the innovated Rizobin, although there are no statistically significant differences between the results. This

Table 4.10. The average grain yield of soybean in pot experiment

Inoculation	Fertilizer	Seed dry matter (g per pot)
Rizobin	Gafsa phosphate	42.2 bcd*
(P-solubilizing rhizobia strains)	Super phosphate	50.2 d
	0	48.7 cd
Rhizobia strains without	Gafsa phosphate	33.6 abc
P-solubilization activity	Super phosphate	44.8 cd
	0	33.2 ab
Control (no inoculation)	Gafsa phosphate	25.6 a
	Super phosphate	22.5 a
	0	23.9 a

*Averages in the column designed by the same letter (a, b, c, d) do not differ significantly (P = 0.05)

result indicated that on inoculation with bacteria capable of releasing phosphorus from hardly soluble compounds, it was even possible to effectively use less soluble fertilizers.

There was no statistically significant difference between the dry matter yields of the soybean in variants with Gafsa phosphate and non-fertilized variants. This fact could be attributed to high content of plant-unavailable phosphorus in European soils. Förster (1983) writes that 1,600-2,000 kg of phosphorus per ha are in a form unavailable to plants. This phosphorus stock partly originated in regular phosphorus fertilization over the last few decades. Up to 90% of supplied fertilizers, however, transformed into hardly soluble forms which are unavailable for plants. Considering that Gafsa phosphate is added to soil at doses equal to 45 kg of phosphorus per ha, it is obvious that the added quantity represents a fragment of the content of insoluble phosphorus in soil. The situation is different where the content of total phosphorus is not increased by regular mineral fertilization (e.g. Asian soils). In Asia, fertilization with rock phosphates and inoculation with P-solubilizing microorganisms are an integral part of the current agricultural practice. For example, in China the positive effect of biological fertilizers is reported between 20 and 30% (Xianghui, 1998). In their study, Menon and Chien (1990) list a few examples from developing countries where expensive and unavailable mineral phosphorus fertilizers are being replaced with rock phosphates from available local sources and acidified with H_2SO_4 before use. An alternative option of acidifying rock phosphates is also possible through microorganisms producing organic acids, as documented in the study by Kapoor (1995), evidencing increased yields of crops fertilized with raw phosphates and simultaneously inoculated with solubilizing microorganisms, based on the results of many Indian and foreign authors.

Besides pot experiments, efficiency of the innovated Rizobin was also verified in field conditions in a small-plot experiment in the Czech

Table 4.11. Soybean small-plot experiment at Tišice

Inoculation	Fertilization	Seed dry matter (g)
Rizobin (P-solubilizing strains)	Fertilzation	131.9
	No fertilzation	112.0
Control (no inoculation)	Fertilzation	112.5
	No fertilzation	105.7

Republic. A variant inoculated with the innovated Rizobin and a control variant without inoculation, were established. Both variants then had plots unfertilized and plots fertilized with superphosphate at a dose equal to 40 kg of P ha^{-1}. At the end of the experiment the value of seed dry matter was evaluated. The results are presented in Table 4.11. (Mikanová, 2000).

Inoculation with the innovated Rhizobin increased the soybean dry matter yields in the non-fertilized plots by about 5%. The difference between the inoculated and non-inoculated variants in the fertilized plots, however, was much higher, i.e. about 17%. There was no difference in soybean dry matter yields between the non-fertilized inoculated plots and non-inoculated fertilized plots, indicating that the inoculation could fully substitute fertilization with superphosphate, 40 kg P ha^{-1}. This result indicates the possibility of decreasing phosphorus fertilization upon inoculation with the innovated Rizobin, which contains strains of the genus *Rhizobium* with P-solubilizing activity.

Bacterial inoculating preparations used in agriculture are, despite all problems, one of the options of resolving some environmental and, to some extent, economic problems.

GENETICS OF PHOSPHATE SOLUBILIZING BACTERIA

Very little is known regarding the genetic regulation governing the mineral phosphate solubilization trait. Since production of organic acids is considered to be the principal mechanism for mineral phosphate solubilization, it could be assumed that any gene involved in organic acid synthesis might have an effect on this character. In a pioneering study, Goldstein and Liu (1987) cloned a gene from P-solubilizing bacterium, *Erwinia herbicola,* whose expression allowed the production of gluconic acid and mineral solubilization activity in *E. coli* HB 101. Similarly, Babu-Khan et al., (1995) cloned a mineral phosphate solubilization gene (*gap Y*) from *Pseudomonas cepacia,* whose expression allowed the induction of the mineral phosphate solubilization phenotype via gluconic acid production in *E. coli* JM 109.

Molecular methods, however, have been mainly used for the identification of the isolated microorganisms, which is a routine procedure, not specific for the phosphate solubilizers (Rodriguez and Fraga, 1999). Mikanová et al., (1997b) isolated several P-solubilizing bacteria, some of them exhibiting repression of this trait under the presence of soluble phosphorus and others showing no repression effect at concentrations up to 50 mM. These data suggest that phosphorus availability could regulate mineral phosphate solubilization in some species, whereas it could have no effect on others. This aspect needs to be further investigated in detail.

CONCLUSION

Future Perspective of Phosphate Solubilizing Microorganisms

Microorganisms play an important role in the integrated soil fertility management and also in sustainable agriculture. The interest in sustainable agriculture is a stimulating research on the rational use of the plant growth-promoting rhizobacteria.

Indeed, microbial phosphorus solubilization may contribute to the sustainability of the farming systems. A better use of the soil microorganisms may improve the plant nutrition with phosphorus and, simultaneously, reduce the input of mineral fertilizers. The efforts in research and technology should continue and be focussed on the good quality of the inoculants, increasing the knowledge on inoculation technology, effective inoculant delivery system and formulation of the policy to exploit biofertilizers successfully (Wani and Lee, 1995).

In some circumstances biofertilizers may not increase the crop yields dramatically, as is the case with mineral fertilizers. From a realistic perspective, one must accept that, in the foreseeable future, chemicals will continue to dominate the market. Only a gradual and modest increase in the use of bacterial inoculants is to be expected. Agriculture in developed countries is definitely the major promoter of microbial inoculants that are "environmentally friendly" (Bashavan, 1998).

Attention should be paid to studies and application of new combinations of phosphate solubilizing bacteria and other plant growth-promoting bacteria for improved results. The mechanisms explaining the synergistic interaction should be a matter of further research to elucidate the biochemical basis of these interactions. Molecular biology and genetic manipulation of phosphate solubilizing bacteria offers a great potential to improve the phosphate solubilizing capabilities of microorganisms (Rodriguez and Fraga, 1999).

Another large field of potential use of the P-solubilizing microorganisms is the unexploited reserves of low-grade rock phosphates, which cannot be economically utilized by the fertilizer industry with the present technologies. Phosphate solubilizing microorganisms have the potential to increase the availability of phosphorus from the native soil phosphorus pool as well as from the rock phosphates. Composting of organic waste matter with the addition of rock phosphates provides an approach for utilization of low-grade rock phosphates (Kapoor, 1995). For improving the effectiveness of low-grade rock phosphate, a granular formulation was developed by immobilizing P-solubilizing bacteria in calcium alginate. This process ensured the required standards of P-solubilizing microorganisms' viability in rock phosphate (Viveganandan, 2002).

In most of the studies focused on P-solubilizers, calcium phosphate has been used predominantly. Research on P-solubilization has thus been limited to neutral and alkaline soils. Just a few studies have demonstrated the presence and the ability of P-solubilizing microorganisms that are able to solubilize Fe or Al phosphates (Antoun, 2002; Barroso and Nahas, 2005). These contributions are particularly important as the low available phosphorus contents are mainly observed in highly weathered and acid soils.

ACKNOWLEDGEMENT

This work was supported by the projects of the Ministry of Education No. 1M6798593901 and Ministry of Agriculture No. MZE 0002700601, Czech Republic.

REFERENCES

Abd-Alla, M.H. (1994). Solubilization of rock phosphate by *Rhizobium* and *Bradyrhizobium*. Folia Microbiol. 39: 53-56.

Adu-Gyamfi, J.J., Fujita, K. and Ogata, S. (1989). Phosphorus absorption and utilization efficiency of pigeon pea [*Cajanus cajan* (L) Millsp.] in relation to dry matter production and dinitrogen fixation. Plant Soil 119: 315-324.

Alagawadi, A.R. and Gaur, A.C. (1988). Associative effect of *Rhizobium* and phosphate-solubilizing bacteria on the yield and nutrient uptake of chickpea. Plant Soil 105: 241-246.

Alexander, M. (1961). Soil microbiology, John Wiley and Sons, Inc., New York, USA, pp. 353-369.

Amijee, F., Tinker, P.D. and Stribley, D.P. (1989). The development of endomycorrhizal systems VII. A detailed study of the effect of soil phosphorus on colonization. New Phytol. 111: 435-446.

Anderson, G. (1980). Assessing organic phosphorus in soils. The role of phosphorus in agriculture (eds.) F.E. Khasawneh, E.C. Sample, E.J. Kamprath, American Society of Agronomy, Madison, Wisconsin, USA, pp. 411-413.

Antoun, H. (2002). Field and greenhouse trials performed with phosphate-solubilizing bacteria and fungi. First International Meeting on Microbial Phosphate Solubilization. Salamanca, Spain, pp. 29-31.

Asea, P.E.A., Kucey, R.M.N. and Stewart, J.W.B. (1988). Inorganic phosphate solubilization by two *Penicillium* species in solution culture and soil. Soil Biol. Biochem. 20: 459-464.

Babenko, Y.S., Tyrygina, G.I., Grigoryev, E.F., Dolgikh, L.M. and Borisova, T.I. (1984). Biological activity and physiologo-biochemical properties of bacteria dissolving phosphates. Microbiology 53: 533-539.

Badr El-Din, S.M.S., Khalafallah, M.A. and Moawad, H. (1986). Response of soybean to dual inoculation with *Rhizobium japonicum* and phosphate dissolving bacteria. Z. Pflanzenern, Bodenk, 149: 130-135.

Babu-Khan, S., Yeo, T.C., Martin, W.L., Duron, M.R., Rogers, R.D. and Goldstein, A.H. (1995). Cloning of a mineral phosphate solubilizing gene from *Pseudomonas cepacia*. Appl. Environ. Microbiol. 61: 972- 978.

Bajpai, P.D. and Rao, W.V.B. (1971). Phosphate solubilising bacteria Part I. Soil Sci. Plant Nutr. 17: 41-43.

Barroso, C.B. and Nahas, E. (2005). The status of soil phosphate fractions and the ability of fungi to dissolve hardly soluble phosphates. Appl. Soil Ecol. 29(1): 73-83.

Bashavan, Y. (1998). Inoculants of plant growth-promoting bacteria for use in agriculture. Biotechnol. Adv. 16: 729-770.

Brady, N.C. (1990). The nature and properties of soil. Cornell University and United States Agency for International Development, USA, pp. 351-368.

Braunberger, P.G., Miller, M.H. and Peterson, R.L. (1991). Effect of phosphorus nutrition on morphological characteristics of vesicular-arbuscular mycorrhizal colonization of maize. New Phytol. 119: 107-113.

Bray, R.H. and Kurtz, L.T. (1945). Determination of total, organic and available forms of phosphorus in soils. Soil Sci. 59: 39-45.

Cassman, K.G., Munns, D.N. and Beck, D.P. (1981). Growth of *Rhizobium* strains at low concentrations of phosphate. Soil Sci. Soc. Am. J. 45: 520-523.

Chhonkar, P.K. and Subba Rao, N.S. (1967). Phosphate solubilization by fungi associated with legume root nodules. Can. J. Microbiol. 13: 749-753.

Cosgrove, D., Irving, G.C.J. and Bromfield, S.M. (1970). Inositol phosphate phosphatase of microbial origin. The isolation of soil bacteria having inositol phosphate phosphatase activity. Aust. J. Biol. Sci. 23: 339- 343.

Datta, M., Banik, S. and Laskar, S. (1992). Effect of inoculation of phosphate-dissolving bacteria on rice (*Oryza sativa*) in acid soil. Indian J. Exp. Biol. 62: 482-485.

Dhingra, K.K., Sekhon, H.S., Sandhu, P.S. and Bhandari, S.C. (1988). Phosphorus- *Rhizobium* interaction studies on biological nitrogen fixation and yield of lentil. J. Agric. Sci. 110: 141-144.

Dinkelaker, B., Romheld, V. and Marschner, H. (1989). Citric acid excretion and precipitation of calcium citrate in the rhizosphere of white lupin (*Lupinus albus* L.). Plant Cell Environ. 12: 285-292.

Domey, S. (1987). Untersuchungen zur mikrobiellen Phosphatmobilisierung in der Rhizosphare von Winterweizen, Winterrogen und Luzerne. (Dizertation), Jena,-IPE. pp. 92

Downey, J. and Kessel, C.H. (1990). Dual inoculation of *Pisum sativum* with *Rhizobium leguminosarum* and *Penicillium bilaji*. Biol. Fertil. Soils 10: 194-196.

Dubey, S.K. (1996). Combined effect of *Bradyrhizobium japonicum* and phosphate-solubilizing *Pseudomonas striata* on nodulation, yield attributes and yield of rainfed soybean (*Glycine max*) under different sources of phosphorus in Vertisols. Indian J. Agric. Sci. 66: 28-32.

Eghball, B., Sander, D.H. and Skopp, J. (1990). Diffusion, adsorption and predicted longevity of banded phosphorus fertilizer in three soils. Soil Sci. Soc. Amer. J. 54: 1161-1165.

Fernandez, M., Cadahia, C., Garate, A. and Esteban, R.M. (1985). The electro- ultrafiltration method for controlling the effect of *Bacillus cereus* on phosphorus mobilization in a calcareous soil. Biol. Fertil. Soils 1: 97-102.

Forster, I. (1983). Mikrobielle Phosphormobilisierung im Boden. Martin Luther Universität, Halle/Saale, Germany, pp. 78.

Gardner, W.K. and Parbery, D.G. (1983). The acquisition of phosphorus by *Lupinus albus* L. I. Some characteristics of the soil/root interface. Plant Soil 68: 19-32.

Gaur, A.C. (1990). Phosphate solubilizing microorganisms as biofertilizers. Omega Sci. Publishers, New Delhi, India, pp. 283.

Goldstein, A.H. (1986). Bacterial solubilization of mineral phosphates: Historical perspective and future prospects. Am. J. Alternative Agric. 1: 51-57.

Goldstein, A.H. and Liu, S.T. (1987). Molecular cloning and regulation of a mineral phosphate solubilizing gene from *Erwinia herbicola*. Biotechnology 5: 72-74.

Goswami, K.P. and Sen, A. (1962). Solubilization of calcium phosphate by three strains of a phosphorbacterium. Indian J. Agric. Sci. 32: 96-101.

Gryndler, M. and Vancura, V. (1991). Vliv dlouhodobého hnojení pùdy fosforem na schopnost vesikuloarbuskulárních mykorrhizních hub kolonizovat koøeny kukuøice. Rostl. Výr. 37: 69-73.

Halder, A.K., Mishra, A.K. and Chakrabartty, P.K. (1991). Solubilization of inorganic phosphates by *Bradyrhizobium*. Ind. J. Exp. Biol. 29: 28-31.

Halder, A.K. and Chakrabartty, P.K. (1993). Solubilization of inorganic phosphate by *Rhizobium*. Folia Microbiol. 38: 325-330.

Hartikainen, H. (1991). Potential mobility of accumulated phosphorus in soil as estimated by the indices of Q/I plots and by extractant. Soil. Sci. 152: 204-209.

Hartikainen, H. and Simojoki, A. (1997). Changes in solid- and solution-phase phosphorus in soil on acidification. Eur. J. Soil Sci. 48: 493-498.

Hoffland, E., Findenegg, G.R. and Nelemans, J.A. (1989a). Solubilization of rock phosphate by rape I. Plant Soil 113: 155-160.

Hoffland, E., Findenegg, G.R. and Nelemans, J.A. (1989b). Solubilization of rock phosphate by rape II. Plant Soil 113: 161-165.

Inskeep, W.P. and Silvertooth, J.C. (1988). Inhibition of hydroxyapatite precipitation in the presence of fulvic, humic, and tannic acids. Soil Sci. Soc. Am. J. 52: 941-946.

Israel, D.W. (1993). Symbiotic dinitrogen fixation and host-plant growth during development of and recovery from phosphorus deficiency. Physiol. Plant. 88: 294-300.

Iswaran, V. and Marwaha, T.S. (1981). Nitrogen fixation by *Azotobacter chroococcum* in the presence of different phosphatic compounds and its solubilization. Zbl. Bakt. II. Abt. 136: 198-200.

Jayachandran, K., Schwab, A.P. and Hetrick, B.A.D. (1989). Mycorrhizal mediation of phosphorus availability: Synthetic iron chelate effects on phosphorus solubilization. Soil Sci. Soc. Am. J. 53: 1701-1706.

Kabesh, M.O., Saber, M.S.M. and Yousry, M. (1975). Effect of phosphate dissolving bacteria and phosphate fertilization on the P-uptake by pea plants cultivated in a calcareous soil. Z. Pflanzenern. Bodenk. 6: 605-611.

Kapoor, K.K. (1995). Phosphate mobilization through soil microorganisms. Proc. Seminar on National Resource Management, Hisar, India, pp. 46-61.

Kaur, S. and Singh, O.S. (1988). Response of ricebean to single and combined inoculation with *Rhizobium* and *Glomus* in a P-deficient sterilized soil. Plant Soil 112: 293-295.

Khurana, A.L. and Dudeja, S.S. (1995). Role of vesicular-arbuscular mycorrhizae (VAM) in sustainable agriculture. Proc. Seminar on National Resource Management, Hisar, India, pp. 37-45.

Kubát, J., Novak, B., Pirkl, J. and Machacek,V. (1985). Rock phosphate solubilisation in cattle slurry. Zbl. fur Mikrobiol. 140: 449-454.

Kubát, J. and Mikanová, O. (1994). P-solubilizing activity of rhizosphere microorganisms and rhizobia. Proc. 8th Symposium. Nyíregyháza-Nancy, France, pp. 23-30.

Kucey, R.M.N. and Leggett, M.E. (1989). Increased yields and phosphorus uptake by westar canola (*Brassica napus* L.) inoculated with a phosphate-solubilizing isolate of *Penicillium bilaji*. Can. J. Soil. Sci. 69 : 425-432.

Kundu, B.S. and Gaur, A.C. (1980). Establishment of nitrogen-fixing and phosphate-solubilising bacteria in rhizosphere and their effect on yield and nutrient uptake of wheat crop. Plant Soil 57: 223-230.

Laheurte, F. and Berthelin, J. (1988). Effect of phosphate solubilizing bacteria on maize growth and root exudation over four levels of labile phosphorus. Plant Soil 105: 11-17.

Lauer, M.J. and Blevins, D.G. (1989). Dry matter accumulation and phosphate distribution in soybean grown on varying levels of phosphate nutrition. J. Plant Nutr. 12: 1045-1060.

Lee, D., Han, X.G. and Jordan, C.F. (1990). Soil phosphorus fractions, aluminium and water retention as affected by microbial activity in an Utisol. Plant Soil 121: 125-136.

Lipton, G.S., Blanchar, R.W. and Blevins, D.G. (1987). Citrate, malate and succinate concentration in exudates from P-sufficient and P-stressed *Medicago sativa* L. seedlings. Plant Physiol. 85: 315-317.

Marinova, R. (1970). Grein yield and quality of soybean as affected by some bacterial manures and microelements. Rast. Nauky 7: 115-123.

Marschner, H., Romheld, V., Horst, W.J. and Martin, P. (1986). Root-induced changes in the rhizosphere: Importance for the mineral nutrition of plant. Z. Pflanzenern. Bodenk. 149: 441-456.

Mendoza, R.E., Canduci, A. and Aprile, C. (1990). Phosphate release from fertilized soils and its effect on the changes of phosphate concentration in soil solution. Fertil. Res. 23: 165-172.

Menge, J.A., Steirle, D., Bagyaraj, D.J., Johnson, F.L.V.and Leonard, R.T. (1978). Phosphorus concentration in plant responsible for inhibition of mycorrhizal infection. New Phytol. 80: 575-578.

Menon, R.G. and Chien, S.H. (1990). Phosphorus availability to maize from partially acidulated phosphate rocks compacted with triple superphosphate. Plant Soil 127: 123-128.

Mikanová, O. and Kubát, J. (1994a). Inoculation by *Bradyrhizobium japonicum* strains with different P-solubilizing activity and its effect on soybean yields. Sci. Agric. Bohemica 1: 59-63.

Mikanová, O. and Kubát, J. (1994b). Phosphorus solubilization from hardly soluble phosphates by soil microflora. Rostl. Výr. 40: 833-840.

Mikanová, O., Kubát, J., Vorisek, K. and Randova, D. (1995a). The capacity of the strains *Rhizobium leguminosarum* to make phosphorus available. Rostl. Výr. 41: 423-425.

Mikanová, O. and Kubát, J. (1995b). The ability to release phosphorus by *Rhizobium* strains. Proc. Seventh International symposium on Microbial Ecology. Santos, Brazil, pp. 138.

Mikanová, O. and Kubát, J. (1996). The role of the P-solubilizing activity of *Rhizobium* and *Bradyrhizobium japonicum* strains in the phosphorus nutrition of soybean. Proc. Inter. Conf. Sustainable Crop Production in Fragile Environments. Hisar, India, p. 86.

Mikanová, O. and Kubát, J. (1997a). The practical utilisation of P-solubilizing microorganisms. Proc. The International Regional Seminar Environment Protection: Modern studies in ecology and microbiology, Uzhgorod, pp. 296-299.

Mikanová, O., Kubát, J., Vorisek, K., Simon, T. and Randova, D. (1997b). Influence of soluble phosphate on P-solubilizing activity of bacteria. Rostl. Výr. 43: 421-424.

Mikanová, O. and Kubát, J. (1997c). The importance P-solubilizing activity of *Bradyrhizobium japonicum* strains for nitrogen fixation. Proc. Inter. Conf. Ecological Agriculture. Chandigarh, India, pp. 119-120.

Mikanová, O. and Kubát, J. (1999). The practical use of the P-solubilizing activity of *Rhizobium* strains. Rostl. Výr. 45: 407-409.

Mikanová, O. (2000). Study of the P-solubilizing microorganisms and their utilization for improving plant nutrition. Ph.D. thesis, University of Agriculture, Prague, Czech Rep., pp. 82.

Mikanová, O. and Novaková, J. (2002). Evaluation of the P-solubilizing activity of soil microorganisms and its sensitivity to soluble phosphate. Rostl. Vyr. 48: 397-400.

Minggang, L., Takuro, S. and Toshiaki, T. (1997). Distribution of exudates of lupin roots in the rhizosphere under phosphorus deficient conditions. Soil Sci. Plant Nutr. 43: 237-245.

Minhoni, M.T.A., Cardoso, E.J. and Eira, A.F. (1991). Effect of five types of organic matter on rock phosphate microbial solubilization. R. Bras. Ci. Solo, Campinas 15: 29-35.

Morton, J.B. and Yarger, J.E. (1990). Soil solution P concentrations necessary for nodulation and nitrogen fixation in mycorrhizal and non-mycorrhizal red clover (*Trifolium pratense* L.). Soil. Biol. Biochem. 22: 127- 129.

Murphy, J.P. and Riley, J.P. (1962). A modified single solution method for the determination of phosphate in natural waters. Anal. Chem. Acta 27: 31-36.

Myskow, W. (1960). The dissolving of calcium phosphate by some soil bacteria. Acta Microbiol. Polonica 9: 367-373.

Natr, L. and Lastuvka, Z. (1980). Nìkolik aspektù pøíjmu a metabolismu fosforu v rostlinách. Agrochem. 20: 8-11.

Piccini, D. and Azcon, R. (1987). Effect of phosphate-solubilizing bacteria and vesicular-arbuscular mycorrhizal fungi on the utilization of Bayovar rock phosphate by alfalfa plants using a sand-vermiculite medium. Plant Soil 101: 45-50.

Pongsakul, P. and Jensen, E.S. (1991). Dinitrogen fixation and soil N uptake by soybean as affected by phosphorus availability. J. Plant Nutr. 14: 809-823.

Pradel, E. and Bouquet, P.L. (1988). Acid phosphatases of *Escherichia coli*: Molecular cloning and analysis of *agp*. The structural gene for a periplasmic acid glucose phosphatase. J. Bacteriol. 170: 4325- 4342.

Ralston, D.B. and McBride, R.P. (1976). Interaction of mineral phosphate-dissolving microbes with red pine seedlings. Plant Soil 45: 493-507.

Raj, J., Bagyaraj, D.J. and Manjunath, A. (1981). Influence of soil inoculation with vesicular-arbuscular mycorrhiza and a phosphate-dissolving bacterium on plant growth and ^{32}P-uptake. Soil. Biol. Biochem. 13: 105-108.

Rea, E. and Tullio, M. (2005). The agro-industrial productions' valorization: The possible role of arbuscular mycorrhizal (AM) fingi. In: Microbial Biotechnology in Agriculture and Aquaculture, Volume I (ed.) R.C. Ray, Science Publisher, Enfield, NH, USA, pp. 101-124.

Reichlova, E. (1972). The utilization of hardly soluble phosphates by the stand of *Rhizobium japonicum*. Rostl.Výr. 18: 205-207.

Robson, A.D., O'Hara, G.W. and Abbott, L.K. (1981). Involvement of phosphorus in nitrogen fixation by subterranean clover. Aust. J. Plant Physiol. 8: 427-436.

Rodriguez, H. and Fraga, R. (1999). Phosphate solubilizing bacteria and their role in plant growth promotion. Biotechnol. Adv. 17: 319-339.

Romer, W., Augustin, J. and Schilling, G. (1988). The relationship between phosphate absorption and root length in nine wheat cultivars. Plant Soil 111: 199-201.

Rydin, E. and Otabbong, E. (1997). Potential release of phosphorus from soil mixed with sewage sludge. J. Environ. Qual. 26: 529-534.

Salih, H.M., Yahya, A.I., Abdul-Rahem, A.M. and Munam, B.H. (1989). Availability of phosphorus in a calcareous soil treated with rock phosphate or superphosphate as affected by phosphate-dissolving fungi. Plant Soil 12: 181-185.

Singal, R., Gupta, R. and Saxena, R.K. (1994). Rock phosphate solubilization under alkaline conditions by *Aspergillus japonicum* and *A. foetidus*. Folia Microbiol. 39: 33-36.

Singh, C.P., Ruhal, D.S. and Singh, M. (1983). Solubilisation of low grade rock phosphate by composting with a farm waste, Pearl-Millet Boobla. Agric. Wastes 8: 17-25.

Singh, C.P. (1985). Preparation of phospho-compost and its effect on the yield of Moong bean and wheat. Biol. Agric. Hortic. 2: 223-229.

Singh, H.P. and Singh, T.A. (1993). The interaction of rockphosphate, *Bradyrhizobium*, vesicular-arbuscular mycorrhizae and phosphate-solubilizing microbes on soybean grown in a Sub-Himalayan mollisol. Mycorrhizae 4: 37-43.

Singh, H.P. (2006). Interactions among beneficial soil microorganism. In: Microbial Biotechnology in Agriculture and Aquaculture, Vol I. (ed.) R.C. Ray, Science Publishers, Enfield, NH, USA, pp. 159-210.

Smith, J.H., Allison, F.E. and Soulides, D.A. (1961). Evaluation of 'Phosphobacterin' as a soil inoculant. Soil Sci. Soc. Am. Proc. 25: 109-111.

Sperber, J.I. (1957). Solution of mineral phosphates by soil bacteria. Nature 180: 994-995.

Sperber, J.I. (1958). Solution of apatite by soil microorganisms producing organic acids. Aust. J. Agric. Res. 9: 782-787.

Sudhakar, P., Singh, B.G. and Rao, L.M. (1989). Effect of *Rhizobium* and phosphorus on growth parameters and yield of blackgram (*Phaseolus mungo*). Indian J. Agric. Sci. 59: 402-404.

Taha, S.M., Mahmoud, S.A.Z., Halim El-Damaty, A. and Abd El-Hafez, A.M. (1969). Activity of phosphate dissolving bacteria in egyptian soils. Plant Soil 31: 149-160.

Thomas, G.V. (1985). Occurrence and ability of phosphate-solubilizing fungi from coconut plantation soils. Plant Soil 87: 357- 364.

Thomson, B.D., Robson, A.D. and Abbot, L.K. (1986). Effects of phosphorus on the formation of mycorrhizas by *Gigaspora calospora* and *Glomus fasciculatum* in relation to root carbohydrates. New Phytol. 103: 751-765.

Tiwari, V.N., Lehri, L.K. and Pathak, A.N. (1989). Effect of inoculating crops with phospho-microbes. Expt. Agric. 25: 47-50.

Vancura, O. (1980). Metabolické interakce rostlin a mikroorganismù v půdě.(Disertace), Praha, Czech Rep., pp. 241-265.

Varma, S. and Mathur, R.S. (1990). The effect of microbial inoculation on the yield of wheat when grown in straw-amended soil. Biol. Wastes 33: 9-16.

Vazquez, P.M. (1996). Bacterias solubilizadoras de fosfatos inorganicos asociadas a la rhizosfera de losmangles : *Avicennia germinans* (L.) L y *Laguncularia racemosa* (L.) Gerth. Tesis para el tifulo de Biologo Marino Univ. Autonoma de Baja California Sur. La Paz, B.C.S.

Vejsadová, H., Hrselova, H., Prikryl, Z. and Vancura, V. (1987). Interrelationships between vesicular-arbuscular mycorrhizal fungi, *Bradyrhizobium japonicum* and soybean plants. Proc. Inter. Symp. Interrelationships between Microorganisms and Plants in Soil. Liblice, Czech Republic, pp. 115-123.

Venkateswarlu, B., Rao, A.V. and Raina, P. (1984). Evaluation of phosphorus solubilisation by microorganisms isolated from aridisols. J. Indian Soc. Soil. Sci. 32: 273-277.

Viveganandan, G. and Jauhri, K.S. (2002). Efficacy of rock phosphate based soil implant formulation of phosphabacteria in soybean. Indian J. Biotechnol. 1: 180-187.

Wani, S.P. and Lee, K.K. (1995). Role of microorganisms in sustainable agriculture. Proc. Seminar on National Resource Management. Hisar, India, pp. 62-88.

Xianghui, L. (1998). Third International Training Course on Biological Fertilizer, Baoding, China.

Yahya, A.I. and Al-Azawi, S.K. (1989). Occurrence of phosphate-solubilizing bacteria in some Iraqi soils. Plant Soil 117: 135-141.

Zaharah, A.R. and Bah, A.R. (1997). Effect of green manures on P-solubilization and uptake from phosphate rocks. Nutrient Cycling in Agroecosystems 48: 247-255.

5

Biotechnology of Biofertilizers for Rice Crop

S. Kannaiyan[1]* *and K. Kumar*[2]

INTRODUCTION

Nitrogen inputs through Biological Nitrogen-Fixation (BNF) support sustainable and environmentally sound agricultural production. BNF conserves soil nitrogen resources. Soil fertility and soil tilth could be improved by practices such as green manuring with dinitrogen (N_2)-fixing legumes, and companion crops or subsequent non-legume crops can also be benefited from BNF (Zahran, 2005, Chapter 3 in Volume I of this series–Microbial Biotechnology in Agriculture and Aquaculture). The selection of crop sequences and the integration of farm inputs for maximal returns are all components of effective soil management. Further, the beneficial interactions/associations between plants and soil microorganisms (Table 5.1) are important in maintaining soil fertility.

To guarantee food for all, the chemical fertilizer consumption has increased from 113 metric tonnes (mt) in 1980 to 307 mt in 2000, most of the deficit noticeably occurring in developing countries where infrastructure facilities for chemical fertilizer production are poor. In contrast, biofertilizers/organic fertilizers are cost-effective, eco-friendly and a renewable source of plant nutrients to supplement chemical fertilizers in sustainable agricultural systems (Emmerling, 2005, Chapter 8 in Volume I of this series–Microbial Biotechnology in Agriculture and Aquaculture). Despite sizable increase in the use of chemical fertilizers over the years, the gap between nutrient removal and replenishment is significantly high in many developing countries. The role of biofertilizers in sustainable

[1]*Chairman, National Biodiversity Authority, Chennai - 600041, India*
[2]*Department of Agricultural Microbiology, Tamil Nadu Agricultural University Coimbatore – 641 003, India*
* *Corresponding author: E-Mail: skannaiyan@hotmail.com*

Table 5.1. Major types of beneficial interactions/associations between plants and soil microorganisms (Subba Rao, 1993)

Nature of interactions/associations	Examples of higher plants involved	Examples of microorganisms involved
Rhizosphere, Rhizoplane and Phyllosphere microflora	All plants with roots and leaves	Bacteria, fungi and Actinomycetes
Ectomycorrhizae	Forest trees – *Pinus*	Mostly basidiomycetous fungi–*Boletus*, *Lactarius* and *Armillaria*
Endomycorrhizae (Arbuscular Mycorrhizal fungal association)	Certain orchids, cereals, grasses and legumes	*Rhizoctonia*, *Endogone* and *Glomus*
Root nodules of nodulating legumes	Soybean, gram, lucerne, etc.	*Rhizobium* spp. and *Bradyrhizobium* spp.
Root nodules of plants other than legumes	*Alnus*, *Myrica*, *Casuarina*, etc.	Actinomycetous endophytes (*Frankia*)
Leaf nodules	*Psychotria*, *Pavetta*	*Klebsiella*
Algal associations with higher plants	*Cycas*, *Zamia*, *Heterozamia*, *Gunnera scabra*, *Azolla*	*Anabaena*, *Nostoc*
Associative symbiosis	Grasses, *Sorghum* and millets	*Azospirillum*

agriculture assumes special significance particularly in the present context of very high cost of chemical fertilizers and environmental hazards associated with its excessive use. The production and application of biofertilizers for leguminous crops, oilseeds, rice, millets and other important crops besides forest nursery plants, are very common in several countries such as Cuba, Thailand, India, China, etc. (Kannaiyan, 2003).

In this volume, application and biotechnology of rhizobial biofertilizers have been discussed by Teamroong and Boonkerd in Chapter 3, and phosphorus (P)-solubilizing biofertilizers by Mikanová and Kubát in Chapter 4. In this chapter, the discussion of various biofertilizers commonly applied to rice agro-ecosystem and the underlying biotechnological principles are explained.

AZOSPIRILLUM BIOFERTILIZER

The concept of "associative symbiosis" has widened our knowledge of free-living N_2-fixing microorganisms. The occurrence of a N_2-fixing bacterium *Spirillum lipoferum* in the roots of digit grass in Brazil was reported for the first time by Dobereiner and Day (1976). The bacterium was found to fix nitrogen in association with roots of various plants. This startling discovery opened new vistas for the world's most staple food crops like rice, maize, sorghum, wheat and millets, which were previously believed to have no potential for biological N_2-fixation.

Azospirillum spp. are of ubiquitous distribution in many parts of the world consisting of tropical, sub-tropical and temperate climatic conditions, and also with various standing crops grown in a variety of soil types. Lakshmikumari et al., (1976) reported the occurrence of N_2-fixing *Spirillum* in the roots of rice, sorghum and maize. Reynders and Vlassak (1976) from Belgium, isolated this organism from more than 50% of soil samples representing a wide range of soil types. Amer et al., (1977) reported that various geographical regions of the world were rich in *Azospirillum*. Neyra and Dobereiner (1977) reported that nearly 10% of the soils and roots from temperate region have *Azospirillum*, while more than 50% tropical soils were found positive for *Azospirillum*.

In India, Rao et al., (1978) reported the presence of *Azospirillum* in different rice soil types (alluvial, laterite, pokkali and kari). The occurrence of *Azospirillum* in the roots of cotton, maize, ragi, kudiraivali, guinea grass, hariyali grass (*Cyanodon dactylon*) (Purushothaman et al., 1980), plantation crops, orchard plants (Subba Rao, 1983), pearl millet, areca nut, banana, cashew, coconut, cocoa, rubber, cardamom (Purushothaman and Govindarajan, 1984; Govindan and Purushothaman, 1985), and upland rice (Purushothaman et al., 1987) has been well documented.

A new acid tolerant sp. i.e. *Azospirillum amazonense*, was reported by Magalhaes et al., (1983) from Amazon region in Brazil. Charyulu and Rao (1980) observed the occurrence of *Azospirillum* with appreciable N_2-fixing ability in highly acid soils. Reinhold-Hurek et al., (1987) reported that *Azospirillum halopraeferens* was characterized by its high salt tolerance. Salt tolerant *Azospirillum* isolates have also been reported by Sundaram et al., (2001). Thiruchenthilvasan and Purushothaman (1999) observed better colonization of sorghum rhizosphere by genetically engineered *Azospirillum* cultures.

So far taxonomists have identified five species in the genus *Azospirillum* i.e. *A.brasilense, A. lipoferum, A. amazonense, A. halopraeferens* and *A. irakense.* They are Gram-negative, vibrioid in shape, motile with a long polar flagellum, and occasionally with peritrichous flagella. Organic acids, such as malate and succinate, are the preferred carbon sources for *Azospirillum.* However, the different species vary in their ability to use dissimilar carbon sources (Okon, 1985). Among the free-living N_2-fixing bacteria *Azospirillum* is considered to be more efficient in nitrogenase activity in comparison with other N_2-fixers. It has been well demonstrated that *Azospirillum* inoculated plants, were able to absorb nutrients from solutions at faster rates than uninoculated plants resulting in accumulation of more dry matter, nitrogen (N), phosphorus (P) and pottasium (K) in the stem and leaves (Okon, 1985).

Azospirillum inoculation has been reported to increase the growth, nitrogen uptake and yield parameters in a number of crops. Field trials conducted in many places with different crops have revealed that biological N_2-fixation by root associated *Azospirillum* contribute significant amounts of nitrogen to the plants, thereby saving valuable nitrogen fertilizers. The results also revealed that the *Azospirillum* inoculation benefits plant growth and increases yield of crops by improving root development, mineral uptake and plant-water relationship. Several experiments have been conducted at different locations in India and all over the world to find out the response of crop plants to *Azospirillum* inoculation under field conditions. A three-year field experiment conducted to assess the response of rice to *Azospirillum* inoculation under graded levels of inorganic nitrogen (0, 25, 50, 75 and 100 kg N ha^{-1}) revealed that *Azospirillum* inoculation at 75 kg N ha^{-1} significantly increased the grain yield (Gopalaswamy et al., 1989). At lower and higher levels of inorganic nitrogen, the inoculation effect was not significant (Table 5.2).

Okon and Labandera-Gonzalez (1994) evaluated worldwide data accumulated over the past 20 years on field inoculation with *Azospirillum*, and concluded that these bacteria are capable of promoting the yield of

Table 5.2. Effect of *Azospirillum* inoculation on the grain yield of rice (var. ADT 36) under graded levels of nitrogen

Treatment	Grain yield (tonnes ha^{-1})			
	1985	1986	1987	Mean
0 kg N ha^{-1}	3.1 d*	3.0 d	3.1 c	3.1 d
0 kg N ha^{-1} + *Azospirillum*	3.1 d	3.4 d	3.5 d	3.3 d
25 kg N ha^{-1}	3.6 d	3.4 d	3.5 d	3.5 d
25 kg N ha^{-1} + *Azospirillum*	3.8 bc	4.0 c	3.9 cd	3.9 c
50 kg N ha^{-1}	4.1 ab	4.3 bc	3.7 cd	4.0 bc
50 kg N ha^{-1} + *Azospirillum*	4.4 a	4.5 ab	4.2 bc	4.4 b
75 kg N ha^{-1}	3.6 cd	4.5ab	3.9 cd	4.0 d
75 kg N ha^{-1}+ *Azospirillum*	4.4 a	4.9 a	5.0 a	4.7 a
100 kg N ha^{-1}	4.3 ab	4.7 ab	4.2 bc	4.3 b
100 kg N ha^{-1}+ *Azospirillum*	4.0 ab	4.7 ab	4.5 ab	4.4 ab
CD (Critical Differences (p = 0.05)	0.5	0.4	0.5	0.4

*In a column means followed by the same letter (a, b, c, d) are not significantly different at 5% level by DMRT (Gopalaswamy et al., 1989)

agriculturally important crops in different soils and climatic regions. The review of the extensive data collected revealed 60-70% success with statistically significant increases in yield in 5-30% of the trials.

AZOLLA BIOFERTILIZER

Azolla is a free-floating water fern, which in symbiotic association with the cyanobacterium–*Anabaena azollae*, contributes a substantial amount of biologically fixed nitrogen to the rice crop. The N_2-fixing water fern, *Azolla* has been used as a biofertilizer for rice crop in India, China, Vietnam, Thailand, Philippines, Korea, Sri Lanka, Bangladesh, Pakistan, Nepal, Burma, Indonesia, Brazil and West Africa (Kannaiyan, 1985, 1987). The technology of utilizing the N_2-fixing water fern *Azolla* has been attempted by these countries, but the full potential of this biological system has not been exploited. The *Azolla–Anabaena* association is a "living floating nitrogen manufacturing factory" which is utilizing the energy from photosynthesis to fix atmospheric nitrogen amounting to 150 to 200 kg N ha^{-1} $year^{-1}$. *Azolla* is used as a biofertilizer for rice crop and it could contribute to 40-60 kg N ha^{-1} per crop (Kannaiyan, 2003). *Azolla* is also used as a biofertilizer in aquaculture/fish culture (Garg and Bhatnagar, 2005, Chapter 9 in Volume I of this series–Microbial Biotechnology in Agriculture and Aquaculture, published in 2005).

The important factor in using *Azolla* as a biofertilizer for rice crop is its quick decomposition in soil and efficient availability of its nitrogen to rice plants (Kannaiyan, 1990a). In tropical rice soils, the applied *Azolla* mineralizes rapidly and its nitrogen is made available to the rice crop in a very short period.

Growth

The growth rate of *Azolla* is very rapid, so much so that it doubles its weight in a mere two-three days. Due to fast relative growth rate, the biomass production is heavy in a given time, hence the biomass is exploited as biofertilizer for rice. *Azolla* has the potential of maintaining an exponential growth rate under optimum conditions. The growth of *Azolla* is initially slow in the growth medium as well as in the rice field followed by rapid growth (Kannaiyan, 1993a). The growth of *Azolla* is subjected to various environmental factors such as temperature, light intensity, season, humidity, wind, pH and salinity (Kannaiyan and Somporn, 1987; Latha et al., 1988; Rajagopal et al., 1994). The response of *Azolla* to phosphorus fertilization is necessary for its growth and N_2-fixation (Kannaiyan et al., 1981; Singh and Singh, 1989). *Azolla* grown in phosphorus deficient solution had a reddish brown discolouration that spread from the centre of the frond to the tip of the body with reduction in size of the frond. Addition of phosphorus is most effective in stimulating the growth of *Azolla* (Singh, 1979). Kannaiyan (1987) reported higher biomass of *Azolla filiculoides* at 20 $\mu g\ g^{-1}$ of phosphorus but found that 5-10 $\mu g\ g^{-1}$ level was adequate for growth and multiplication. Singh (1977a) recommended 4-6 kg P_2O_5 ha^{-1} per week for large-scale multiplication of *Azolla pinnata* under field conditions. Kannaiyan et al., (1982) have demonstrated the effectiveness of split application of phosphorus for *Azolla* growth and multiplication.

Mass Production of *Azolla* Biomass

A simple *Azolla* nursery method for large-scale multiplication of *Azolla microphylla* in the fields has been evolved for easy adoption by the farmers (Kannaiyan, 2003). The field selected for *Azolla* nursery must be thoroughly prepared and levelled uniformly. The field is divided into plots (20 m x 2 m) by providing suitable bunds and irrigation channels. Water is maintained at a depth of 10 cm. Ten kg of fresh cattle dung mixed in 20 litres of water is sprinkled in each plot and *A. microphylla* SK-TNAU inoculum at 8 kg is inoculated into each plot. Super phosphate (100 g) is applied in three split doses at four days interval as top dressing fertilizer for *A. microphylla* SK-TNAU (Plate 5.1). For insect pest control, Furadan^{R} [(carbofuran) (2,3–dihydro–2,2-dimethybenzofuran- 7-yl methyl carbamate; Rallis India Ltd.)] granules at 100 g $plot^{-1}$ (40 m^2 area) are applied seven days after inoculation. Fifteen days after inoculation, *A.*

microphylla SK-TNAU is harvested and introduced into the main field as a source of primary inoculum. From one harvest, 100-150 kg fresh *Azolla* is obtained from each plot. Again, *Azolla* is inoculated in the same plot and cattle dung, super phosphate and furadan granules are applied and harvested. Four 40 m^2 nursery plots are required for raising *A. microphylla* SK-TNAU for inoculation to one hectare main field.

Azolla Spore Production Technology

Sporulation in different *Azolla* species is not well understood, and the factors responsible for sporulation and spore germination are not well defined. *Azolla microphylla* SK-TNAU, a sporulating culture, was selected for spore inoculum development. Fresh young fronds of *A. microphylla* SK-TNAU were inoculated in a puddled and well prepared field under flooded condition. A basal nutrition of 10 kg fresh cattle dung was mixed in 20 litres of water and applied in 40 m^2. After a weak of inoculation of *A. microphylla* SK-TNAU, the growth regulator gibberelic acid (GA_3) at 100 $\mu g\ ml^{-1}$ was applied as foliar spray. Foliar spray with GA_3 induced the growth and stimulated the production of sporocarps. These sporulated fronds were harvested after four weeks and formed a heap with a flat surface. One hundred kg of sporulated fresh fronds of *A. microphylla* SK-TNAU formed a heap. The heap was then covered with a thin layer of clay soil slurry (20%) and allowed to undergo partial decomposition for a period of 21 days. Subsequently, the heaps were removed and well-dried under shade for a week. The decomposed and well-dried frond materials are called "Frond based spore inoculum" of *Azolla* (Kannaiyan, 1990b). The viability of the spore inoculum was found to be good for 10-15 months (Shanmugasundaram and Kannaiyan, 1992a).

Mahesh and Kannaiyan (1994) have investigated the influence of growth regulators i.e. GA_3, naphthyl acetic acid (NAA), 2,4-dichlorophenoxy acetic acid (2,4-D) and benzyl amino purine (BAP), at various concentrations on the germination and fertilization of the megasporocarps of *A. microphylla* and found that GA_3 and BAP at 20 to 80 $\mu g\ l^{-1}$ stimulated the germination and fertilization of megasporocarps. The growth regulators also enhanced the survival of the sporelings and the heterocyst frequency of the algal symbiont in the sporelings (Shanmugasundaram and Kannaiyan, 1992b).

Inoculation of Fresh Biomass of *Azolla*

Fresh biomass of *Azolla* is broadcasted in the main field 7-10 days after transplanting rice. Inoculation of the fresh biomass of *A. microphylla* at 200 kg ha^{-1} could multiply faster and cover the rice fields as a green mat in a period of two-three weeks, with 15-25 tonnes biomass accumulation.

However, inoculation of even 50-100 kg biomass of various *Azolla* spp., i.e. *A. microphylla, A. pinnata, A. caroliniana* and *A. filiculoides* per hectare will multiply and establish in rice fields, but it takes four to five weeks for full coverage in the transplanted rice. Under optimum conditions with cloudy days and low temperature in day/night cycle of 25±1°C/18±1°C, the inoculated *Azolla* spreads very fast in transplanted rice. *Azolla* technology is very efficient in terms of N_2-fixation and biomass accumulation during wet season due to the better environmental conditions prevailing during the second rice season in India, particularly cloudy days coupled with low temperature favours its vegetative multiplication (Kannaiyan, 2003). The most suitable season for *Azolla* utilization as dual crop with rice is wet season. It can also be used for late dry season crop. Season is very critical for its multiplication and biomass accumulation. *A. microphylla* is known to survive even in the summer season, but its full potential as biofertilizer could be exploited only during the second wet rice season crop in tropical conditions. Inoculation of *Azolla* in transplanted rice crop supports the growth and multiplication. Growing *Azolla* along with rice crop is called dual culture (Plate 5.2) and it does not affect the growth of rice crop in any way, however, it suppresses the weed growth in wetland rice (Kannaiyan et al., 1984).

Application of *Azolla* Biofertilizer for Rice Crop

Srinivasan (1980) reported that *Azolla* as green manure could reduce the nitrogen input by 25-30% in lowland rice. *Azolla* incorporation as green manure increased the yield of rice by 9-38% (Singh, 1977a), 35.8% (Kannaiyan, 1981) and 44% (Baratharkur and Talukdar, 1983). Subramani and Kannaiyan (1987) reported that application of gypsum at 5 tonnes ha^{-1} with 60 kg N ha^{-1} as urea and *Azolla* inoculation as dual crop recorded the highest grain yield of rice. Jayanthi et al., (1994) have reported that incorporation of 10 tonnes ha^{-1} fresh *Azolla* biomass along with 50 kg N ha^{-1} resulted in increased nitrogen uptake, available nitrogen in soil, and rice yield. Gopalaswamy et al., (1994) have investigated the potential of *Azolla* hybrids and reported that inoculation of *Azolla* hybrids significantly improved the soil organic carbon status besides increasing the rice yield.

In wetland rice ecosystem, the productivity of rice could be increased by raising the efficiency of applied nitrogen either by natural means or by chemical fertilization. Latha et al., (1988) have established that the inoculation of *Azolla* coupled with Urea Super Granule (USG) application, has recorded significantly higher grain yield of rice than Prilled Urea (PU) application. Thangaraju and Kannaiyan (1990) reported that *A. microphylla* and *A. filiculoides* coupled with USG application, significantly increased the grain yield of rice. Kannaiyan (1992a) reported increased rice yield by the combined application of USG and *A. microphylla*. Shanmugasundaram and Kannaiyan (1994) have reported that inoculation of *Azolla* hybrid RS-SK-TNAU-1 with PU as well as USG, recorded higher grain yield and nitrogen uptake in rice. The N_2-fixing green manure plants i.e. *Azolla* and *Sesbania rostrata*, could be effectively utilized by incorporating USG to increase the rice production in lowland rice ecosystem (Kumar et al., 1995). The nitrogen use efficiency in lowland rice system could be increased substantially by adopting the integrated nutrient management system using N_2-fixing green manures or *Azolla* biofertilizer.

Azolla inoculation has considerably increased the activity of both the soil microbial community and enzymes (Kannaiyan, 1994; Kumar, 1994). Zhang Zhuang et al., (1987) have reported that incorporation of 22.5-37.5 tonnes (t) fresh *Azolla* ha^{-1} biomass before planting increased the rice yield by 9.6-13.0%. Patil and Mandhare (1991) reported that incorporation of 5 t *Azolla* ha^{-1} with 100% recommended nitrogen level yielded 6.7 t ha^{-1} which is more or less equal to the incorporation of 5 t *Azolla* ha^{-1} with 75% of the recommended dose of nitrogen. Alexander et al., (1992) have observed maximum rice yield increase with an output ratio recovery of 2.07 when 7.5 t fresh *Azolla* ha^{-1} biomass is incorporated along with 90 kg N ha^{-1}. An increase in the grain yield of rice by the application of *Azolla* as green manure, dual crop or both with rice was equivalent to 30-40 kg N ha^{-1} and this has been well documented by several investigators in various countries such as China (Liu Chung Chu, 1987), The Philippines (Mabbayad, 1987; Watanabe, 1987), India (Singh, 1977b; Kannaiyan, 1981), Thailand (Loudhapasitiporn and Kanareugsa, 1987), Sri Lanka (Kulasooriya et al., 1987), USA (Talley et al., 1977) and West Africa (Diara et al., 1987). The use of *Azolla* in rice fields has considerably increased the soil microbial as well as enzymes activity (Kannaiyan, 1994; Kumar, 1994), which in turn maintained soil fertility. It is an important low cost input, which plays a vital role in improving soil quality management in sustainable rice farming (Kannaiyan, 1992b, 1993b, 1995). Dual cropping of *Azolla* is a most effective, ecofriendly practice, which plays a major role in sustainable rice farming (Kannaiyan, 1992b, 1993b, 1995). Though *Azolla-Anabaena* symbiosis is the only plant-cyanobacterial symbiosis used as a rice biofertilizer for maintaining soil health and productivity, this system could still be improved physiologically by using the growth regulator GA_3 (Gopalaswamy and Kannaiyan, 1996). *Azolla* hybrids were found to undergo rapid decomposition and to increase the floodwater ammoniacal nitrogen content, which is very vital for the rice crop. By carefully screening different *Azolla* cultures, *Azolla* hybrid AH-C2 was found to be saline tolerant. Rice production could be significantly increased with *Azolla* hybrids in combination with 75 kg N ha^{-1} USG (Gopalaswamy and Kannaiyan, 2000).

Kumar and Kannaiyan (2000) developed a mutant, *A. microphylla*–KK-SK-E20 that exhibited higher nitrogenease and ammonia assimilating enzyme activity than the wild parent. Variation among N_2-fixing *Anabaena azollae* strains isolated from four different *Azolla* cultures were characterized based on their total protein profile and RAPD (Random Amplified Polymorphic DNA) profile. Considerable variation in protein banding at 40-70 kDa (Dalton) protein subunits was shown by different isolates and specific polymorphic bands were generated for various isolates with different primers (Subhashini et al., 2000).

BLUE-GREEN ALGAE (BGA)

The rice field ecosystem provides an environment favourable for the growth of blue-green algae with respect to their requirements for light, water, high temperature and nutrient availability. Algal biofertilizers constitute a perpetual source of nutrients and they do not contaminate ground water and deplete the resources. In addition to contributing 25-30 kg N ha^{-1} of biologically fixed nitrogen, they also add organic matter to the soil, excrete growth promoting substances, solubilize insoluble phosphates and amend the physical and chemical properties of the soil (Venkataraman, 1972; Kannaiyan, 1978; Goyal, 1982).

Distribution of BGA in Indian Soils

Under submerged soil in tropical conditions in rice fields of India, BGA form the most promising biological system–in addition to contributing nitrogen, they benefit the crop in many other ways (Goyal, 1982). The occurrence and distribution of BGA in Indian rice fields have been reported by several workers (Mitra, 1951; Pandey, 1965; Venkataraman, 1972; Bendra and Kumar, 1975; Sinha and Mukherjee, 1975; Kannaiyan, 1978; Prasad et al., 1978; Bongale and Bharathi, 1980; Sardeshpandey and Goyal, 1981a; Kannaiyan, 1993a). The available information shows that while some forms like *Nostoc, Anabaena, Calothrix, Aulosira* and *Plectonema* are ubiquitous in Indian soils, others like *Haplosiphon, Scytonema* and *Cylindrospermum* have localized distribution (Venkataraman, 1972; Kannaiyan, 2003).

The beneficial effect of algal inoculation on the grain yield of rice has been well documented at a number of locations (Venkataraman, 1972; Kannaiyan, 1978). Algal inoculation to rice crop was effective in different agro climatic conditions and soil types (Goyal and Venkataraman, 1971; Kannaiyan, 2000, 2003). Blue-green algal inoculation with composite cultures was found to be more effective than single culture inoculation (Kannaiyan, 1978). A technology for mass scale production of composite culture of BGA under rice field condition was developed by Kannaiyan (1978) and the soil based BGA inoculum could survive for more than two years (Kannaiyan, 1990c). At many sites where algal inoculation was used for three to four consecutive cropping seasons, the inoculated algae established well and the effect persisted over subsequent rice crop (Kannaiyan, 1990b). Technologies for utilizing N_2-fixing organisms in lowland rice were reported (Venkataraman, 1972; Roger and Watanabe, 1986) and the beneficial role of blue-green algal inoculation in rice soils of Tamil Nadu province of India was reported by Kannaiyan (2003).

N_2-Fixing BGA

The species of BGA which are known to fix atmospheric nitrogen are classified into three groups: (i) heterocystous - aerobic forms (ii) aerobic unicellular forms, and (iii) non-heterocystous, filamentous, microaerophilic forms. The blue-green algal culture's composite inoculum consisting of *Nostoc, Anabaena, Calothrix, Tolypothrix, Plectonema, Aphanothece, Gloeocapasa, Oscillatoria, Cylindrospermum, Aulosira* and *Scytonema* have been used for inoculation in rice (Kannaiyan, 2003). These N_2-fixing cultures have better adaptability in wetland rice.

Mass Scale Multiplication Under Field Conditions

Mass production of dried soil based algal flakes under rice field condition was developed by Kannaiyan (1978). The field selected for mass multiplication of BGA is prepared well and leveled uniformly. The field is divided into small-plots size of 20 m × 2 m (40 m^2), by providing suitable bunds and irrigation channels. Water is maintained at a height of 10 cm. Two kg of super phosphate added to the plot and inoculated with 5 kg of starter culture of BGA. To prevent the activities of soil insects, 200 g of carbofuran (furadan) is also added. Water is let in at periodical intervals, so that the height of water level is always maintained at 10 cm. The growth of algae is rapid and in about 10 days, a thick mat is formed which floats on the surface of the water. After 15 days of inoculation, the plots are allowed to dry in the sun and the algal flakes are collected and stored. From each harvest, 30-40 kg of dry algal flakes are obtained per plot.

Algal Inoculation to Rice Crop

Inoculation of BGA is highly suitable during dry season. Application of dried soil based blue-green algal flakes at the rate of 10 kg ha^{-1} is recommended for rice crop. At the time of inoculation of BGA, 500 g of single super phosphate may be mixed with 10 kg of soil based BGA which encourages better growth and establishment in rice field. A composite starter culture consisting of more than one species or genera is more preferable. Fast growing species, which can tolerate a certain amount of extremities, are more suitable than highly susceptible ones. The algal species should have a fair amount of tolerance to agricultural chemicals such as pesticides and weedicides. The algal inoculant is to be applied 10 days after transplantation of rice crop. Continuous application of blue-green algae to the rice fields for at least three or four consecutive seasons sustains the high crop yield at reduced levels of commercial nitrogen fertilizers input in ensuing cropping seasons (Kannaiyan, 1993a).

Residual Effect of BGA Inoculation

Blue-green algae are living biosystems and once they establish, their biological activity is a continuous process. Inoculation of blue-green algae for three or four cropping seasons continuously results in an appreciable population build up and consequently their beneficial effect can be seen in the subsequent years without any further inoculation unless some unfavourable soil ecological conditions operate. The results of several field trials conducted in Tamil Nadu province of India have shown the positive residual effect of algal inoculation on rice yield (Kannaiyan, 1978).

Induction of Native BGA

Phosphorus nutrition is known to induce the native BGA (Kannaiyan, 2003). Top dressing (5 kg ha^{-1}) of single super phosphate (SSP) 10-15 days after transplanting induces the native N_2- fixing algal flora. Top dressing of 5 kg of SSP not only stimulates the growth of native BGA but also encourages the growth of inoculated BGA. Environmental conditions prevailing during dry season are more favourable for multiplication and establishment of BGA. It has been shown that the incorporation of various types of green manure in rice fields have induced the N_2-fixing and non N_2-fixing algal flora, and certain N_2-fixing algae exhibit heterotrophic characteristics in rice fields (Kannaiyan, 1992a). Algal populations were able to colonize well in rice straw. Addition of 2 kg of rice straw bits per 40 m^2 plot induces the populations of BGA in algal multiplication plots.

Blue-green algae are ideal biological systems under tropical situations for better N_2-fixation and conservation in rice soil. Blue-green algae are renewable biofertilizers and they represent a self-supporting biological agent because they can photosynthetically provide energy for N_2-fixation. Rice cropping system provides favourable aquatic environment for the growth, multiplication and N_2-fixation by utilizing the solar energy. Thus, these potential groups of N_2-fixing algae may be better exploited for harnessing atmospheric nitrogen so as to increase the productivity of rice.

Stress Tolerant Cyanobacteria

In India, out of a total of 157 million ha of cultivable area, 49 million ha of land are acidic in nature. Moore (1963) reported the occurrence of many cyanophyceae in acid soils. Many cyanobacterial cultures were isolated from acid soils of Kerala (India) having a pH of 3.8 (Aiyer, 1965; Madhusoodanan and Dominic, 1995). The survival of cyanobacteria under acid soil rice field condition was studied by Jurgensen and Davey (1968). Later, Sardeshpandey and Goyal (1981b) observed that the native isolates survived better under acid soils than the introduced cultures. Singh et al., (1997) isolated many acid tolerant strains of cyanobacteria from acid soils of Nagaland (India). The effect of low pH (4-6) on growth and biomass production of different cyanobacterial cultures was examined and compared with a normal non-acid soil cyanobacterium *Anabaena* – NS-TGK-A1. *Westiellopsis* – AT-TGK-W1 showed maximum tolerance to acidity upto pH 4 followed by *Westiellopsis* – AT-TGK-W2 and *Anabaena* – AT-TGK–A1 (Gopalaswamy et al., 2002).

Cyanobacteria react to salt stress either through osmotic adjustment or by secretion of excess polysaccharide (Flowers et al., 1977; Saxena and Kaushik, 1991). Salt tolerance in cyanobacteria is mediated by an osmo-regulatory mechanism, which involves the accumulation of carbohydrates, amino acids and quarternary ammonium compounds (Reed et al., 1986). Salt adaptation strategy of cyanobacteria includes two main processes: (i) maintenance of low internal contents of inorganic ions by active export mechanism, and (ii) synthesis and accumulation of osmoprotective compounds corresponding to the osmotic potential of the surrounding medium.

The first process is achieved by Na^+/H^+ antiporters that are coupled to cytochrome oxidase (Militor et al., 1986). Several compounds like sucrose, trehalose, glucosyl glycerol and glycine acting in the second process were identified in salt loaded cyanobacteria (Reed and Stewart, 1988). The physiological basis of salt tolerance may be due to extrusion of Na^+ and maintenance of low intracellular Na^+, accumulation of K^+ or ionic balance between K^+ and Na^+ (Miller et al., 1976; Apte and Thomas, 1986; Jha and Kaushik, 1988).

There is increasing evidence that respiration plays a primary role in the adaptation of cyanobacteria to higher concentration of salt (Erber et al., 1986; Fry et al., 1986; Wastyn et al., 1987). In *Anacystis nidulans* increasing salt concentration led to decreased photosynthesis and respiration (Vonshak et al., 1988). When cyanobacterial cells were exposed to salt stress, a short term increase in the cellular sodium concentration was noticed, which was due to a transient increase in the permeability of plasma membrane during the first few seconds of exposure to high salt and that led to the immediate inhibition of the photosynthetic and respiratory systems (Reed et al., 1986). Erdmann et al., (1992) have reported decreased photosynthesis and chlorophyll contents in *Microcystis firma* and *Synechocystis* sp. PCC 6803 during the first few days after a salt shock.

With increased sodium chloride (NaCl) concentration, the filaments of *Anabaena doliolum*, a salt sensitive freshwater cyanobacterium, was shorter with less number of heterocysts, which was an indirect evidence of reduced N_2-fixation, decrease in relative protein and chlorophyll content (Rai and Abraham, 1993). Schubert et al., (1993) observed that *Synechocystis* sp. PCC 6803 when grown in different salt concentrations, the optimum growth was found to be around 34.2 mM NaCl and at this salinity, photosynthesis and chlorophyll and phycocyanin contents showed the highest values. Selvakumar et al., (2002) analyzed pigment content in butachlor (N – butoxymethyl-2 chloro-2, 6 diethyl acetanilide) tolerant cyanobacteria and showed that the C- phycoerythrin content of *Anabaena* HT-SGK-A_2, *Westiellopsis* – HT-SGK-W1 and *Westiellopsis* – HT-SGK-W_2 was not affected even at 12 $\mu g\ ml^{-1}$ level of butachlor. Interestingly, butachlor was found to be stimulatory to *Anabaena* – HT-SGK-A_2 at 9 $\mu g\ ml^{-1}$ concentration.

Anand et al., (1994) have reported the following salt tolerance levels of different cyanobacteria: *Chroococcus minor* and *Oscillatoria saline* grew upto 3% salinity, *Nostoc piscinale, Tolypothrix tenuis* and *Gloeocapsa polydesmatica* failed to grow beyond 1.5% salinity. The cells of *Anabaena* strain L-31, when exposed to 350 mM sucrose and 10 mM KNO_3, osmotic stress induced proteins (OSPS) were stimulated which resulted in significant osmoprotection (Iyer et al., 1994).

Immobilization of Cyanobacteria

Cyanobacteria could be immobilized into the pores of polyurethane foam (PUF). It has been well documented that N_2-fixing cyanobacteria are known to colonize outside PU foams as well as immobilize into pores of the foam (Kannaiyan et al., 1994). In PUF, colonization on the foam surface followed by penetration into the pores by the cyanobacteria, was reported earlier by Brouers et al., (1988). *Anabaena azollae* and *A.variabilis*, when immobilized in polyurethane and polyvinyl foams and calcium alginate beads, have been shown to release ammonia extracellularly (Kerby et al., 1986; Brouers et al., 1988). The algal symbiont *A. azollae* (Plate 5.3) is known to colonize and immobilize well in polyvinyl foam (PVF) (Kannaiyan et al., 1994). Kannaiyan et al., (1994) have shown ammonia excretion by *A. azollae* (AS-DS) and *Nostoc muscorum* immobilized in PVF at different foam densities. Significantly higher quantities of ammonia were excreted by

Anabaena sp. immobilized in PUF in the presence of the glutamine analogue MSX (L-methionine DL- sulphoximine) (Brouers and Hall, 1986; Shi et al., 1987a; Kannaiyan et al., 1994).

Boussiba (1997) reported that N_2-fixing activity of the algal symbionts was 6-10 times higher in immobilized condition than the activity of free-living cyanobacteria, and that the independent isolate of *Anabaena azollae* from *Azolla filiculoides* registered higher heterocyst frequency and nitrogenase activity. An in vitro study with *A. azollae*–AS-DS has recorded higher nitrogenase activity followed by *A. variabilis*-SA_0. Interestingly, the immobilized cyanobacterial cultures treated with the systemic fungicide Bavistin (methyl benzimidazol –2- yl carbamate; BASF [India] Ltd.,) at 5 μg ml^{-1} level stimulated the nitrogenase activity (Table 5.3).

Table 5.3. Effect of Bavistin (methyl benzimidazol –2- yl carbamate) on nitrogenase activity of cyanobacteria immobilized on PUF (Kannaiyan et al., 1997)

Cynobacterial cultures	Nitrogenase activity (n moles ethylene produced H (hydrogen)$^{-1}$g^{-1} dry wt.)	
	Control	Bavistin (5 μg ml^{-1})
Anabaena variabilis SA_0	2678	3086
A. variabilis SA_1	1904	2226
A. azollae AS-DS	3571	3874
N. muscorum SK_6	1309	1454
N. muscorum SK	1492	1651
N. muscorum DOH	2237	2417
SED (Standard Error of Deviations)	9.55	
CD (Critical Difference) at 5%	19.8	

Kannaiyan et al., (1994) have shown that the systemic fungicide Benlate (methyl-1-butyl-carbamoyl-2 benzimidazole carbamate) stimulated ammonia production by *A. azollae* in the immobilized state in a photobioreactor system. The N_2-fixing cyanobacterial cultures were used for immobilization in PUF and investigated for their performance on the ammonia production under laboratory conditions. Hall et al., (1997) immobilized two cyanobacterial cultures i.e. *A. azollae*-AS-DS-SK and *A. variabilis*-SA_0 in PUF and inoculated them in transplanted rice. They assessed the ammonia production in the floodwater of rice and found a positive influence on rice crop.

N_2-fixing BGA, both symbiotic and free-living types, are known to contribute nitrogen to rice plants in terms of ammonia under natural ecological conditions. The algal symbionts *A. azollae*-MPK-SK-AF-38, *A. azollae*-MPK-SK-AM-24 and *A. azollae*-MPK-SK-AM-27 have excreted significantly higher levels of ammonia into the medium and the same cyanobacterial cultures when immobilized in PUF further increased the ammonia production (Table 5.4 and 5.5). Higher ammonia excretion by *A. azollae* immobilized in sodium alginate and PUF was reported (Kannaiyan, 1994).

Table 5.4. Ammonia excretion by the algal symbionts under free-living conditions at different intervals (Kannaiyan et al., 1997)

Algal symbionts	Ammonia excretion (n moles ml^{-1})			
	7th day	14th day	21st day	28th day
Anabaena azollae-MPK-SK-AM-4	510.5	516.8	562.8	489.2
A. azollae-MPK-SK-AM-7	455.6	475.5	518.5	535.7
A. azollae-MPK-SK-AM-24	625.7	657.9	735.6	677.4
A. azollae-MPK-SK-AM-27	556.3	610.8	712.9	646.8
A. azollae-MPK-SK-AF-38	599.2	571.1	687.1	605.3
	SED		CD	
Algal symbionts	6.26		12.66	
Days	5.60		11.32	
Algal symbionts x Days	12.5		25.29	

Table 5.5. Ammonia excretion by the algal symbionts under PUF immobilized condition at different intervals (Kannaiyan et al., 1999)

Algal symbionts	Ammonia excretion (n moles ml^{-1})			
	7th day	14th day	21st day	28th day
Anabaena azollae-MPK-SK-AM-4	622.8	598.3	718.9	671.4
A. azollae-MPK-SK-AM-7	575.5	656.1	635.4	602.1
A. azollae-MPK-SK-AM-24	732.9	766.1	843.1	771.8
A. azollae-MPK-SK-AM-27	672.4	691.7	761.7	723.6
A. azollae-MPK-SK-AF-38	705.1	745.4	809.2	756.3
	SED		CD	
Algal symbionts	8.36		16.89	
Days	7.47		15.09	
Algal symbionts x Days	16.70		33.75	

Systemic Fungicides on Ammonia Production

The treatment of the PUF immobilized cyanobacterial symbionts with the systemic fungicides Bavistin and Vitavax (5,6 dihydro-2 methyl – N - phenyl- 1,4 –oxathin – 3- carboxamide) induced ammonia production at significantly higher rates. The cyanobacterial culture, *A. azollae*-MPK-SK-AM-24, immobilized on PUF excreted maximum amount of ammonia on the 21st day (Kannaiyan et al., 1997). They also noticed that the immobilization of the algal symbionts on PUF and treatment with the systemic fungicide Bavistin greatly decreased the activity of glutamine synthetase (GS). Of the algal symbionts, *A. azollae*-MPK-SK-AM-24 registered very low GS activity and resulted in higher ammonia production. Paul and Kannaiyan (1994) developed mutants of *A. azollae*-AS-DS using colchicine and these mutants have recorded higher excretion of ammonia when treated with the herbicide butachlor. It was shown that incorporation of the granular insecticide carbofuran at 5 $\mu g\ ml^{-1}$ into nitrogon-free medium stimulated ammonia excretion by the cyanobacterial cultures immobilized in PUF. This may be due to the membrane permeability of the cells by the immobilization process (Shi et al., 1987b), which could allow a greater accessibility of substrates like inhibitors of ammonia assimilating enzymes, and thereby facilitating excretion of ammonia. Kannaiyan et al., (1994) found that Benlate (methyl - 1 – [butylcarbamoyl] benzimidazole – 2 – yl carbamate; Dupont, USA) due to its systemic action, is likely to diffuse into the cyanobacterial cells where it could, like methyl sulphoxamine (MSX), inhibit both ammonia assimilating enzyme and glutamine synthetase, and thereby increasing ammonia production. It is apparent that the systemic fungicide Bavistin could induce significant amounts of ammonia production by the algal symbionts under immobilized condition (Kannaiyan et al., 1997).

Nitrogen Regulation In Rice By Immobilized Cyanobacteria

Fertilizer nitrogen, usually from an inorganic source, as well as nitrogen fixed by biological sources/nitrogen derived from mineralization of organic matter are subjected to a series of physical, chemical and biological changes, which ultimately determine its fate in flooded rice soil and subsequent availablity to rice crop. Integrated use of organics and fertilizer nitrogen apparently increases the nitrogen utilization in rice crop. In lowland rice soil ecosystem, nitrogen could be efficiently supplied by the use of free-living cyanobacterial and symbiotic *Azolla* biofertilizers. The nitrogen excreted as ammoniacal form by the immobilized cyanobacteria or due to mineralization of both symbiotic as well as free-living cyanobacterial biomass, is utilized by the rice plants at all its critical stages of growth, like tillering, panicle initiation and grain filling. Hall et al., (1997) have shown that the application of immobilized N_2-fixing cyanobacteria increased the growth as well as grain and straw yield of rice. Kannaiyan (1990a) has shown significant increase in grain yield of rice when *Azolla* biofertilizer was used along with USG. Recently, Kannaiyan (1995) has established that the use of neem cake (*Azadirachta indica*) at 200 kg ha^{-1} along with *Azolla*, or neem cake-coated urea (NCU) i.e. 20% NCU with *Azolla* biofertilizer, significantly increased the nitrogen use efficiency in rice. The alkaloid "azadirachtin" present in neem cake (NC) inhibits nitrifying bacteria, which in turn inhibits the ammonia oxidation in rice fields and thereby the accumulation of ammonia occurs. The inhibitory action of NC on efficient utilization of nitrogen in rice crop is well documented by Kannaiyan (1995) by developing a pathway, which reflects in increasing the grain and straw yield of rice.

Carrier Based Immobilized Cyanobacterial Inoculum

The symbiotic N_2-fixing cyanobacterium i.e. *Anabaena azollae* – AS-DS and the free-living N_2-fixing cyanobacterium i.e. *Anabaena variabilis* – SA_0, were immobilized in solid matrices such as PUF and PVF, and it was found that the ammonia excretion was stimulated as a result of the immobilization of cyanobacterial cells. The effect of inoculation of immobilized cyanobacterial inoculum in the transplanted rice was quite significant. Similarly, inoculation of sugarcane waste immobilized strains of *A. azollae*-AS-DS and *A. variabilis*–SA_0 in transplanted rice field significantly increased the flood water ammonia content and also increased the grain and straw yield of rice. Among the different treatments tested, the inoculation of *A. azollae* immobilized on sugarcane waste along with amendment of 60 kg N ha^{-1} recorded maximum grain and straw yield of rice (var. ADT 36). Hence, a carrier based immobilized cyanobacterial inoculant was developed to maintain better quality. The PUF, sugarcane waste and paper waste immobilized cyanobacteria were the components of the inoculant. Rice husk and sterilized soil were used in equal proportion as a carrier material for preparing this inoculant. The carrier based immobilized cyanobacterial inoculum may be applied for rice crop at 2 kg ha^{-1} one week after transplanting.

A field experiment was conducted to study the effect of inoculation of immobilized cyanobacteria alone with fertilizer nitrogen on the floodwater ammonia of rice. Higher ammonia level was recorded in the floodwater of the experimental plots due to inoculation with immobilized cyanobacteria than uninoculated control (Table 5.6). It is seen from the data that the PUF and sugarcane waste immobilized cyanobacteria could produce ammonia in the floodwater of rice crop. Both *A. azollae*– AS-DS-SK and *A. variabilis*–SA_0 immobilized in PUF and sugarcane waste have excreted significant amounts of ammonia in experimental plots and the results have clearly indicated that the immobilized cyanobacterial inoculum in rice field have produced ammonia in floodwater.

SESBANIA ROSTRATA

Legumes that nodulate on aboveground stems are referred to as stem nodulating legumes. The term "Stem Nodule" was first reported in

Table 5.6. Effect of inoculation of immobilized cyanobacteria on ammonia production of rice floodwater

Treatments	Ammonia production (n moles ml^{-1})				
	2nd day	8th day	11th day	14th day	28th day
Control	318	341	428	727	311
Anabaena azollae (PUF)	890	1137	1027	942	341
A. variabilis (PUF)	850	1559	1267	1195	350
A. azollae (SCW)	1013	1234	1169	1199	331
A. variabilis (SCW)	1072	1364	1445	1120	360
A. azollae (PUF) + 60 kg N ha^{-1}	1653	1624	1462	1608	389
A. variabilis (PUF) + 60 kg N ha^{-1}	1234	1786	1364	1624	370
A. azollae (SCW) + 60 kg N ha^{-1}	1150	1786	1559	1494	380
A. variabilis + kg + 60 N ha^{-1}	1267	1591	1656	1786	389
60 kg N ha^{-1}	357	357	584	890	321
SED (Standard Error of Deviations)	5.950	2.654	1.431	3.446	0.995
CD (Critical Difference) at 5%	11.890	5.307	3.005	6.893	1.909

PUF: Polyurethane foam; SCW: sugarcane waste

histological studies on *Aeschynomene aspera* by Hagerup (1928). Since then, several workers reported stem nodulation in *Aeschynomene paniculata* (Suessenguth and Beyerle, 1936), *Neptunia oleracea* (Schaede, 1940), *Aeschynomene indica* (Arora, 1954), *A. elaphroxylon* (Jenik and Kubikova, 1969), *A. evenia* and *A. villosa* (Barrios and Gonzales, 1971). However, the importance of stem nodulation was recognized only when Dreyfus and Dommergues (1981) reported profuse stem nodulation on the fast growing leguminous plant *Sesbania rostrata*. Until now, 26 different stem nodulating legume species from various genera i.e. *Aeschynomene, Neptunia* and *Sesbania* have been reported. However, fast growing, extensively nodulating, erect leguminous plants such as *S. rostrata, Aeschynomene afraspera* and *A. nilotica* have been reported as most promising N_2-fixing green manures for lowland rice (Rinaudo et al., 1983; Alazard and Becker, 1987; Becker et al., 1990; Kannaiyan, 1993a).

The most distinctive characteristic of stem nodulating legumes is the presence of predetermined nodulation sites on the stems, which are independent of infection with rhizobia (Dreyfus et al., 1984). They comprise primordia of adventitious roots that are able to grow out under waterlogged conditions (Tsien et al., 1983) and they are arranged in straight vertical rows (*Sesbania*) or in spiral shaped rows winding around the stem (*Aeschynomene*). Ladha et al., (1998) defined "stem nodules" as those formed on aerial portions of the plant, whose location prevents contact with soil and whose formation is initiated on a primordium and not on an outgrowing adventitious root (Plate 5.4).

Among the leguminous green manure crops, *S. rostrata* was found to be highly sensitive to photoperiod (Visperas et al., 1987). Low biomass production and low nitrogen content of *S. rostrata* in The Philippines between October and February 1998 have been ascribed to the day length of less than 12 hours (Becker et al., 1990). *S. rostrata* flowered after 37 and 125 days when planted during December and April respectively, while *A. afraspera* took 54 and 85 days to flower respectively, in the same season in The Philippines (Becker et al., 1990).

The specific acetylene reduction activities in stem nodules of *S. rostrata* and *Aeschynomene scabra* were 1.5 and 9 times higher respectively, than in root nodules. Stem nodules have both higher enzyme activities and metabolic content than root nodules, suggesting that they fix N_2 with greater energy efficiency, which was correlated to photosynthesis in the cortex of stem nodules (Hungria et al., 1992). Sadasivam et al., (1993) found a close relationship between nodule phosphoenolpyruvate carboxylase and nitrogenase activities, and suggested that nodule CO_2-fixation contributed directly to nitrogen assimilation in stem nodules of *S. rostrata*. Moreover, nitrogenase component II was localized in bacteriods within the outermost layers of infected cells, suggesting that a low CO_2 was maintained, despite the chloroplasts being nearby (James et al., 1996).

Characteristics of the Symbiont *Azorhizobium caulinodans*

Rhizobia isolated from root and stem nodules of *Sesbania* and *Aeschynomene* were able to produce nodules in both the stem and root (Yatazawa and Yoshida, 1979; Legocki et al., 1983; Alazard, 1985; Dreyfus et al., 1988; Subba Rao, 1988). Dreyfus et al., (1984) isolated two types of strains; one that fixes atmospheric N_2 under free-living and nodulates both the root and stem of *S. rostrata*, while the other does not fix N_2 in culture and nodulates only roots of *S. rostrata*. Rhizobial strains from stem nodules of *A. indica, A. aspera* and *S. rostrata* were characterized and the strains were found to be of intermediate type, sharing characteristics of both fast and slow growers (Stowers and Eaglesham, 1983; Chakravarty et al., 1986).

Dreyfus et al., (1983) observed free-living N_2-fixation and growth of *Azorhizobium* with N_2 as the sole nitrogen source. Gebhardt et al., (1984) investigated the important characteristics of free-living N_2-fixation by *Azorhizobium* and noticed higher tolerance to oxygen. Urban et al., (1986) reported that *Azorhizobium* strain ORS 571 fixes nitrogen at a relatively high oxygen concentration (3%) and temperature (37°C). Free-living growth and N_2-fixation have also been shown for isolates of *A. indica, A. sensitiva, A. protensis* and *A. afraspera* (Ladha et al., 1998).

Kalidurai and Kannaiyan (1988) have shown that *S. rostrata* stem isolates recorded relatively more in vitro nitrogenase activity in arabinose and mannitol, followed by arabinose and sucrose in semi-solid agar medium. The role of azorhizobia as a free-living N_2-fixer in rice soil was demonstrated by Ladha et al., (1998). Free-living N_2-fixation by *A. caulinodans* strain ORS 571 was 10 times higher than that fixed by *Bradyrhizobium* strain CB 756 (Alazard, 1990).

Spermatospheric presence of *Azorhizobium* in seeds of *S. rostrata* was noticed by Ladha et al., (1988) and Balasubramani and Kannaiyan (1991a) and when released from stem nodules can survive and grow in flooded soils (Ladha et al., 1998). The epiphytic occurrence of *A. caulinodans* in leaves and flowers of *S. rostrata* was reported (Adebayo et al., 1989; Balasubramani and Kannaiyan, 1991b). The blister beetle (*Mylabris pustulata*) generally seen during the flowering season of *S. rostrata* was found to transmit *A. caulinodans* from one plant to another (Balasubramani and Kannaiyan, 1991a). The isolates obtained from *S. rostrata* leaves, seeds, flowers and from the insect-blister beetle were found to nodulate *S. rostrata* (Table 5.7). The internal movement of *A. caulinodans* on *S. rostrata* stems was also reported (Table 5.8) (Chitra, 1992). Robertson and Alexander (1994) suggested that rain and wind blown soil are important means of transmitting *Azorhizobium* for stem nodulation in *S. rostrata*. The occurrence and distribution of *Azorhizobium* in four vegetation zones of Senegal was determined by Robertson et al., (1995).

Based on whole cell protein analysis, DNA-DNA and DNA-rRNA hybridization, Dreyfus et al., (1988) proposed a new genus and species *Azorhizobium caulinodans* for the stem nodulating isolates of *S. rostrata*.

Table 5.7. Studies on stem by *Azorhizobium* isolated from insect, seed, leaf and flower of *S. rostrata*

Azorhizobium isolates	Stem nodule number plant^{-1}		
	30 DAS	40 DAS	60 DAS
Insect isolate	33	42	50
Seed isolate	42	53	67
Leaf isolate	34	47	60
Flower isolate	24	31	43

DAS: Days after sowing

Table 5.8. Internal movement of *A. caulinodans* in seedlings of *S. rostrata*

Inoculation	Recovery (%)			
	Root bits		Stem bits	
	IRRI $^{(+)}$ NO_3	IRRI $^{(-)}$ NO_3	IRRI $^{(+)}$ NO_3	IRRI $^{(-)}$ NO_3
Inoculated	20.0	100.0	40.0	100.0
Uninoculated	-	30.0	30.0	20.0

IRRI: International Rice Research Institute

Rinaudo et al., (1991) studied DNA homology among members of the genus *Azorhizobium* and other stem and root nodulating rhizobia isolated from *S. rostrata* and found that most isolates belonged to the genus *Azorhizobium* with G+C content ranging from 66 to 68%. The sequence surrounding the *A. caulinodans* regulatory *nod* D gene was analyzed by Geelen et al., (1995). Tomekpe et al., (1996) reported that the greater infectivity of *A. caulinodans* strain ORS 571 was due to more efficient production of *nod* factors at infection sites located on the stem.

Seed/Foliar Application of *S. rostrata*

The primary objective of seed inoculation is to introduce rhizobia in soil in such a way that the inoculum will remain viable until the host seedlings would be able to accept an infection and compete with native rhizobia for optimum nodulation and higher N_2-fixation. In general, the seed coat of different leguminous green manure seeds is very thick which affects the seed germination. Pre-soaking the seeds of *Sesbania* spp. and subsequent inoculation with *Rhizobium*/*Azorhizobium* have substantially increased the colonization and survival of *Rhizobium*/*Azorhizobium* and higher population load in pre-soaked seeds than unsoaked seeds.

Though the nodules on the stem appear spontaneously on the preformed infection sites, inoculation by foliar spraying with *A. caulinodans* has been found to activate the nodule initiation. During summer season when the temperature is high (37±1°C), the infection by *A. caulinodans* on *S. rostrata* stem is very low. However, the external supply of *A. caulinodans* inoculation by foliar spraying triggers the infection and nodule formation. Eaglesham and Szalay (1983) showed that a single inoculation of the stem of *Aeschynomene* plant with rhizobial strain BTAI-1 induced the formation of stem nodules within eight days.

Rinaudo et al., (1982) reported that stem inoculation of *S. rostrata* in Northern Senegal resulted in increased N_2-fixing activity. The positive effect of stem inoculation on the growth and N_2-fixing activity of *S. rostrata* was reported by Ladha et al., (1998). Chitra and Kannaiyan (1995) found that foliar inoculation of *A. caulinodans* significantly increased the fresh weight, nodule number and nutrient content of *S. rostrata* (Table 5.9). Kalidurai and Kannaiyan (1991) have also shown that foliar inoculation with *A. caulinodans* favoured stem nodulation in *S. rostrata*.

Stem inoculation seems to be appropriate particularly in acid soil where root infection of rhizobia is not possible (Ladha et al., 1998). Foliar inoculation of *A. caulinodans* by foliar spraying increases the population load considerably on the stem surface of *S. rostrata*. It is also presumed that the inoculated *A. caulinodans* might survive on the plant parts such as leaves and flowers, and later might spread to the stem portion of *S. rostrata* for increasing the colonization, infection and nodulation.

Table 5.9. Effect of foliar inoculation of *A. caulinodans* on nodulation, biomass production and nutrient content of *S. rostrata*

Foliar Inoculation	Nodule numbers plant^{-1}		Fresh weight (g)	Nutrient content (%)		
	Root	Stem		N	P	K
Sprayed	34.5	1173.5	972.59	3.31	0.42	2.00
Unsprayed	29.2	87.7	510.00	2.06	0.40	1.90
SED	1.02	95.94	92.84	0.05	0.01	0.01
CD at 5%	2.06	201.56	195.72	0.12	0.02	0.02

SED: Standard Error of Deviations; CD: Critical Difference

Ladha et al., (1998) have demonstrated that both seed and stem inoculation have improved the growth and N_2-fixation in *S. rostrata*. Seed inoculation coupled with foliar spraying, has substantially increased the stem nodule numbers per plant (Chitra, 1992). Seed inoculation and foliar spraying have also increased the nitrogen, phosphorus and potassium content of *S. rostrata*.

Foliar spraying with appropriate strains of *A. caulinodans* on the stems of two to seven week old plants have been shown to produce a greater number of stem nodules (Ladha et al., 1998). Chitra (1992) has reported that foliar inoculation of 20-25 day old plants of *S. rostrata* recorded a maximum number of stem nodules per plant (Table 5.10). Foliar inoculation both in the morning and evening recorded maximum number of nodules per plant. However, foliar inoculation in the evening resulted in more number of stem nodules than the morning spray.

Stem Nodulation and N_2-Fixation in *S. rostrata*

The tolerance of leguminous green manures to flooding is an important trait for species used in wetland rice. Eaglesham and Szalay (1983) studied the effect of flooding on root and stem nodulation of *Aeschynomene scabra* and found that the number of root nodules decreased to almost zero compared with 1,000 stem nodules per plant in non-flooded condition. *S. rostrata* was found to tolerate water logging and produced about

Table 5.10. Influence of waterlogging condition on stem nodule formation in *S. rostrata*

Azorhizobium isolates	Stem nodule number plant^{-1}					
	Water logging			Saturation		
	35 DAS	45 DAS	60 DAS	35 DAS	45 DAS	60 DAS
SRS2	104	118	134	48	55	62
SRR1	75	87	98	35	43	54

DAS: Days after sowing

25 t ha^{-1} biomass (Plate 5.5 and 5.6) after approximately 45 days growth (Kalidurai and Kannaiyan, 1988; Ndoye and Dreyfus, 1988). Becker et al., (1990) reported higher biomass production and N_2-fixation in 40-50 day old *S. rostrata* and *A. afraspera*. Stem nodulation in *S. rostrata* was twice in flooded condition (Table 5.10) compared to water saturated condition (Balasubramani and Kannaiyan, 1993). Becker and George (1995) found maximum nitrogen accumulation of 275 mg $plant^{-1}$ by *S. rostrata* which was 20% more than *S. cannabina*, in flooded soils. Flooding the soil increased the relative humidity of the crop microenvironment by 4 to 11% and induced early appearance of stem nodules in *S. rostrata* (Patel et al., 1996).

Nutritional factors are responsible for the wide variation in the amount of N_2 fixed by legume-*Rhizobium* symbiosis (O'Hara et al., 1988). The growth of *S. rostrata* and *A. afraspera* was vigorous on both fertile and relatively poor sandy soils (Becker et al., 1990).

The requirement of phosphorus for efficient nodulation and maximum nodule activity was much greater in nodules than the host plant growth (De Mooy et al., 1973). Balasubramani and Kannaiyan (1991a) have shown enhanced nodule induction and biomass production in *S. rostrata* by phosphorus nutrition. Becker et al., (1991) showed that the application of mineral phosphorus and potassium fertilizers stimulated nodulation and acetylene reduction activity (ARA) in *S. rostrata* and *A. afraspera,* which led to a 40% increase in nitrogen content. Potassium alone enhanced plant growth, whereas phosphorus activated it. A positive effect of phosphorus and potassium fertilizers on biomass production of *S. rostrata* has also been reported by Meelu and Morris (1988). Becker et al., (1991) recommended the application of phosphorus and potassium fertilizers to green manure crops for higher biomass production.

The type and concentration of mineral nitrogen influenced both nodulation and N_2-fixation in root nodulating legumes (Eaglesham and Szalay, 1983). However, the N_2-fixation by stem nodulating legumes is reported to be unaffected to some extent by mineral nitrogen (Dreyfus and Dommergues, 1981; Eaglesham and Szalay, 1983; Becker et al., 1986; 1989; 1991; Kannaiyan, 1993a). Rinaudo et al., (1983) have reported complete inhibition of root and stem nodule ARA in hydroponically grown *S. rostrata* where the ammonium nitrate concentrations were constantly maintained at 1.5 -3.0 mM and 6 mM, respectively.

Kwon and Beevers (1993) found that 0-9.0 mM NO_3 had no adverse effect on stem nodulation in *S. rostrata*. However, when the NO_3 concentration was raised above 20 mM, the growth of nodules and ARA were strongly inhibited. Balasubramani and Kannaiyan (1990) have reported that the dry weight of stem nodule and nitrogenase activity of root and stem nodules of *S. rostrata*, were not affected adversely upto 90 kg N ha^{-1}, but interestingly 30 to 90 kg N ha^{-1} increased the fresh and dry weight of nodules. Reduction in the amount of BNF due to applied mineral nitrogen was greater in aerobic soil than in flooded soil. The N_2-fixing activity of stem nodules was less sensitive to NH_4^+ concentration in flooded soils than root nodules and resulted in more biologically fixed nitrogen in *S. rostrata* (Becker and George, 1995).

Ghai et al., (1988) evaluated 36 *Sesbania* spp. in the wet season and recorded an average contribution of 63 kg N ha^{-1} in 30-45 days. Under flooded conditions, the nitrogen accumulation potential ranged from 41 kg N ha^{-1} in *A. indica* (Crozat and Sangchyosawat, 1985) to more than 200 kg N ha^{-1} in *S. rostrata* (Morris et al., 1987; Becker et al., 1990). Ladha et al., (1992) found that 50 day old *S. rostrata* could accumulate 190-267 kg N ha^{-1}. Pareek et al., (1990) have estimated N_2-fixation of *S. rostrata* by ^{15}N technique and found *S. rostrata* to be the fastest N_2-fixing plant known, fixing 70-95% of the total plant nitrogen within 45-55 days. They also found that *S. rostrata* could accumulate more nitrogen than *S. cannabina* during dry season.

Stem nodulation has been reported to be more efficient than root nodulation in N_2-fixing system. Eaglesham and Szalay (1983) reported activities of over 160 μM C_2H_4 h^{-1} $plant^{-1}$ of aboveground portions of *A. scabra*. Dreyfus et al., (1985) reported nitrogenase activity of about 600 μM $C_2H_4h^{-1}$ $plant^{-1}$ in a three month old *S. rostrata* and most of the nitrogenase activity was attributed to stem nodules. Negligible ARA has also been recorded in *S. rostrata* root nodules (Rinaudo et al., 1982; Becker et al., 1990). The nitrogenase activity and the nodule dry weight were higher in stem nodules than in root nodules (Balasubramani et al., 1992; Saraswati et al., 1992).

Xylem Colonization by *A. caulinodans* in *S. rostrata*

Rhizobia are bacteria that invade legumes after activating cell division in the plant root cortex and initiating a new organ, the N_2-fixing nodule (Long, 1996). Since the nodule is generally considered as the only endophytic destination of invading rhizobia, most previous studies have focussed on the rhizobial invasion pathway into and within the cortex (Kijne, 1992). Many non-rhizobia endophytic bacteria colonize the vascular system of plants, sometimes, but not always leading to plant disease (Bell et al., 1995). For example, the xylem of healthy alfalfa, a legume, is colonized by non-rhizobial endophytes (Gagne et al., 1987). Agrobacteria and plant pathogens are closely related taxonomically to rhizobia, and they invade the xylem of several species including *S. rostrata* (Vlachova et al., 1987). Inoculation of *S. rostrata* with *A. caulinodans* ORS

571 (Dreyfus et al., 1988) results in production of nodules on roots (Ndoye et al., 1994) and stems (Tsien et al., 1983). Interestingly, strain ORS 571 is able to establish itself endophytically in the roots of rice (Christiansen-Weniger, 1996) and wheat (Sabry et al., 1997). The presence of azorhizobia in the xylem was confirmed by light and electron microscopy. The recent availability of various genetic tools and reporter genes, based on fusions with *lacZ* and *gus*A, is having a major impact on studying the xylem colonization (Vande Broek et al., 1993; Arsene et al., 1994; Gough et al., 1996). Bacteria re-isolated from inoculated plants and plated onto selective media containing X-Gal, produced blue staining colonies, confirming that some azorhizobia still retained pXLGD4. Re-isolated bacteria also exhibited typical azorhizobial colony morphology on semi-solidified TY (Tryptone-Yeast extract) medium and failed to take up Congo red dye (a diagnostic test for rhizobia) (Somasegaran and Hoben, 1994). Recently it was found that the strain ORS 571 colonizes xylem elements, in addition to inducing and invading nodules in the root cortex of *S. rostrata*. However, xylem colonization is not regulated in the same way as nodulation. Thus, for the first time, a species of legume nodule bacteria has been found to colonize, reproducibly, regions of the host plant other than nodules. This novel endophytic interaction will be of interest to phytopathologists, to researchers investigating legume – rhizobia interactions, and to workers attempting to extend endophytic rhizobial N_2-fixation to non-legumes (O'Callaghan et al., 1997).

A. caulinodans

N_2 fixing nodules are usually found on the roots of the host leguminous plants. Only some legume species belonging to the genera *Neptunia, Aeschynomene* and *Sesbania* bear nodules on both the root and stem. Dreyfus et al., (1988) found that the tropical legume *S. rostrata* was symbiotically associated with *A. caulinodans.*

The cells of *A. caulinodans* are Gram-negative, small rods (0.5-0.6 μm by 0.5-2.5 μm) and motile. The cells have peritrichous flagella on solid medium and one lateral flagellum in liquid medium (Dreyfus et al., 1988). On agar medium, the colonies appear circular, viscid and cream coloured, and fix N_2 in microaerophilic condition. Bacteria growing within the para-nodule displayed variable morphology, differing in size and cell contents. Electron transparent regions within some bacterial cells of these sections may have accumulations of poly-β-hydroxybutyrate. Considerable amounts of indeterminate material were found together with these bacteria. This was surprising, since the bacteria were extracellular in the middle lamellar space or in rare cases, present in dead cells. *A. caulinodans* ORS 571 induces the formation of root and stem nodules on the tropical legume, *S. rostrata*. *Azorhizobium* (can use nitrogen while free-living) and organic acids such as lactate and succinate are the favourite carbon substrates for both NH_4^+ and N_2-dependent growth.

This bacterium contains essential nodulation genes as that of *nod*ABC of rhizobia (Goethals et al., 1989). Hence it is possible to elucidate the *nod* genes in the biosynthesis of the Nod factor induced by the host plant exudate. In view of the reports of associative N_2-fixation by *Azorhizobium* in rice fields, it will be interesting to study this behaviour and identify functions that contribute to the interactions between *Azorhizobium* and rice plants. The induction of para-nodules and the invasion of rice by rhizobia, either spontaneously or after the treatment of the roots with cell wall degrading enzymes, have been reported (Cocking et al., 1990). The nodule like structures induced by 2,4-D can be inhabited by diazotrophs (Nie, 1983). It is possible that 2,4-D has at least two effects:

(1) an auxin-like activity, controlling development of nodular structure, and
(2) an effect enabling increased frequency of infection of plant root cells and colonization of nodular structures with certain bacteria.

The choice of the bacterial strain *A. caulinodans* ORS 571 was based on three characteristics (de Bruijn, et al., 1995). First, as it is unlikely that there will be bacteroid differentiation in non-legume plants, strain ORS 571 has the unusual ability among the rhizobia to fix nitrogen in free-living conditions without differentiation into bacteroids. Second, as it is also unlikely that non-legumes will be able to reproduce the sophisticated oxygen barriers which exist in legume nodules to protect the nitrogenase enzyme complex from inhibitory oxygen levels, strain ORS 571 can tolerate upto 12 μM dissolved oxygen while fixing nitrogen in culture. Third, strain ORS 571 forms nodules on its host plant following crack entry infection i.e. inter-cellularly between adjacent cells. It has been proved that the *A. caulinodans* strain ORS 571 is able to enter the hosts (rice, wheat, sorghum, oil seed rape and *S. rostrata*) through crack entry at the point of emergence of bacterial roots, their colonization and invasion were significantly stimulated by the flavonoid naringenin. Though in rice, *A. caulinodans* was found to stimulate the formation of root-hairs, no root-hair curling was observed. However, abnormal lateral root development was observed within two weeks after inoculation. More than 80% of the rice roots exhibited thick, short lateral roots.

Flavonoids as Signal Molecules for Endophytic Colonization by *A. caulinodans*

Flavonoids are a class of phenolics secreted by plant roots as part of the regulatory system for colonization by *A. caulinodans*. Four families of flavonoids, such as flavones, flavonols, flavonones and iso-flavonones, have been shown to have *nod*-gene inducing activity. Naringenin–a specific flavonoid, is known to stimulate the colonization of *A. caulinodans* in rice roots. It is a signal molecule for attracting the population of *A. caulinodans*. However, rice root did not produce the flavonoid naringenin. The flavonoid naringenin is considered as a potential signal molecule to activate the colonization of *A. caulinodans* in rice roots and, therefore, a specific gene could be introduced into rice plants for synthesis of naringenin. Flavonoids enhance the growth rate of certain rhizobia, and phenolic compounds in general, and could induce expression of a variety of genes in plant-associated bacteria. The flavonoids naringenin and diadzein, at low concentration were found to significantly stimulate the frequency of lateral root colonization (LRC) and inter-cellular colonization by *A. caulinodans* in non-legumes by acting as chemical stimulants (Gough et al., 1996). Flavonoids have been shown to accumulate in many plants under infection. However, there are few reports in cereal crops showing such accumulation. Naringenin was detected in rice leaves exposed to UV (ultra violet) irradiation. Kumar et al., (1996), while studying the effect of flavonoid naringenin on the endophytic colonization by *A. caulinodans* in rice, reported favourable stimulation of lateral root formation by naringenin in rice (var. ADT36). They also reported that the main mode of entry of *A. caulinodans* in rice roots is by crack entry through the emerging lateral roots. The effect of naringenin (3.0 $\mu g\ ml^{-1}$), enzyme mixture and *A. caulinodans* on induction of para-nodules in rice and maize was reported (Buvana and Kannaiyan, 2002; Senthil Kumar, 2000). It was found that addition of naringenin enhanced the short thickened lateral root formation. The seedlings treated with naringenin, enzyme mixture and *A. caulinodans* recorded maximum number of lateral rootlets, rhizoplane population, nodule like structures and nitrogenase activity. Lateral root development of rice and colonization of lateral root cracks by bacteria were shown to be stimulated by the flavonoid naringenin (Table 5.11). The effect of flavonoid on root morphology of different rice varieties was studied and it was found that flavonoid naringenin at 5×10^{-5} M concentration significantly stimulated the lateral root formation in rice varieties (Gopalaswamy et al., 2000). The symbiont *A. caulinodans* ORS 571 upon inoculation, stimulated the lateral root development as well as formation of nodular like structures in the roots of rice plants (Kannaiyan et al., 2001).

Endophytic Colonization by *A. caulinodans*

"Endophytic N_2-fixation" is a process where by the N_2-fixing bacteria enter the plant roots through cracks or fissures in the lateral roots, colonize

Table 5.11. Effect of naringenin on LRC (lateral root colonization) on four varieties of *Oryza sativa* inoculated with *A. caulinodans* ORS 571 (pXLGD4)

Variety	Treatment	Percentage of LRCs colonized per plant (±S.D.)	
		Seminal root	Adventitious roots
Lemont	–	9.4 ± 0.7	5.7 ± 1.2
	+ naringenin	20.9 ± 1.5	10.7 ± 0.6
CO43	–	8.3 ± 0.6	11.0 ± 0.4
	+ naringenin	10.2 ± 1.3	16.1 ± 1.0
ADT36	–	6.7 ± 1.1	14.5 ± 1.5
	+ naringenin	9.3 ± 0.6	37.6 ± 3.1
CR 1009	–	5.3 ±0.6	11.3 ± 1.1
	+ naringenin	8.8 ± 0.9	16.5 ± 1.2

– Without naringenin
± S.D. (Standard Deviations)

the inter-cellular spaces (cortex, within few cells of epidermis, inner cortex to epidermis, vascular tissues and xylem), where without differentiating into bacteroids, fix atmospheric nitrogen micro-aerobically, with the help of the enzyme "nitrogenase" (Kennedy et al., 1997). In the earlier days, it was thought that only by inducing nodular structures to provide a place for the organism specifically, BNF could be extended to cereals (Arora, 1954). However, recent researches have vividly reduced the difficulties and increased the scope for N_2-fixation in cereals by means of endophytic N_2-fixation (Stoltzfus et al., 1997; Kannaiyan, 2000).

The extension of biological N_2-fixation to non-legume crops such as rice, maize etc. would be of enormous economical and environmental impact. The discovery of the fact that bacteria of the genera *Rhizobium* and *Bradyrhizobium* can form effective N_2-fixing nodules on the non-legume *Parasponia* (Davey et al., 1997), provided added impetus to this approach. Many of these nodular structures have to be artificially induced by adding 2,4-D (Tchan et al., 1991; Kennedy et al., 1997) or cell wall degrading enzymes (Al- Mallah et al., 1989; Cocking et al., 1990; Tchan et al., 1991). Cocking et al., (1993) reported that *Rhizobium* of *Parasponia* and *Aeschynomene* produce nodule like structures on the emerging lateral roots of rice and maize without any enzyme treatment. Kennedy and Tchan (1992) studied various approaches for the induction of nodule like structures on the roots of rice and other cereals.

Zeman et al., (1992) reported that the exogenous application of the synthetic auxin 2,4-D to rice induces modified root outgrowths (MROs), which result from the induction of meristem. It has been reported that certain rhizobia can enter rice plant at low frequency (Cocking et al., 1994). Christiansen-Weniger (1996) reported the formation of a nodule like structure in rice roots treated with 2,4-D and inoculated with *A. caulinodans*. It is recommended to have a thorough knowledge about the plant genes involved in nodulation and symbiotic N_2-fixation and the function of their gene products before the development of true nodulation of cereals (de Bruijn et al., 1995; Kennedy et al., 1997; Webster et al., 1997). Views have consequently changed, and presently the emphasis is to try to establish stable endophytic associations between diazotrophic bacteria and non-legume crops.

Two major approaches are currently being undertaken to try to extend biological N_2-fixation to non-legume crops. The first is to look for natural endophytic diazotrophs of non-legume crops and to assess such endophytic N_2-fixation. This is being exploited with sugarcane (Boddey et al., 1995) and other crops such as rice and wheat (Sabry et al., 1997). The second approach is to determine whether the rhizobia can internally colonize and ultimately fix nitrogen in non-legumes. The first step is to study the possibility of inter-cellular colonization of diazotrophs with non-legumes including entry, development and the mechanism involved. A sensitive and quantifiable system is needed for this. Microscopic observation of colonization is greatly facilitated by using a genetically modified bacterial strain tagged with a reporter gene (Webster et al., 1997). The efficacy of endophytic colonization by diazotrophs *Azospirillum* and *Azorhizobium* tagged with *lacZ* marker genes in rice was compared. It was observed that the colonization by diazotrophs was mainly through crack entry. In the case of *Azospirillum*, the bacteria entered the lateral cracks and their invasion was not so deep and very much limited to the endodermis and cortex, whereas, *Azorhizobium* exhibited deep colonization. They first got entry through lateral cracks and traversed intercellularly and colonized xylem of rice var. ADT 36 (Gopalaswamy et al., 2000).

Pre-requisites for an Endophytic Relationship

There are several prerequisites for the establishment of an endophytic situation (Quispel, 1974):

(1) Hostile reactions from the host, like hypersensitive reactions and production of phytoalexins, which have to be prevented.

(2) The bacteria must find an inter- or intra-cellular niche, where sufficiently reducing conditions are provided and the pO_2 (partial pressure of oxygen) is regulated by diffusion barriers and/or leghaemoglobins.

(3) Membrane systems around the endosymbiont of microbial and host origin must be adapted in such a way that a bi-directional transport of substances is possible.

Recently, Chaintreuil et al., (2000) have investigated the presence of endophytic rhizobia within the roots of the wetland wild rice *Oryza breviligulata*, which is the ancestor of the African cultivated rice *O. glaberrima*. This primitive rice species grows in the same wetland sites as *Aeshynomene sensitiva*, an aquatic stem-nodulated legume associated with photosynthetic strains of *Bradyrhizobium*. Twenty endophytic and aquatic isolates were obtained at three different sites in West Africa (Senegal and Guinea) from nodal roots of *O. breviligulata* and surrounding water by using *A. sensitiva* as a trap legume. Most endophytic and aquatic isolates were photosynthetic and belonged to the same phylogenetic *Bradyhizobium, Blastobacter* subgroup as the typical photosynthetic *Bradyrhizobium* strains previously isolated from *Aeschynomene* stem nodules. N_2-fixing activity, measured by acetylene reduction, was detected in the rice plants inoculated with endophytic isolates. A 20% increase in the shoot growth and grain yield of *O. breviligulata* grown in a greenhouse was also observed upon inoculation with one endophytic strain and one *Aeschynomene* photosynthetic strain. The photosynthetic *Bradyrhizobium* sp. strain ORS 278 extensively colonized the root's surface, followed by intercellular and rarely intracellular, bacterial invasion of the rice root, which was determined with a *lacZ*-tagged mutant of ORS 278. The discovery of the fact that photosynthetic *Bradyrhizobium* strains, which are usually known to induce N_2-fixing nodules on stem of the legume *Aeschynomene*, are also natural true endophytes of the primitive rice *Oryza brevigulata*, could significantly enhance cultivated rice production.

Para-nodulation by *A. caulinodans*

Recently, de Bruijn et al., (1995) reviewed the potential and pitfalls of extending symbiotic interactions between N_2-fixing organisms and cereals such as rice. They considered chemical signalling between plant and microbe, nodulation and the prospects for symbiotic N_2-fixation with such systems. The results of experiments carried out in China on the induction of "nodule-like structures" on rice roots by rhizobia were highlighted. They concluded that the formation of hypertrophies on rice roots infected and colonized with microbes had been confirmed, but that little evidence supported their designation as "nodules" or even "nodule-like". The frequency of formation of the root structures was low, but non-saprophytic colonization of the rice-root endorhizosphere had clearly been observed and deserved further study in this important area of research.

During the past few years, the status of biological N_2-fixation in non-leguminous field crops were reviewed (Kennedy, 1994, 1997; Kattupitiya et al., 1995a, 1995b; Yu and Kennedy, 1995; Kennedy et al., 1997). The approach is based on the initial observation of Nie (1983) at Shandong University, Thailand that 2,4-D acting as an auxin could stimulate colonization by rhizobia in 'nodular structures' modified from the lateral roots of many plants species.

Although the use of synthetic auxin such as 2,4-D presents difficulties for field application, it is considered that the use of this procedure is justified as a laboratory model, providing improved colonization and N_2-fixation that allow the controlling factors to be studied. It may be possible to provide similar outcomes to the use of 2,4-D biologically, such as by using mutants producing higher amounts of indole acetic acid (IAA). However, it should be recognized that many systems of agricultural monoculture already apply very stringent chemical treatments, sometimes with dramatic effects on plant development (e.g. cotton production using hormones and chemical defoliants).

Para-nodular Bacteria: Mode of Entry and Colonization

Para-nodules are nodular structures produced by modification from the lateral roots of many plant species due to the stimulating effect of auxins such as 2,4-D (Tchan and Kennedy, 1989). These structures inhabited by diazotrophs may be quite dissimilar to legume nodules (Kennedy and Tchan, 1992).

The invasion of root cells by endophytic bacteria and their colonization has attracted the interest of many workers. The bacteria are attracted by certain root exudates. In cereals like rice and maize, since the exopolysaccharide production and the production of flavonoids like naringenin, luteoline and maiden are low, by supplying them artificially at very low concentrations invasion can be enhanced. (Kennedy et al., 1997; Sabry et al., 1997). The bacteria search for cracks or fissures through which they can enter the roots. The bacteria enter through cracks in the lateral roots. They move through the intercellular spaces and find a site of low redox potential, possibly better suited for N_2-fixation. A large number of bacteria was also found associated with the para-nodules (Kennedy et al., 1997).

Structure of Para-nodules

Para-nodular structures were formed when seedlings were treated with 2,4-D. The para-nodules were spheroidal in shape, and less than a millimeter in diameter. The frequency of para-nodulation on the young roots was 1-4 nodules per cm of root length. Sections of para-nodules show the presence of central vascular tissue, connected to the root stele. Several layers of cortical cells, usually without recognizable cytoplasmic contents, surrounded this rudimentary stele. This is in contrast to the organization of vascular tissue in legume nodules, in which stele surrounds the infected cell. In the early stages, root-cap like structures were found on the para-nodules. In the later stages, outer cells in the region showed swelling and sloughed off frequently. Occasionally, bacteria were observed inside the outer 'empty' cortical cells but never in large numbers. In contrast, extensive bacterial infection was confined to the inter-cellular spaces. Serial sectioning of para-nodules showed that the bacteria and associated material were present in large, irregular channels open to the exterior. Examination of roots by phase-contract and light microscopy indicated extensive colonization of the root surface and root hair by the endophytic bacteria. Electron microscopical observation of the para-nodules revealed the presence of large numbers of bacteria in inter-cellular spaces. These irregular spaces or channels were developed by splitting of the middle lamella, during bacterial colonization. These channels were surrounded by shrunken and almost empty cortical cells. Only cytoplasmic debris was observed in these empty cells.

Effect of Growth Regulators and Cell Wall Degrading Enzymes on Induction of Para-nodules in Rice

The inoculation of the bacterium *A. caulinodans* with growth regulators produces lateral rootlets formation in rice. During the development of lateral root emergence, a minute opening is formed on both sides of the lateral root by which the bacterium *A. caulinodans* could colonize well and invade into the roots through crack entry and gradually reach to the inner region of the rice root tissues. Seven day old seedlings of rice (var. ADT-36), when inoculated with the symbiotic N_2-fixing bacterium, *A. caulinodans* along with the growth regulators such as 2,4-D, NAA, kinetin could induce the formation of nodular structures. Buvana and Kannaiyan (2001) studied the effect of growth regulator NAA and cell wall degrading enzymes (cellulase and pectinase) on induction of para-nodules in rice by *A. caulinodans*. Inoculation of *A. caulinodans* with NAA and cell wall degrading enzyme mixture (cellulase + pectinase) decreased the root growth of seedlings compared to inoculation of *A. caulinodans* alone. Inoculation of *A. caulinodans* and cell wall degrading enzyme mixture developed lateral rootlets which were short and thickened. The number of nodule like structures and total nitrogen content was maximum with inoculation of *A. caulinodans* strains in combination with cell wall degrading enzyme mixture and NAA. Inoculation of *A. caulinodans* was found to increase the root and shoot growth, total biomass and lateral rootlets of rice seedlings over uninoculated control. The growth regulator kinetin was used at 3.0, 5.0 and 7.0 $\mu g\ ml^{-1}$ concentration, and the levels above 3.0 $\mu g\ ml^{-1}$concentration were found to be inhibitory to the seedlings by recording decreased root and shoot growth, total biomass and lateral rootlets. The inoculation of *A. caulinodans* could nullify the adverse effect of kinetin to some extent compared to the incorporation of growth regulator treatment alone. Increased concentrations of kinetin with inoculation of *A. caulinodans* increased the number of nodule like structures in the roots of young rice seedlings. The number of nodule like structure were relatively more in rice variety TKM-9 compared to the rice variety ADT-36. It is interesting to note that the increased concentrations of kinetin with inoculation of *A. caulinodans* increased the number of nodule like structures in rice seedlings. However, there is no nodule like structure formation in rice roots when *A. caulinodans* and growth regulator kinetin were treated in the seedlings separately (Amutha and Kannaiyan, 1999; 2000).

Xylem Colonization by *A. caulinodans* in Rice

The diazotroph *A. caulinodans* ORS 571 colonizes xylem of roots of the legume *S. rostrata* (O' Callaghan et al., 1997) in addition to colonizing nodules in that legume, which led us to investigate whether this rhizobial strain, might nevertheless be able to colonize the xylem of rice roots. Colonization of the xylem of rice by *A. caulinodans* ORS 571 is of interest because it might provide a suitable nutritional and environmental niche for N_2-fixation, comparable to the situation in sugar cane in which endophytic diazotrophs internally colonize the plant, including the xylem, and are probably major contributors to the observed high levels of N_2-fixation (Boddey et al., 1995).

A large number of plant species exhibit xylem colonization by bacteria without disease symptoms and xylem colonization is increasingly being seen as a common aspect of plant-microbe interactions (Hallmann et al., 1997). Moreover, it is becoming increasingly realized that the xylem may be more robust structurally and physiologically than previously envisaged, and that it should no longer be regarded as a vulnerable pipeline on the edge of disaster (Canny, 1998). The non-rhizobial diazotrophic *Acetobacter diazotrophicus* has been shown to penetrate sugarcane roots intercellularly at the root tip and at cracks in lateral root junctions, and to colonize xylem vessels following the inoculation of aseptically grown plants (James et al., 1994). In the presence of the flavanoid naringenin, *A. caulinodans* ORS 571 entered the rice through lateral root cracks and was present in large numbers in cortex (Webster et al., 1997). Reddy et al., (1997) have also observed that azorhizobia invade the outer cracks intercellularly. Little is known, however, as to how exactly bacteria reach and invade the xylem. It has been suggested that xylem elements are possible sites of N_2-fixation by diazotrophs, since the xylem elements could provide the low pO_2 and a site for exchange of metabolites necessary for N_2-fixation (James et al., 1994; Rolfe et al., 1998). Recently, Gopalaswamy et al., (2000) performed microscopic analyses of sections of the junction regions of lateral and primary roots of rice inoculated in the presence of naringenin with ORS 571 ($pXLGD_4$) carrying the *lacZ* reporter gene, and for the first time demonstrated the xylem colonization of rice roots by *A. caulinodans*. In fact, the number of plant species found to exhibit xylem colonization by bacteria without plant symptoms has increased dramatically in recent years, suggesting the xylem colonization is a common aspect of plant – microbe interactions (Kloepper et al., 1992). Since *A. caulinodans* ORS-571 is able to fix nitrogen in the free-living state without differentiation into bacteriods upto 3% oxygen (Kitts and Ludwing, 1994), this first report of the colonization of xylem elements of rice by azorhizobia may be of significance in this respect. The pO_2 in rice roots is likely to be influenced by waterlogging of the rice plant and more extensive invasion will probably be required to have an impact on the nitrogen balance.

Ecological and Economic Aspects

It has to be considered whether under field conditions, these non-legumes show the properties that could be expected on the basis of greenhouse experiments. As it happens in the case of legumes, it is now being realized that these endophytes also aid in improving soil properties (Dobereiner et al., 1995). The crop residues and stubbles are a major route for the transfer of fixed nitrogen to the soil. The decomposition of dead roots and nodules, and a possible excretion of nitrogen have also to be considered (Rinaudo et al., 1983). It is thought that these organisms have an ability to act as pioneers in nitrogen poor soils and subsequently raise the level of soil nitrogen (Jain and Patriquinn, 1984).

In summary, there is reasonable evidence that this endophytic N_2-fixation can attain a magnitude, which is of definite ecological and economic significance. Additionally, it should be noted that these endophytes constitute a group of species, which can tolerate adverse conditions of growth such as low temperature, or waterlogged and acid soils. Thus, the group is in a position to fulfill a role of nitrogen providers under very varied circumstances.

PHOSPHOBACTERIAL BIOFERTILIZER

Phosphate solubilizing bacteria, otherwise called phosphobacteria, play a major role in the solubilization and uptake of native and applied soil phosphorus (Mikanová and Kubát, Chapter 4 in this Volume). The phosphorus requirement of crops is often met by the addition of phosphatic fertilizers, but the efficiency of applied phosphorus used is very low due to its fixation either in the form of aluminium or iron phosphates in acid soils, or in the form of calcium phosphate in neutral or alkaline soils. The introduction of efficient P-solubilizers in the rhizosphere of crops and in soils increases the availability of phosphorus. Phosphobacteria bring about dissolution of bound forms of phosphates in soil, and hence advocated as biofertilizers to crop plants. Several soil bacteria, fungi and rhizosphere microorganisms, notably species of *Bacillus*, *Pseudomonas*, *Aspergillus* and *Penicillium*, secrete organic acids and lower the pH in their vicinity to bring about solubilization of bound phosphates in soils (Sundara Rao and Sinha, 1963; Rodriguez and Fraga, 1999).

Beneficial effects due to inoculation with phosphate solubilizing organisms had been reported for different crops under diverse agro climatic conditions (Gaur, 1990; Mikanová and Kubát, Chapter 4 in this Volume). Phosphobacteria inoculation with rock phosphate has shown increased benefits in terms of crop yield (Rodriguez and Fraga, 1999). Significant increase in grain yield of wheat and paddy was obtained in field inoculations with *Pseudomonas striata* and rock phosphate application to soil (Table 5.12).

POTENTIAL FOR FUTURE RESEARCH

Extending endophytic N_2-fixation to all cereal crops of economic significance may be tried by the following two methods:

Genetic Modification of Host Plant

- To avoid hypersensitive reactions of the host plant.
- To avoid production of phyto-alexins.

Table 5.12. Effect of phosphate solubilizing microorganisms in yield of wheat and paddy

Treatment	Yield (kg ha^{-1})	
	Wheat	Paddy
Control	4975	4866
Bacillus polymyxa - no P	5000	4966
B. polymyxa with MRP*	5000	5300
Pseudomonas striata - no P	5066	5083
P. striata with MRP	5466	5200
Aspergillus awamori - no P	5166	4783
A. awamori with MRP	4966	5000
Super phosphate at 50 kg P_2O_5 ha^{-1}	5500	5083
Super phosphate at 100 kgP_2O_5 ha^{-1}	5633	5600
MRP at 100 kg P_2O_5 ha^{-1}	5250	5166

*MRP – Mussoorie rock phosphate (Gaur, 1990)

- To increase host compatibility to accommodate endophytic N_2-fixers.
- To provide areas of low redox potential for the establishment of endophytes by means of diffusion barriers or leghaemoglobins.
- To make the membrane systems around the endosymbiont facilitate bi-directional transport of substances.
- To study *Nif* gene transfer.

Genetic Modification of the Microorganisms

- To induce production of auxins such as 2,4-D, NAA, etc.
- To produce cellulases and pectinases, which can degrade the cell walls.
- To make the endophytes more tolerant to oxygen concentrations.
- To increase the ability of endophytes to colonize both intra- and inter-cellularly.
- To increase the span of microbial life in vegetative phase.
- To make the microorganisms use the host materials as an energy source but without any phyto-pathogenic effect.
- To make microorganisms produce sufficient amounts of oxyen dismutase.
- Selection of promising strains possessing high nitrogenase activity.
- To screen germplasm for diazotrophic colonization.
- To investigate root morphogenesis in relation to invasion.
- To study in detail the interactive responses between flavonoids and nod-factor.
- To investigate the metabolic programming of cereals for attachment of microorganisms and protection of nitrogenase from oxygen inactivation.

CONCLUSION

The management of native soil nitrogen, which is the principal nitrogen source for rice and biological N_2-fixation, may play a role in sustaining the soil nitrogen under intensive rice cropping. Biological N_2-fixation is an important process in rice farming systems because it is an inexpensive source of nitrogen for increasing the productivity of crops. The N_2-fixing bio-systems such as *Azolla*, cyanobacteria and legume green manures are able to adopt well under wetland rice field ecological conditions and fix considerable amounts of nitrogen. The biomass of these biosystems decomposes rapidly in rice soil and supplies the additional nitrogen. These biofertilizers or biomanures also contribute significant amounts of phosphorus, potassium, sulphur, zinc, iron, molybdenum and other micronutrients. The organic acids released during the mineralization process of the biomass of biomanures accelerate the phosphorus availability in rice soil. These biological fertilizers or biomanures are known to influence the total soil microbial population such as bacteria, fungi, and actinomycetes, and also substantially activate the population dynamics of N_2-fixing free-living bacteria like *Azotobacter*, *Azospirillum* and also the symbiotic bacterium such as *Rhizobium*. It is evidently clear now that the application of biological fertilizers is greatly involved in the accumulation of soil enzymes, which directly reflects on soil fertility index (Fioretto and Fuggio, 1995, Chapter 2, in Volume I of this series). The use of biofertilizers would reduce the cost of chemical fertilizers involved in crop production. The effective utilization of biofertilizers for crops not only provides economic benefits to the farmers, but also improves and maintains the soil fertility and sustainability on natural soil ecosystem. Since these biosources represent a great diversity in chemical, physical and biological characteristics, their efficient use depends upon the particular agro-ecological environment and local availability. Therefore, it is extremely essential to develop a strong, workable and compatible package of nutrient management through organic sources for various crops based on scientific facts, local conditions and economic viability.

REFERENCES

Adebayo, A., Watanabe I, and Ladha J.K. (1989). Epiphytic occurrence of *Azorhizobium caulinodans* and other rhizobia on the host and non-host legumes. Appl. Environ. Microbiol. 55(9): 2407-2409.

Aiyer, R.S. (1965). Comparative algological studies in rice fields of Kerala state. Agric. Res. J. Kerala. India, 3(1): 100-104.

Alazard, D. (1985). Stem and root nodulation in *Aeschynomene* spp. Appl. Environ. Microbiol. 50: 732-734.

Alazard, D. and Becker, M. (1987). *Aeschynomene* as green manure for rice. Plant Soil 101: 141-143.

Alazard, D. (1990). Nitrogen fixation in pure culture by rhizobia isolated from stem nodules of tropical *Aeschynomene* species. FEMS Microbiol. Lett. 68(1-2): 177-182.

Alexander, D., Sadanandan, N. and Karunakaran, K. (1992). Evaluation of *Azolla* as an organic matter for rice production. In: New Trends in Biotechnology (eds.) N.S. Subba Rao, C. Balagopalan and S.V. Ramakrishnan, Oxford & IBH Publ. Co., New Delhi, India, pp. 255-262.

Al-Mallah, M. K., Davey, M.R. and Cocking, E.C. (1989). A new approach to the nodulation of non-legumes by rhizobia and the transformation of cereals by agrobacteria using enzymatic treatment of root hairs. Int. J. Gene Manip. Plants 5: 1-7.

Amer H., Hegazi, N.A. and Nonib, M. (1977). A survey on the occurrence of nitrogen fixing *Spirillum* in Egypt. Non-symbiotic nitrogen fixation Newsletter 5: 101-102.

Amutha, K. and Kannaiyan, S. (1999). Effect of 2,4-D, NAA and inoculation with *Azorhizobium caulinodans* on induction of lateral rootlets and para nodules in rice. Ind. J. Microbiol. 39: 125-128.

Amutha, K. and Kannaiyan, S. (2000). Effect of kinetin and inoculation of *Azorhizobium caulinodans* on induction of lateral rootlets and para nodules in rice. J. Microb. World. 2(1): 1-7.

Anand, N., Hopper, R.S.S. and Jagatheswari, G. (1994). Response of certain blue-green algae (cyanobacteria) in salinity. In: Recent Advances in Phycology (eds.) A.K. Kashyap and H.D. Kumar, Rastogi Publications, Varanasi, India, pp. 125-129.

Apte, S.K. and Thomas, J. (1986). Membrane electrogenesis and sodium transport in filamentous nitrogen fixing cyanobacteria. Eur. J. Biochem. 154: 395-401.

Arora, N. (1954). Morphological development of the root and stem nodules of *Aeschynomene indica* L. Phytomorphology 4: 211-216.

Arsene, F., Katupitiya, S., Kennedy, I.R. and Elmerich, C. (1994). Use of *lac*Z fusions to study the expression of *nif* genes of *Azospirillum brasilense* in association with plants. Mol. Plant Microbe Interact. 7(6): 748-757.

Balasubramani, G. and Kannaiyan, S. (1990). Studies on the effect of fertilizer nitrogen on nodulation and nitrogen fixation and biomass production in *Sesbanai rostrata*. Paper presented in International Network on Soil Fertility and Sustainable Rice Farming, Sub-Network Planning Meeting, Int. Rice Res, Inst., Manila, The Philippines.

Balasubramani, G. and Kannaiyan, S. (1991a). Effect of phosphorus on nodulation and biomass production in *Sesbania rostrata*. INSURF Netwatch 1: 2-3.

Balasubramani, G. and Kannaiyan, S. (1991b). Epiphytic occurrence of *Azorhizobium caulinodans* in leaves and flowers of *Sesbania rostrata*. INSURF Netwatch 1: 4-5.

Balasubramani, G. and Kannaiyan, S. (1993). Influence of water logging condition on stem nodule formation in *Sesbania*. Madras Agric. J., India 80(11): 657-658.

Balasubramani, G., Kumar, K. and Kannaiyan, S. (1992). Studies on nitrogen fixation activity in stem nodulating *Sesbania rostrata*. In: Biological Nitrogen Fixation and Biogas Technology (eds.). S. Kannaiyan, K. Ramasamy, K. Ilamurugu, and K. Kumar, Tamil Nadu Agric. Univ., Coimbatore, India, pp. 84-89.

Baratharkur, H.B. and Talukdar, H. (1983). Use of *Azolla* and commercial nitrogen fertilizer in Jorhat, India. Int. Rice Res. Newslett. 8(2): 20.

Barrios, S. and Gonzales, V. (1971). Rhizobial symbiosis on Venezuelan savannas. Plant Soil 34: 707-719.

Becker, M., Alazard, D. and Ottow, J.C.G. (1986). Mineral nitrogen effect on nodulation and nitrogen fixation of the stem nodulating legume *Aeschynomene afraspera*. Z. Pflanzenernaehr Boden. 149: 485-491.

Becker, M., Pareek, R.P., Ladha, J.K. and Ottow, J.C.G. (1989). Biofertilizer production of stem cut planted and seeded *Sesbania rostrata*. Int. Rice Res. Newslett. 14(2): 30-31.

Becker, M., Ladha, J.K. and Ottow, J.C.G. (1990). Growth and N_2 fixation of two stem nodulating legumes and their effect as green manure on lowland rice. Soil Biol. Biochem. 22: 1109-1119.

Becker, M., Diekmann, K.H., Ladha, J.K., De Datta, S.K. and Ottow, J.C.G. (1991). Effect of N, P, K on growth and nitrogen fixation of *Sesbania rostrata* as green manure for lowland rice (*Oryza sativa* L.). Plant Soil 132: 149-158.

Becker, M. and George, T. (1995). Nitrogen fixation response of stem and root nodulating *Sesbania* species to flooding and mineral nitrogen. Plant Soil 175(2): 189-196.

Bell, C.R., Dickle, G.A., Harvey, W.L.G. and Chan, J.W.Y. (1995). Endophytic bacteria in grapevine. Can. J. Microbiol. 41: 46-53.

Bendra, A.M. and Kumar, S. (1975). Cyanophyceae of Meerut. Phykos 14(1-2): 1-7.

Boddey, R.M., de Oliveira, O.C., Urquiaga, S., Reis, V.M., de Olivares, F.L., Baldani V.L.D. and Dobereiner, J. (1995). Biological nitrogen fixation associated with sugarcane and rice: Contributions and prospects for improvement. Plant Soil 174: 195-209.

Bongale, U.D. and Bharathi, S.G. (1980). On the algal flora of cultivated soils of Karnataka State, India. Phykos 19(1): 95-113.

Boussiba, S. (1997). Ammonia assimilation and its biotechnological aspects in cyanobacteria. In: Cyanobacterial Nitrogen Metabolism and Environmental Biotechnology (ed.) A.K. Rai, Narosa Publishing House, New Delhi, India, pp. 36-62

Brouers, M. and Hall, D.O. (1986). Ammonia and hydrogen production by immobilized cyanobacteria. J. Biotechnol. 3: 307-321.

Brouers, M., Jong, H.D., Shi, D.I. and Hall, D.O. (1988). Immobilized cyanobacteria for sustained ammonia production. In: Algal Biotechnology (eds.). T. Stadler, J. Mollion, M.C. Verdus, Y. Karamanos, H. Morran and D. Christian, Elsevier Applied Science Publishers, Essex, UK pp. 265-275.

Buvana, R. and Kannaiyan S. (2001). Influence of cell wall degrading enzymes on colonization of N_2-fixing bacterium *Azorhizobium caulinodans* in rice. Indian J. Exp. Biol. 40: 369-372.

Buvana, R. and Kannaiyan, S. (2002). Effect of 2,4-D and naringenin on the induction of paranodules by *Azorhizobium caulinodans* in cereals. Indian J. Microbiol. 42: 111-115.

Canny, M.J. (1998). Applications of the compensating pressure theory of water transport. Am. J. Bot. 85: 897-909.

Chaintreuil, C., Prin, J., Lourquin, A.B., Gillis, M. de, Lajudie, P. and Dreyfus, B.L. (2000). Photosynthetic bradyrhizobia are natural endophytes of the African wild rice *Oryza breviligulata* Appl. Environ. Microbiol. 66: 5437-5447.

Chakravarty, S.K., Mishra, A.K. and Chakravarty, P.K. (1986). A note on physiological characteristics and genetic relatedness of a fast growing *Rhizobium* from stem nodules of *Aeschynomene aspera*. L. J. Appl. Bacteriol. 60: 463-468.

Charyulu, P.B.B.N. and Rao, V.R. (1980). Influence of various soil factors on nitrogen fixation by *Azospirillum* spp. Soil Biol. Biochem. 12: 343-346.

Chitra, C. (1992). Studies on seed coating with certain carbon sources on the survival of *Azorhizobium caulinodans* in seeds and foliar inoculation on nodulation, nitrogen fixation and biomass production in *Sesbania rostrata*. M.Sc. (Ag.) Thesis, Tamil Nadu Agric. Univ., Coimbatore, India, pp. 230.

Chitra, C. and Kannaiyan, S. (1995). Application of stem nodulating *Sesbania rostrata* for rice production. In: Rice Management Biotechnology (ed.) S. Kannaiyan, Associate Publishing Co., New Delhi, India, pp. 217-236.

Christiansen-Weniger, G. (1996). Endophytic establishment of *Azorhizobium caulinodans* through auxin-induced root tumors of rice (*Oryza sativa* L.). Biol. Fertil. Soils 21: 293-302.

Cocking, E.C., Al Mallah, M.K., Benson, E. and Davey, M.R. (1990). Nodulation of non legumes by rhizobia In: Nitrogen fixation: Achievements and Objectives (eds.) P.M. Gresshoff, E.C. Roth, G. Stacey and W.E. Newton, Chapman and Hall, New York, USA, pp. 813-823.

Cocking, E.C., Davey, M.R., Kothari, S.L, Srivastava, M., Jing, Y.X., Ridge, R.W. and Rolfe, B.G. (1993). Altering the specificity control of the interaction between rhizobia and plants. Symbiosis. 14: 123-130.

Cocking, E.C., Webster, G., Kothari, S.L., Pawlowski, K., Jones, J., Batchelor, J.P., Jotham, P., Jain, S. and Davey, M.R. (1994). Nodulation and nitrogen fixation in non-legume *Parasponia* species: Dues for non-legume crops. In: Proceedings of the 1st European

Nitrogen Fixation Conference (eds.) G.B. Kiss and G. Endre, Officina Press, Szeged, Hungary, pp. 215-219.

Crozat, I. and Sangchyosawat, C. (1985). Evaluation of different green manures on rice yield in Sonkla Laka Basin, Sonklanadarin. J. Sci. Technol. 7: 391-397.

Davey, M.R., Webster, G., Manders, G., Ringrose, F.L., Power, J.B. and Cocking, E.C. (1997). Effective nodulation of micropropagated shoots of the non-legume *Parasponia andersonii* by *Bradyrhizobium*. J. Exp. Bot. 44: 863-867.

de Bruijn, F.J., Jing, Y. and Dazzo, F.B. (1995). Potential and pitfalls of trying to extend symbiotic interactions of nitrogen-fixing organisms to presently non-nodulated plants, such as rice. Plant Soil 174: 225-240.

De Mooy, J.C., Pesek, J. and Spaldon, E. (1973). Mineral nutrition. In: Soybean Improvement, Production and Uses (ed.) B.E. Caldwell, Am. Soc. Agron. Madison, USA, pp. 267-352.

Diara, H.F., van Brandt, H., Diop, A.M and van Hove, C. (1987). *Azolla* and its use in rice culture in West Africa. In: *Azolla* Utilization, Proc. Workshop on *Azolla* use, Fuzhou, Fujian, China, Int. Rice Res. Inst., Los Banos, Philippines, pp. 147-152.

Dobereiner, J. and Day, J.M. (1976). Associative symbiosis in tropical grasses: Characterization of microorganisms and nitrogen fixing sites. In: Proceedings of the First International Symposium on Nitrogen Fixation (ed.) W.E. Newton, Washington State University Press, Pullman, USA, pp. 518-536.

Dobereiner, J., Baldani, V.L.D. and Reis, V.M. (1995). Endophytic occurrence of diazotrophic bacteria in non-leguminous crops. In: NATO ASI Series, Vol. G37 (*Azospirillum* and related microorganisms) (eds.) I. Fedrik, M. del Gallo, J. Vanderleyden and M. de Zamoroczy, Springer Verlag, Berlin, Heidelberg, Germany, pp. 3-14.

Dreyfus, B.L. and Dommergues. Y.R. (1981). Stem nodules on the tropical legume *Sesbania rostrata*. In: Current Perspective in Nitrogen Fixation (ed.) A.H. Gibson, Aust. Acad. Sci. Canberra, Australia, p. 615.

Dreyfus, B.L., Rinaudo, G. and Dommergues, Y. (1983). Use of *Sesbania rostrata* as a green manure in paddy fields. ORSTOM, Dakar, Senegal, pp. 18-19.

Dreyfus, B.L., Alazard, D. and Dommergues, Y.R. (1984). Stem nodulating rhizobia In: Current Perspectives of Microbial Ecology (eds.) M.G. Klug and C.E. Reddy, American Society of Microbiologists, Washington DC, USA, pp.161-169.

Dreyfus, B.L., Rinando, G. and Dommergues, Y.R. (1985). Observations on the use of *Sesbania rostrata* as green manure in paddy fields. Mircen. J. Microbiol. 1: 111-121.

Dreyfus, B.L., Garcia, J.L and Gillis, M. (1988). Characterization of *Azorhizobium caulinodans*, a stem, nodulating nitrogen fixing bacterium isolated from *Sesbania rostrata*. Int. J. Syst. Bacteriol. 38: 89-98.

Eaglesham, A.R.J. and Szalay, A.A. (1983). Aerial stem nodules on *Aeschynomene* sp. Plant Sci. Lett. 29: 265-272.

Emmerling, C. (2005). Soil microbial biomass, and activity and the lasting impact on agricultural deintensification. In: Microbial Biotechnology in Agriculture and Aquaculture, Volume I (ed.) R.C. Ray, Science Publishers, Enfield, New Hampshire, USA, pp. 239-260.

Erber, W.W.A., Nitschmann, W.H., Munchl, R. and Peschek, G.A. (1986). Endogenous energy supply to the plasma membrane of dark aerobic cyanobacterium, *Anacystis nidulans* ATPase-independent efflux of H^+ and Na^+ from respiring cells. Arch. Biochem. Biophys. 247: 28-39.

Erdmann, N., Fulda, S. and Hagemann, M. (1992). Glucosylglycerol accumulation during salt accumulation of two unicellular cyanobacteria. J. Gen. Microbiol. 138: 363-368.

Fioretto, A. and Fuggio, A. (2005). Biotechnology of soil enzymes. In: Microbial Biotechnology in Agriculture and Aquaculture, Volume I (ed.) R.C. Ray, Science Publishers, Volume I, Enfield, New Hampshire, USA, pp. 31-70.

Flowers, T.J., Troke, P.K. and Yeo, A.R. (1977). The mechanism of salt tolerance in halophytes. Annu. Rev. Plant Physiol. 28: 89-121.

Fry, I.V., Huflejt, M.A., Erber, W.W., Peschek, G.A.and Packer, L. (1986). The role of respiration during adaptation of the fresh water cyanobacterium, *Synechococcus*-6311 to salinity. Arch. Biochem. Biophys. 244: 686-691.

Gagne, S., Richard, C., Rosseau, H. and Antoun, H. (1987). Xylem-residing bacteria in alfalfa roots. Can. J. Microbiol. 33: 996-1000.

Garg, S.K. and Bhatnagar, A. (2005). Use of microbial biofertilizers for sustainable aquaculture/fish culture. In: Microbial Biotechnology in Agriculture and Aquaculture, Volume I (ed.) R.C. Ray, Science Publishers, Enfield, New Hampshire, USA, pp. 261-278.

Gaur, A.C. (1990). Phosphate solubilizing microorganisms as biofertilizers. Omega Scientific Publishers, New Delhi, India, pp. 176.

Gebhardt, C., Turner, C.L., Gibson, A.H., Dreyfus, B.L. and Bergerson, F.J. (1984). Nitrogen fixing growth in continuous culture of a strain of *Rhizobium* sp. isolated from stem nodule of *S. rostrata*. J. Gen. Microbiol., 130: 843-848.

Geelen, D., Goethals, K., Montagu, M.V. and Holsters, M. (1995). The *nod*D locus from *Azorhizobium caulinodans* is flanked by two repetitive elements. Gene 164(l): 107-111.

Ghai, S.K., Rao, D.L.N. and Batra, L. (1988). Nitrogen contribution to wetland rice by green manuring with *Sesbania* sp. in an alkaline soil. Biol. Fertil. Soils 6: 22-25.

Goethals, K., Gao, M., Tomekpe, K., van Montagu, M. and Holsters, M. (1989). Common *nod*ABC genes in *Nod* locus 1 of *Azorhizobium caulinodans*: Nucleotide sequence and plant-inducible expression. Mol. Gen. Genet. 219: 289-298.

Gopalaswamy, G., Vidhyasekaran, P. and Chelliah, S. (1989). Effect of *Azospirillum lipoferum* inoculation and inorganic nitrogen on wetland rice. Oryza 26: 378-380.

Gopalaswamy, G., Anthoni Raj, S. and Abdul Kareem, A. (1994). Interaction of herbicides with *Azolla* and soil microbes. Indian. J. Weed Sci. 26 (3& 4): 28-34.

Gopalaswamy, G. and Kannaiyan, S. (1996). Effect of growth regulator GA_3 on *Azolla*. Phykos 35 (1&2): 183-188.

Gopalaswamy, G. and Kannaiyan S. (2000). Changes in soil microbial and enzyme activities during decomposition of biomass of *Azolla* hybrids. J. Microb. World 2(2): 1-8.

Gopalaswamy, G., Kannaiyan, S., O' Callagham K.J., Davey M.R. and Cocking E.C. (2000). The xylem of rice (*Oryza sativa*) is colonized by *Azorhizobium caulinodans*. Proc. Royal Soc. Lond. 267: 103-107.

Gopalaswamy, G., Tamil Selvam, B. and Kannaiyan, S. (2002). Biomass production and biochemical composition of acid tolerant cyanobacteria. Indian J. Microbiol. 42: 121-124.

Gough, C., Webster, G., Vasse, J., Galera, C., Batchelor, C., O' Callaghan, K.J. and Davey, M.R. (1996). Specific flavonoids stimulate intercellular colonization of non-legumes by *Azorhizobium caulinodans*. In: Biology of Plant Microbe Interactions (eds.) G. Stacey, B. Mulin and P.M. Gresshoff. Proc. 8th Int. Cong. Molec. Plant Microbe Interactions, 14-19 July, 1996, Knoxville, USA, pp. 409-415.

Govindan, M. and Purushothaman, D. (1985). Association of nitrogen fixing bacteria with certain plantation crops. Natl. Acad. Sci. Lett. India 8: 1631-1665.

Goyal S.K. and Venkataraman, G.S. (1971). Response of high yielding rice varieties to algalization. II. Interaction of soil types with algal inoculation. Phykos 10(1-2): 32-33.

Goyal, S.K. (1982). Blue-green algae and rice cultivation. Proc. Natl. Sym. Biol. Nitrogen Fixation, Indian Agric. Res. Inst., New Delhi, India, pp. 346-376.

Hagerup, O. (1928). En hygrotil Bael-g-plante (*Aeschynomene aspera* L.) med bakterieknolde paa staengelen. Dan Bot. 5: 19.

Hall, D.O., Kannaiyan, S. and Vander Leij, M. (1997). Ammonia biofertilizer production in rice paddies using immobilized cyanobacteria. In: Asian Paddy Fields (eds.) M. Oshirra, Y.E. Spratt and J.W.B. Stewart, Univ. Saskatchewan, Canada, p. 218.

Hallman, J., Quadt, A.M.W.E. and Klopper, J.W. (1997). Bacterial endophytic in agricultural crops. Can. J. Microbiol. 43: 895-914.

Hungria, M., Eaglesham, A.R.J. and Hardy, R.W.F. (1992). Physiological comparisons of root and stem nodules of *Aeschynomene scabra* and *Sesbania rostrata*. Plant Soil 139(l): 7-13.

Iyer, V., Fernandes, T. and Apte, S.K. (1994). A role for osmotic stress induced proteins in the osmotolerance of nitrogen fixing cyanobacterium *Anabaena* sp. strain L-31. J. Bacteriol. 176(18): 5868-5870.

Jain, D. and Patriquin, D. (1984). Root hair deformation, bacterial attachment and plant growth in wheat – *Azospirillum* interaction. Appl. Environ. Microbiol. 48: 1208-1213.

James, E.K., Reis, V.M., Olivares, F.L., Baldani, J.I. and Dobereiner, J. (1994). Infection of sugarcane by the nitrogen fixing bacterium *Acetobacter diazotrophicus*. J. Exp. Bot. 45: 757-766.

James, E.K., Fannetta, P.P.M., Nixon, P.J., Whiston, A.J., Peat, L., Crawford, R.M.M., Sprent, J.I. and Brewin, N.J. (1996). Photosystem II and oxygen regulation in *Sesbania rostrata* stem nodules. Plant Cell Environ. 19(8): 895-910.

Jayanthi, M., Rajagopal, V. and Shantaram, M.V. (1994). Studies on conjunctive use of *Azolla* and fertilizer in rice cultivation. Natl. Seminar on *Azolla* and Algal Biofertilizers for Rice, Tamil Nadu Agricultural University, Coimbatore, India, (abstr.), p. 23.

Jenik, J. and Kubikova, J. (1969). Root systems of tropical trees. Peslia, Prague, Czech. Rep. 14: 59-65.

Jha, M.N. and Kaushik, B.D. (1988). Response of *Westiellopsis prolifica* and *Anabaena* sp. to salt stress II. Uptake of Na^+ in the presence of K^+ as chloride, nitrate and phosphate. Curr. Sci. 57: 667-668.

Jurgensen, M.F. and Davey, C.B. (1968). Nitrogen fixing blue green algae in acid forest and nursery soils. Can. J. Microbiol. 14: 1179-1183.

Kalidurai, M. and Kannaiyan, S. (1988). Studies on biomass production, nodulation and nitrogen fixation in stem nodulating *Sesbania rostrata*. In: Int. Symp. Biological Nitrogen Fixation in Rice Soil. Central Rice Res. Inst., Cuttack, India, (abstr.) p. 18.

Kalidurai, M. and Kannaiyan, S. (1991). *Sesbania* as a biofertilizer for rice. Biores. Technol. 36: 141-145.

Kannaiyan, S. (1978). A note on *Azolla* and blue-green algae. Special Bull., Dy. Director Agric., Chennai, India, p. 18.

Kannaiyan, S. (1981). *Azolla* biofertilizer for rice. Int. Rice Res. Inst., Los Banos, Manila, Philippines, pp. 1-11.

Kannaiyan, S., Thangaraju, M. and Oblisami, G. (1981). Mineral nutrition of *Azolla*. Tamil Nadu Agric. Univ. Newslett., India, 8: 2.

Kannaiyan, S., Thangaraju, M. and Oblisami, G. (1982). Studies on the multiplication and utilization of *Azolla* biofertilizer for rice. In: Proc. National Symp. Biological Nitrogen Fixation, Indian Agri. Res. Inst., New Delhi, India, pp. 451-460.

Kannaiyan, S., Thangaraju, M. and Oblisami, G. (1984). Effect of *Azolla* application on grain yield of rice. Agric Res. J. Kerala, India, 22: 183-185.

Kannaiyan, S. (1985). Potentiality of *Azolla* biofertilizer for rice. In: Soil Biology (eds.) M.M. Mishra and K.K. Kapoor, Haryana Agric. Univ., Hisar, India, pp. 259-265.

Kannaiyan, S. (1987). Sheath blight disease of rice. In: Advances in Rice Pathology (ed.) S. Kannaiyan, Tamil Nadu Agri. Univ., Coimbatore, India, pp. 210-233.

Kannaiyan, S. and Somporn, C. (1987). Studies on sporulation, biomass production and nitrogen fixing potential of eleven cultures of the aquatic fern *Azolla*. Indian J. Microbiol. 27: 22-25.

Kannaiyan, S. (1990a). Biotechnology of Biofertilizer for Rice Crop. Tamil Nadu Agric. Univ., Coimbatore, India, p. 212.

Kannaiyan, S. (1990b). *Azolla-Anabaena* complex as biofertilizer for rice. Proc. Current Trends in Biotechnol., Cochin Univ. Sci. Technol., Cochin, India, pp. 16-27.

Kannaiyan, S. (1990c). Blue green algal biofertilizers. In: Biotechnology of Biofertilizers For Rice Crop. (ed.) S. Kannaiyan, Tamil Nadu Agric. Univ. Publication, Coimbatore, India, p. 225.

Kannaiyan, S. (1992a). Potentiality of *Azolla* as biofertilzer in sustainable rice farming. In: Biodiversity Implications for Global Food Security (eds.) M.S. Swaminathan and S. Jana, MS Swaminathan Res. Foundation, Chennai, India, pp. 221-233.

Kannaiyan, S. (1992b). *Azolla* Biofertilizer Technology for Rice Crop. Tamil Nadu Agric. Univ., Coimbatore, India, p. 54.

Kannaiyan, S. (1993a). Biofertilizers For Rice. Tamil Nadu Agric. Univ., Coimbatore, India, p. 60.

Kannaiyan. S. (1993b). Nitrogen contribution by *Azolla* to rice crop. Proc. Indian Natl. Sci. Acad. 59(5&6): 21-31.

Kannaiyan, S. (1994). Ammonia production by immobilized cyanobacteria for its use as biofertilizer for rice crop. Final Technical Report, Tamil Nadu Agric. Univ., Coimbatore, India.

Kannaiyan, S., Rao, K.K. and Hall, D.O. (1994). Immobilization of *Anabaena azollae* from *Azolla filiculoides* in polyvinyl foam for ammonia production in photobioreactor system. World J. Microbiol. Biotechnol. 10: 55-58.

Kannaiyan, S. (1995). Biofertilizer for rice: status and scope. In: Rice Management Biotechnology (ed.) S. Kannaiyan, Associate Publishing Co. New Delhi, India, pp. 237-265.

Kannaiyan, S., Aruna, S.J., Merina, P.K.S. and Hall, D.O. (1997). Immobilized cyanobacteria as a biofertilizer for rice crops. J.Appl. Phycol. 9: 167-174.

Kannaiyan, S. (2000). Biofertilizer Technology and Quality Control. Publication Directorate, Tamil Nadu Agric. Univ., Coimbatore, India, pp. 256.

Kannaiyan, S., Kumar, K. and Gopalaswamy, G. (2001). Stem nodulating *Sesbania rostrata* as a bioresource for rice production, induction of nodulation and endophytic N_2 fixation in rice. Madras. Agric. J. India 88(10-12): 547-560.

Kannaiyan, S. (2003). Inoculant production in developing countries: Problems, potentials and success. In: Maximising the Use of Biological Nitrogen Fixation in Agriculture (eds.) G. Hardarson and W.J. Broughton, Kluwer Academic Publishers, Dordrecht, Netherlands, pp. 187-198.

Kattupitiya, S., New, P.B., Elmerich, C. and Kennedy, I.R. (1995a). Improved nitrogen fixation in 2,4-D treated wheat roots associated with *Azospirillun lipoferum*: Colonization using reporter genes. Soil Biol. Biochem. 27: 447-452.

Kattupitiya, S., Millet, J., Vesk, M., Viccars, L., Seman, A., Zhao, L., Elmerich, C. and Kennedy, I.R. (1995b). A mutant of *Azospirillum brasilense* Sp7 impaired in flocculation with modified colonization and superior nitrogen fixation in association with wheat. Appl. Environ. Microbiol. 61: 1987-1995.

Kennedy, I.R. and Tchan, Y.T. (1992). Biological nitrogen fixation in non-leguminous field crops: Recent advances. Plant Soil 141: 93-118.

Kennedy, I.R. (1994). Auxin–induced nitrogen–fixing associations between *Azospirillum brasilense* and wheat. In: Nitrogen Fixation with Non-Legumes (eds.) N.A. Hegazi, M. Fayez and M. Monib, Cairo Press, Cairo, Egypt, pp. 513-523.

Kennedy, I.R. (1997). The establishment of new non-nodular endophytic systems. In: Biological Nitrogen Fixation: The Global Challenge and Future Needs (eds.) I.R. Kennedy and E.C. Cocking. The Rockefeller Foundation, Bellagio Conference Centre, Lake Como, Italy, April 8-12, 1997, pp. 65-67.

Kennedy, I.R., Lily, L., Pereg-Gerk, C., Wood, R., Deaker, K. and Katupitiya, S. (1997). Biological nitrogen fixation in non-leguminous field crops: Facilitating the evolution of an effective association between *Azospirillum* and wheat. Plant Soil 194: 65-79.

Kerby, N.W., Musgrave, S.C., Shestakov, S.V. and Stewart, W.D.P. (1986). Photoproduction of ammonium by immobilized mutants strains of *Anabaena variabilis*. Appl. Microbiol. Biotechnol. 24: 42-45.

Kijne, J.W. (1992). The *Rhizobium* infection process. In: Biological Nitrogen Fixation (eds.) G.Stacey, R. H. Burris and H.J. Evans, Chapman and Hall, New York, USA, pp. 349-398.

Kitts, C.L. and Ludwing, R.A. (1994). *Azorhizobium caulinodans* respires with at least four terminal oxidases. J. Bacteriol. 176: 886-895.

Kloepper, J.W., Schippers, B. and Bakker, P.A.H.M. (1992). Proposed elimination of the term endorhizosphere. Phytopathology 82: 726-727.

Kulasooriya, S.A., Hirimburegama, S.A. and Abeysekava, S.W. (1987). Use of *Azolla* in Sri Lanka. In: *Azolla* Utilization. Proc. Workshop on *Azolla* Use, Fuzhou, Fujian, China, Int. Rice Res. Inst., Los Banos, Philippines, p. 140.

Kumar, K. (1994). Studies on the development of *Azolla* mutants and their nitrogen fixing potential, microbial decomposition and biofertilizer value for rice. Ph.D thesis, Tamil Nadu Agric. Univ., Coimbatore, India, pp. 323.

Kumar, K., Thangaraju, M. and Kannaiyan, S. (1995). Use of nitrification inhibitors and nitrogen use efficiency in lowland rice soil. In: Rice Management Biotechnology (ed.) S. Kannaiyan, Associate Publishing Co., New Delhi, India, pp. 101-106.

Kumar, K., Kannaiyan, S. and Cocking, E.C. (1996). Studies on the colonization of *Azorhizobium caulinodans* in rice roots and their role in symbiotic nitrogen fixation. Paper presented in the 37th Annual conference of the Association of Microbiologists of India, IIT, Chennai, India, 4-6, December, 1996, (Abstr.) p. 118.

Kumar, K. and Kannaiyan. S. (2000). Selection of mutants of *Azolla* for higher nitrogen potential, ammonia assimilation and nutrient content. J. Microb. World 2(1): 35-41.

Kwon, D.K. and Beevers H. (1993). Adverse effects of nitrate on stem nodules of *Sesbania rostrata* Brem. New phytol. 125(2): 345-350.

Ladha, J.K. Pareek, R.P. and Becker, M. (1992). Stem-nodulating legume-*Rhizobium* symbiosis and its agronomic use in lowland rice. Adv. Soil Sci. 20: 147-192.

Ladha, J.K., Garcia, M., Miyan, S., Padre, A.T. and Watanabe, I. (1998). Survival of *Azorhizobium caulinodans* in the soil and rhizosphere of wetland rice under *Sesbania rostrata*. Appl. Environ. Microbiol. 55: 454-460.

Lakshmikumari, M., Kavimandan, S.K. and Subba Rao, N.S. (1976). Occurrence of nitrogen fixing *Spirillum* in roots of rice, sorghum, maize and other plants. Indian J. Expt. Biol. 14: 638-639.

Latha, K.R., Kannaiyan, S. and Subramanian, S. (1988). Uptake of nitrogen in different growth phases of rice through *Azolla* biofertilizer. Acta Agron. Hungarica 37(3-4): 239-243.

Legocki, R.P., Eaglesham, A.R.J. and Szalay, A.A. (1983). Stem nodulation in *Aeschynomene*: A model system for bacterium- plant interaction. In: Molecular Genetics of the Bacteria-Plant Interaction (ed.) A. Puhler, Springer Verlag, Berlin, Heidelberg, Germany, pp. 210-219.

Liu Chung Chu. (1987). Re-evaluation of *Azolla* utilization in agricultural production. In: *Azolla* Utilization. Proc. on the Workshop on *Azolla* use. Fuzhou, Fujian, China, Int. Rice Res. Instt., Los Banos, Manila, Philippines, pp. 67-76.

Long, S.R. (1996). *Rhizobium* symbiosis- Nod factors in perspective. The Plant Cell 8: 1885-1898.

Loudhapasitiporn, L. and Kanareugsa. C. (1987). *Azolla* use in Thailand. In: *Azolla* Utilization Proc. Workshop on *Azolla* Use, Fuzhou, Fujian, China, Int. Rice Res. Instt., Los Banos, Manila, Philippines, pp. 119-122.

Mabbayad, B.B. (1987). The *Azolla* programme of the Philippines. In: *Azolla* Utilization. Proc. Workshop *Azolla* Use, Fuzhou, Fujian, China, Int. Rice Res. Instt., Los Banos, Manila, Philippines, pp. 101-108.

Madhusoodanan, P.V. and Dominic, T.K. (1995). On the acid tolerance of certain strains of cyanobacteria from Kerala. Natl. Symp. Algal Biofertilizer, Tamil Nadu Agric. Univ. Coimbatore, India, (abstr.) p. 19.

Magalhaes, F.M., Baldani, J.I., Santo, J. Kuykendall, J.R. and Dobereiner, J. (1983) A new acid tolerant *Azospirillum* species. Ann. Acad. Brar. Clienc. 55: 417-430.

Mahesh, G. and Kannaiyan, S. (1994). Influence of certain growth regulators on the germination and fertilization of sporocarps *Azolla microphylla*. Paper presented in the National Seminar on *Azolla* and Algal Biofertilizer for Rice. Tamil Nadu Agric. Univ., Coimbatore, India, 1-3, March, 1994 (abstr.) p. 16.

Meelu, O.P. and Morris. R.A. (1988). Green manure management in rice-based cropping systems. In: Sustainable Agriculture: Green Manure Rice Farming. Int. Rice Res. Inst., Los Banos, Manila, Philippines, pp. 209-222.

Militor, V., Erber, W.W.A. and Peschek, G.A. (1986). Increased levels of cytochrome C oxidase and sodium antiporter in the plasma membrane of *Anaystis nidulans* after growth in sodium enriched media. FEMS Microbiol. Lett. 204: 151-255.

Miller, D.M., Johnes, J.V. Yopp, J.H. Tindall, D.R. and Schmid, W.D. (1976). Ion metabolism in a halophilic blue green alga *Aphanothece halophytica*. Arch. Microbiol. 111: 145-149.

Mitra, A.K. (1951). The algal flora of certain Indian soils. Indian J. Agric. Sci. 21: 375.

Moore, A.W. (1963). Occurrence of non-symbiotic nitrogen fixing microorganisms in nitrogen soils. Plant Soil. 19: 385-395.

Morris, R.A., Furoc, R.E., Rajbandari, R., Marqueses E.P. and Dizon, M.A. (1987). Nitrogen accumulation in water logging tolerant green manure. Agron. J. 79(5): 937-945.

Ndoye, I. and Dreyfus, B.L. (1988). N_2 fixation by *Sesbania rostrata* and *Sesbania sesban* estimated using ^{15}N and total N difference methods. Soil Biol. Biochem. 20: 209-213.

Ndoye, I., De Billy, F., Vasse, J., Dreyfus, B. and Trucjet, G. (1994). Root nodulation of *Sesbania rostrata*. J. Bacteriol. 176: 106-1068.

Neyra, C.A. and Dobereiner, J. (1977). Nitrogen fixation in grasses. Adv. Agron. 29: 1-38.

Nie, Y.E. (1983). Studies on induced nodulation in plants lacking root nodules. Nature. 6: 326-360.

O' Callaghan, K.J., Davey, M.R. and Cocking, E.C. (1997). Xylem colonization of the legume *Sesbania rostrata* by *Azorhizobium caulinodans*. Proc. Royal. Soc. Lond. B 264: 1821-1826.

O'Hara, G.W., Boonkerd, N. and Dilworth, M.J. (1988). Mineral constraints to nitrogen fixation. Plant Soil, 108: 93-110.

Okon, Y. (1985). *Azospirillum* as a potential inoculant for agriculture. TIBTECH. 3(9): 223-228.

Okon, Y. and Labandera-Gonzalez, C.A. (1994). Agronomic applications of *Azospirillum*: An evaluation of 20 years worldwide field inoculation. Soil Biol. Biochem. 26: 1591-1601.

Pandey. D.C. (1965). A study of the algae from paddy soils of Ballia and Ghazipur districts of Uttar Pradesh, India. 1. Cultural and ecological considerations. Nova. Hedwigia 9: 299-334.

Pareek, R.P., Ladha, J.K. and Watanabe, I. (1990). Estimating N_2 fixation by *Sesbania rostrata* and *S. cannabina* (syn *S. aculeata*) in lowland rice soil by ^{15}N dilution method. Biol. Fertil. Soils 10: 77-88.

Patel, L.B., Sidhu, B.S. and Beri, V. (1996). Symbiotic efficiency of *Sesbania rostrata* and *S. cannabina* as affected by agronomic practices. Biol. Fertil. Soils 21(3): 149-151.

Patil, P.L. and Mandhare V.K. (1991). Effect of direct incorporation of *Azolla* with nitrogen levels on yield of rice. 31st Annu. Conf. Assoc. Microbiol. India, Tamil Nadu Agric. Univ., Coimbatore, India, (abstr.), p. 135.

Paul, J.V. and Kannaiyan, S. (1994). Effect of the herbicide Butachlor on ammonia excretion by the colchicine mutants of *Anabaena azollae* (AS-DS). Paper presented in the National Seminar on *Azolla* and Algal Biofertilizer for Rice. Tamil Nadu Agric. Univ., Coimbatore, India, 1-3, March, 1994, (abstr.), p. 3.

Prasad, B.N., Mehrotra, R.K. and Singh, Y. (1978). On pH tolerance of some soil blue-green algae. Acta Bot. Indica 6(2): 130-138.

Purushothaman, D. and Govindarajan, K. (1984). Rhizosphere ecology and nitrogen fixation of *Azospirillum* in pearl millet (*Pennisetum americanum* (L.) Leeke). In: Working Group Meeting on Cereal Nitrogen Fixation. 9-12, October, 1984, Int. Crop. Res. Inst. Semi-Arid Tropics, Patencheru, Andhra Pradesh, India, (abstr.), p. 16.

Purushothaman, D., Gunasekaran, S. and Oblisami, G. (1980). Nitrogen fixation by *Azospirillum* in some tropical plants. Proc. Ind. Natl. Sci. Acad. 46: 713-717.

Purushothaman, D., Jayanthamurali, K. and Gunasekaran, S. (1987). Association of *Azospirillum* with the roots of upland rice. In: Biofertilizers: Potentials and Problems. (eds.) S.P. Sen and P. Palit. Naya Prakasan, Calcutta, India, pp. 171-182.

Quispel, A. (1974). The endophyte of root nodules in non-leguminous plants. In: The Biology of Nitrogen Fixation (ed.) A. Quispel, North – Holland Publishing Company, Amsterdam, The Netherlands, pp. 520-599.

Rai, A.K. and Abraham, G. (1993). Salinity tolerance and growth analysis of the cyanobacterium, *Anabaena doliolum*. Bull. Environ. Cont. and Toxicol. 51(5): 724-731.

Rajagopal, V., Jayanthi, M. and Shantaram, M.V. (1994). Growth studies on *Azolla* under Telangana conditions of Andra Pradesh, India. Natl. Seminar on *Azolla* and Algal Biofertilizer for Rice, Tamil Nadu Agric. Univ., Coimbatore, India, (abstr.), p. 15.

Rao, V.R., Charyulu, P.B.B.N., Nayak, D.N. and Ramakrishna, C. (1978). Nitrogen fixation by free-living organisms in tropical rice soils. In: Limitations and Potentials for Biological Nitrogen Fixation in the Tropics. (ed.). J. Dobereiner, R.H. Burris and A.L. Hollander, Plenum Press, New York, USA, pp. 354-368.

Reddy, P.M., Ladha, J.K., So, R.B., Hernandez, R., Ramos, M.C., Angeles, O.R., Dazzo, E.B. and de Bruijn, E.J. (1997). Rhizobial communication with rice roots: Induction of phenotypic changes, mode of invasion and extent of colonization. Plant Soil 194: 81-98.

Reed, R.H., Borowitzka L.J., Mackay M.A., Foster J.A., Warr S.R.C., Moore D.J. and Stewart W.D.P. (1986). Organic solute accumulation in osmotically stressed cyanobacteria. FEMS Microbiol. Lett. 2: 83-86.

Reed, R.H. and Stewart, W.D.P. (1988). The responses of cyanobacteria to salt stress. In: Biochemistry of the Algae and Cyanobacteria (eds.) L.J. Rogers and J.R. Gallon, Clarendon Press, Oxford, UK, pp. 217-237.

Reinhold-Hurek, B., Hurek, T., Fendrik, I., Pot, B., Gillis, M., Kersters, K., Thielemans, S. and DeLey, J. (1987). *Azospirillum halopraeferens* sp. nov., a nitrogen-fixing organism associated with roots of kallar grass (*Leptochola fusca*) (L.) kunth.). Int. J. Syst. Bacteriol. 37: 43-51.

Reynders, L. and Vlassak, K. (1976). Nitrogen fixing *Spirillum* species in Belgium soils. Agriculture (Belgium) 24: 329-336.

Rinaudo, G., Dreyfus, B.L. and Dommergues, Y.R. (1982). *Sesbania rostrata* as a green manure for rice in West Africa. BNF Technol. Trop. Agric. 441-445.

Rinaudo, G., Dreyfus, B.L. and Dommergues, Y. (1983). Influence of *Sesbania rostrata* green manure on the nitrogen content of rice crop and soil. Soil Biol. Biochem. 15: 111-113.

Rinaudo, G., Orenga, S., Fernandez, M.P., Meugnier, H. and Bardin, R.A.D. (1991). DNA homologies among members of the genus *Azorhizobium* and other stem and root nodulating bacteria isolated from the tropical legume *Sesbania rostrata*. Int. J. Syst. Bacteriol. 41(l): 114-120.

Robertson, B.K. and Alexander, M. (1994). Mode of dispersal of stem nodulating bacterium. Soil Biol. Biochem. 26(11): 1535-1540.

Robertson, B.K., Dreyfus, B.L. and Alexander, M. (1995). Ecology of stem nodulating *Rhizobium* and *Azorhizobium* in four vegetation zones of Senegal. Microb. Ecol. 29: 71-81.

Rodriguez, H. and Fraga, R. (1999). Phosphate solubilizing bacteria and their role in plant growth promotion. Biotechnol. Adv. 17 : 319-339.

Roger, R.A. and Watanabe, I. (1986). Technologies for using biological nitrogen fixation in wetland rice: Potentialities, current stage and limiting factors. Fertil. Res. 9: 39-77.

Rolfe, B.G. et al. (1998). Round table: Agriculture 2020: Eight billion people. In: Current Plant Science and Biotechnology in Agriculture: Biological Nitrogen Fixation for the 21st Century. (eds.) C. Elmerich, A. Kondorosi and W.E. Newton, Kluwer Academic Publishers, Dordrecht, The Netherlands, pp. 685-692.

Sabry, S.R.S., Saleh, S.A., Batchelor, A. Jones, J., Jotham, J., Webster, G., Kothari, S.L., Devey, M.R. and Cocking, E.C. (1997). Endophytic establishment of *Azohizobium caulinodans* in wheat. Proc. Soc. Lond. B 264: 341-346.

Sadasivam, S., Lakshmi, P.H. and Kannaiyan, S. (1993). Phosphoenolpyruvate carboxylase activity, fixation of ^{14}C in amino acids and nitrogen transport in stem nodules of *Sesbania rostrata*. Biol. Plant. 35(2): 303-306.

Saraswati, R., Matoh, T. and Sekiva, J. (1992). Nitrogen fixation of *Sesbania rostrata*: contribution of stem nodules to nitrogen acquisition. Soil Sci. Plant Nutr. 38(4): 775-780.

Sardeshpandey, J.S. and Goyal, S.K. (1981a). Distributional pattern of blue green algae in rice field soils of Konkan region of Maharastra State. Phykos 20 (1&2): 102-106.

Sardeshpandey, J.S. and Goyal, S.K. (1981b). Effect of pH on growth and nitrogen fixation by blue green algae. Phykos 20(1-2): 107-113.

Saxena, S. and Kaushik, B.D. (1991). Polysaccharides (biopolymers) from halotolerant cyanobacteria. Indian J. Exp. Biol. 30: 433-434.

Schaede, R. (1940). Die Knollchen der adventiven Wasser Wurzeln Von *Neptunia oleracae* and ihre Bakteriensymbiose. Planta 31: 1-21.

Schubert, H., Fulda, S. and Hagemann, M. (1993). Effects of adaptation of different salt concentration on photosynthesis and pigmentation of the cyanobacterium, *Synechocystis* sp. PCC 6803. J. Plant Physiol. 142: 291-295.

Selvakumar, G., Gopalaswamy, G. and Kannaiyan, S. (2002). Pigment analysis and ammonia excretion in herbicide tolerant cyanobacterial. Indian J. Exp. Biol. 40: 934-940.

Senthil Kumar, M. (2000). Studies on the induction of para nodules by the stem nodulating *Azorhizobium caulinodans* and its nitrogen fixing activity in rice and maize. M.Sc. (Ag.) thesis, Tamil Nadu Agricultural University, Coimbatore, India, pp. 73-92.

Shanmugasundaram, R. and Kannaiyan, S. (1992a). Studies on the sporocarp germination of *Azolla microphylla* and isolation of algal symbionts from megasporocarp. In: Biological Nitrogen Fixation and Biogas Technology (eds.) S. Kannaiyan, K. Ramasamy, K. Kumar, K. Ilamurugu, Tamil Nadu Agrl. Univ. Coimbatore, India, pp. 44-48.

Shanmugasundaram, R. and Kannaiyan, S. (1992b). Effect of growth regulator and fungicide combination on germination, fertilization and viability of the dried sporocarps of *Azolla microphylla*. Phykos 31: 61-67.

Shanmugasundaram, R. and Kannaiyan, S. (1994). Studies on the development of *Azolla* and algal biofertilizer for rice. Tamil Nadu Agric. Univ., Coimbatore, India, (abstr.), p. 17.

Shi, D.J., Brouers, M. Hall, D.O. and Robins, R.J. (1987a). The effects of immobilization on the biochemical, physiological and morphological features of *Anabaena azollae*. Planta 172: 298-308.

Shi, D.J., Hall, D.O. and Tang, P.S. (1987b). Photosynthesis, nitrogen fixation, ammonia photoproduction and structure of *Anabaena azollae* immobilized in natural and artificial systems. In: Progress in Photosynthesis Research, Vol.II (ed.) J. Biggens, Martinus Nijhoff Publishers, Dordrecht, The Netherlands, pp. 641-644.

Singh, A.L. and Singh, P.K. (1989). N_2 Fixation in Indian Rice Field (*Azolla* and Blue green algae). Agrobotanical Publisher, Bikaner, India, pp. 236.

Singh, N.I., Dorycanta, H., Devi, G.A. Singh, N.S. and Singh, M.S. (1997). Blue green algae from rice field soils of Nagaland. Phykos 36(182): 115-120.

Singh, P.K. (1977a). Multiplication and utilization of fern *Azolla* containing nitrogen fixing algal symbiont, a green manure in rice cultivation. Il Riso 26: 125-137.

Singh, P.K. (1977b). The use of *Azolla pinnata* as green manure for rice. Int. Rice Res. Newsletter 2(7): 7.

Singh, P.K. (1979). Use of *Azolla* in rice production in India. In: Nitrogen and Rice, Int. Rice Res. Inst., Los Banos, Manila, Philippines, pp. 407-418.

Sinha, J.P. and Mukherjee, D. (1975). Blue-green algae from the paddy fields of Bankura districts of West Bengal. Phykos 14(1-2): 117-118.

Somasegaran, P. and Hoben, H.J. (1994). Handbook for Rhizobia-Methods in Legume-*Rhizobium* Technology. Springer-Verlag, New York, USA, pp. 450.

Srinivasan, G. (1980). *Azolla* manuring and grain yield of rice. Int. Rice Res. Newslett. 6: 25.

Stoltzfus, J.R., So, R., Malarvithi, P.P., Ladha, J.K. and de Bruijn, F.J. (1997). Isolation of endophytic bacteria from rice and assessment of their potential for supplying rice with biologically fixed nitrogen. Plant Soil 194: 25-36.

Stowers, M.P. and Eaglesham, A.R.J. (1983). A stem-nodulating *Rhizobium* with physiological characteristics of both fast and slow growers. J. Gen. Microbiol. 129: 3651-3655.

Subba Rao, N.S. (1983). Nitrogen fixing bacteria associated with plantation and orchard plants. Can. J. Microbiol. 29: 863-866.

Subba Rao, N.S. (1988). Microbiological aspects of green manure in lowland rice soil. In: Sustainable Agriculture: Green Manure in Rice Farming. Int. Rice Res. Inst., Los Banos, Manila, Philippines, pp. 132-149.

Subba Rao, N.S. (1993). Biofertilizers in Agriculture and Forestry, III edn, Oxford & IBH Publications, New Delhi, India.

Subhashini, R. Kumar, K. and Kannaiyan, S. (2000). Variation revealed by RAPD – PCR profiles of microsymbiont *Anabaena azollae* strains from four N_2 fixing *Azolla* cultures. Acta Biotechnnol. 20 (2): 169-174.

Subramani, S. and Kannaiyan, S. (1987). Effect of urea on decomposition of *Azolla*. Int. Rice Res. Newslett 12(4): 57.

Suessenguth, K. and Beyerle, R. (1936). Uber Bakterienknollchen am sprob von *Aeschynomene paniculata* wild. Nova Hedwigia 75: 234-237.

Sundara Rao, W.V.B. and Sinha, M.K. (1963). Phosphate dissolving organisms in soil and rhizosphere. Indian J. Agric. Sci. 33: 272-278.

Sundaram, S.P., Kannaiyan, S. and Muthukumarasamy, S. (2001). Transformation studies in *Azospirillum halopraeferens*. http://www.microbiologyou. Com, p. 6.

Talley, S.N., Talley, B.J. and Rains, D.W. (1977). Nitrogen fixation by *Azolla* in rice fields. In: Genetic Engineering for Nitrogen Fixation (ed.) Al. Hollander, Plenum Publishing Company, New York, USA, pp. 259-281.

Tchan, Y.T. and Kennedy, I.R. (1989). Possible N_2-fixing root nodules induced in non-legumes. Agricultural Science (Melbourne New Series) 2: 57-59.

Tchan, Y.T., Zeman, A.M.M. and Kennedy, I.R. (1991). Nitrogen fixation in para-nodules of wheat roots by introduced free-living diazotrophs. Plant Soil 137: 43-47.

Thangaraju, M. and Kannaiyan, S. (1990). Influence of *Azolla, Sesbania* and urea super granules (USG) on rice yield and nitrogen uptake. Int. Rice Res. Newslett. 15(1): 25.

Thiruchenthilvasan, S. and Purushothaman, D. (1999). Rhizosphere colonization of genetically engineered *Azospirillum* strains and response of sorghum to inoculation. XI Southern Conference on Microbial Inoculants. 19th and 20th Feb. 1999. Guntur, India (abstr.), p. 1.

Tomekpe, K., Dreyfus, B.L. and Holsters, M. (1996). Root nodulation of *Sesbania rostrata* suppresses stem nodulation by *Sinorhizobium teranga* but not *Azorhizobium caulinodans*. Can. J. Microbiol. 42(2): 187-190.

Tsien, H.C., Dreyfus, B.L. and Schmidtt, E.L. (1983). Initial stages in the morphogenesis of nitrogen-fixing stem nodules of *Sesbania rostrata*. J. Bacteriol. 150: 888-897.

Urban, J.S., Davis, L.C. and Brown, S.J. (1986). *Rhizobium trifolii* - 0403 is capable of growth in the absence of combined nitrogen. Appl. Environ. Microbiol. 52: 1060-1067.

Vande Broek, A., Michiels, J., Van Gool, A. and Vanderleyden, J. (1993). Spatial-temporal colonization patterns of *Azospirillum brasilense* on the wheat root surface and expression of the bacterial *nif* H gene during association. Mol. Plant Microbe Interact. 6: 592-600.

Venkataraman, G.S. (1972). Algal Biofertilizers and Rice Cultivation. Today and Tomorrow's Printers and Publishers, New Delhi, India.

Visperas, R.M., Furoc, R.E., Morris, R.A., Vergara, B.S. and Patna, G. (1987). Flowering response of *Sesbania rostrata* to photoperiod. Philipp. J. Crop Sci. 12: 147-149.

Vlachova, M., Metz., B.A., Schell, J. and de Bruijn, E.J. (1987). The tropical legume *Sesbania rostrata* - Tissue culture, plant regeneration and infection with *Agrobacterium tumefaciens* and *rhizogenes* strains. Plant Soil 50: 213-223.

Vonshak, A., Guy, R. and Guy, M. (1988). The response of the filamentous cyanobacterium, *Spirulina platensis* to salt stress. Arch. Microbiol. 150: 417-420.

Wastyn, M., Achatz, A. Trnka, M. and Peschek, G.A. (1987). Immunological and special characterisation of partly purified cytochrome oxidase from the cyanobacterium, *Synechocystis* 6714. Biochem. Biophys. Res. Commun. 149: 102-111.

Watanabe, I. (1987). Fertility and fertilizer evaluation for rice. In: *Azolla* Utilization. Proc. Workshop on *Azolla* use Fuzhou Fujian China, Int. Rice Res. Instt. Los Banos, Philippines, pp. 208.

Webster, G., Cough, C. Vasse, J. Batchelor, C.A. O'Callaghan, K.J. Kothari, S.L. Davey, M.R. Denarie, J. and Cocking, E.C. (1997). Interactions of rhizobia with rice and wheat. Plant Soil 194: 115-122.

Yatazawa, M. and Yoshida, S. (1979). Stem nodules in *Aeschynomene indica* and their capacity of nitrogen fixation. Plant Physiol. 45: 293-295.

Yu, D. and Kennedy, I.R. (1995). Nitrogenase activity ($C_2 H_2$ reduction) of *Azorhizobium* in 2, 4-D induced root structures of wheat. Soil Biol. Biochem. 27: 459-462.

Zahran, H.H. (2005). Rhizobial nitrogen fixation in agriculture: biotechnological perspectives. In: Microbial Biotechnology in Agriculture and Aquaculture, Volume I (ed.) R.C. Ray, Science Publishers, Enfield, New Hampshire, USA, pp. 71-100.

Plate 5.1. Close-up view of *Azolla microphylla* (SK-TNAU)

Plate 5.2. Dual crop of *Azolla* and rice

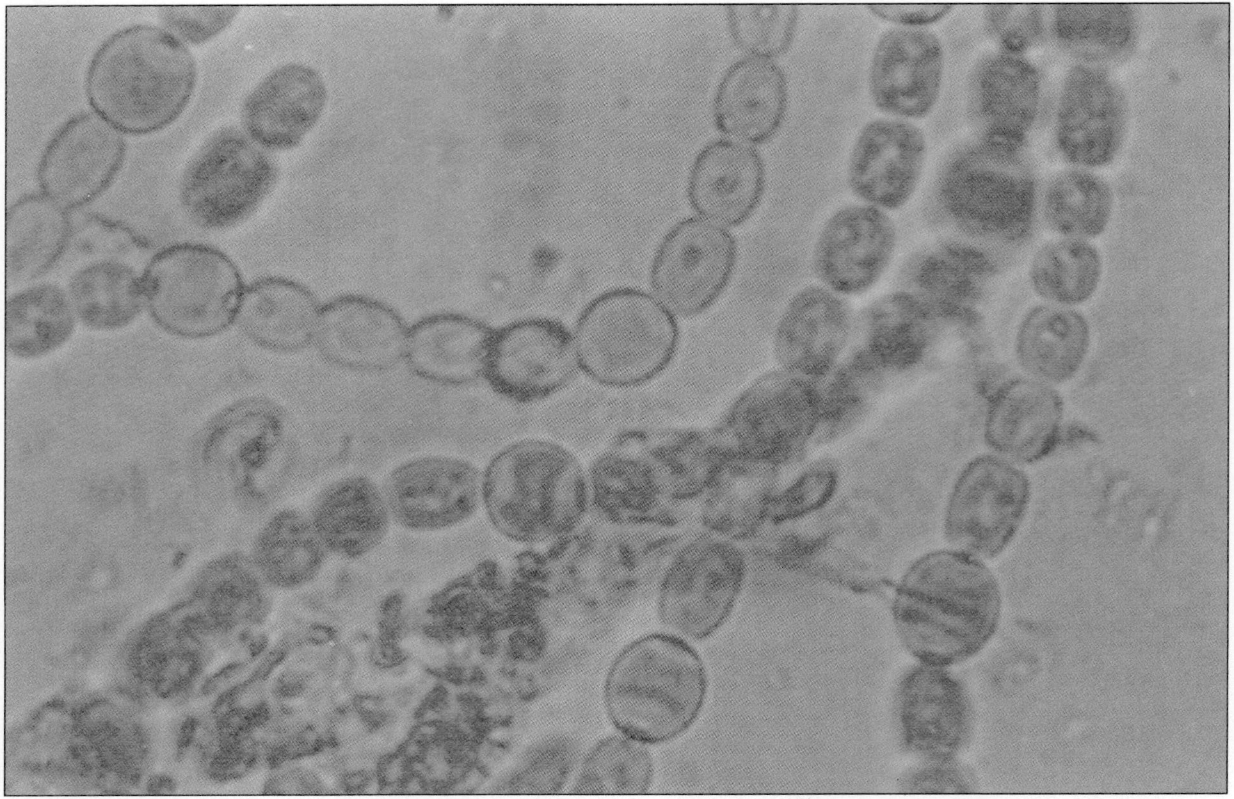

Plate 5.3. Algal symbiont - *Anabaena azollae*

Plate 5.4. Stem nodules of *Sesbania rostrata*

Plate 5.5. Biomass of *Sesbania rostrata*

Plate 5.6. Incorporation of *Sesbania rostrata*

Zeman A.M., Tchan, Y.T., Elmerich, C. and Kennedy, I.R. (1992). Nitrogenase activity in wheat seedlings bearing para- nodules induced by 2,4 - dichlorophenoxyacetic acid (2,4-D) and inoculated with *Azospirillum*. Res. Microbiol. 143: 847-855.

Zhang Zhuang Ta, Ke Yu-Si, Ling De-Quan, Duan Bing-Yuan and Liu Xi-Lian. (1987). Utilization of *Azolla* in agricultural production in Guandong Province, China. In: *Azolla* Utilization. Proc. Workshop on *Azolla* Use. Fuzhou, Fujian, China, Int. Rice Res. Instt., Los Banos, Manila, The Philippines, pp. 141-145.

6

Physiological and Genetic Effects of Bacterial ACC Deaminase on Plants

*Saleema Saleh-Lakha, Nikos Hontzeas and Bernard R. Glick**

INTRODUCTION

Ethylene mediates a wide range of plant responses and developmental steps. It is involved in seed germination, tissue differentiation, formation of root and shoot primordia, root elongation, lateral bud development, flowering initiation, anthocyanin synthesis, flower opening and senescence, fruit ripening and degreening, production of volatile organic compounds responsible for aroma formation in fruits, storage product hydrolysis, programmed cell death, leaf and fruit abscission, and the response of plants to biotic and abiotic stresses (Abeles et al., 1992; Arshad and Frankenberger, 2002; van Loon and Glick, 2004; Saleh and Glick, 2005).

Stress ethylene, which refers to an increased level of ethylene produced in response to temperature stress, pathogen attack, chemicals, insect damage and disease, ultraviolet light, mechanical wounding and water stress, can cause a variety of symptoms in plants such as epinastic curvature, leaf abscission, inhibition of root growth, formation of aerenchyma and the onset of senescence. High levels of ethylene, essential for fruit ripening, are also the primary cause of the rotting of fruits and senescence of plants (Stearns and Glick, 2003). At the same time, low levels of ethylene are required for cell wall strengthening, production of phytoalexins and the synthesis of defensive proteins (van Loon and Glick,

Department of Biology, University of Waterloo, Waterloo, ON, Canada N2L 3G1

**Corresponding author: E-mail: glick@sciborg.uwaterloo.ca*

2004; Saleh and Glick, 2006). Hence, it is essential to balance the ethylene level so that it provides beneficial effects and the deleterious symptoms are minimized.

Chemicals have been used to control ethylene levels in plants. The application of compounds such as rhizobitoxin (Yasuta et al., 1999), an amino acid secreted by several strains of bacteria, and its synthetic analog, aminoethoxyvinylglycine (AVG), can inhibit ethylene biosynthesis; silver thiosulfate can inhibit ethylene action, and 2-chloroethylphosphoric acid (ethephon), regarded by some researchers as "liquid ethylene", can release ethylene (Abeles et al., 1992). In addition, Sisler and Serek (1997) discovered that cyclopropenes can block ethylene perception and are potentially useful for extending the life span of cut flowers and the display life of potted plants. In addition, tropolone compounds, isolated from wood (Mizutani et al., 1998), which can inhibit the growth of wood rotting fungi, inhibit the biosynthesis of ethylene in excised peach pits. Unfortunately, many of these chemicals are potentially harmful to the environment: AVG and silver thiosulfate are highly toxic in food, and silver thiosulfate causes blackspotting in flowers (Liao et al., 2000).

ETHYLENE BIOSYNTHESIS

Ethylene that is synthesized in response to a variety of environmental stresses appears in two peaks (Abeles et al., 1992). The first peak is small and often difficult to detect, and generally occurs within a few hours after the stress. This peak of ethylene is thought to act as a signal to turn on the transcription of genes, which encode proteins that help to protect the plant against environmental stress such as a pathogen (Melchers and Stuiver, 2000). The second peak is much larger and generally occurs several days after the stress, often concomitant with the appearance of visible damage to the plant. Ideally, it would be advantageous to engineer plants to synthesize the first but not the second peak of ethylene in response to an environmental stress such as infection by a fungal pathogen. One way to do this is by using plant growth-promoting bacteria that contain the enzyme 1-aminocyclopropane-1-carboxylate (ACC) deaminase. This enzyme can decrease the concentration of ACC in plants, where ACC is the immediate precursor of ethylene (Glick et al., 1998). In addition, since (at least in some strains of bacteria) ACC deaminase is an induced enzyme, the bacterial expression of ACC deaminase does not occur for several hours after the appearance of increased amounts of ACC.

In higher plants, ethylene is produced from L-methionine via the intermediates, S-adenosyl-L-methionine (SAM) and ACC (Yang and Hoffman, 1984). The enzymes involved in this metabolic sequence are

SAM synthetase, which catalyzes the conversion of methionine to SAM (Giovanelli et al., 1980); ACC synthase, which is responsible for the hydrolysis of SAM to ACC and 5'-methylthioadenosine (Kende, 1989); and ACC oxidase which further metabolizes ACC to ethylene, carbon dioxide, and cyanide (John, 1997).

In most plants, ACC synthase is encoded by medium-sized gene families. In addition, ACC synthase genes are differentially regulated by various developmental, hormonal and environmental signals (Bleecker and Kende, 2000). Plants treated with exogenous ethylene develop the so-called triple response, which consists of agravitropic (horizontal) growth, inhibition of stem elongation and stem thickening (Bleecker and Kende, 2000). A number of *Arabidopsis* mutants have been isolated that are unable to generate the triple response leading to the identification of ethylene insensitive mutants (Stepanova and Ecker, 2000). Similarly, plants that exhibit the triple response in the absence of exogenous ethylene have been identified and characterized as ethylene overproducing or constitutive-signaling mutants. Based on these mutants, numerous genes that are implicated in the ethylene-signaling cascade have been identified (Chang and Schockey, 1999).

An ethylene response pathway has been identified starting from ethylene perception and leading to changes in gene expression. Ethylene is perceived by receptors, referred to collectively as the ethylene-receptor family, which is similar to bacterial two-component histidine kinase receptors (Urao et al., 2000). Numerous two-component regulators from various plants have been identified. They primarily consist of histidine kinases, response regulators and phospho-relay intermediates made up of a histidine-containing phosphotransfer domain (HPt). In *Arabidopsis*, at least five integral membrane receptor family members have been identified. They are: ethylene receptor1 (ETR1), ETR2, ethylene insensitive4 (EIN4), ethylene response sensor1 (ERS1), and ERS2 (Stepanova and Ecker, 2000).

At the next step in the ethylene biosynthesis pathway there are several alternatives. The ACC produced by ACC synthase, can either: (1) be converted to ethylene, via the action of ACC oxidase, (2) it can be converted to an inactive storage compound such as glutamyl ACC or malonyl ACC, or (3) the ACC can be exuded into environment such as soil and rhizosphere and taken up by plant growth as promoting bacteria (Glick et al., 1999). The latter route is the subject of discussion for the remaining part of this chapter.

PLANT GROWTH-PROMOTING BACTERIA AND ACC DEAMINASE

Plant growth-promoting bacteria are defined as organisms that can directly, or indirectly enhance plant growth, following binding to either roots (rhizosphere bacteria) or leaves (phyllosphere bacteria) (Glick et al., 1999). The highest concentrations of these microorganisms normally exist around the roots, in the rhizosphere, most probably due to the high levels of nutrients (amino acids, sugars and organic acids) that are exuded from the roots of most plants (Whipps, 1990). The bacteria can then utilize these nutrients to support their growth (Glick, 2003; Penrose and Glick, 2003). Numerous plant growth-promoting rhizobacteria have been isolated to date, each containing one or more mechanisms for enhancement of plant growth (Burdman et al., 2000).

Direct promotion of plant growth may involve nitrogen fixation (Boddey and Dobereiner, 1994; James and Olivares, 1998; Burdman et al., 2000), enzymatic lowering of ethylene levels that are otherwise an impediment to plant growth (Glick et al., 1998; 1999), solubilization of minerals such as phosphorus (Rodriguez and Fraga, 1999; Mikanová and Kubát, Chapter 4 in this Volume), sequestration of iron by siderophores (low molecular weight iron-binding compounds) (Buchenauer, 1998; Bhatia et al., 2005, Chapter 12 in the Volume I of this series–Microbial Biotechnology in Agriculture and Aquaculture, published in 2005), or production of phytohormones such as auxins and cytokinins (Glick et al., 1999, Burdman et al., 2000). A particular strain of bacteria may possess one or more of these plant growth promoting mechanisms, and moreover may utilize them at different times during the life cycle of the plant. In this chapter, the focus is on lowering the concentration of ethylene in plants, via the action of the bacterial enzyme, ACC deaminase.

Among the first ACC deaminase enzymes isolated were those from *Pseudomonas* sp. strain ACP, and from the yeast, *Hansenula saturnus* (Honma and Shimomura, 1978; Minami et al., 1998). Since then, ACC deaminase has been detected in the fungus, *Penicillium citrinum*, (Honma, 1993) and in a number of other bacterial strains (Klee and Kishore, 1992; Jacobson et al., 1994; Glick et al., 1995; Campbell and Thomson 1996; Ma et al., 2003b), all of which originated in the soil. Many of these microorganisms were identified by their ability to grow on minimal media containing ACC as its sole nitrogen source. They include *Pseudomonas*, *Enterobacter*, and more recently, *Bacillus* and *Rhizobium* species (Honma and Shimomura, 1978; Klee et al., 1991; Honma, 1993; Jacobson et al., 1994; Glick et al., 1995; Campbell and Thomson, 1996; Burd et al., 1998; Belimov et al., 2001, 2005; Ghosh et al., 2003; Ma et al., 2003b).

Enzymatic activity of ACC deaminase is assayed by monitoring the production of either ammonia or α-ketobutyrate, the products of ACC hydrolysis (Honma and Shimomura, 1978). To date, ACC deaminase has been found only in microorganisms, and there are no microorganisms that synthesize ethylene via ACC (Fukuda et al., 1993). However, there is strong evidence that the fungus, *Penicillium citrinum*, produces ACC from SAM via ACC synthase, one of the enzymes of plant ethylene biosynthesis, and degrades the ACC by ACC deaminase. In this case, it appears that the ACC, which accumulates in the intracellular spaces, can induce ACC deaminase (Jia et al., 2000).

Genes encoding ACC deaminase have been cloned from a number of different soil bacteria including *Pseudomonas* sp. strain ACP (Sheehy et al., 1991), *Pseudomonas* sp. strains 6G5 and 3F2 (Klee et al., 1991; Klee and Kishore, 1992), *Pseudomonas* sp. strain 17 (Campbell and Thomson, 1996), *Enterobacter cloacae* strains CAL2 and UW4 (Shah et al., 1998), and *Rhizobium leguminosarum* bv. *viciae* 128C53K (Ma et al., 2003b); yeast, *Hansenula saturnus* (Minami et al., 1998); and fungus, *Penicillium citrinum* (Jia et al., 2000).

The ACC deaminase genes from *Pseudomonas* sp. strains 6G5 and F17, and *Enterobacter cloacae* strains UW4 and CAL2, all have an open reading frame (ORF) of 1014 nucleotides that encodes a protein containing 338 amino acids with a calculated molecular weight of approximately 36.8 kDa (Klee et al., 1991; Campbell and Thomson, 1996; Shah et al., 1998). The genes from these strains are highly homologous to each other: at the nucleotide level 6G5, F17, UW4 and CAL2 are 85-95% identical to each other (Campbell and Thomson, 1996; Shah et al., 1998) and most of the dissimilarities are in the wobble position (Shah et al., 1998). However, the DNA sequences from strains UW4 and CAL2 show only about 74% homology with the sequence of the ACC deaminase gene from *Pseudomonas* sp. strain ACP (Sheehy et al., 1991; Shah et al., 1998).

Sequence data indicate that strain UW4 contains a DNA region similar to that of the anaerobic transcription regulator, FNR (fumarate and nitrate regulator), at positions –39 to –49 (Grichko and Glick, 2001a, b). This region has been shown to be involved in regulating the transcription of ACC deaminase in this bacterial strain (Grichko and Glick, 2000b; Li and Glick, 2001). Moreover, the ACC deaminase gene promoter in strain UW4 is also under the transcriptional control of a nearby gene that has a DNA sequence similar to a leucine responsive regulatory protein (LRP) and the LRP-like gene is transcriptionally regulated by ACC (Grichko and Glick, 2001b; Li and Glick, 2001). In a separate study with *Rhizobium leguminosarum* bv. *viciae* 128C53K, insertional mutants in the regulatory leucine-responsive protein-like gene (*lrp*L) were unable to synthesize ACC deaminase and also demonstrated a significantly decreased nodulation efficiency in pea plants compared to the parental non-transformed strain (Ma et al., 2003a). These results reflect the fact that the ACC deaminase gene in *Rhizobium leguminosarum* bv. *viciae* 128C53K is also under the control of LRP.

A model has been proposed (Glick et al., 1998) in which plant growth-promoting bacteria can lower ethylene levels and in turn stimulate plant growth. It was hypothesized that following the attachment of bacteria to the surface of plant roots or seeds, and in response to tryptophan and other small molecules found in root and seed exudates (Whipps, 1990), the auxin indole-3-acetic acid (IAA) is synthesized and secreted. This IAA, together with endogenous plant IAA, can function to stimulate plant cell proliferation or plant cell elongation, and it can induce the synthesis of the plant enzyme ACC synthase, which converts SAM to ACC (Kende, 1993). Some of this ACC is exuded from seeds or roots, and taken up by the plant growth-promoting bacteria bound to the plant. The ACC is then converted to ammonia and α-ketobutyrate by ACC deaminase. In effect, the ACC deaminase-containing bacteria act as a sink for ACC and thereby lower the level of ethylene that could otherwise result from various environmental stresses or during the course of plant development.

Plants that have been treated with microorganisms that produce ACC deaminase, exhibit longer roots and can better resist the inhibitory effects of stress ethylene on plant growth that is typically produced in response to flooding (Grichko and Glick, 2001a, b), heavy metals (Burd et al., 2000), drought, salt and pathogens (Wang et al., 2000; Mayak et al., 2004a, b). It has been suggested that to protect plants from a variety of environmental stresses, the use of ACC deaminase-containing plant growth-promoting bacteria may be an effective alternative to genetically engineered plants (Glick et al., 1999). Moreover, engineering plant growth-promoting bacteria to express ACC deaminase may increase both the fitness and the utility of the bacterial strain. For example, when *Azospirillum brasilense* Cd was transformed with an ACC deaminase gene from *Enterobacter cloacae* UW4, it was found that the genetically modified strain had improved fitness and growth-promoting capability, in terms of increased IAA synthesis and greater bacterial growth rate, as well as an increased ability to survive on the surface of tomato leaves and promote the growth of tomato seedlings (Holguin and Glick, 2003).

Flower Wilting

Ethylene is a key signal in the initiation of senescence of flowers in most plants. For example, carnation flowers produce minute amounts of ethylene until there is an endogenous rise (climacteric burst) in the level of this hormone. This rise in endogenous ethylene concentration is responsible for flower senescence (Mol et al., 1995), which in carnations, is characterized by in-curling of the petals.

However, ethylene does not cause senescence in all flower families, and even the features of senescence that are caused by ethylene, differ from plant to plant (Woltering and van Doorne, 1988): for example, Caryophyllaceae (e.g. carnations) show ethylene mediated wilting of their petals whereas ethylene causes petal abscission in Rosaceae (e.g. roses), and does not cause any senescence of petals in Compositae (e.g. sunflowers).

Since ethylene is a key element in the senescence of flower petals, a reduction in endogenous ACC should lower the amount of ethylene synthesized by the flower and delay the senescence of the petals. Many cut flowers (e.g. carnations and lilies), sold commercially, are routinely treated with the ethylene inhibitor, silver thiosulfate, which in high concentrations is potentially phytotoxic and environmentally hazardous (Liao et al., 2000). More recently, cut flowers have been treated with the compound 1-methylcyclopropene, an inhibitor of ethylene perception (Watkins and Miller, 2003), as a means of delaying senescence. In addition, the use of ACC deaminase-containing plant growth-promoting bacteria could be an environmentally friendly method of lowering ACC levels in cut flowers. As the first step towards determining the feasibility of this approach, carnation petals were treated with ACC deaminase-containing plant growth-promoting bacteria; petal senescence was delayed by several days when compared with untreated flower petals (Nayani et al., 1998). The treatment of these petals with a chemical ethylene inhibitor, namely AVG, had a similar effect in delaying the senescence of petals.

Fruit Ripening

The process of fruit ripening is highly regulated by the expression of a large number of gene products. Enzymes involved in the degradation of cell walls and complex carbohydrates, as well as other macromolecules, must be coordinately expressed with enzymes that impart more flavor and color to the fruit. Studies to decipher the role of various ripening-related genes have led to the construction of several mutants. It is now widely established that ethylene synthesis is one of the early events in the process of fruit ripening. It is involved in turning on the transcription of genes involved in degreening, flavor and color development, as well as softening of several climacteric fruits including melons, bananas, kiwi and tomatoes (Table 6.1). Synthesis of ethylene has been significantly reduced in transgenic tomato plants by expression of antisense constructs of ACC oxidase or ACC synthase cDNAs (Oeller et al., 1991). In addition,

Table 6.1. Genes involved in ripening that are regulated by ethylene

Plant	Functions Regulated by Ethylene	Reference
Melon	Increasing the activity of galactanase, α-arabinosidase, β-galactosidase, edo-polygalacturonase and ACC synthase in early stages of ripening	Pech et al., 1999
Tomato	Turn on the ethylene-responsive genes corresponding to regulatory genes involved in signal transduction, or transcriptional and post-translational modifications	Zehzouti et al., 1999
Oil palm fruit	Initiation of inner tissue signaling and gene expression cascades in ripening and abscission	Henderson and Osborne, 1999
Citrus	Promotion of flavor of fruit and maturation of peel	Cubella-Martinez et al., 1999
Snapdragon	Regulation of gravitropic response in cut flowers	Philosoph-Hadas et al., 1999
Tobacco	Triggering of early stages of female sporogenesis	De Martinis et al., 1999
Citrus sinensis	Activation of esterase and hydroxynitril lyase causing abscission of *Citrus sinensis* leaves	Goren et al., 2003
Sunflower	Stimulation of seed germination	Corbineau and Come, 2003
Apricot	Regulates carotenoid biosynthesis	Bureau et al., 2003

insertion of a sense version of these cDNAs, also causes a decrease in ethylene levels, because expression of SAM or ACC is co-suppressed by extra copies of sense mRNA. Another enzyme, whose expression has been altered using this kind of genetic approach, is SAM hydrolase, which degrades SAM, and hence decreases the amount of SAM available for ethylene biosynthesis (Stearns and Glick, 2003). In all these cases, a delay in ripening of fruits was observed.

Klee (1993) and Klee et al., (1991) measured the levels of ethylene in tomato (*Lycopersicon esculentum*) and determined that a higher ethylene content is present in detached fruit compared to attached fruit. This higher level of ethylene also promotes faster ripening. Studies with transgenic plants revealed that tomato plants expressing ACC deaminase under the transcriptional control of the constitutive 35S promoter ripened much later and had lower levels of ethylene than control plants (Klee, 1993). Another study with bananas investigated the characteristics of ethylene biosynthesis and the levels of mRNA transcripts encoding ACC synthase and ACC oxidase, the two key enzymes in the ethylene biosynthesis pathway. It was determined that at the onset of the climacteric period, there was an increase in ethylene production, and a corresponding increase in ACC synthase mRNA. This was followed by ripening, where the activity of ACC oxidase was enhanced. Since ACC oxidase converts ACC to ethylene, this observation reinforces the conclusion that ethylene plays a major role in ripening (Liu et al., 1999). Moreover, in this case both ACC synthase and ACC oxidase must be involved.

Nodulation

Ethylene inhibits nodulation of various legumes by Rhizobia (Ma et al., 2003a; Teamroong and Boonkerd, Chapter 3 in this Volume). Thus, if the primary strain of bacteria that causes the formation of nodules, namely, various *Rhizobium* species, could also lower the amount of ethylene in plants, then one could optimize the number and size of nodules formed, and hence increase the amount of fixed nitrogen available to the plants. In order to test this hypothesis, an ACC deaminase-containing strain of *Rhizobium leguminosarum* was isolated and specifically mutagenized to knock out the rhizobial ACC deaminase gene. Then both the mutant and wild-type strains were tested for the ability to form nodules on their host plant (peas). The mutant strain demonstrated a significant decrease in nodulation efficiency compared to the wild-type strain (Ma et al., 2003a). The nodulation efficiency of the mutant was restored when the mutant was transformed with an exogenous rhizobial ACC deaminase gene suggesting that a non-functional ACC deaminase enzyme was unable to break down ACC, resulting in an increase in the ethylene concentration and hence a greater inhibition of nodulation compared to the wild-type strain. Wild-type strains that possessed ACC deaminase activity enhanced the nodulation process, presumably by modulating localized ethylene levels in plant roots during the early stages of nodule development (Ma et al., 2003a).

PHYSIOLOGICAL EFFECTS OF ACC DEAMINASE ON PLANTS

There are a large number of situations in which the manipulation of ACC deaminase genes could be used to improve agricultural and horticultural practices. Organisms containing these genes may find use in promoting early root development from either seeds or cuttings, increasing the life of cut flowers, protecting plants against a wide range of environmental stresses, facilitating the production of volatile organic compounds responsible for aroma formation and phytoremediation of contaminated soils.

Currently, many consumers worldwide are reluctant to embrace the use of genetically modified plants as sources of foods (Watrud, 2000; Arvanitoyannis, 2003 and Chapter 2 in this Volume). Thus, it may be advantageous to use either natural or genetically engineered plant growth-promoting bacteria as a means of lowering plant ethylene levels rather than genetically modifying the plant itself to achieve the same result. Moreover, given the large number of different plants, the various cultivars of those plants and the multiplicity of genes that would need to be introduced into plants, it is not feasible to genetically engineer all plants to be resistant to all types of pathogens and environmental stresses (Murphy, 2003; Minocha and Page, Chapter 1 in this Volume). Rather, it makes a lot of sense to engineer plant growth-promoting bacteria to do this job, and the first step in this direction could well be the introduction of appropriately regulated ACC deaminase genes (Ray and Edison, 2005, Chapter 1 in Volume I of this series–Microbial Biotechnology in Agriculture and Aquaculture).

A lot of research has been carried out an plant growth-promoting bacteria that produce ACC deaminase (Klee et al., 1991; Klee, 1993). Furthermore, the effect of ACC deaminase on plant physiology, especially under stress conditions, has evinced a great -deal of interest. Several different plants, including canola, carnation, Indian mustard, lettuce, mungbean, oats, pepper, rye, tall fescue, tobacco, tomato, and wheat have either been treated with plant growth-promoting bacteria that produce ACC deaminase, or transgenic lines of these crops have been produced that have one or more copies of ACC deaminase in the plant genome

(Bansal et al., 1999; Cushman and Bohnert, 2000). In general, a greater effect is observed with ethylene sensitive plants such as dicots, compared to monocots, which are less ethylene sensitive (Hall et al., 1996). These plants have been subjected to various types of stress factors, and their behavior monitored. Each of these factors is reviewed, and the major findings are presented below:

Flooding

Flooding is a common stress that affects many plants, often several times during the same growing season. Plant roots suffer a lack of oxygen as a consequence of flooding, and hence, the creation of an anaerobic environment. This in turn causes deleterious effects such as epinasty, leaf chlorosis, necrosis and reduced fruit yield. Two of the ACC synthase genes are rapidly induced in the roots of flooded tomato plants (Olson et al., 1995) eventually resulting in an increase in the amount of ACC in the plant. However, since ACC oxidase-catalyzed ethylene synthesis cannot occur in the anaerobic environment of flooded roots, ACC is transported into the aerobic shoots where it is converted to ethylene (Else and Jackson, 1998). Treatment of tomato plants with ACC deaminase-containing plant growth-promoting bacteria significantly decreases the damage suffered by these plants as a result of the deleterious effects of ethylene that normally occurs as a consequence of flooding (Grichko and Glick, 2001a). These ACC deaminase-containing plant growth-promoting bacteria can act as a sink for ACC, thereby lowering the level of ethylene that can be formed in the shoots and thus protecting the tomato plants against some of the damage that would otherwise be caused by flooding.

Grichko and Glick (2001b) also conducted a study where they used transgenic tomato plants expressing a bacterial ACC deaminase gene, and compared them to non-transformed strains in their response to flooding stress. Three lines of transgenic plants were used: the first contained the ACC deaminase gene under the control of the constitutive 35S promoter from cauliflower mosaic virus; the second line had the ACC deaminase gene under the control of the *rol*D promoter which is root specific, and in the third line, the ACC deaminase gene was under the control of the pathogenesis-related, *PRB-1b*, promoter. Shoot length, fresh and dry weight, epinasty as well as ACC deaminase activity, leaf chlorophyll and protein content were compared. All of the transgenic tomato plants were found to be more tolerant to flooding stress and were less subject to the deleterious effects of root hypoxia on plant growth than were the non-transformed plants. A transgenic tomato line that had the ACC deaminase gene under the control of the *rol*D promoter demonstrated the most enhanced effects, or the most protection under flooding stress. Transgenic plants were healthier, greener, had larger shoots and total chlorophyll amounts compared to non-transgenic plants (Grichko and Glick, 2001b). Interestingly, the *rol*D transgenic plants may be regarded as almost identified of the three types of transgenic plants to the situation in which non-transformed plants are treated with an ACC deaminase-containing soil bacterium.

Xenobiotics

Phytoremediation refers to the use of plants to remove, destroy or sequester hazardous substances from the environment (Cunningham et al., 1995). A number of different types of plants are effective at stimulating the degradation of organic molecules in the rhizosphere. Typically, these plants have extensive fibrous roots, which form an extended rhizosphere. These plants include many common grasses as well as beans, corn, peas, soybean and wheat. While plants grown on highly contaminated soils might be able to withstand some of the inhibitory effects of high concentrations of organic material within a plant, the environmental stress can not only induce the production of ethylene in the plant, but also decrease the availability of iron in the environment. Plant growth-promoting bacteria may be able to relieve these problems, and thus enhance the uptake and degradation of hazardous organic substances by the plants.

Typically, phytoremediation has been the predominant approach for removal of recalcitrant substances such as polyaromatic hydrocarbons, polychlorinated biphenyls and total petroleum hydrocarbons. Bioremediation, which refers to the addition of contaminant-degrading bacteria to the soil, has also been effective to a certain extent. However, recent studies on a multi-purpose phytoremediation system, which is composed of a combination of land-farming (aeration and light exposure), contaminant degrading bacteria, plant growth-promoting bacteria and the use of contaminant-tolerant plants, has produced much more promising results. Huang et al., (2004a, b), first described this multi-system approach in the removal of polyaromatic hydrocarbons from soil. The application of this multi-process method resulted in an overall bioremediation process that was much more effective than phytoremediation, land-farming or bioremediation alone. The choice of plant species used also plays a role in enhancing the uptake of contaminant from the environment. The physiological characteristics that are important for phytoremediation include increased water content, accelerated root growth and maintenance of chlorophyll content in plant tissue. Addition of ACC deaminase-containing plant growth-promoting bacteria can significantly enhance root growth (and hence shoot growth) in the presence of these compounds, which are of significant stress to the plant, and the increased root biomass can in turn lead to a greater remediation potential (Huang et al., 2004b).

The most recent application of the multi-process phytoremediation system was in the removal of total petroleum hydrocarbons (TPH) from soils. It was found that this approach was on an average twice as effective as land-farming in removal of persistent TPHs, 50% more effective than bioremediation alone, and 45% more effective than phytoremediation alone. Moreover, this approach was especially successful (in marked contrast to the other procedures) in removing the high molecular weight TPHs from soil over an eight-month period (Huang et al., 2004a).

Metals

ACC deaminase-containing plant growth-promoting bacteria that facilitate the proliferation of various plants under environmentally stressful conditions, such as the presence of high concentrations of metals, can lower the amount of stress ethylene produced by a plant and also provide the plant with iron from the soil. The net result of adding these bacteria to plants is a significant increase in both the number of seeds that germinate and the amount of biomass that the plants are able to attain, making phytoremediation in the presence of ACC deaminae-containing plant growth-promoting bacteria an attractive possibility.

It was recently observed that transgenic canola plants that express *Enterobacter cloacae* UW4 ACC deaminase could proliferate and accumulate metal in the presence of high levels of arsenate in the soil. While arsenate-resistant plant growth-promoting bacteria did not significantly protect plants from arsenate inhibition, transgenic canola plants expressing a bacterial ACC deaminase were significantly more resistant to the toxic effects of arsenate than were non-transformed plants (Nie et al., 2002). Regardless of the presence or absence of plant growth-promoting bacteria, a maximum of 25% of non-transformed plants germinated in arsenic contaminated soil. In contrast, about 70% of the transgenic canola seeds germinated under similar toxic conditions. Although a small ethylene pulse is imperative for breaking seed dormancy in many plants, too much ethylene can inhibit plant seed germination. In the presence of arsenate, ACC deaminase may hydrolyze excess ACC, hence lowering inhibitory levels of ethylene produced. In addition, each transgenic plant was found to accumulate approximately four times as much arsenate as non-transformed canola plants.

Although the presence of arsenate leads to a significant decrease in leaf chlorophyll content, transgenic plants were less affected than non-transformed plants. The total biomass of transgenic canola grown in the presence of arsenate was significantly greater than non-transformed plants so that on a per plant basis, transgenic plants took up two to three times as much arsenate as non-transformed plants. In addition, a greater number of transgenic seeds germinated in arsenate contaminated soil, hence, the total amount of arsenate taken up per 100 seeds planted was approximately seven times greater for transgenic plants (Nie et al., 2002).

Transgenic tomato plants were also stimulated in their ability to concentrate and uptake metals such as cadmium, cobalt, copper, magnesium, nickel, lead or zinc. In general, transgenic tomato plants expressing ACC deaminase acquired a greater amount of metal within the plant tissues, and were less subjected to the deleterious effects of these metals on plant growth, than were non-transgenic plants (Grichko et al., 2000).

Canola seeds inoculated with the ACC deaminase-containing and siderophore-producing, *Kluyvera ascorbata* SUD165 in the presence of high concentrations of nickel chloride were partially protected against nickel toxicity. This was evident under gnotobiotic (i.e. in the absence of other microorganisms) conditions, as well as in pot experiments. The presence of this bacterium had no effect on the amount of nickel accumulated per milligram of either roots or shoots of canola plants. Rather, its plant growth-promoting attributes enable it to lower the level of stress ethylene induced by nickel, and hence promote the health of the plant. Moreover, part of the toxic effect of nickel in plants is an iron deficiency, which *K. ascorbata* is able to offset. When a siderophore overproducer of this strain was used, it was even more effective at protecting plants against the inhibitory effects of high concentration of nickel, lead and zinc. This was indicated by higher fresh and dry weights, and protein and chlorophyll content in plant leaves (Burd et al., 1998; 2000).

Studies of transgenic tobacco plants under nickel toxicity conditions demonstrated that a transgenic tobacco line with a bacterial ACC deaminase gene under transcriptional control of the *rol*D promoter was significantly more effective in improving the biomass of plants when fresh and dry weights were compared to non-transformed plants (Li et al., unpublished results). In addition, the *rol*D transgenic tobacco line was most effective in offsetting the inhibitory effects of nickel, and this protective effect was enhanced when the ACC deaminase-containing plant growth-promoting bacterium *E. cloacae* UW4 was added to the plants. Furthermore, chlorophyll levels were also increased by this treatment. This indicates that ACC deaminase enhances the growth and development of the tobacco plants by lowering the amount of stress ethylene that is induced due to the presence of nickel, and by offsetting the deleterious effects of nickel and stress ethylene, it imparts health and added vigor to the plant (Li et al., unpublished results). The enhanced growth effect observed upon the addition of *Pseudomonas putida* UW4 is likely the consequence of siderophores and/or indole acetic acid produced by the bacterium.

In a parallel study, canola plants transformed with an ACC deaminase gene under the control of the *rol*D promoter accumulated ten times as much nickel in roots, compared to non-transformed canola plants. Preliminary microarray analysis of leaf mRNAs produced in *rol*D versus non-transformed canola plants suggests an interaction between the ethylene and auxin pathways due to an upregulation of auxin response factors in transgenic canola plants, compared to non-transformed plants (Stearns and Glick, unpublished results).

Salt

The bacterium *Achromobacter piechaudii* ARV8 was isolated from a saline soil environment in the southern portion of the Negev desert in Israel. It was able to grow on minimal media with ACC as the sole nitrogen source and hence, found to possess ACC deaminase activity. When applied to two-week old tomato seedlings, the bacterial strain significantly increased the fresh and dry weight of plants grown in the presence of 172 mM NaCl salt. It also reduced the production of ethylene, which results from the salt stress, in these tomato seedlings. In addition, this bacterium increased the uptake of phosphorus and potassium by plants. Hence, the presence of ACC deaminase-containing *A. piechaudii* ARV8 significantly enhanced the growth and vigor of tomato plants grown under high salt conditions (Mayak et al., 2004b)

Studies of the effects of high salt were undertaken with transgenic canola plants where the ACC deaminase gene was inserted into the plant genome, under the transcriptional control of the *rol*D promoter. These plants were assayed for parameters such as protein content, fresh weight, dry weight and chlorophyll content and it was found that the *rol*D transgenic line was able to sustain and grow much better in salty environments compared to non-transformed canola (Sergeeva et al., 2006). This study reinforces the inferred role of ACC deaminase in conferring additional immunity to plants that are grown under stress, most probably by decreasing the amount of stress ethylene that is produced by the plant.

Drought

Mayak et al., (2004a) conducted a study with the above-mentioned ACC deainase-containing plant growth-promoting bacterium *A. peichaudii* ARV8 in which tomato and pepper plants were subjected to prolonged periods of drought and then re-watered. This bacterial strain was found to increase the fresh and dry weight of tomato and pepper seedlings that were exposed to drought conditions. In addition, the bacterium reduced the production of ethylene by tomato seedlings following water stress, presumably a consequence of the presence of ACC deaminase. When the plants were subjected to drought conditions, the bacterium did not influence the reduction in relative water content; however, it significantly improved recovery of plants when watering was resumed. This study was a pioneering effort in determining the role of ACC deaminase on plant growth and survival in arid environments and bodes well for the use of ACC deaminase-containing bacteria as an adjunct to agricultural practice in arid environments.

Pathogens

Agricultural production is wrought by plant disease. In 1997, it was estimated that worldwide crop losses attributed to various plant diseases accounted for losses of more than 100 billion US$ (Yang et al., 1997). The application of fungicides and pesticides has helped tremendously in alleviating this loss; however, in the long term, these chemicals are likely to have a huge negative impact on the environment. Hence, plant researchers have focused on unraveling and exploiting the complex defense mechanisms used by plants, as a more natural alternative to fighting plant disease. Plants are constantly under attack from microbes, viruses, invertebrates and other plants. Since plants lack an immune system, this necessitates each plant cell to have an induced plant defense mechanism (Hammond-Kosack and Jones, 1997). Several plant resistance genes against pathogens (*R* genes) such as fungi, viruses and bacteria have been isolated in the last decade. Plant resistance requires complementary pairs of dominant genes, pathogen avirulence (*Avr*) and plant *R* genes. This is referred to as the "gene-for-gene" model (Hammond-Kosack and Jones, 1997; Yu et al., 1998).

One of the main defense mechanisms triggered by the gene-for-gene resistance pathway is programmed cell death at the infection site. This is referred to as the hypersensitive response (HR). A hypersensitive response (HR) involves rapid death of cells at the site of pathogen infection and is thought to limit pathogen growth throughout the plant. For example, on infection with *Xanthomonas campestris* bv. *vesicatora*, an upregulation of ethylene receptor genes was observed. An antisense orientation of the ethylene receptor gene in the plant caused increased sensitivity to ethylene on infection with a pathogen, and hence, an increase in the HR response. Treatment with 1-methylcyclopropene (1-MCP), an ethylene action inhibitor, alleviates the enhanced HR response in antisense plants, denoting that the changes in pathogen response are a result at least in part from increased ethylene sensitivity (Ciardi et al., 2001).

Many of the symptoms of a diseased plant arise as a direct result of the stress imposed by pathogen infection. In other words, much of the damage sustained by plants infected with fungal phytopathogens occurs as a result of the response of the plant to the increased levels of stress ethylene (van Loon, 1984). It has also been observed that exogenous ethylene often increases the severity of a fungal infection and, in addition, ethylene synthesis inhibitors significantly decrease the severity of a fungal infection (Ray and Ravi, 2005). In a study with over 60 different cultivars and breeding lines of wheat, ethylene production increased as a result of infection with the fungal phytopathogen, *Septoria nodorum*, and was correlated with increased plant disease susceptibility (Hyodo, 1991). The damage caused by the fungal phytopathogen, *Alternaria*, decreased in cotton plants by treating them with chemical inhibitors of ethylene synthesis (Bashan, 1994). The levels of both ethylene and disease severity decreased in melon plants infected by the fungal phytopathogen, *Fusarium oxysporum*, following treatment of the plants with ethylene inhibitors (Cohen et al., 1986). Fungal disease development increased in both cucumber plants infected with *Colletotrichum lagenarium* (Biles et al., 1990) and in tomato plants infected with *Verticillium dahliae* (Cronshaw and Pegg, 1976) when the plants were pretreated with ethylene. Treatment with ethylene inhibitors decreased disease severity in roses, carnations, tomatoes, pepper, French beans and cucumbers infected with the fungus, *Botrytis cinerea* (Elad, 1990).

Several biocontrol strains of bacteria were transformed with the *Enterobacter cloacae* UW4 ACC deaminase gene and the effect of the transformation was assessed by using the cucumber-*Pythium ultimum* system (Wang et al., 2000). The results of the experiments indicated that ACC deaminase-containing biocontrol bacterial strains were significantly more effective than biocontrol strains that lacked this enzyme. Yet another example of pathogen resistance by the ACC deaminase gene was illustrated in a study where the biocontrol bacterium *Pseudomonas fluorescens* strain CHA0, transformed with the ACC deaminase gene had a vastly improved ability to protect potato tubers against Erwinia soft rot (Wang et al., 2000).

Moreover, transgenic tomato plants that express ACC deaminase are also protected, to a significant extent, against phytopathogen-mediated damage from several different phytopathogens (Lund et al., 1998; Robison et al., 2001a, b). In effect, ACC deaminase acts synergistically with other mechanisms of biocontrol, such as the production of antibiotics or pathogenesis-related proteins, to prevent phytopathogens from damaging plants. As with other types of stress, it is assumed that ACC deaminase can act to prevent the accumulation of ACC that would otherwise occur as a result of environmental stress.

Ethylene has been observed to both inhibit and promote Verticillium wilt in tomato plants. Application of AVG (aminoethoxyvinylglycine), an ethylene inhibitor, along with ACC, a precursor of ethylene, minimized disease symptoms (Robison et al., 2001a, b). AVG treated plants, either treated prior to or at inoculation, had reduced disease symptoms. Addition of ACC to AVG-treated plants, at the time of pathogen inoculation, reduced symptom severity to a greater extent than AVG alone. Furthermore, addition of greater amounts of ethylene to AVG-treated plants led to greater reductions in disease severity. Based on these and other observations, it was hypothesized that a small peak of ethylene is rapidly produced after inoculation of plants (Stearns and Glick, 2003). This small ethylene peak could function to induce resistance responses and defense proteins. Conversely, production of large quantities of ethylene after successful pathogen attack are known to promote disease symptoms by inducing senescence-like effects such as chlorosis, wilting and abscission (van Loon and Glick, 2004). Reduction of Verticillium wilt severity in tomato may be possible either by enhancing the smaller peak of ethylene that is produced upon inoculation, or by reducing the much larger amount of ethylene that is produced (usually a few days) after successful pathogen attack.

Three transgenic tomato lines, with an ACC deaminase gene under the control of the 35S promoter (constitutive expression), the *rol*D promoter (which limits expression to roots that are the site of infection), and the *PRB-1b* promoter (which limits expression to certain environmental situations such as disease infection), were created. Significant reductions in the symptoms of Verticillium wilt were obtained for the *rol*D and *PRB-1b*, but not for the 35S transformants. This result is similar to the flooding, high salt and metal stress experiments where *rol*D plants demonstrated the greatest enhancement in growth and the smallest decrease in stress damage compared to the *PRB-1b* and 35S transgenic lines of tomato or canola. Collectively, these results indicate that while it may be beneficial to the plant to lower ethylene levels in roots, lowering ethylene levels throughout the plant is not as beneficial as might be imagined. This suggests that low levels of ethylene may be required for the optimal functioning of the plant.

GENOMICS IN PLANT DISEASE AND POSSIBLE APPLICATIONS

New technologies are permitting experimentation on an unimaginable scale. For example, DNA microarrays are furnishing massive amounts of data providing new insights as well as new challenges. It is thought that these new approaches will affect most areas of biology, including an understanding of plant stress. Genomics involves the discovery and study of a large number of genes simultaneously on a genome-wide scale. Genomics can be divided into the three fields of structural genomics, comparative genomics and functional genomics. Comparative genomics involves the molecular basis of differences between organisms at a variety of taxonomic levels, while functional genomics focuses on the function of genes (Michelmore, 2000). Preliminary functional genomic analysis has been performed in an effort to assess some of the ways in which canola plants respond to plant growth-promoting bacteria in general, and to ACC deaminase in particular. In one study, RNA arbitrarily primed PCR (RAP-PCR) was used to identify the differential expression of genes of canola plants when subjected to treatment with plant growth-promoting bacteria that contain ACC deaminase. RAP-PCR is a technique that uses longer primers than usual (18- to 20-oligomers) for both reverse transcription and PCR amplification, which allows for increased reproducibility of band display patterns and direct sequencing of PCR products. Total RNA from the roots of canola plants was first isolated, and RAP-PCR performed to study the band display patterns of different treatments of plant growth-promoting bacteria on canola plants. The bands that were upregulated or downregulated were identified, cloned and sequenced. Plants that were treated with bacteria containing ACC deaminase as compared to an ACC deaminase knockout mutant of the same bacterium displayed an upregulation of genes involved in metabolism and protein synthesis, which translated physiologically into longer root lengths. In addition, canola treated with ACC deaminase-containing bacteria showed a downregulation of an identified RNA binding protein that could be involved in plant stress response. A decrease in plant ethylene levels attributed to the presence of ACC deaminase could account for the downregulation of genes involved in plant stress response. This also explains why plants treated with the ACC deaminase knockout mutant demonstrated an upregulation of genes involved in plant defense response (Hontzeas et al., 2004).

Another technique that falls under the classification of functional genomics is DNA microarrays. This approach involves using mRNA to probe membranes (or DNA chips) that have the entire genome of an organism spotted onto the surface. Recently, Stearns and Glick (2003) attempted to study the differential expression of canola genes from plants that were subjected to nickel stress. Several genes were upregulated, many of which were involved in the indole acetic acid (IAA biosynthesis pathway). In parallel, genes involved in stress response were also differentially expressed, compared to control plants. This preliminary study suggests a possible crosstalk between the IAA and ethylene biosynthesis pathways, and, at the very least, it gives an idea of how complex gene regulation really is and how functional genomics can aid tremendously in determining functions of various classes of genes (Stearns et al., 2003).

CONCLUSION

It is clear that a large number of beneficial soil microorganisms contain ACC deaminase, and that these organisms can facilitate plant growth under a variety of conditions. As we develop a more profound understanding of some of the mechanisms employed by these organisms, we should be able to find additional use for these microorganisms and new ways in which they can be used to promote plant growth, especially under stressful conditions. Finally, despite the checkered history of using plant growth-promoting bacteria in the field, the more in-depth understanding of the mechanisms employed by these organisms that now exist should enable workers to isolate, construct and utilize organisms that are appropriate for a variety of crops and soil types.

ACKNOWLEDGEMENT

We are grateful to the Natural Science and Engineering Research Council of Canada for their ongoing support for research in our laboratory.

REFERENCES

Abeles, F.B., Morgan, P.W. and Saltveit, M.E. (1992). Ethylene in Plant Biology, 2nd ed., Academic Press, San Diego, USA.

Arshad, M. and Frankenberger, W.T. Jr. (2002). Ethylene from Agricultural Sources and Applications. Kluwer Academic/Plenum Publishers, New York, USA.

Arvanitoyannis, I.S. (2003). Genetically engineered/modified organisms in foods. Appl. Biotechnol. Food Sci. Policy 1(1): 3-12.

Bansal, K.C., Barthakur, S. and Ray, S.K. (1999). Postharvest biotechnology in fruits and vegetables. In: Biotechnology and its Application in Horticulture (ed.) S.P. Ghosh, Narosa Publishing House, New Delhi, India, pp. 165-182.

Bashan, Y. (1994). Symptom expression and ethylene production in leaf blight of cotton caused by *Alternaria macrospora* and *Alternaria alternata* alone and combined. Can. J. Bot. 72: 1574-1579.

Belimov, A.A., Safranova, V.I., Sergeyeva, T.A., Egorova, T.N., Matveyeva, V.A., Tsyganov, V.E., Borisov, A.Y., Tikhonovich, I.A., Kluge, C., Preisfeld, A., Dietz, K.J. and Stepanok, V.V. (2001). Characterization of plant growth promoting rhizobacteria isolated from polluted soils containing 1-aminocyclopropane-1-carboxylate deaminase. Can. J. Microbiol. 47: 642-652.

Belimov, A.A., Hontzeas, N., Safranova, V.I., Dodd, I.C., Demchinskaya, S.V., Piluzza, G., Bullitta, S., Davies, W.J. and Glick, B.R. (2005). Cadmium tolerant root growth-promoting bacteria of the rhizoplane of Indian mustard (*Brassica juncea* L. Czern.). Soil Biol. Biochem. 37: 241-250.

Bhatia, S., Dubey, R.C. and Maheshwari, D.K. (2005). Control of soil-borne sclerotial pathogens using fluorescent pseudomonads. In: Microbial Biotechnology in Agriculture and Aquaculture, Volume I (ed.) R.C. Ray, Science Publishers, Enfield, New Hampshire, USA, pp. 377-409.

Biles, C.L., Abeles, F.B. and Wilson, C.L. (1990). The role of ethylene in anthracnose of cucumber, *Cucumis sativus*, caused by *Colletotrichum lagenarium*. Phytopathol. 80: 732-736.

Bleecker, A.B. and Kende, A. (2000). Ethylene, a gaseous signal molecule in plants. Annu. Rev. Cell. Dev. Biol. 16: 1-18.

Boddey, R.M. and Dobereiner, J. (1994). Biological nitrogen fixation associated with graminaceous plants. In: *Azospirillum*/Plant Associations (ed.) Y. Okon, CRC Press, Boca Raton, Florida, USA, pp. 119-135.

Buchenauer, H. (1998). Biological control of soil-borne diseases by rhizobacteria. Z. Pflanzenkrankh. Pflanzenschutz 105: 329-348.

Burd, G.L., Dixon, D.G. and Glick, B.R. (1998). A plant growth-promoting bacterium that decreases nickel toxicity in seedlings. Appl. Environ. Microbiol. 64: 3663-3668.

Burd, G.L., Dixon, D.G. and Glick, B.R. (2000). Plant growth-promoting bacteria that decrease heavy metal toxicity in plants. Can. J. Microbiol. 46: 237-245.

Burdman, S., Jurkevitch, E. and Okon, Y. (2000). Recent advances in the use of plant growth promoting rhizobacteria (PGPR) in agriculture. In: Microbial Interactions in Agriculture and Forestry (eds.) N.S. Subba Rao and Y.R. Dommergues, Science Publishers, Enfield, New Hampshire, USA, pp. 231-250.

Bureau, S., Gouble, B., Reich, M., Audergon, J.M. and Albagnac, M. (2003). Regulation by ethylene of carotenoid biosynthesis in fruits of two contrasted apricot varieties. In: Biology and Biotechnology of the Plant Hormone Ethylene III (eds.). M. Vendrell, H. Klee, J.M. Pech and F. Romojaro, International Organizational Services B.V. Press, Burke, Virginia, USA, pp. 233-234.

Campbell, B.G. and Thomson, J.A. (1996). 1-aminocyclopropane-1-carboxylate-deaminase genes from *Pseudomonas* strains. FEMS Micobiol. Lett. 138: 207-210.

Chang, C. and Schockey, J.A. (1999). The ethylene response pathway: signal perception to gene regulation. Curr. Opin. Plant Biol. 2: 352-358.

Ciardi, J.A., Tieman, D.M., Jones, J.B. and Klee, H.J. (2001). Reduced expression of the tomato ethylene receptor gene LeETR4 enhances the hypersensitive response to *Xanthomonas campestris* bv. *vesicatoria*. Mol. Plant. Microb. 14: 487-495.

Cohen, R., Riov, J., Lisker, N. and Katan, J. (1986). Involvement of ethylene in herbicide-induced resistance to *Fusarium oxysporum* f. sp. *Melonis*. Phytopathol. 76: 1281-1285.

Corbineau, F. and Come, D. (2003). Germination of sunflower seeds as related to ethylene synthesis and sensitivity - an overview. In: Biology and Biotechnology of the Plant Hormone Ethylene III (eds.) M. Vendrell, H. Klee, J.M. Pech and F. Romojaro, International Organizational Services B.V. Press, Burke, Virginia, USA, pp. 216-221.

Cronshaw, R. and Pegg, G.F. (1976). Ethylene as a toxin synergist in *Verticillium* wilt of tomato. Physiol Plant Pathol. 9: 33-38.

Cubella-Martinez, X., Alonso, J.M., Sanchez-Ballesta, M.T. and Granell, A. (1999). Ethylene perception and response in Citrus fruit. In: Biology and Biotechnology of the Plant Hormone Ethylene II (eds.) A.K. Kanellis, C. Chang, H. Klee, A.B. Bleecker, J.C. Pech and D. Greirson, Kluwer Academic Publishers, Dordrecht, The Netherlands, pp. 137-144.

Cunningham, S.D., Berti, W.R. and Huang, J.W. (1995). Phytoremediation of contaminated soils. Trends Biotechnol. 13: 393-397.

Cushman, J.C. and Bohnert, H.J. (2000). Genomic approaches to plant stress tolerance. Curr. Opin. Plant Biol. 3: 117-124.

De Martinis, D., Haenen, I., Pezzotti, M., Benvenuto, E. and Mariani, C. (1999). Ethylene and flower development in tobacco plants. In: Biology and Biotechnology of the Plant Hormone Ethylene II (eds.) A.K. Kanellis, C. Chang, H. Klee, A.B. Bleecker, J.C. Pech and D. Greirson, Kluwer Academic Publishers, Dordrecht, The Netherlands, pp. 157-164.

Elad, Y. (1990). Production of ethylene in tissues of tomato, pepper, French bean and cucumber in response to infection by *Botrytis cinerea*. Physiol. Mol. Plant Pathol. 36: 277-287.

Else, M.A. and Jackson, M.B. (1998). Transport of 1-aminocyclopropane-1-carboxylic acid (ACC) in the transpiration stream of tomato (*Lycopersicon esculentum*) in relation to foliar ethylene production and petiole epinasty. Aust. J. Plant Physiol. 25: 453-458.

Fukuda, H., Ogawa, T. and Tanase, S. (1993). Ethylene production by microorganisms. Adv. Microb. Physiol. 35: 275-306.

Ghosh, S., Penterman, J.N., Little, R.D., Chavez, R. and Glick, B.R. (2003). Three newly isolated plant growth-promoting bacilli facilitate the growth of canola seedlings. Plant Physiol. Biochem. 41: 277-281.

Giovanelli, J., Mudd, S.H. and Datko, A.H. (1980). Sulfur amino acids in plants. In: The Biochemistry of Plants, Vol. 5, (ed.) B.J. Miflin, Academic Press, New York, USA, pp. 453-505.

Glick, B.R., Karaturovic, D.M. and Newell, P.C. (1995). A novel procedure for rapid isolation of plant growth-promoting pseudomonads. Can. J. Microbiol. 41: 533-536.

Glick, B.R., Penrose, D.M. and Li, J. (1998). A model for the lowering of plant ethylene concentrations by plant growth-promoting bacteria. J. Theor. Biol. 190: 63-68.

Glick, B.R., Patten, C.L., Holguin, G. and Penrose, D.M. (1999). Biochemical and genetic mechanisms used by plant growth-promoting bacteria. Imperial College Press, London, UK.

Glick, B.R. (2003). Phytoremediation: synergistic use of plants and bacteria to clean up the environment. Biotechnol. Adv. 21: 383-393.

Goren, R., Zhong, G.Y., Huberman, M., Feng, X., Holland, D. and Sisler, E.C. (2003). Effect of 1-methylcyclopropene on ethylene-induced abscission of *Citrus sinensis* leaves. . In: Biology and Biotechnology of the Plant Hormone Ethylene III (eds.) M. Vendrell, H. Klee, J.M. Pech and F. Romojaro, International Organizational Services B.V. Press, Burke, Virginia, USA, pp. 198-204.

Grichko, V.P., Filby, B. and Glick, B.R. (2000). Increased ability of transgenic plants expressing the bacterial enzyme ACC deaminase to accumulate Cd, Co, Cu, Ni, Pb, and Zn. J. Biotechnol. 81: 45-53.

Grichko, V.P. and Glick, B.R. (2001a). Amelioration of flooding stress by ACC deaminase-containing plant growth-promoting bacteria. Plant Physiol. Biochem. 39: 11-17.

Grichko, V.P. and Glick, B.R. (2001b). Flooding tolerance of transgenic tomato plants expressing the bacterial enzyme ACC deaminase controlled by the 35S, *rolD* or PRB-1*b* promoter. Plant Physiol. Biochem. 39: 19-25.

Hall, K.A., Peirson, D.G., Ghosh, S. and Glick, B.R. (1996). Root elongation in various agronomic crops by the plant growth promoting rhizobacterium *Pseudomonas putida* GR12-2. Israel J. Plant Sci. 44: 37-42.

Hammond-Kosack, K.E. and Jones, J.D.G. (1997). Plant disease resistance genes. Annu. Rev. Plant Physiol. Mol. Biol. 48: 575-607.

Henderson, J. and Osborne, D.J. (1999). Ethylene as the initiator of the inter-tissue signaling and gene expression cascades in ripening and abscission of oil palm fruit. In: Biology and Biotechnology of the Plant Hormone Ethylene II (eds.) A.K. Kanellis, C. Chang, H. Klee,

A.B. Bleecker, J.C. Pech and D. Greirson, Kluwer Academic Publishers, Dordrecht, The Netherlands, pp. 129-136.

Holguin, G. and Glick, B.R. (2003). Transformation of *Azospirillum brasilense* Cd with an ACC deaminase gene from *Enterobacter cloacae* UW4 fused to the Tet^r gene promoter improves its fitness and plant growth promoting ability. Microb. Ecol. 46: 122-133.

Honma, M. and Shimomura, T. (1978). Metabolism of 1-aminocyclopropane-1-carboxylic acid. Agric. Biol. Chem. 42: 1825-1831.

Honma, M. (1993). Stereospecific reaction of 1-aminocyclopropane-1-carboxylate deaminase. In: Cellular and Molecular Aspects of the Plant Hormone Ethylene (eds.) J.C. Pech, A. Latche and C. Balague, Kluwer Academic Publishers, Dordrecht, The Netherlands, pp. 111-116.

Hontzeas, N., Saleh, S.S. and Glick, B.R. (2004). Changes in gene expression in canola roots induced by ACC-deaminase containing plant growth promoting bacteria. Mol. Plant Microb. 17: 865-871.

Huang, X.D., El-Alawi, Y., Penrose, D.M., Glick, B.R. and Greenberg, B. (2004a). A multi-process phytoremediation system for removal of polycyclic aromatic hydrocarbons from contaminated soils. Environ. Pollut. 130: 465-476.

Huang, X.D, El-Alawi, Y., Penrose, D.M., Glick, B.R. and Greenberg, B. (2004b). Responses of plants to creosote during phytoremediation and their significance for remediation process. Environ. Pollut. 130: 453-463.

Hyodo H. (1991). Stress/wound ethylene. In: The Plant Hormone Ethylene (eds.) A. K. Matoo and J.C. Suttle, CRC Press, Boca Raton, Fl., US, pp. 65-80.

Jacobson, C.B., Pasternak, J.J. and Glick, B.R. (1994). Partial purification and characterization of ACC deaminase from the plant growth-promoting rhizobacterium *Pseudomonas putida* GR12-2. Can. J. Microbiol. 40: 1019-1025.

James, E.K. and Olivares, F.L. (1998). Infection and colonization of sugarcane and other graminaceous plants by endophytic diazotrophs. Crit. Rev. Plant Sci. 17: 77-119.

Jia, Y.J., Matsui, H. and Honma, M. (2000). 1-Aminocyclopropane-1-carboxylate (ACC) deaminase induced by ACC synthesized and accumulated in *Penicillium citrinum* intracellular spaces. Biosci. Biotechnol. Biochem. 64: 299-305.

John, P. (1997). Ethylene biosynthesis: the role of 1-aminocyclopropane-1-carboxylate (ACC) oxidase, and its possible evolutionary origin. Physiol. Plant. 100: 583-592.

Kende, H. (1993). Ethylene biosynthesis. Annu. Rev. Plant. Physiol., Plant Mol. Biol. 44: 283-307.

Klee, H.J., Hayford, M.B., Kretzmer, K.A., Barry, G.F. and Kishore, G.M. (1991). Control of ethylene synthesis by expression of a bacterial enzyme in transgenic tomato plants. Plant Cell. 3: 1187-1193.

Klee, H. J. and Kishore, G.M. (1992). Control of fruit ripening and senescence in plants. United States Patent Number: 5,702,933.

Klee, H.J. (1993). Ripening Physiology of fruit from transgenic tomato (*Lycopersicon esculentum*) plants with reduced ethylene synthesis. Plant Physiol. 102: 911-916.

Li, J. and Glick, B.R. (2001). Transcriptional regulation of the *Enterobacter cloacae* UW4 1-aminocyclopropane-1-carboxylate (ACC) deaminase gene (*acdS*). Can. J. Microbiol. 47: 359-367.

Liao, L.J., Lin, Y.H., Huang, K.L., Chen, W.S. and Cheng, Y.M. (2000). Postharvest life of cut rose flowers as affected by silver thiosulfate and sucrose. Bot. Bull. Acad. Sin. 41: 299-303.

Liu, X., Shiomi, S., Makatsuka, A., Kubo, Y., Nakamura, R. and Inaba, A. (1999). Characterization of ethylene biosynthesis associated with ripening in banana fruit. Plant Physiol. 121: 1257-1265.

Lund, S.T., Stall, R.E. and Klee, H.J. (1998). Ethylene regulates the susceptible response to pathogen infection in tomato. Plant Cell. 10: 371-382.

Ma, W., Guinel, F.C. and Glick, B.R. (2003a). *Rhizobium leguminosarum* biovar *viciae* 1-aminocyclopropane-1-carboxylate deaminase promotes nodulation of pea plants. Appl. Environ. Microbiol. 69: 4396-4402.

Ma, W., Sebestianova, S.B., Sebestian, J., Burd, G.I., Guinel, F.C. and Glick, B.R. (2003b). Prevalence of 1-aminocyclopropane-1-carboxylate deaminase in *Rhizobium* spp. Anton. van Leeuwenhoek. J. Microbiol. 83: 285-291.

Mayak, S., Tirosh, T. and Glick, B.R. (2004a). Plant growth-promoting bacteria that confer resistance to water stress in tomatoes and peppers. Plant Sci. 166: 525-530.

Mayak, S., Tirosh, T. and Glick, B.R. (2004b). Plant growth-promoting bacteria confer resistance in tomato plants to salt stress. Plant Physiol Biochem. 42: 565-572.

Melchers, L.S. and Stuiver, M.H. (2000). Novel genes for disease-resistance breeding. Curr. Opin. Plant Biol. 3: 147-152.

Michelmore, R. (2000). Genomic approaches to plant disease resistance. Curr. Opin. Plant Biol. 3: 125-131.

Minami, R., Uchiyama, K., Murakami, T., Kawai, J., Mikami, K., Yamada, T., Yokoi, D., Ito, H., Matsui, H. and Honma, M. (1998). Properties, sequence, and synthesis in *Escherichia coli* of 1-aminocyclopropane-1-carboxylate deaminase from *Hansenula saturnus*. J. Biochem. 123: 1112-1118.

Mizutani, F., Rabbany, A.B.M.G. and Akiyoshi, H. (1998) Inhibition of ethylene production by tropolone compounds in young excised peach seeds. J. Japan Soc. Hort. Sci. 67: 166-169.

Mol, J.N.M., Holton, T.A. and Koes, R.E. (1995). Floriculture: Genetic engineering of commercial traits. Trends Biotechnol. 13 (9): 350-356.

Murphy, D.J. (2003). Agricultural biotechnology and oil crops: current uncertainty and future potentials. Appl. Biotechnol. Food Sci. Policy 1(1): 25-38.

Nayani, S., Mayak, S. and Glick, B.R. (1998). The effect of plant growth promoting rhizobacteria on the senescence of flower petals. Ind. J. Exp. Biol. 36: 836-839.

Nie, L., Shah, S., Burd, G.I., Dixon, D.G. and Glick, B.R. (2002). Phytoremediation of arsenate contaminated soil by transgenic canola and the plant growth-promoting bacterium *Enterobacter cloacae* CAL2. Plant Physiol. Biochem. 40: 355-361.

Oeller, P.W., Min-Wong, L., Taylor, L.P., Pike, D.A. and Theologis, A. (1991). Reversible inhibition of tomato fruit senescence by antisense RNA. Science 254: 437-439.

Olson, D.C., Oetiker, J.H. and Yang, S.F. (1995). Analysis of *LE-ACS3*, a 1-aminocyclopropane-1-carboxylic acid synthase gene expressed during flooding in the roots of tomato plants. J. Biol. Chem. 270: 14056-14061.

Pech, J.C., Guis, M., Botondi, R., Ayub, R., Bouzayen, M., Lelievre, J.M., El Yahyaoui, F. and Latche, A. (1999). Ethylene-dependent and ethylene-independent pathways in a climacteric fruit, the melon. In: Biology and Biotechnology of the Plant Hormone Ethylene II (eds.) A.K. Kanellis, C. Chang, H. Klee, A.B. Bleecker, J.C. Pech and D. Greirson, Kluwer Academic Publishers, Dordrecht, The Netherlands, pp. 105-110.

Penrose, D.M. and Glick, B.R. (2003). Methods for isolating and characterizing ACC deaminase-containing plant growth-promoting rhizobacteria. Physiol. Plant. 118: 10-15.

Philosoph-Hadas, S., Freidman, H., Berkovitz-Simantov, R., Rosenberger, I., Woltering, E.J., Halvey, A.H. and Meir, S. (1999). Involvement of ethylene biosynthesis and action in regulation of the gravitropic response of cut flowers. In: Biology and Biotechnology of the Plant Hormone Ethylene II (eds.) A.K. Kanellis, C. Chang, H. Klee, A.B. Bleecker, J.C. Pech and D. Greirson, Kluwer Academic Publishers, Dordrecht, The Netherlands, pp. 151-156.

Ray, R.C. and Edison, S. (2005). Microbial biotechnology in agriculture and aquaculture- an overview. In: Microbial Biotechnology in Agriculture and Aquaculture, Volume 1 (ed.) R.C. Ray, Science Publishers, Enfield, New Hampshire, USA, pp. 1-30.

Ray, R.C. and Ravi, V. (2005). Post harvest spoilage of sweet potato in the tropics and control measures. Crit. Rev. Food Sci. Nutr. 45: 623-644.

Robison, M.M., Shah, S., Tamot, B., Pauls, K.P., Moffatt, B.A. and Glick, B.R. (2001a). Reduced symptoms of *Verticillium* wilt in transgenic tomato expressing a bacterial ACC deaminase. Mol. Plant Pathol. 2: 135-145.

Robison, M.M., Griffith, M., Pauls, K.P. and Glick, B.R. (2001b). Dual role for ethylene in susceptibility of tomato to *Verticillium* wilt. J. Phytopathol. 149: 385-388.

Rodriguez, H. and Fraga, R. (1999). Phosphate solubilizing bacteria and their role in plant growth promotion. Biotechnol. Adv. 17: 319-339.

Saleh, S. and Glick, B.R. (2006). A review of plant growth-promoting bacteria. In: Modern Soil Microbiology (eds.) J. Trevors and A. van Elsas, Marcel Dekker Inc., New York, USA, in press.

Sergeeve, E., Shah, S. and Glick, B.R. (2006). Growth of transgenic canola (*Brassica napus* cv. *Westrar*) expressing a bacterial 1-aminocyclopropane-1-carboxylase (ACC) deaminase gene on high concentrations of salt. World J. Microbiol Biotechnol. 22: 277-282.

Shah, S., Li, J., Moffat, B.A. and Glick, B.R. (1998). Isolation and characterization of ACC deaminase genes from two different plant growth-promoting rhizobacteria. Can. J. Microbiol. 44: 833-843.

Sheehy, R.E., Honma, M., Yamada, M., Sasaki, T., Martineau, B. and Hiatt, W.R. (1991). Isolation, sequence, and expression in *Escherichia coli* of the *Pseudomonas* sp. Strain ACP gene encoding 1-aminocyclopropane-1-carboxylate deaminase. J. Bacteriol. 173: 5260-5265.

Sisler, E.C. and Serek, M. (1997). Inhibitors of ethylene responses in plants at the receptor level: Recent developments. Physiol. Plant. 100: 577-582.

Stearns, J.C. and Glick, B.R. (2003). Transgenic plants with altered ethylene biosynthesis or perception. Biotech. Adv. 21: 193-210.

Stearns, J.C., Shah, S. and Glick, B.R. (2003). Transgenic canola containing ACC deaminases have increased tolerance to nickel. Biological effects of pollutants: Proteomics and Genomics, 22nd conference of the European Society for Comparative Physiology and Biochemistry, Alessandria, Italy, 14-18, December 2003.

Stepanova, A.N. and Ecker, J.R. (2000). Ethylene signaling: from mutants to molecules. Curr. Opin. Plant Biol. 3: 353-360.

Urao, T., Yamaguchi-Shinozaki, K. and Shinozaki, K. (2000). Two-component systems in plant signal transduction. Trends Plant Sci. 5: 67-74.

van Loon, L.C. (1984). Regulation of pathogenesis and symptom expression in diseased plants by ethylene. In: Ethylene: Biochemical, Physiological and Applied Aspects (eds.) Y. Fuchs and E. Chalutz, Martinus Nijhoff/Dr W. Junk, The Hague, The Netherlands, pp. 171-180.

van Loon, L.C. and Glick, B.R. (2004). Increased Plant Fitness by Rhizobacteria. In: Molecular Ecotoxicology of Plants (ed.) H. Sandermann, Springer-Verlag, Berlin, Heidelberg, Germany, pp. 177-207.

Wang, C., Knill, E., Glick, B.R. and Degafo, G. (2000). Effect of transferring 1-aminocyclopropane-1-carboxylic acid (ACC) deaminase genes into *Pseudomonas fluorescens* strain CHA0 and its gacA derivative CHA96 on their growth-promoting and disease-suppressive capacities. Can. J. Microbiol. 46: 898-907.

Watkins, C.B. and Miller, W.B. (2003). Implications of 1-methylcyclopropene registration for use on horticultural products. In: Biology and Biotechnology of the Plant Hormone

Ethylene III (eds.) M. Vendrell, H. Kee, J.C. Pech and F. Romojaro, International Organizational Services B.V. Press, Burke, Virginia, US, Amsterdam, The Netherlands, pp. 385-390.

Watrud, L.S. (2000). Genetically engineered plants in the environment- applications and issues. In: Microbial Interactions in Agriculture and Forestry (eds.) N.S. Subba Rao and Y.R. Dommergues, Science Publishers, Enfield, New Hampshire, USA, pp. 61-82.

Whipps, J.M. (1990). Carbon utilization. In: The Rhizosphere (ed.) JM Lynch,Wiley Interscience, Chichester, UK, pp. 59-97.

Woltering, E.J. and van Doorne, W.G. (1988). Role of ethylene in senescence of petals-morphological and taxonomical relationships. J. Exp. Bot. 39: 1605-1616.

Yang, S.F. and Hoffman, N.E. (1984). Ethylene biosynthesis and its regulation in higher plants. Annu. Rev. Plant Physiol. 35: 155-189.

Yang, Y., Shah, J. and Klessig, D.F. (1997). Signal perception and transduction in plant defense responses. Genes Develop. 11: 1621-1639.

Yasuta, T., Satoh, S. and Minamisawa, K. (1999). New assay for rhizobitoxine based on inhibition of 1-aminocyclopropane-1-carboxylate synthase. Appl. Environ. Microbiol. 65: 849-852.

Yu, I.C., Parker, J. and Bent, A.F. (1998). Gene-for-gene disease resistance without the hypersensitive response in *Arabidopsis dnd1* mutant. Proc. Natl. Acad. Sci. US, 95: 7819-7824.

Zehzouti, H., Jones, B., Tournier, B., Leclercq, J., Bernadac, A. and Bouzayen, M. (1999). Isolation and characterization of novel ethylene-responsive cDNA clones involved in signal transduction, transcription and mRNA isolation. In: Biology and Biotechnology of the Plant Hormone Ethylene II (eds.) A.K. Kanellis, C. Chang, H. Klee, A.B. Bleecker, J.C. Pech and D. Greirson, Kluwer Academic Publishers, Dordrecht, The Netherlands, pp. 111-118.

7

Rhizobial Strain Improvement: Genetic Analysis and Modification

*Marta Laranjo and Solange Oliveira**

INTRODUCTION

The symbiosis between rhizobia and legumes is a special example of an association between prokaryotic and eukaryotic organisms (Perret et al., 2000) that plays an important role in agriculture and has thus been intensively studied over the recent years. The rhizobia-legume symbiosis is the most important biological mechanism for providing nitrogen to the soil/plant system, reducing the need of plant crops for chemical fertilizers. This justifies the ecological importance of legume crops not only in the improvement of nitrogen nutrition to other crops, such as pasture or cereals, but also in land remediation.

The main advantage of biological N_2-fixation, in grain and forage legumes, is the increase in yield potential without the use of chemical nitrogen fertilizers and with the consequent decrease in pollution (Freiberg et al., 1997; Stephens and Rask, 2000). Furthermore, the use of rhizobial inoculants has permitted the introduction of legumes outside their native regions.

Rhizobia are soil bacteria, which are able to infect the roots of leguminous plants, promoting the formation of N_2-fixing nodules. These are produced as a result of complex signal exchanges between the symbiotic partners. This chemical dialogue is responsible for the specificity of the symbiosis: microsymbionts can have a narrow or broad host range and, similarly, host legumes are either specific or promiscuous (Fisher and Long, 1992; Spaink, 1995; van Rhijn and Vanderleyden, 1995; Denarié et al., 1996). Table 7.1 shows the rhizobial species that usually

Department of Biology, University of Évora, Apartado 94, 7002-554 Évora, Portugal
**Corresponding author: E-mail: ismo@uevora.pt*

Table 7.1. Rhizobial species that commonly nodulate legumes used for food and forage

Legume			
Common Name	Species	Rhizobial Species	Reference
Chickpea	*Cicer arietinum*	*Mesorhizobium ciceri*	Nour et al., (1994)
		M. mediterraneum	Nour et al., (1995)
Common bean	*Phaseolus vulgaris*	*Rhizobium etli*	Segovia et al., (1993)
		R. gallicum	Amarger et al., (1997)
		R. giardinii	Amarger et al., (1997)
		R. leguminosarum bv. *phaseoli*	Frank (1889)
		R. mongolense	van Berkum et al., (1998)
		R. tropici	Martínez-Romero et al., (1991)
Faba bean	*Vicia faba*	*R. leguminosarum* bv. *viciae*	Frank (1889)
Soybean	*Glycine max*	*Bradyrhizobium elkanii*	Kuykendall et al., (1992)
		B. japonicum	Jordan (1982)
		B. liaoningense	Xu et al., (1995)
		S. fredii	de Lajudie et al., (1994)
		Sinorhizobium xinjiangense	Peng et al., (2002)
Pea	*Pisum sativum*	*R. leguminosarum* bv. *viciae*	Frank (1889)
Alfalfa	*Medicago* spp.	*S. medicae*	Rome et al., (1996)
		S. meliloti	
			de Lajudie et al., (1994)
Clover	*Trifolium* spp.	*R. leguminosarum* bv. *trifolii*	Frank (1889)

nodulate some of the most cultivated legumes in the world. The rhizobia described so far belong to four different 16S rDNA branches in the α-Proteobacteria: *Azorhizobium*; *Bradyrhizobium*; *Mesorhizobium-Rhizobium-Sinorhizobium* (Young et al., 2001); and *Methylobacterium* (Ibekwe et al., 1997; Sy et al., 2001). More recently, rhizobia from the β-Proteobacteria have also been described (Moulin et al., 2001), changing the current rhizobial taxonomy.

N_2-fixation depends not only on the physiological state of the host plant, but also on the effectiveness and environmental fitness of the microsymbiont and on the interaction between the two partners.

Rhizobia strains used as inoculants must be competitive with the indigenous populations for nodule occupancy and be efficient in N_2-fixation afterwards (Carter et al., 1995; Stephens and Rask, 2000; Bloem and Law, 2001; Okogun and Sanginga, 2003). Inoculant strains must be selected for individual cultivars of a given legume, and also preferably for particular soil and environmental conditions; they should show high tolerance to climatic and edaphic stress conditions, such as extreme temperatures, adverse pH, and drought (Stephens and Rask, 2000). Another fundamental characteristic is its persistence in the soil over time (Carter et al., 1995; Howieson, 1995).

At the recommended inoculation rates (Somasegaran and Hoben, 1994; Brockwell and Bottomley, 1995), inoculant strains often dominate in nodulation in the first year of a newly introduced crop (Singleton and Tavares, 1986). Furthermore, on nodule senescence, a large number of viable rhizobia are released in the soil (McDermott et al., 1987). Several studies have shown that inoculant strains still predominate in nodules 5 to 15 years after initial inoculation (Lindström et al., 1990), confirming that they are effective soil saprophytes persisting in soil for many years in the absence of their host (Sanginga et al., 1994).

The main objectives of strain modification for inoculant development are the increase of symbiotic efficiency, the introduction of tolerance traits to adverse environmental and soil conditions and, eventually, the extension of nodulation ability to other hosts.

GENETIC ANALYSIS

In order to plan strain improvement strategies for the development of rhizobial inoculants, molecular analysis of genes involved in symbiosis, competitiveness and stress response is of major importance.

Analysis of Complete Genomes

The analysis of complete genomes of rhizobia has improved knowledge on symbiotic and stress related genes (Downie and Young, 2001).

The genes involved in symbiosis can be located on symbiotic plasmids (p*Sym*), as in *S. meliloti* (Barloy-Hubler et al., 2000; Finan et al., 2001), *R. etli* (González et al., 2003) and *Mesorhizobium amorphae* (Wang et al., 1999), or on a "symbiotic island" of low G+C content in the chromosome (Downie and Young, 2001), like in *Mesorhizobium loti* (Sullivan et al., 1995) and *B. japonicum* (Kaneko et al., 2002b). The chromosomal location has also been suggested for chickpea rhizobia (Cadahía et al., 1986).

In evolutionary terms, plasmid encoded symbiotic genes have the advantage of being more easily exchangeable within populations, than genes located in the chromosome (Downie, 1997). Acquisition and loss of *Sym* plasmids are a dynamic and continuous process, which seems to play a major role in the evolution of rhizobia. Contrary to symbiotic genes, stress related genes are usually located on the chromosome (Nakamura et al., 2004). There are currently two major databases that gather most information available on complete genomes of rhizobia and related bacteria: RhizoBase (http://www.kazusa.or.jp/rhizobase) and RhizoBD (http://www.rhizo.bham.ac.uk).

The first completely sequenced rhizobial genome was that of *M. loti* strain MAFF303099 (Kaneko et al., 2000a, b) (http://www.kazusa.or.jp/rhizobase/), which comprises one chromosome (7.0 Mb) and two plasmids (352 and 208 kb). Many of the symbiotic genes are located in the chromosome on a "symbiotic island" of 611 kb (Hirsch et al., 2001).

Immediately afterwards, the sequence of *S. meliloti* strain 1021 was published by a French consortium (Capela et al., 1999; Barloy-Hubler et al., 2000; Finan et al., 2001; Galibert et al., 2001). This genome is composed of a circular chromosome (3.6 Mb) and two megaplasmids: pSymA (1354 kb) and pSymB (1683 kb) (http://www.bioinfo.genopole-toulouse.prd.fr/annotation/iANT/bacteria/rhime/). Most symbiotic genes are located in the two *Sym* plasmids (Nakamura et al., 2004).

The complete genome of *B. japonicum* strain USDA110, which was released in 2002 (Barloy-Hubler et al., 2001; Kaneko et al., 2002a, b), consists of a single very large circular chromosome (9.1 Mb) (http://www.kazusa.or.jp/rhizobase/Bradyrhizobium/). The complete genome of the *Rhizobium etli* type strain CFN 42^T has been published in March 2006 (González et al., 2006) (http://www.cifn.unam.mx/retlidb/). This genome is composed by seven replicons: one chromosome and six plasmids, ranging from about 180 to 640 kb. The total genome size is approximately 6.5 Mb. The symbiotic plasmid, which had been sequenced before (González et al., 2003), is p42d (371 kb).

The genome sequence of *R. leguminosarum* bv. *viciae* strain 3841 is available from the Sanger Institute webpage (http://www.sanger.ac.uk/Projects/R_leguminosarum/). The total genome size is over 7.7 Mb. It is composed of one circular chromosome and six plasmids, ranging from about 150 to 870 kb each. The symbiotic plasmid is pRL10JI (488 kb).

All strains of *R. leguminosarum* have several large plasmids, but the numbers and sizes of plasmids vary among strains. These large numbers of plasmids distinguish *R. leguminosarum* from the other rhizobia for which complete genome sequences have been determined.

Several symbiotic plasmids and "islands" from different rhizobial species have also been sequenced: *Rhizobium* sp. NGR234 symbiotic plasmid pNGR234a (536 kb) (Freiberg et al., 1997) and pNGR234b (594 kb) (Streit et al., 2004) (http://www.kazusa.or.jp/rhizobase/NGR234/); *M. loti* strain R7A symbiosis island (500 kb) (Sullivan et al., 2002) (http://www.kazusa.or.jp/rhizobase/R7A).

Furthermore, the following sequencing projects are currently ongoing: *Rhizobium* sp. strain NGR234 (ANU 265) (http://www.g2l.bio.uni-goettingen.de/projects/f_projects.html); *Rhizobium tropici* strain PRF 81 (http://www.bnf.lncc.br/rht/index.php); and *Bradyrhizobium* sp. strain BTAi (http://www.genome.jgi-psf.org/draft_microbes/bra_b/bra_b.home.html). Another member of the Rhizobiaceae is the plant-pathogenic bacterium *Agrobacterium tumefaciens*, whose full sequence of its type strain C58 has also been determined (Allardet-Servent et al., 1993; Wood et al., 2001) (http://www.depts.washington.edu/agro/). Its genome is composed of one circular (2.8 Mb) and one linear chromosome (2.1 Mb), the tumour inducing plasmid pTiC58 (214 kb) and the pAtC58 (543 kb). *Agrobacterium radiobacter* bv. II strain K84 and *Agrobacterium vitis* bv. III strain S4 is also being sequenced by the same laboratory (http://www.depts.washington.edu/agro/).

Rhodopseudomonas palustris is a purple non-sulphur phototrophic bacterium that bears a close relationship to *Bradyrhizobium*. The sequences and annotation of strain CGA 009 were released in 2004 (Larimer et al., 2004) (http://www.genome.jgi-psf.org/finished_microbes/rhopa/rhopa.home.html). This genome is composed of one chromosome (5.4 Mb) and one plasmid (8.4 kb). Other available complete genomes of related bacteria are those of *Brucella melitensis* (Del Vecchio et al., 2002) strain 16M and *B. suis* (Paulsen et al., 2002) strain 1330, which are animal pathogens that have fundamental similarities with plant pathogens and symbionts such as rhizobia. They both have two small chromosomes (2.1 and 1.2 Mb). The larger chromosome shows extensive synteny with the chromosomes of *Sinorhizobium meliloti* and *Agrobacterium tumefaciens*.

The availability of complete genome sequences has demonstrated that lateral gene transfer and recombination are essential components in the evolution of bacterial genomes (van Berkum et al., 2003). Furthermore, the ability of bacteria to exploit new environments and to respond to different selective pressures is more easily explained by the acquisition of new genes by horizontal gene transfer than by gene modification through accumulation of point mutations (Francia et al., 2004).

Work on whole genome sequence analysis of *S. meliloti* 1021 (Guo et al., 2003) and *Rhizobium* sp. NGR234 (Mavingui et al., 2002) has shown that the structure of rhizobial genomes can be highly dynamic generating different genomic rearrangements, through homologous recombination between reiterated DNA sequences.

The transfer of symbiotic information in soil and the formation of new symbiotic plasmids by genetic recombination could result in a heterogeneous population with high adaptive capacity to interact and nodulate different legumes (Soberón-Chávez et al., 1991).

It is now possible to analyze whole genomes of fully sequenced strains by DNA microarrays (Brown and Botstein, 1999; Bontemps et al., 2005) in order to obtain more information on the expression pattern and function of rhizobial genes, namely symbiotic genes.

Symbiotic Genes

The symbiotic genes, which are responsible for the establishment of an effective symbiosis, can be divided into two categories: N_2-fixation genes (*nif* and *fix*) (Kaminski et al., 1998) that are involved in atmospheric N_2-fixation, and nodulation genes (*nod, nol* and *noe*) (Downie, 1998), which are responsible for nodulation (van Rhijn and Vanderleyden, 1995; Freiberg et al., 1997). Other genes involved in the process of symbiosis are listed in Table 7.2.

Several studies have shown that relatively few genes are required for nodulation of legumes (Long, 1989; van Rhijn and Vanderleyden, 1995).

The different rhizobial lineages probably diverged before the appearance of legumes (Turner and Young, 2000). Thus nodulation capacity may have been acquired in a single lineage after the diversification of bacteria and, consequently, symbiotic genes may have spread by lateral gene transfer among the different rhizobia (Hirsch et al., 2001). This hypothesis is supported by recent data obtained from the fully sequenced plasmids and genomes (Sullivan and Ronson, 1998; Kaneko et al., 2000b; Galibert et al., 2001; Sullivan et al., 2002).

Complex events of gene recruitment, combined with duplication and lateral gene transfer, contributed to the evolution and spread of symbiotic abilities among rhizobia (Moulin et al., 2004).

Plant genes involved in symbiosis have been identified by Expressed Sequence Tags (ESTs) sequencing in *Medicago truncatula* (Journet et al., 2002) providing the means to study the interaction between macro- and microsymbiont, at the level of gene networks, in the symbiotic N_2-fixation process.

Table 7.2. Other rhizobial genes involved in symbiotic N_2-fixation

Genes	Function	Species	Reference
exo	Exopolyssacharide production	*Sinorhizobium meliloti*	Glazebrook and Walker (1989)
hup	Hydrogen uptake	*Bradyrhizobium japonicum*	Maier (1986)
gln	Glutamine synthase	*Rhizobium phaseoli*	Carlson et al., (1987)
dct	Dicarboxylate transport	*R. leguminosarum* *S. meliloti*	Jiang et al., (1989)
nfe	Nodulation efficiency	*S. meliloti*	Sanjuan and Olivares (1989)
ndv	β-1,2 glucans synthesis	*S. meliloti*	Breedveld et al., (1994)
lps	Lipopolysaccharide production	*R. phaseoli*	Carlson et al., (1987)

Nodulation Genes

Nodule formation requires a complex and dynamic molecular interaction between the two symbiotic partners. The onset of symbiosis is mediated by an exchange of diffusible signals (Caetano-Anollés, 1997).

Flavonoids seem to be the most important plant exudates for the symbiosis (Perret et al., 2000) since they direct the bacteria to the rhizosphere by chemotaxis and enable the rhizobia to distinguish their host (Hirsch et al., 2001). The flavonoids or isoflavonoids induce the expression of nodulation genes together with the NodD protein (van Rhijn and Vanderleyden, 1995).

The genetic control of nodulation by rhizobia involves nodulation genes that determine the production of lipo-chitooligosaccharides (LCOs) known as Nod factors. Nod factors act as specific morphogenic signal molecules on legume hosts and induce early host responses such as root hair deformation and cortical cell division (Cullimore et al., 2001). They are indispensable for nodulation and are seemingly recognized by high-affinity legume receptors (Cullimore et al., 2001).

Structural nodulation genes, which encode Nod factors, can be classified into two groups, (1) the common, and (2) the host-specific genes (*hsn*). Additionally, genotype-specific nodulation (*gsn*) genes have also been described (Sadowsky et al., 1991). These genes are responsible for the nodulation of specific plant genotypes within a single legume species.

The common *nod*ABC genes are structurally and functionally conserved and are usually part of a single operon (van Rhijn and Vanderleyden, 1995). The lipo-chitooligosaccharide backbone is synthesized under the control of *nod*A (acyl transferase), *nod*B (deacetylase) and *nod*C (N-acetylglucosaminyl transferase), all of which are present as single copy genes in most known rhizobia (Rastogi, 1992; Downie, 1998).

*Nod*A varies in its fatty acid specificity, thus contributing to the rhizobial host range (Roche et al., 1996). *Nod*I and *nod*J are located downstream of *nod*C and seem to be part of the same operon (van Rhijn and Vanderleyden, 1995). They encode two proteins belonging to a bacterial inner membrane transport system of small molecules (Vázquez et al., 1993).

The *hsn* genes are responsible for the host range and most mutations result in alteration or extension of the host range (van Rhijn and Vanderleyden, 1995). For example, *nod*F and *nod*G encode acyl carrier proteins, *nod*E encodes a β-acetoacethylsynthase (Sharma et al., 1993) and *nod*P, *nod*Q and *nod*H are involved in the sulfation of the Nod factor reducing sugar in *S. meliloti* (Roche et al., 1991).

Examples of *gsn* genes are *nol*C, which are involved in cultivar-specific nodulation of soybean and share homology with a heat-shock gene

(Krishnan and Pueppke, 1991) and *nol*BTUVW that regulate the nodulation of *Glycine max* in a cultivar-specific manner (Meinhardt et al., 1993).

Most *nod* genes share a conserved promoter sequence called the *nod* box (Freiberg et al., 1997).

The regulatory NodD protein is crucial in the activation of other *nod* genes. *Nod*D is the only nodulation gene that is constitutively expressed even in non-symbiotic bacteria (Long, 1989). The spectrum of flavonoid specificity of NodD correlates with the broadness of the host range (van Rhijn and Vanderleyden, 1995).

Phylogeny based on common nodulation genes does not correlate with the 16S phylogeny of rhizobia; instead it reflects the phylogeny of host plants (Dobert et al., 1994).

N_2-fixation Genes

There are two main types of N_2-fixation genes: *nif* and *fix* genes. Rhizobial *nif* genes are structurally homologous to those of *Klebsiella pneumoniae*, a free-living diazotrophic bacteria (Drummond, 1984; Arnold et al., 1988). *Fix* genes are essential to N_2-fixation, but do not have homologous equivalents in *K. pneumoniae* (Fischer, 1994).

The nitrogenase complex, which has the catalytic function of N_2-fixation, is encoded by the structural genes *nif*HDK. *Nif*D and *nif*K code for the α- and β-subunits of the dinitrogenase or Mo-Fe protein subunit. *Nif*H codes for the nitrogenase reductase or Fe protein subunit of the enzyme complex. Phylogenetic studies have preferably focused on this N_2-fixation gene. A Mo-Fe cofactor is required for the activation of the Mo-Fe protein and is assembled from the *nif*B, V, N and E gene products (Freiberg et al., 1997).

The gene products of *nif*A and *nif*L control the regulation of all other *nif* genes: NifA is the positive activator of transcription of *nif* operons and NifL is the negative activator (Morett et al., 1990).

Nif gene expression is regulated by oxygen and ammonia levels in the soil. Ammonia causes *nif*L to act as a negative control and prevents the activator function of *nif*A. This has been called the nitrogen (N) control system (Sadowsky and Graham, 2000). Symbiotic N_2-fixation is influenced by the presence of mineral nitrogen in the soil: the nodulation process may be promoted by relatively low levels of available nitrate or ammonia, higher concentrations of which usually depress nodulation (Davidson and Robson, 1986a, b; Eaglesham, 1989). Furthermore, soil NO_3^- inhibits root infection, nodule development and nitrogenase activity (Zahran, 1999).

Nitrogenase is rapidly inactivated by oxygen, thus the enzyme must be protected from O_2. In the root nodules of legumes, protection is provided by the symbiotic synthesis of the red pigmented leghemoglobin, an oxygen binding protein (Quispel, 1988). The globin portion of the protein is synthesized by the plant in response to bacterial infection, while the heme group is synthesized by the rhizobia (Appleby, 1984). Leghemoglobin has a high affinity to O_2, keeping its concentration low enough to protect the nitrogenase, while providing O_2 transport for the aerobic rhizobia.

The N_2-fixation turnover is strongly associated with the concentration of leghemoglobin (Werner et al., 1981).

Leghemoglobins are interesting proteins as they are the only globins synthesized by plants, having genetic and structural homologies with animal oxyhemoproteins, like hemoglobin (Appleby, 1984). They occur as multigene families in legumes, and several protein forms are expressed in nodules (Long, 1989).

Rhizobia have developed an independent enzymatic complex, *Hup* (Hydrogen uptake) system with nitrogenase activity, able to oxidize all or part of the hydrogen produced by nitrogenase, allowing the recycling of hydrogen (Baginsky et al., 2002).

The *fix*ABCX genes are usually located on a single operon, except in *B. japonicum*, and are directly involved in the electron transfer to nitrogenase (Fischer, 1994). The FixNOPQ proteins constitute a membrane-bound cytochrome C-oxidase complex, which supports bacteroid respiration under low oxygen conditions inside the nodule (Preisig et al., 1993). The *fix*GHIS genes encode transmembrane proteins (Fischer, 1994). Other *fix* genes are associated to the development and metabolism of bacteroids (Fischer, 1994).

Most species of rhizobia carry large plasmids that may vary in number and size. These are not always associated with symbiosis; they may contain essential or housekeeping genes (Barnett et al., 2001; Downie and Young, 2001; Finan et al., 2001; Galibert et al., 2001) and some of them are cryptic (Mercado-Blanco and Toro, 1996).

A negative correlation has been detected between symbiotic effectiveness (SE) and plasmid number (Fig. 7.1) (Thurman et al., 1985; Harrison et al., 1988; Laranjo et al., 2001, 2002). In fact, symbiotic effectiveness is usually higher in rhizobial isolates carrying one plasmid, which suggests that additional plasmids may confer some competitive advantages not directly related to the symbiosis (Thurman et al., 1985).

Stress Genes

Environmental conditions can act as limiting factors to the establishment of the symbiosis and to N_2-fixation. Different limiting or stress factors can be distinguished: salt and osmotic stress; soil moisture deficiency;

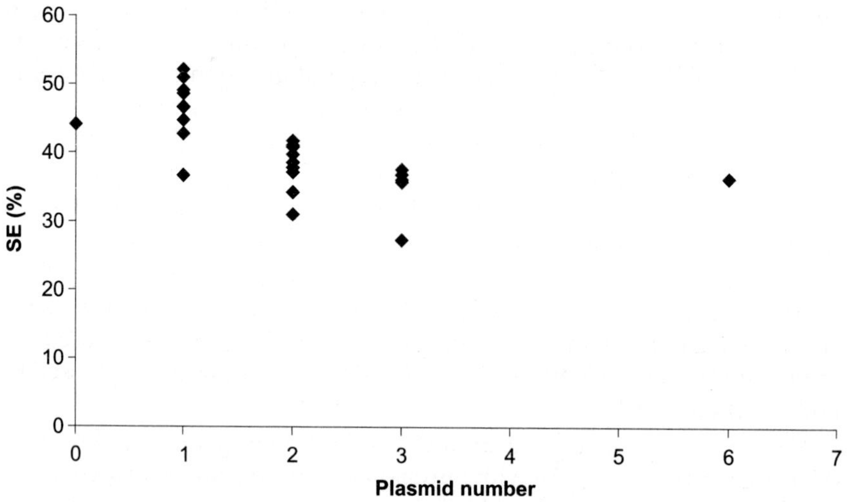

Fig. 7.1. Correlation between plasmid number and symbiotic effectiveness (SE) of chickpea rhizobia isolates from Portuguese soils (Laranjo et al., 2001)

temperature extremes and stress; soil acidity and sodicity; nutrient deficiency; soil amendment and amelioration (Zahran, 1999; 2005).

Temperature stress

When subjected to a sudden increase in growth temperature, rhizobia, like other bacteria, show an induced synthesis of several heat shock proteins (Hsps), such as chaperons, small Hsps (sHsps) and proteases (Table 7.3) (Nakamura et al., 2004). Chaperons are involved in the proper folding of denaturated proteins; proteases are required for the degradation of irreversibly damaged proteins. These two types of Hsps are also important during normal growth. On the contrary, sHsps are only active during heat stress, maintaining denaturated proteins in a folding competent state (Münchbach et al., 1999; Nocker et al., 2001).

Most bacteria apparently have a small number of Hsps (Münchbach et al., 1999) according to a survey on the microbial genomes sequenced so far. Rhizobia can contain multiple *sHsps* genes (Michiels et al., 1994), like, for instance, *B. japonicum* that has twelve (Münchbach et al., 1999). It is interesting to note that a heat-sensitive bean nodulating *Rhizobium* strain presented several *Hsps*, but a heat-tolerant one enclosed only two (Michiels et al., 1994), as detected by two-dimensional gel analysis.

The GroEL has been one of the first and most intensively studied stress related proteins (Fayet et al., 1989). In rhizobia, the *gro*EL gene is expressed constitutively and it encodes a chaperonin that is co-regulated with

Table 7.3. Some stress related genes identified in rhizobia (Nakamura et al., 2004). Most genes are located in the chromosome

Gene	Function	Organism
*clp*B	Heat shock protein	*S. meliloti*
*cop*B	Copper tolerance protein	*B. japonicum*
*cop*C	Copper tolerance protein	*B. japonicum*
*csp*A	Cold shock protein	*B. japonicum*
*dna*J	Heat shock protein	*M. loti*
*dna*K	Heat shock protein	*M. loti*
*gro*EL	Chaperonin	*B. japonicum*
*gro*ES	Heat shock protein	*B. japonicum* *M. loti*
*grp*E	Heat shock protein	*M. loti*
*hsl*O	Heat shock protein	*B. japonicum*
*hsl*U	Small heat shock protein	*B. japonicum* *M. loti* *S. meliloti*
*hsl*V	Heat shock protein	*B. japonicum* *M. loti* *S. meliloti*
*hsp*33	Heat shock protein	*M. loti*
*hsp*A	Small heat shock protein	*B. japonicum*
*hsp*B	Small heat shock protein	*B. japonicum*
*hsp*C	Small heat shock protein	*B. japonicum* *M. loti* *S. meliloti*
*hsp*D	Small heat shock protein	*B. japonicum*
*hsp*E	Small heat shock protein	*B. japonicum*
*hsp*F	Small heat shock protein	*B. japonicum*
*hsp*H	Small heat shock protein	*B. japonicum* *M. loti*
*htp*G	Heat shock protein	*M. loti* *S. meliloti*
*htp*X	Heat shock protein	*B. japonicum* *M. loti* *S. meliloti*
*htr*A	Heat shock protein	*B. japonicum* *M. loti* *R. etli*
*ibp*A	Small heat shock protein	*S. meliloti*
*nol*C	Heat shock protein	*R. etli*
ohr	Putative stress induced protein	*B. japonicum* *S. meliloti*

symbiotic N_2-fixation genes (Govezensky et al., 1991; Fischer et al., 1993). This protein is required for the assembly of a functional nitrogenase (Govezensky et al., 1991; Fischer et al., 1999) and for the correct folding of the regulatory protein NodD (Govezensky et al., 1991; Fischer et al., 1993, 1999; Ribbe and Burgess, 2001; Yeh et al., 2002).

NolC seems to be a heat shock protein with strong sequence homology to *Escherichia coli*'s *DnaJ*, although it does not directly function in the heat-shock response (Krishnan and Pueppke, 1991).

Acidity Stress

Rhizobia exhibit an adaptive acid tolerance response (ATR) that is influenced by calcium concentration (Dilworth et al., 2001). The ATR has been shown to require de novo protein synthesis during adaptation to acid stress (O'Hara and Glenn, 1994).

Genes essential for growth at low pH, such as *act*A, *act*P, *exo*R, *act*R and *act*S have been identified in *S. meliloti*, using Tn5 mutagenesis, *gus*A fusions and proteome analysis (Glenn et al., 1999). *Act*A encodes a lipid acyl transferase, *act*R a cognate regulator and *act*S a histidine protein kinase "sensor" (Reeve et al., 1998). These genes seem to be involved in a signal transduction system and are constitutively expressed (Tiwari et al., 1996a, b).

Such an adaptive response may contribute to the effective nodulation and persistence of rhizobial strains in soil microenvironments, which show marked differences in pH (Rickert et al., 2000). However, different species may vary in their response to low pH (Rickert et al., 2000).

An important practical implication of the ATR might be in the preparation of inoculants for acid soils: cultures that have been exposed to sub lethal acid conditions will have better survival chances in soils with low pH, than cultures that have been grown at neutral pH (Rickert et al., 2000).

Salt Stress

Under salt stress, an osmoadaptive response is induced in rhizobia, which is mainly characterized by the repression of the flagella and pilin biosynthesis genes (*fla*A, *fla*B, *fla*C, *fla*D, *pil*A), as well as chemotaxis genes (*che*Y1, *che*W3, *mcp*X, *mcp*Z) and by the induction of genes involved in the surface polysaccharide biosynthesis (*exo*N, *exo*Y, *exs*I) (Rüberg et al., 2003).

The *act*A gene, involved in lipid metabolism, plays a role in tolerance to heavy metals (Reeve et al., 1998; Dilworth et al., 2001). The *phr*R gene in *S. meliloti* is induced by high concentrations of Zn^{2+}, Cu^{2+}, H_2O_2 and ethanol (Reeve et al., 1998).

Most stress related genes are usually found in the chromosome of rhizobia. There are, however, a few exceptions, like the *gro*EL copies in both p*Sym*A and p*Sym*B of *S. meliloti* strain 1021 and in pMLa of *M. loti* strain MAFF303099 (Nakamura et al., 2004).

Stress-induced gene expression in bacteria has been examined primarily using 2D-PAGE of differentially expressed gene products (Spector et al., 1986; Givskov et al., 1994). These studies have revealed sets of proteins that are specifically induced or enhanced under stress conditions. Stress genes have been identified also by random insertional mutagenesis (Tn5) (Priefer et al., 2001) and by cDNA (complementary DNA)-AFLP (Amplifed Fragment Length Polymorphism) technique (Simões-Araújo et al., 2002).

Djordjevic and co-workers (Djordjevic et al., 2003) used a combination of two-dimensional gel electrophoresis and peptide mass fingerprinting to look into the protein patterns of rhizobia in response to different stress conditions.

Based on the complete *S. meliloti* strain 1021 genome sequence, DNA microarrays have been established as a comprehensive tool for systematic genome-wide gene expression analysis in the osmoadaptive response (Rüberg et al., 2003). Nowadays, cDNA array technology is seen as a powerful tool for monitoring global changes in gene expression patterns following subjection to several stress conditions (Fan et al., 2002).

MOLECULAR METHODS FOR STRAIN IDENTIFICATION

The evaluation of rhizobial diversity is essential for the identification and selection of appropriate inoculant strains for a given crop. Before the emergence of modern molecular methods, the main criteria used in the assignment of rhizobia into different species were host range, together with growth rate, morphology and metabolic tests.

Strain Identification

Presently, the identification of rhizobia is often based on a polyphasic approach using several different techniques to describe new genera and species. The Sub-committee on *Rhizobium* and *Agrobacterium* of the International Committee on Systematic Bacteriology of the International Union of the Microbiological Societies has recommended a minimal set of standard criteria for identifying new genera and species of nodule bacteria (Graham et al., 1991; Stackebrandt et al., 2002). These criteria include cultural, biochemical, symbiotic and molecular data. The most common approaches are 16S rDNA Restriction Fragment Length Polymorphism (RFLP) (Laguerre et al., 1992; 1994) and sequencing (Young et al., 1991;

Willems and Collins, 1993; Yanagi and Yamasato, 1993; Young, 1996; Young and Haukka, 1996); DNA-DNA hybridization (Willems et al., 2001a, b), Fatty Acids Methyl Esters (FAME) analysis (So et al., 1994; Jarvis et al., 1996) and Multilocus Enzyme Electrophoresis (MLEE) (Strain et al., 1995).

16S rRNA gene sequencing is fundamental for the identification of species and its full sequence is required for the description of a new species. However, occasionally recombination events between species have caused some incoherencies between phylogenies based on different parts of the gene (Young and Haukka, 1996). Transfer of parts of the 16S rDNA sequences between species has been suggested and there is indeed evidence for its recombination (van Berkum et al., 2003). Thus, the 16S rDNA sequence should be only one of the many characters used for bacterial taxonomy, instead of the predominant one (van Berkum et al., 2003).

As an example of a 16S rDNA based phylogeny, Fig. 7.2 represents a tree showing isolates able to nodulate chickpea, probably belonging to five species of rhizobia.

DNA-DNA hybridization has been required for the description of a new species (Stackebrandt et al., 2002; Kämpfer et al., 2003). Recently, a new approach, which combines DNA-DNA hybridization with DNA microarrays, was described for the identification of bacteria (Cho and Tiedje, 2001). Bacterial total DNA is randomly cut into 1 kb pieces and representative fragments are spotted on microarrays and hybridized to type genomes. This method has the advantage that a database of hybridization profiles can be established.

Genetic Diversity and Strain Differentiation

The study of rhizobial population diversity can be achieved by using both phenotypic and molecular approaches. Several methods have been used for strain differentiation, which evaluate rhizobia at various levels, namely DNA, RNA, phenotypic traits, protein expression and functions, fatty acids and quimiotaxonomic markers. The great variety of available techniques has led to the proposal of a polyphasic approach of bacterial taxonomy that include results obtained with the different methods (Stackebrandt et al., 2002).

Some of the most commonly used differentiation approaches include phenotypic methods like assimilation, biochemical and enzymatic tests (de Lajudie et al., 1994), intrinsic antibiotic resistance (IAR) patterns (Abaidoo et al., 2002), along with molecular methods, such as MLEE (Strain et al., 1995) and SDS-PAGE (Sodium Dodecyl Sulphate-Polyacrylamide Gel

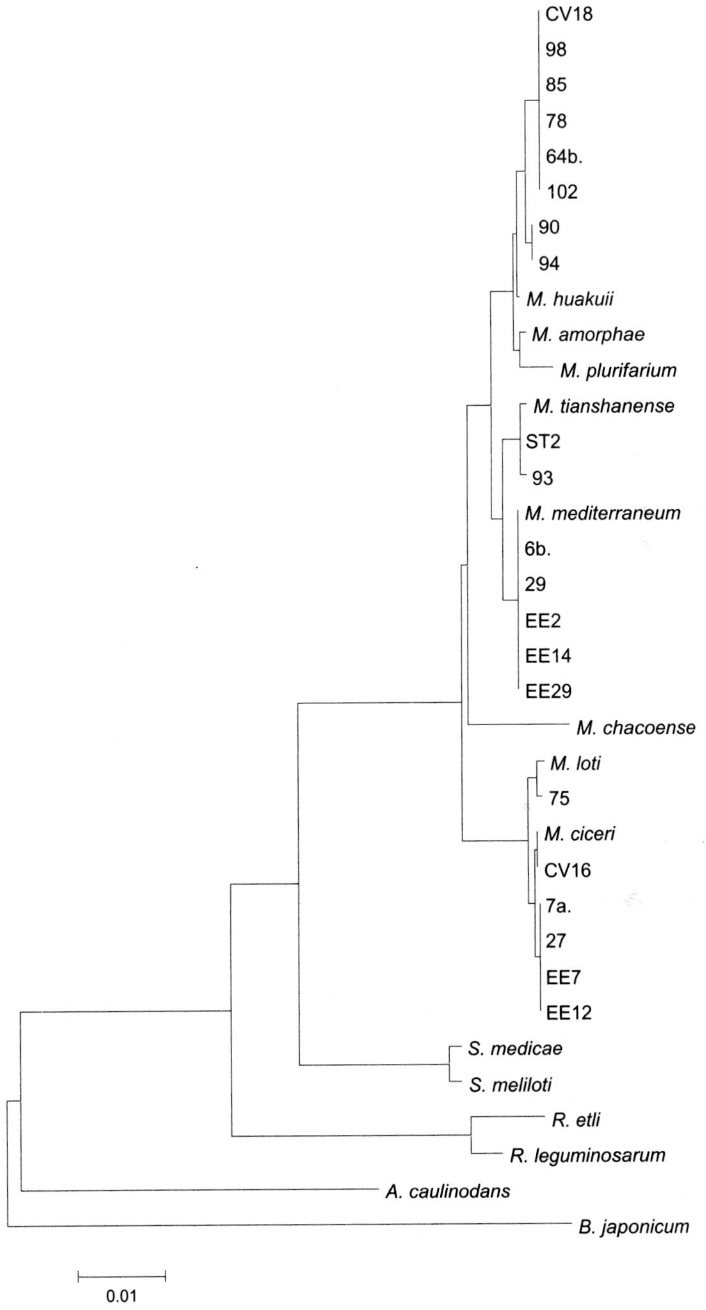

Fig. 7.2. 16S rDNA-based phylogeny of chickpea rhizobia isolates, adapted from (Laranjo et al., 2004).

Electrophoresis) analysis of total cell proteins (de Lajudie et al., 1994; Dupuy et al., 1994).

DNA fingerprinting techniques have been widely used to study biodiversity of rhizobial populations (Demezas, 1998). Restriction enzyme digestion of total genomic DNA (Glynn et al., 1985) was certainly one of the first approaches with resolution up to the species level (Laguerre et al., 1992). DNA restriction profiles with rare-cutting endonucleases provide less complex patterns, but the generated fragments are too large and need to be separated by Pulsed Field Gel Electrophoresis (PFGE) (Haukka and Lindström, 1994).

Plasmid profile analysis through an in situ lysis method (Eckhardt, 1978) is also a commonly used DNA fingerprinting technique, which has a discriminating power at the same level as total DNA restriction profiles (Laguerre et al., 1992). This approach assumes a particular importance in rhizobia, whose genome is composed, in most cases, by several plasmids and megaplasmids.

RNA fingerprinting techniques include the recently described Low Molecular Weight RNA (LMW RNA) that allows the separation of ribosomal and transfer RNAs (Velázquez et al., 1998).

Due to the improvement in technology of PCR in the last 20 years, numerous PCR-based techniques have been developed for evaluation of rhizobial diversity and strain differentiation. Some of these methods consist of the amplification of repeated regions in the genome, like REP (Repetitive Extragenic Palindromic) - PCR (de Bruijn, 1992; Judd et al., 1993), ERIC (Enterobacterial Repetitive Intergeneric Consensus) - PCR (de Bruijn, 1992; Niemann et al., 1999) and BOX (Box element) - PCR (Vinuesa et al., 1998), that can be used to distinguish between closely related strains of rhizobia. The obtained patterns are specific and reproducible and can be used in bacterial taxonomy at the strain level (Nick et al., 1999).

Another type of PCR-based technique consists of the random amplification of genome fragments and this kind of approach has been used to differentiate among rhizobial strains (Kishinevsky et al., 1996). Random Amplified Polymorphic DNA (RAPD) (Harrison et al., 1992; Dooley et al., 1993; Richardson et al., 1995; van Rossum et al., 1995; Selenska-Pobell et al., 1996) is a PCR-based approach that uses one random primer of about 10 nucleotides combined with a low annealing temperature to generate strain specific profiles. However, this technique has been proved to have some reproducibility constraints (Selenska-Pobell et al., 1995).

TP (Two Primers) - RAPD is another technique in which two primers, homologous to highly conserved regions of the 16S rDNA, are used (Rivas et al., 2001). The obtained profiles seem subspecies characteristic and not strain dependent.

Direct Amplified Polymorphic DNA (DAPD) is a recently described PCR-based DNA profiling method useful for assessing biodiversity and genetic relatedness among rhizobia (Laranjo et al., 2002, 2004) (Fig. 7.3). DAPD profiles are highly reproducible probably due to the stringent amplification conditions and the use of a specific 20-nucleotide primer, in contrast with RAPD analysis.

GENE EXPRESSION ANALYSIS

The availability of complete genome sequences has changed the whole basis on which rhizobia can be studied. The way is now opened to the functional genomics of rhizobia: genomics, transcriptomics and proteomics. Gene expression analyses include monitoring changes in the transcriptome and the proteome. In the post-genomic era, experimental approaches have moved from the targeted investigation of individual

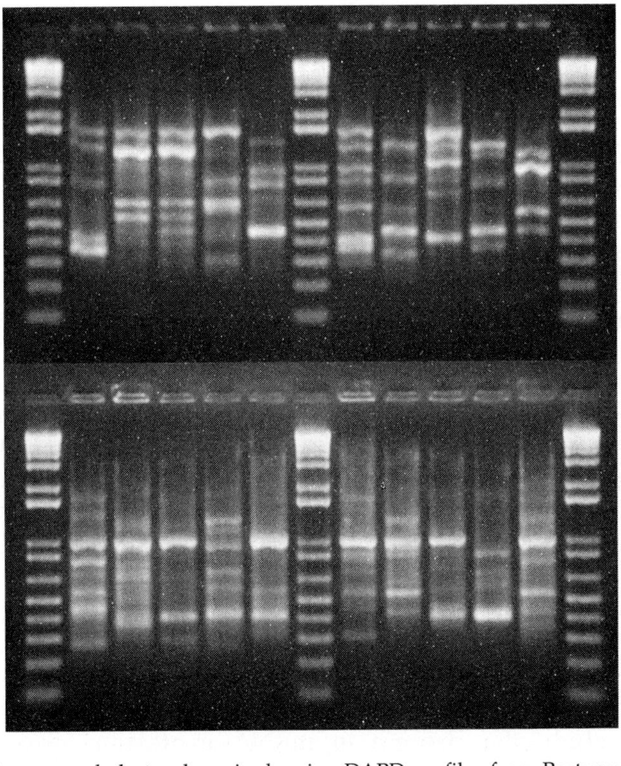

Fig. 7.3. Agarose gel electrophoresis showing DAPD profiles from Portuguese chickpea rhizobia isolates. (M-1 Kb Plus)

genes to the study of thousands of genes simultaneously by array techniques (Pühler et al., 2004), giving the possibility to study rhizobial genes involved in symbiosis, nodule development, stress response, etc.

Prior to the DNA array techniques for gene expression profiling, several methods were available for detecting and quantifying gene expression levels, including Northern blot analysis (Alwine et al., 1977), S1 nuclease protection (Berk and Sharp, 1977), sequencing of cDNA libraries (Adams et al., 1991; Okubo et al., 1992) and Serial Analysis of Gene Expression (SAGE), a method for comprehensive study of gene expression patterns that allows the quantitative and simultaneous analysis of a large number of transcripts (Velculescu et al., 1995).

Other approaches such as transposon mutagenesis (Milcamps et al., 1998), RNA fingerprinting (Cabanes et al., 2000) and promoter probe assays (Oke and Long, 1999) were also used for the analysis of gene function and expression patterns.

Insertional mutagenesis with promoter probe elements has been used to isolate and characterize genes activated under particular conditions (Milcamps et al., 1998). The use of Tn5 has been shown to be an extremely useful tool for random mutagenesis in *S. meliloti* (Honeycutt et al., 1993; Reeve et al., 1998).

Commonly used RNA fingerprinting methods include RNA-fingerprinting by arbitrarily primed PCR (RAP-PCR) (Welsh et al., 1992) and RT-PCR differential display (DD/RT-PCR) (Liang and Pardee, 1992). Recent advances in the cDNA-AFLP technique have made this method to be highly sensitive and reproducible for detecting differentially expressed transcripts. This approach may overcome the limitations of traditional RNA fingerprinting methods (Bachem et al., 1998).

In recent years, DNA micro- and macroarrays have developed into powerful tools for large-scale expression analysis at the level of transcription. DNA microarrays provide a simple and natural vehicle for exploring the genome in a way that is both systematic and comprehensive. There are several features in this technology that can be termed as being advantages: it is relatively cheap, universal, fast and user-friendly (Brown and Botstein, 1999). cDNA microarrays can profile gene expression patterns of tens of thousands of genes in a single experiment (Duggan et al., 1999).

Pilot micro- and macroarrays for *S. meliloti* strain 1021 have already been designed (Ampe et al., 2003; Bergès et al., 2003; Rüberg et al., 2003). These pilot experiments allowed the definition of the conditions that can be used to design whole-genome arrays and have validated the use of arrays to study the changes in mRNA expression in *S. meliloti*.

Furthermore, a number of candidate genes involved in the osmoadaptive and oxygen limitation stresses were identified.

GENETIC MODIFICATION

For many years, it has been a challenge for researchers working with rhizobia to develop an inoculant that can promote higher levels of N_2-fixation in practical field conditions. This is the final goal of strain improvement and, nowadays, this might be achieved by molecular techniques within our reach.

Since the work of Johnston and co-workers on conjugation (Johnston et al., 1978), several authors have shown the transfer of *Sym* plasmids between rhizobial strains in vitro. Nevertheless, not all types of chromosomal backgrounds can sustain all plasmids (Young, 1992). Many studies have demonstrated the importance of the genomic background on symbiosis (Hynes et al., 1986; Martínez et al., 1987; Brom et al., 1988). The interaction between chromosome and plasmids has been shown to influence the conjugation frequency even between closely related strains (Geniaux and Amarger, 1993).

The recent boom of complete genome sequences has made it possible to design pathways of rearrangements that lead to different genomic structures in *Rhizobium* with potential biological implications (Flores et al., 2000). However, there is evidence of a strong selective pressure imposed by the host, which favours the maintenance of certain chromosome-plasmid associations (Silva et al., 2003).

Studies in different organisms have shown that recombination and replication are closely related. Furthermore, additional replication can induce recombination through either homologous or heterologous events. Recombination enhancement by replication (RER) is the induction of θ-type replication from a supernumerary origin. This phenomenon has been described in *R. etli* and provokes a 1,000-fold increase in the formation of deletions in its p*Sym* (Valencia-Morales and Romero, 2000).

An important prerequisite for genetic improvement of any bacterial species is the availability of a highly efficient gene transfer system. Traditional transformation methods have been developed for rhizobia with little success (Selvaraj and Iyer, 1981; Bullerjahn and Benzinger, 1982; Kiss and Kálmán, 1982; Courtois et al., 1988). The referred methods are all time consuming and restricted to plasmid vectors carrying the *mob* gene (Garg et al., 1999). Thus, gene transfer to rhizobia has mainly been possible through conjugation with *Escherichia coli* (Johnston et al., 1978).

More recently, electrotransformation protocols have been developed for rhizobia (Garg et al., 1999; Hayashi et al., 2000). Compared to the conventional methods, electroporation yields much higher transformation efficiencies. However, the optimal electroporation conditions vary from strain to strain, making it hard to establish an efficient transformation system for rhizobial strains.

Rhizobia genetically modified to be easily traceable and/or to be improved in their expression of beneficial traits have been constructed and released with plants in a number of experimental field plots (Selbitschka et al., 1995). With these releases, it has been possible to monitor the modified inoculant bacteria after their introduction in field ecosystems and to assess their impact on the resident population (Amarger, 2002).

Marker genes that have been introduced in rhizobia to study the survival, persistence and spread of genetically modified bacteria in the field, include transposons or genes conferring antibiotic or mercury resistance, which can be used with reporter genes, such as *gus*A, *luc* or *lac*Z (Amarger, 2002).

Genetic modifications that have been introduced in rhizobia aimed at improving the strains used as inoculants on traits such as competitiveness, with indigenous strains and N_2-fixation capacity.

Generation of *Rhizobium* strains with improved symbiotic properties has been achieved by random DNA amplification (RDA), a new strategy for the improvement of bacterial strains through the selection of desired properties without the need to identify the genes involved in the process (Mavingui et al., 1997).

To increase the nodulation competitiveness under field conditions, two approaches have been used: (1) the introduction of genes involved in the biosynthesis of trifolitoxin, an antibiotic to which native rhizobia are susceptible, in *R. etli*, the microsymbiont of common bean (Robleto et al., 1998); and (2) the overexpression of *put*A, a metabolic gene involved in root surface colonization, in *S. meliloti*, associated with alfalfa (van Dillewijn et al., 2001, 2002).

Two strategies have also been developed to improve the N_2-fixing capacity of rhizobial inoculants: modifications of the expression of (1) *nif*A, one of the *nif* genes regulatory protein, and (2) dicarboxylate transport (*dct*) genes, involved in the supply of carbon and energy to the N_2-fixation process. *S. meliloti* strains with an extra copy of *nif*A and *dct* genes were constructed to inoculate alfalfa leading to an increase in plant yield under field conditions of nitrogen limitation, low endogenous rhizobial competitors and sufficient moisture (Bosworth et al., 1994). Nevertheless, other authors (Ronson et al., 1990) have shown that inoculation of two varieties of soybean with strains of *B. japonicum*, having an extra copy of *nif*A, did not increase soybean yield.

CONCLUSION

Overall, the studies indicated above show that it may be possible to improve symbiotic effectiveness through genetic modification of rhizobial strains. However, despite the facts already known about genetics of symbiotic N_2-fixation and strain improvement, most commercial inoculants are native selected strains, usually best fitted, rather than modified strains.

Another possible strategy to improve the efficiency of the symbiotic N_2-fixation process involves the genetic modification of the plant instead of the rhizobia. However, legumes have generally been considered difficult to transform (Atkins and Smith, 2000).

Genes regulating nodule development have been identified by isolating and characterizing plant mutants, and using positional cloning to uncover the role of the mutated gene in the symbiosis (Caetano-Anollés, 1997). The cloning of plant nodulation genes will provide insight into the role of the plant in the symbiotic process. It will also reveal whether functional gene homologues exist in other legumes and non-legumes. This will help in identifying the evolutionary relationship of the symbiotic genes to other genes used for normal cellular processes.

The understanding of the molecular differences between plants capable or incapable of entering into symbiosis may open up new avenues to the eventual extension of nodulation to other plants by genetic engineering (Caetano-Anollés, 1997).

Indeed, a long-standing goal of biological N_2-fixation research has been to extend the N_2-fixing symbiosis to non-nodulated cereal plants such as rice. A genetic predisposition of rice for symbiosis with rhizobia has been described and induction of nodulation and infection of rice with *Rhizobium* has been accomplished (Reddy et al., 1997; 1998). Nevertheless, the final goal is to make self-sufficient plants, such as rice, replenish in nitrogen content, which might be achieved by the transfer of the *nif* gene cluster to its genome. Some advances have been made with model legumes, however "making cereals fix their own nitrogen" still seems unattainable (de Bruijn et al., 1995).

Upto now several transgenic plants have already been constructed, such as *M. truncatula* (Cook, 1999; Bell et al., 2001) and *Lotus japonicus* (Handberg and Stougaard, 1992; Hayashi et al., 2001). The next step will be to modify these plants in order to express N_2-fixation genes.

The engineering of N_2-fixing cereal crops has been a long-term pioneering and ambitious project with several barriers to its success. Recent advances in plant molecular biology and technical developments in cereal transformation have made it possible to introduce multiple genes into important food crops, such as rice (Kohli et al., 1998).

The recently described oxygen-tolerant nitrogenase from *Streptomyces thermoautotrophicus* (Ribbe et al., 1997) might be helpful in the development of N_2-fixing plants, as long as the plant tolerates the generation of the superoxide radical required as an electron donor to the enzyme.

From the limited results obtained so far in engineering N_2-fixing plants, it seems that the improvement of the N_2-fixation process at the level of the microsymbiont symbiotic abilities may still be the best strategy.

The opportunity to carry out large-scale analyses in functional genomics now opens the way to the identification of large sets of co-regulated genes involved in biological processes at the level of gene networks instead of individual genes (Journet et al., 2002). This kind of approach is striking for a better global understanding of complex developmental programmes such as those activated during interactions between plants and microorganisms.

The importance of symbiotic N_2-fixation for developing sustainable agricultural systems is enormous and current genome projects include a diversity of legumes and rhizobial species. The application of new genomic tools to identify symbiotic genes both in model macro- and microsymbionts makes way to apply these findings to improve legume and other crops that are relevant to world agriculture.

ACKNOWLEDGEMENT

This work was supported by projects PRAXIS/P/AGR/11129/1998 and POCTI/BME/44140/2002 from Fundação para a Ciência e a Tecnologia, co-financed by EU-FEDER. Marta Laranjo acknowledges a PhD fellowship (SFRH/BD/6772/2001) from Fundação para a Ciência e a Tecnologia.

REFERENCES

Abaidoo, R.C., Keyser, H.H., Singleton, P.W. and Borthakur, D. (2002). Comparison of molecular and antibiotic resistance profile methods for the population analysis of *radyrhizobium* spp. (TGx) isolates that nodulate the new TGx soybean cultivars in Africa. J. Appl. Microbiol. 92: 109-117.

Adams, M.D., Kelley, J.M., Gocayne, J.D., Dubnick, M., Polymeropoulos, M.H., Xiao, H., Merril, C.R., Wu, A., Olde, B., Moreno, R.F., Kerlavage, A.R., McCombie, W.R. and Venter, J.C. (1991). Complementary-DNA sequencing - expressed sequence tags and human genome project. Science 252: 1651-1656.

Allardet-Servent, A., Michaux-Charachon, S., Jumas-Bilak, E., Karayan, L. and Ramuz, M. (1993). Presence of one linear and one circular chromosome in the *Agrobacterium tumefaciens* C58 genome. J. Bacteriol. 175: 7869-7874.

Alwine, J.C., Kemp, D.J. and Stark, G.R. (1977). Method for detection of specific RNAs in agarose gels by transfer to diazobenzyloxymethyl-paper and hybridization with DNA probes. Proc. Natl. Acad. Sci., USA 74: 5350-5354.

Amarger, N. (2002). Genetically modified bacteria in agriculture. Biochimie 84: 1061-1072.

Amarger, N., Macheret, V. and Laguerre, G. (1997). *Rhizobium gallicum* sp. nov. and *Rhizobium giardinii* sp. nov., from *Phaseolus vulgaris* nodules. Int. J. Syst. Bacteriol. 47: 996-1006.

Ampe, F., Kiss, E., Sabourdy, F. and Batut, J. (2003). Transcriptome analysis of *Sinorhizobium meliloti* during symbiosis. Genome Biol. 4: R15.

Appleby, C.A. (1984). Leghemoglobin and *Rhizobium* respiration. Annu. Rev. Plant Physiol. Plant Molec. Biol. 35: 443-478.

Arnold, W., Rump, A., Klipp, W., Priefer, U.B. and Pühler, A. (1988). Nucleotide sequence of a 24,206-base-pair DNA fragment carrying the entire nitrogen fixation gene cluster of *Klebsiella pneumoniae*. J. Mol. Biol. 203: 715-738.

Atkins, C.A. and Smith, P.M.C. (2000). Genetic transformation of legumes. In: Nitrogen Fixation: From Molecules to Crop Productivity (ed.) F.O. Pedrosa, Kluwer Academic Publishers, Dordrecht, The Netherlands pp. 647-652.

Bachem, C.W.B., Oomen, R. and Visser, R.G.F. (1998). Transcript imaging with cDNA-AFLP: a step-by-step protocol. Plant Mol. Biol. Rep. 16: 157-173.

Baginsky, C., Brito, B., Imperial, J., Palacios, J.M. and Ruiz-Argüeso, T. (2002). Diversity and evolution of hydrogenase systems in rhizobia. Appl. Environ. Microbiol. 68: 4915-4924.

Barloy-Hubler, F., Capela, D., Barnett, M.J., Kalman, S., Federspiel, N.A., Long, S.R. and Galibert, F. (2000). High-resolution physical map of the *Sinorhizobium meliloti* 1021 pSyma megaplasmid. J. Bacteriol. 182: 1185-1189.

Barloy-Hubler, F., Lelaure, V. and Galibert, F. (2001). Ribosomal protein gene cluster analysis in eubacterium genomics: homology between *Sinorhizobium meliloti* strain 1021 and *Bacillus subtilis*. Nucleic Acids Res. 29: 2747-2756.

Barnett, M.J., Fisher, R.F., Jones, T., Komp, C., Abola, A.P., Barloy-Hubler, F., Bowser, L., Capela, D., Galibert, F., Gouzy, J., Gurjal, M., Hong, A., Huizar, L., Hyman, R.W., Kahn, D., Kahn, M.L., Kalman, S., Keating, D.H., Palm, C., Peck, M.C., Surzycki, R., Wells, D.H., Yeh, K.C., Davis, R.W., Federspiel, N.A. and Long, S.R. (2001). Nucleotide sequence and predicted functions of the entire *Sinorhizobium meliloti* p*Sym*A megaplasmid. Proc. Natl. Acad. Sci. USA 98: 9883-9888.

Bell, C.J., Dixon, R.A., Farmer, A.D., Flores, R., Inman, J., Gonzales, R.A., Harrison, M.J., Paiva, N.L., Scott, A.D., Weller, J.W. and May, G.D. (2001). The *Medicago* genome initiative: a model legume database. Nucleic Acids Res. 29: 114-117.

Bergès, H., Lauber, E., Liebe, C., Batut, J., Kahn, D., de Bruijn, F.J. and Ampe, F. (2003). Development of *Sinorhizobium meliloti* pilot macroarrays for transcriptome analysis. Appl. Environ. Microbiol. 69: 1214-1219.

Berk, A.J. and Sharp, P.A. (1977). Sizing and mapping of early Adenovirus messenger-RNAs by Gel-electrophoresis of S1 endonuclease-digested hybrids. Cell 12: 721-732.

Bloem, J.F. and Law, I.J. (2001). Determination of competitive abilities of *Bradyrhizobium japonicum* strains in soils from soybean production regions in South Africa. Biol. Fertil. Soils 33: 181-189.

Bontemps, C., Golfier, G., Gris-Liebe, C., Carrere, S., Talini, L. and Boivin-Masson, C. (2005). Microarray-based detection and typing of the *Rhizobium* nodulation gene nodC: Potential of DNA arrays to diagnose biological functions of interest. Appl. Environ. Microbiol. 71: 8042-8048.

Bosworth, A.H., Williams, M.K., Albrecht, K.A., Kwiatkowski, R., Beynon, J., Hankinson, T.R., Ronson, C.W., Cannon, F., Wacek, T.J. and Triplett, E.W. (1994). Alfalfa yield response to inoculation with recombinant strains of *Rhizobium meliloti* with an extra copy of *dct*ABD and/or modified *nif*A expression. Appl. Environ. Microbiol. 60: 3815-3832.

Breedveld, M.W., Yoo, J.S., Reinhold, V.N. and Miller, K.J. (1994). Synthesis of glycerophosphorylated cyclic Beta-(1,2)-glucans by *Rhizobium meliloti ndv* mutants. J. Bacteriol. 176: 1047-1051.

Brockwell, J. and Bottomley, P.J. (1995). Recent advances in inoculant technology and prospects for the future. Soil Biol. Biochem. 27: 683-697.

Brom, S., Martinez, E., Dávila, G. and Palacios, R. (1988). Narrow- and broad-host-range symbiotic plasmids of *Rhizobium* spp. strains that nodulate *Phaseolus vulgaris*. Appl. Environ. Microbiol. 54: 1280-1283.

Brown, P.O. and Botstein, D. (1999). Exploring the new world of the genome with DNA microarrays. Nat. Genet. 21: 33-37.

Bullerjahn, G.S. and Benzinger, R.H. (1982). Genetic transformation of *Rhizobium leguminosarum* by plasmid DNA. J. Bacteriol. 150: 421-424.

Cabanes, D., Boistard, P. and Batut, J. (2000). Identification of *Sinorhizobium meliloti* genes regulated during symbiosis. J. Bacteriol. 182: 3632-3637.

Cadahía, E., Leyva, A. and Ruiz-Argüeso, T. (1986). Indigenous plasmids and cultural characteristics of rhizobia nodulating Chickpeas (*Cicer arietinum* L.). Arch. Microbiol. 146: 239-244.

Caetano-Anollés, G. (1997). Molecular dissection and improvement of the nodule symbiosis in legumes. Field Crop. Res. 53: 47-68.

Capela, D., Barloy-Hubler, F., Gatius, M.T., Gouzy, J. and Galibert, F. (1999). A high-density physical map of *Sinorhizobium meliloti* 1021 chromosome derived from bacterial artificial chromosome library. Proc Natl. Acad. Sci. USA 96: 9357-9362.

Carlson, R.W., Kalembasa, S., Turowski, D., Pachori, P. and Noel, K.D. (1987). Characterization of the lipopolysaccharide from a *Rhizobium phaseoli* mutant that is defective in infection thread development. J. Bacteriol. 169: 4923-4928.

Carter, J.M., Tieman, J.S. and Gibson, A.H. (1995). Competitiveness and persistence of strains of rhizobia for Faba Bean in acid and alkaline soils. Soil Biol. Biochem. 27: 617-623.

Cho, J.-C. and Tiedje, J.M. (2001). Bacterial species determination from DNA-DNA hybridization by using genome fragments and DNA microarrays. Appl. Environ. Microbiol. 67: 3677-3682.

Cook, D.R. (1999). *Medicago truncatula*-a model in the making! Curr. Opin. Plant Biol. 2: 301-304.

Courtois, J., Courtois, B. and Guillaume, J. (1988). High-frequency transformation of *Rhizobium meliloti*. J. Bacteriol. 170: 5925-5927.

Cullimore, J.V., Ranjeva, R. and Bono, J.-J. (2001). Perception of lipo-chitooligosaccharide Nod factors in legumes. Trends Plant Sci. 6: 24-30.

Davidson, I.A. and Robson, M.J. (1986a). Effect of contrasting patterns of nitrate application on the nitrate uptake, N_2-fixation, nodulation and growth of White Clover. Ann. Bot. 57: 331-338.

Davidson, I.A. and Robson, M.J. (1986b). Interactions between nitrate uptake and N_2 -fixation in White Clover. Plant Soil 91: 401-404.

de Bruijn, F.J. (1992). Use of repetitive (repetitive extragenic palindromic and enterobacterial repetitive intergeneric consensus) sequences and the polymerase chain reaction to fingerprint the genomes of *Rhizobium meliloti* isolates and other soil bacteria. Appl. Environ. Microbiol. 58: 2180-2187.

de Bruijn, F.J., Jing, Y. and Dazzo, F.B. (1995). Potential and pitfalls of trying to extend symbiotic interactions of nitrogen-fixing organisms to presently non-nodulated plants, such as rice. Plant Soil 174: 225-240.

de Lajudie, P., Willems, A., Pot, B., Dewettinck, D., Maestrojuan, G., Neyra, M., Collins, M.D., Dreyfus, B., Kersters, K. and Gillis, M. (1994). Polyphasic taxonomy of rhizobia - emendation of the genus *Sinorhizobium* and description of *Sinorhizobium meliloti* comb nov, *Sinorhizobium saheli* sp nov, and *Sinorhizobium teranga* sp nov. Int. J. Syst. Bacteriol. 44: 715-733.

Del Vecchio, V. G., Kapatral, V., Redkar, R. J., Patra, G., Mujer, C., Los, T., Ivanova, N., Anderson, I., Bhattacharyya, A., Lykidis, A., Reznik, G., Jablonski, L., Larsen, N., D'Souza, M., Bernal, A., Mazur, M., Goltsman, E., Selkov, E., Elzer, P. H., Hagius, S., O'Callaghan, D., Letesson, J.-J., Haselkorn, R., Kyrpides, N. and Overbeek, R. (2002). The genome sequence of the facultative intracellular pathogen *Brucella melitensis* 10.1073/ pnas.221575398. PNAS 99: 443-448.

Demezas, D.H. (1998). Fingerprinting bacterial genomes using restriction fragment length polymorphisms. In: Bacterial Genomes: Physical Structure and Analysis (eds.) F.J. de Bruijn, J.R. Lupski and G.M. Weinstock, Chapman and Hall, New York, USA, pp 383-398.

Denarié, J., Debellé, F. and Promé, J.C. (1996). *Rhizobium* lipo-chitooligosaccharide nodulation factors: signaling molecules mediating recognition and morphogenesis. Annu. Rev. Biochem. 65: 503-535.

Dilworth, M.J., Howieson, J.G., Reeve, W.G., Tiwari, R.P. and Glenn, A.R. (2001). Acid tolerance in legume root nodule bacteria and selecting for it. Aust. J. Exp. Agric. 41: 435-446.

Djordjevic, M.A., Chen, H.C., Natera, S., van Noorden, G., Menzel, C., Taylor, S., Renard, C., Geiger, O. and Weiller, G.F. (2003). A global analysis of protein expression profiles in *Sinorhizobium meliloti*: Discovery of new genes for nodule occupancy and stress adaptation. Mol. Plant Microbe Interact. 16: 508-524.

Dobert, R.C., Brei, B.T. and Triplett, E.W. (1994). DNA sequence of the common nodulation genes of *bradyrhizobium elkanii* and their phylogenetic relationship to those of other nodulating bacteria. Mol. Plant-Microbe Interact. 7: 564-572.

Dooley, J.J., Harrison, S.P., Mytton, L.R., Dye, M., Cresswell, A., Skøt, L. and Beeching, J.R. (1993). Phylogenetic grouping and identification of *Rhizobium* isolates on the basis of random amplified polymorphic DNA profiles. Can. J. Microbiol. 39: 665-673.

Downie, A. (1997). Fixing a symbiotic circle. Nature 387: 352-354.

Downie, J.A. (1998). Functions of rhizobial nodulation genes. In: The Rhizobiaceae: Molecular Biology of Model Plant-Associated Bacteria. (eds.) H.P. Spaink, A. Kondorosi and P.J.J. Hooykaas , Kluwer Academic Publishers, Dordrecht, The Netherlands, pp 387-402.

Downie, J.A. and Young, J.P.W. (2001). The ABC of symbiosis. Nature 412: 597-598.

Drummond, M.H. (1984). The nitrogen fixation genes of *Klebsiella pneumoniae* - a model system. Microbiol. Sci. 1: 30-33.

Duggan, D.J., Bittner, M., Chen, Y., Meltzer, P. and Trent, J.M. (1999). Expression profiling using cDNA microarrays. Nat. Genet. 21: 10-14.

Dupuy, N., Willems, A., Pot, B., Dewettinck, D., Vandenbruaene, I., Maestrojuan, G., Dreyfus, B., Kersters, K., Collins, M.D. and Gillis, M. (1994). Phenotypic and genotypic characterization of bradyrhizobia nodulating the leguminous tree *Acacia albida*. Int. J. Syst. Bacteriol. 44: 461-473.

Eaglesham, A.R.J. (1989). Nitrate inhibition of root-nodule symbiosis in doubly rooted soybean plants. Crop Sci. 29: 115-119.

Eckhardt, T. (1978). Rapid method for identification of plasmid deoxyribonucleic acid in bacteria. Plasmid 1: 584-588.

Fan, J.S., Yang, X.L., Wang, W.G., Wood, W.H., Becker, K.G. and Gorospe, M. (2002). Global analysis of stress-regulated mRNA turnover by using cDNA arrays. Proc. Natl. Acad. Sci. USA, 99: 10611-10616.

Fayet, O., Ziegelhoffer, T. and Georgopoulos, C. (1989). The GroES and GroEL heat shock gene products of *Escherichia coli* are essential for bacterial growth at all temperatures. J. Bacteriol. 171: 1379-1385.

Finan, T.M., Weidner, S., Wong, K., Buhrmester, J., Chain, P., Vorhölter, F.J., Hernandez-Lucas, I., Becker, A., Cowie, A., Gouzy, J., Golding, B. and Pühler, A. (2001). The complete sequence of the 1,683-kb pSymB megaplasmid from the N_2-fixing endosymbiont *Sinorhizobium meliloti*. Proc. Natl. Acad. Sci. USA, 98: 9889-9894.

Fischer, H.M., Babst, M., Kaspar, T., Acuña, G., Arigoni, F. and Hennecke, H. (1993). One member of a *gro*ESL-like chaperonin multigene family in *Bradyrhizobium japonicum* is co-regulated with symbiotic nitrogen fixation genes. EMBO J. 12: 2901-2912.

Fischer, H.M. (1994). Genetic regulation of nitrogen fixation in rhizobia. Microbiol. Rev. 58: 352-386.

Fischer, H.M., Schneider, K., Babst, M. and Hennecke, H. (1999). GroEL chaperonins are required for the formation of a functional nitrogenase in *Bradyrhizobium japonicum*. Arch. Microbiol. 171: 279-289.

Fisher, R.F. and Long, S.R. (1992). *Rhizobium*-plant signal exchange. Nature 357: 655-660.

Flores, M., Mavingui, P., Perret, X., Broughton, W.J., Romero, D., Hernández, G., Dávila, G. and Palacios, R. (2000). Prediction, identification, and artificial selection of DNA rearrangements in *Rhizobium*: toward a natural genomic design. Proc. Natl. Acad. Sci., USA, 97: 9138-9143.

Francia, M.V., Varsaki, A., Garcillán-Barcia, M.P., Latorre, A., Drainas, C. and de la Cruz, F. (2004). A classification scheme for mobilization regions of bacterial plasmids. FEMS Microbiol. Rev. 28: 79-100.

Frank, B. (1889). Über die pilzsymbiose der leguminosen. Berichte der Deutschen Botanischen Gesellschaft 7: 332-346.

Freiberg, C., Fellay, R., Bairoch, A., Broughton, W.J., Rosenthal, A. and Perret, X. (1997). Molecular basis of symbiosis between *Rhizobium* and legumes. Nature 387: 394-401.

Galibert, F., Finan, T.M., Long, S.R., Pühler, A., Abola, P., Ampe, F., Barloy-Hubler, F., Barnett, M.J., Becker, A., Boistard, P., Bothe, G., Boutry, M., Bowser, L., Buhrmester, J., Cadieu, E., Capela, D., Chain, P., Cowie, A., Davis, R.W., Dréano, S., Federspiel, N.A., Fisher, R.F., Gloux, S., Godrie, T., Goffeau, A., Golding, B., Gouzy, J., Gurjal, M., Hernandez-Lucas, I., Hong, A., Huizar, L., Hyman, R.W., Jones, T., Kahn, D., Kahn, M.L., Kalman, S., Keating, D.H., Kiss, E., Komp, C., Lalaure, V., Masuy, D., Palm, C., Peck, M.C., Pohl, T.M., Portetelle, D., Purnelle, B., Ramsperger, U., Surzycki, R., Thébault, P., Vandenbol, M., Vorhölter, F.J., Weidner, S., Wells, D.H., Wong, K., Yeh, K.C. and Batut, J. (2001). The composite genome of the legume symbiont *Sinorhizobium meliloti*. Science 293: 668-672.

Garg, B., Dogra, R.C. and Sharma, P.K. (1999). High-efficiency transformation of *Rhizobium leguminosarum* by electroporation. Appl. Environ. Microbiol. 65: 2802-2804.

Geniaux, E. and Amarger, N. (1993). Diversity and stability of plasmid transfer in isolates from a single field population of *Rhizobium leguminosarum* bv. *viciae*. FEMS Microbiol. Ecol. 102: 251-260.

Givskov, M., Eberl, L. and Molin, S. (1994). Responses to nutrient starvation in *Pseudomonas putida* kt2442: two-dimensional electrophoretic analysis of starvation- and stress-induced proteins. J. Bacteriol. 176: 4816-4824.

Glazebrook, J. and Walker, G.C. (1989). A novel exopolysaccharide can function in place of the calcofluor-binding exopolysaccharide in nodulation of alfalfa by *Rhizobium meliloti*. Cell 56. 58(4): 661-672.

Glenn, A.R., Reeve, W.G., Tiwari, R.P., Dilworth, M.J., Cook, G.M., Booth, I.R., Poole, R.K., Foster, J.W., Slonczewski, J.L., Padan, E., Epstein, W., Skulachev, V., Matin, A. and Fillingame, R.H. (1999). Acid tolerance in root nodule bacteria. In: Bacterial Response to pH, John Wiley & Sons Ltd, Chichester, UK, pp. 112-130.

Glynn, P., Higgins, P., Squartini, A. and O'Gara, F. (1985). Strain identification in *Rhizobium trifolii* using DNA restriction analysis, plasmid DNA profiles and intrinsic antibiotic resistances. FEMS Microbiol. Lett. 30: 177-182.

González, V., Bustos, P., Ramírez-Romero, M.A., Medrano-Soto, A., Salgado, H., Hernández-González, I., Hernández-Celis, J.C., Quintero, V., Moreno-Hagelsieb, G., Girard, L., Rodríguez, O., Flores, M., Cevallos, M.A., Collado-Vides, J., Romero, D. and Dávila, G. (2003). The mosaic structure of the symbiotic plasmid of *Rhizobium etli* CFN42 and its relation to other symbiotic genome compartments. Genome Biol. 4: R36.

González, V., Santamaria, R. I., Bustos, P., Hernandez-Gonzalez, I., Medrano-Soto, A., Moreno-Hagelsieb, G., Janga, S. C., Ramirez, M. A., Jimenez-Jacinto, V., Collado-Vides, J. and Davila, G. (2006). The partitioned *Rhizobium etli* genome: Genetic and metabolic redundancy in seven interacting replicons 10.1073/pnas.0508502103. PNAS 103: 3834-3839.

Govezensky, D., Greener, T., Segal, G. and Zamir, A. (1991). Involvement of GroEL in *nif* gene regulation and nitrogenase assembly. J. Bacteriol. 173: 6339-6346.

Graham, P.H., Sadowsky, M.J., Keyser, H.H., Barnet, Y.M., Bradley, R.S., Cooper, J.E., Deley, D.J., Jarvis, B.D.W., Roslycky, E.B., Strijdom, B.W. and Young, J.P.W. (1991). Proposed minimal standards for the description of new genera and species of root- and stem-nodulating bacteria. Int. J. Syst. Bacteriol. 41: 582-587.

Guo, X.W., Flores, M., Mavingui, P., Fuentes, S.I., Hernández, G., Dávila, G. and Palacios, R. (2003). Natural genomic design in *Sinorhizobium meliloti*: novel genomic architectures. Genome Res. 13: 1810-1817.

Handberg, K. and Stougaard, J. (1992). *Lotus japonicus*, an autogamous, diploid legume species for classical and molecular genetics. Plant J. 2: 487-496.

Harrison, S.P., Jones, D.G., Schunmann, P.H.D., Forster, J.W. and Young, J.P.W. (1988). Variation in *Rhizobium leguminosarum* biovar *trifolii* Sym plasmids and the association with effectiveness of nitrogen fixation. J. Gen. Microbiol. 134: 2721-2730.

Harrison, S.P., Mytton, L.R., Skøt, L., Dye, M. and Cresswell, A. (1992). Characterisation of *Rhizobium* isolates by amplification of DNA polymorphisms using random primers. Can. J. Microbiol. 38: 1009-1015.

Haukka, K. and Lindström, K. (1994). Pulsed-field gel electrophoresis for genotypic comparison of *Rhizobium* bacteria that nodulate leguminous trees. FEMS Microbiol. Lett. 119: 215-220.

Hayashi, M., Maeda, Y., Hashimoto, Y. and Murooka, Y. (2000). Efficient transformation of *Mesorhizobium huakuii* subsp *rengei* and *Rhizobium* species. J. Biosci. Bioeng. 89: 550-553.

Hayashi, M., Miyahara, A., Sato, S., Kato, T., Yoshikawa, M., Taketa, M., Pedrosa, A., Onda, R., Imaizumi-Anraku, H., Bachmair, A., Sandal, N., Stougaard, J., Murooka, Y., Tabata, S., Kawasaki, S., Kawaguchi, M. and Harada, K. (2001). Construction of a genetic linkage map of the model legume *Lotus japonicus* using an intraspecific F-2 population. DNA Res. 8: 301-310.

Hirsch, A.M., Lum, M.R. and Downie, J.A. (2001). What makes the rhizobia-legume symbiosis so special? Plant Physiol. 127: 1484-1492.

Honeycutt, R.J., McClelland, M. and Sobral, B.W. (1993). Physical map of the genome of *Rhizobium meliloti* 1021. J. Bacteriol. 175: 6945-6952.

Howieson, J.G. (1995). Rhizobial persistence and its role in the development of sustainable agricultural systems in mediterranean environments. Soil Biol. Biochem. 27: 603-610.

Hynes, M.F., Simon, R., Müller, P., Niehaus, K., Labes, M. and Pühler, A. (1986). The two megaplasmids of *Rhizobium meliloti* are involved in the effective nodulation of alfalfa. Mol. Gen. Genet. 202: 356-362.

Ibekwe, A.M., Angle, J.S., Chaney, R.L. and van Berkum, P. (1997). Differentiation of clover *Rhizobium* isolated from biosolids-amended soils with varying pH. Soil Sci. Soc. Am. J. 61: 1679-1685.

Jarvis, B.D.W., Sivakumaran, S., Tighe, S.W. and Gillis, M. (1996). Identification of *Agrobacterium* and *Rhizobium* species based on cellular fatty acid composition. Plant Soi 184: 143-158.

Jiang, J., Gu, B., Albright, L.M. and Nixon, B.T. (1989). Conservation between coding and regulatory elements of *Rhizobium meliloti* and *Rhizobium leguminosarum dct* genes. J. Bacteriol. 171: 5244-5253.

Johnston, A.W.B., Beynon, J.L., Buchanan-Wollaston, A.V., Setchell, S.M., Hirsch, P.R. and Beringer, J.E. (1978). High frequency transfer of nodulating ability between strains and species of *Rhizobium*. Nature 276: 634-636.

Jordan, D.C. (1982). Transfer of *Rhizobium japonicum* Buchanan 1980 to *Bradyrhizobium* gen. nov., a genus of slow-growing, root nodule bacteria from leguminous plants. Int. J. Syst. Bacteriol. 32: 136-139.

Journet, E.P., van Tuinen, D., Gouzy, J., Crespeau, H., Carreau, V., Farmer, M.J., Niebel, A., Schiex, T., Jaillon, O., Chatagnier, O., Godiard, L., Micheli, F., Kahn, D., Gianinazzi-Pearson, V. and Gamas, P. (2002). Exploring root symbiotic programs in the model legume *Medicago truncatula* using EST analysis. Nucleic Acids Res. 30: 5579-5592.

Judd, A.K., Schneider, M., Sadowsky, M.J. and de Bruijn, F.J. (1993). Use of repetitive sequences and the polymerase chain reaction technique to classify genetically related *Bradyrhizobium japonicum* serocluster 123 strains. Appl. Environ. Microbiol. 59: 1702-1708.

Kaminski, P.A., Batut, J. and Boistard, P. (1998). A survey of symbiotic nitrogen fixation by rhizobia. In: The Rhizobiaceae: Molecular Biology of Model Plant-Associated Bacteria, (eds.) H.P. Spaink, A. Kondorosi and P.J.J. Hooykaas, Kluwer Academic Publishers, Dordrecht, The Netherlands, pp. 432-460.

Kämpfer, P., Buczolits, S., Albrecht, A., Busse, H.J. and Stackebrandt, E. (2003). Towards a standardized format for the description of a novel species (of an established genus): *Ochrobactrum gallinifaecis* sp. nov. Int. J. Syst. Evol. Microbiol. 53: 893-896.

Kaneko, T., Nakamura, Y., Sato, S., Asamizu, E., Kato, T., Sasamoto, S., Watanabe, A., Idesawa, K., Ishikawa, A., Kawashima, K., Kimura, T., Kishida, Y., Kiyokawa, C., Kohara, H., Matsumoto, M., Matsuno, A., Mochizuki, Y., Nakayama, S., Nakazaki, N., Shimpo, S., Sugimoto, M., Takeuchi, C., Yamada, M. and Tabata, S. (2000a). Complete genome structure of the nitrogen-fixing symbiotic bacterium *Mesorhizobium loti* (supplement). DNA Res. 7: 381-406.

Kaneko, T., Nakamura, Y., Sato, S., Asamizu, E., Kato, T., Sasamoto, S., Watanabe, A., Idesawa, K., Ishikawa, A., Kawashima, K., Kimura, T., Kishida, Y., Kiyokawa, C., Kohara, M., Matsumoto, M., Matsuno, A., Mochizuki, Y., Nakayama, S., Nakazaki, N., Shimpo, S., Sugimoto, M., Takeuchi, C., Yamada, M. and Tabata, S. (2000b). Complete genome structure of the nitrogen-fixing symbiotic bacterium *Mesorhizobium loti*. DNA Res. 7: 331-338.

Kaneko, T., Nakamura, Y., Sato, S., Minamisawa, K., Uchiumi, T., Sasamoto, S., Watanabe, A., Idesawa, K., Iriguchi, M., Kawashima, K., Kohara, M., Matsumoto, M., Shimpo, S., Tsuruoka, H., Wada, T., Yamada, M. and Tabata, S. (2002a). Complete genomic sequence of nitrogen-fixing symbiotic bacterium *Bradyrhizobium japonicum* USDA110 (supplement). DNA Res. 9: 225-256.

Kaneko, T., Nakamura, Y., Sato, S., Minamisawa, K., Uchiumi, T., Sasamoto, S., Watanabe, A., Idesawa, K., Iriguchi, M., Kawashima, K., Kohara, M., Matsumoto, M., Shimpo, S., Tsuruoka, H., Wada, T., Yamada, M. and Tabata, S. (2002b). Complete genomic sequence of nitrogen-fixing symbiotic bacterium *Bradyrhizobium japonicum* USDA110. DNA Res. 9: 189-197.

Kishinevsky, B.D., Sen, D. and Yang, G. (1996). Diversity of rhizobia isolated from various *Hedysarum* species. Plant Soil 186: 21-28.

Kiss, G.B. and Kálmán, Z. (1982). Transformation of *Rhizobium meliloti* 41 with plasmid DNA. J. Bacteriol. 150: 465-470.

Kohli, A., Leech, M., Vain, P., Laurie, D.A. and Christou, P. (1998). Transgene organization in rice engineered through direct DNA transfer supports a two-phase integration mechanism mediated by the establishment of integration hot spots. Proc. Natl. Acad. Sci. USA 95: 7203-7208.

Krishnan, H.B. and Pueppke, S.G. (1991). *nolC*, a *Rhizobium fredii* gene involved in cultivar-specific nodulation of soybean, shares homology with a heat-shock gene. Mol. Microbiol. 5: 737-745.

Kuykendall, L.D., Saxena, B., Devine, T.E. and Udell, S.E. (1992). Genetic diversity in *Bradyrhizobium japonicum* Jordan 1982 and a proposal for *Bradyrhizobium elkanii* sp. nov. Can. J. Microbiol. 38: 501-505.

Laguerre, G., Mazurier, S.I. and Amarger, N. (1992). Plasmid profiles and restriction-fragment-length-polymorphism of *Rhizobium leguminosarum* bv. *viciae* in field populations. FEMS Microbiol. Ecol. 101: 17-26.

Laguerre, G., Allard, M.-R., Revoy, F. and Amarger, N. (1994). Rapid identification of rhizobia by restriction fragment length polymorphism analysis of PCR-amplified 16s rRNA genes. Appl. Environ. Microbiol. 60: 56-63.

Laranjo, M., Rodrigues, R., Alho, L. and Oliveira, S. (2001). Rhizobia of chickpea from southern Portugal: Symbiotic efficiency and genetic diversity. J. Appl. Microbiol. 90: 662-667.

Laranjo, M., Branco, C., Soares, R., Alho, L., Carvalho, M.D. and Oliveira, S. (2002). Comparison of chickpea rhizobia isolates from diverse Portuguese natural populations based on symbiotic effectiveness and DNA fingerprint. J. Appl. Microbiol. 92: 1043-1050.

Laranjo, M., Machado, J., Young, J.P.W. and Oliveira, S. (2004). High diversity of chickpea *Mesorhizobium* species isolated in a Portuguese agricultural region. FEMS Microbiol. Ecol. 48: 101-107.

Larimer, F.W., Chain, P., Hauser, L., Lamerdin, J., Malfatti, S., Do, L., Land, M.L., Pelletier, D.A., Beatty, J.T., Lang, A.S., Tabita, F.R., Gibson, J.L., Hanson, T.E., Bobst, C., Torres, J.L., Peres, C., Harrison, F.H., Gibson, J. and Harwood, C.S. (2004). Complete genome sequence of the metabolically versatile photosynthetic bacterium *Rhodopseudomonas palustris*. Nat. Biotechnol. 22: 55-61.

Liang, P. and Pardee, A.B. (1992). Differential display of eukaryotic messenger-RNA by means of the polymerase chain reaction. Science 257: 967-971.

Lindström, K., Lipsanen, P. and Kaijalainen, S. (1990). Stability of markers used for identification of two *Rhizobium galegae* inoculant strains after five years in the field. Appl. Environ. Microbiol. 56: 444-450.

Long, S.R. (1989). *Rhizobium*-legume nodulation: Life together in the underground. Cell 56: 203-214.

Maier, R.J. (1986). Biochemistry regulation and genetics of hydrogen oxidation in *Rhizobium*. Crit. Rev. Biotechnol. 3: 17-38.

Martínez, E., Palacios, R. and Sánchez, F. (1987). Nitrogen-fixing nodules induced by *Agrobacterium tumefaciens* harboring *Rhizobium phaseoli* plasmids. J. Bacteriol. 169: 2828-2834.

Martínez-Romero, E., Segovia, L., Mercante, F.M., Franco, A.A., Graham, P. and Pardo, M.A. (1991). *Rhizobium tropici*, a novel species nodulating *Phaseolus vulgaris* L. beans and *Leucaena* sp. trees. Int. J. Syst. Bacteriol. 41: 417-426.

Mavingui, P., Flores, M., Romero, D., Martínez-Romero, E. and Palacios, R. (1997). Generation of *Rhizobium* strains with improved symbiotic properties by random DNA amplification (RDA). Nat. Biotechnol. 15: 564-569.

Mavingui, P., Flores, M., Guo, X., Dávila, G., Perret, X., Broughton, W.J. and Palacios, R. (2002). Dynamics of genome architecture in *Rhizobium* sp. strain NGR234. J. Bacteriol. 184: 171-176.

McDermott, T.R., Graham, P.H. and Brandwein, D.H. (1987). Viability of *Bradyrhizobium japonicum* bacteroids. Arch. Microbiol. 148: 100-106.

Meinhardt, L.W., Krishnan, H.B., Balatti, P.A. and Pueppke, S.G. (1993). Molecular cloning and characterization of a sym plasmid locus that regulates cultivar-specific nodulation of soybean by *Rhizobium fredii* USDA 257. Mol. Microbiol. 9: 17-29.

Mercado-Blanco, J. and Toro, N. (1996). Plasmids in rhizobia: The role of nonsymbiotic plasmids. Mol. Plant-Microbe Interact. 9: 535-545.

Michiels, J., Verreth, C. and Vanderleyden, J. (1994). Effects of temperature stress on bean-nodulating *Rhizobium* Strains. Appl. Environ. Microbiol. 60: 1206-1212.

Milcamps, A., Ragatz, D.M., Lim, P., Berger, K.A. and de Bruijn, F.J. (1998). Isolation of carbon- and nitrogen-deprivation-induced loci of *Sinorhizobium meliloti* 1021 by Tn5-*lux*AB mutagenesis. Microbiology 144: 3205-3218.

Morett, E., Kreutzer, R., Cannon, W. and Buck, M. (1990). The influence of the *Klebsiella pneumoniae* regulatory gene *nif*L upon the transcriptional activator protein *nif*A. Mol. Microbiol. 4: 1253-1258.

Moulin, L., Munive, A., Dreyfus, B. and Boivin-Masson, C. (2001). Nodulation of legumes by members of the beta-subclass of Proteobacteria. Nature 411: 948-950.

Moulin, L., Béna, G., Boivin-Masson, C. and Stepkowski, T. (2004). Phylogenetic analyses of symbiotic nodulation genes support vertical and lateral gene co-transfer within the *Bradyrhizobium* genus. Mol. Phylogenet. Evol. 30: 720-732.

Münchbach, M., Nocker, A. and Narberhaus, F. (1999). Multiple small heat shock proteins in rhizobia. J. Bacteriol. 181: 83-90.

Nakamura, Y., Kaneko, T. and Tabata, S. (2004). RhizoBase: The Genome Database for Rhizobia (http://www.kazusa.or.jp/rhizobase/index.html). In: Kasuza DNA Research Institute, Japan.

Nick, G., Jussila, M., Hoste, B., Niemi, R.M., Kaijalainen, S., de Lajudie, P., Gillis, M., de Bruijn, F.J. and Lindström, K. (1999). Rhizobia isolated from root nodules of tropical leguminous trees characterized using DNA-DNA dot-blot hybridisation and rep-PCR genomic fingerprinting. Syst. Appl. Microbiol. 22: 287-299.

Niemann, S., Dammann-Kalinowski, T., Nagel, A., Pühler, A. and Selbitschka, W. (1999). Genetic basis of enterobacterial repetitive intergenic consensus (ERIC)-PCR fingerprint pattern in *Sinorhizobium meliloti* and identification of *S. meliloti* employing PCR primers derived from an ERIC-PCR fragment. Arch. Microbiol. 172: 22-30.

Nocker, A., Krstulovic, N.P., Perret, X. and Narberhaus, F. (2001). ROSE elements occur in disparate rhizobia and are functionally interchangeable between species. Arch. Microbiol. 176: 44-51.

Nour, S.M., Fernandez, M.P., Normand, P. and Cleyet-Marel, J.-C. (1994). *Rhizobium ciceri* sp. nov., consisting of strains that nodulate chickpeas (*Cicer arietinum* L.). Int. J. Syst. Bacteriol. 44: 511-522.

Nour, S.M., Cleyet-Marel, J.-C., Normand, P. and Fernandez, M.P. (1995). Genomic heterogeneity of strains nodulating chickpeas (*Cicer arietinum* L.) and description of *Rhizobium mediterraneum* sp. nov. Int. J. Syst. Bacteriol. 45: 640-648.

O'Hara, G.W. and Glenn, A.R. (1994). The adaptive acid tolerance response in root-nodule bacteria and *Escherichia coli*. Arch. Microbiol. 161: 286-292.

Oke, V. and Long, S.R. (1999). Bacterial genes induced within the nodule during the *Rhizobium*-legume symbiosis. Mol. Microbiol. 32: 837-849.

Okogun, J.A. and Sanginga, N. (2003). Can introduced and indigenous rhizobial strains compete for nodule formation by promiscuous soybean in the moist savanna agroecological zone of Nigeria? Biol. Fertil. Soils 38: 26-31.

Okubo, K., Hori, N., Matoba, R., Niiyama, T., Fukushima, A., Kojima, Y. and Matsubara, K. (1992). Large-scale cDNA sequencing for analysis of quantitative and qualitative aspects of gene expression. Nat. Genet. 2: 173-179.

Paulsen, I. T., Seshadri, R., Nelson, K. E., Eisen, J. A., Heidelberg, J. F., Read, T. D., Dodson, R. J., Umayam, L., Brinkac, L. M., Beanan, M. J., Daugherty, S. C., Deboy, R. T., Durkin, A. S., Kolonay, J. F., Madupu, R., Nelson, W. C., Ayodeji, B., Kraul, M., Shetty, J., Malek, J., Van Aken, S. E., Riedmuller, S., Tettelin, H., Gill, S. R., White, O., Salzberg, S. L., Hoover, D. L., Lindler, L. E., Halling, S. M., Boyle, S. M. and Fraser, C. M. (2002). The *Brucella suis* genome reveals fundamental similarities between animal and plant pathogens and symbionts 10.1073/pnas.192319099. PNAS 99: 13148-13153.

Peng, G.X., Tan, Z.Y., Wang, E.T., Reinhold-Hurek, B., Chen, W.F. and Chen, W.X. (2002). Identification of isolates from soybean nodules in Xinjiang Region as *Sinorhizobium xinjiangense* and genetic differentiation of *S. xinjiangense* from *Sinorhizobium fredii*. Int. J. Syst. Evol. Microbiol. 52: 457-462.

Perret, X., Staehelin, C. and Broughton, W.J. (2000). Molecular basis of symbiotic promiscuity. Microbiol Mol. Biol. Rev. 64: 180-201.

Preisig, O., Anthamatten, D. and Hennecke, H. (1993). Genes for a microaerobically induced oxidase complex in *Bradyrhizobium japonicum* are essential for a nitrogen-fixing endosymbiosis. Proc. Natl. Acad. Sci. USA 90: 3309-3313.

Priefer, U.B., Aurag, J., Boesten, B., Bouhmouch, I., Defez, R., Filali-Maltouf, A., Miklis, M., Moawad, H., Mouhsine, B., Prell, J., Schlüter, A. and Senatore, B. (2001). Characterisation of *Phaseolus* symbionts isolated from Mediterranean soils and analysis of genetic factors related to pH tolerance. J. Biotechnol. 91: 223-236.

Pühler, A., Arlat, M., Becker, A., Gottfert, M., Morrissey, J.P. and O'Gara, F. (2004). What can bacterial genome research teach us about bacteria-plant interactions? Curr. Opin. Plant Biol. 7: 137-147.

Quispel, A. (1988). Bacteria-plant interactions in symbiotic nitrogen-fixation. Physiol. Plant. 74: 783-790.

Rastogi, V. K., Bromfield, E. S. P., Whitwill, S. T. and Barran, L. R. (1992). A cryptic plasmid of indigenous *Rhizobium meliloti* possesses reiterated *nod*C and *nif*E genes and undergoes DNA rearrangement. Can. J. Microbiol. 38: 563-568.

Reddy, P.M., Ladha, J.K., So, R.B., Hernandez, R.J., Ramos, M.C., Angeles, O.R., Dazzo, F.B. and de Bruijn, F.J. (1997). Rhizobial communication with rice roots: Induction of phenotypic changes, mode of invasion and extent of colonization. Plant Soil 194: 81-98.

Reddy, P.M., Ladha, J.K., Ramos, M.C., Maillet, F., Hernandez, R.J., Torrizo, L.B., Oliva, N.P., Datta, S.K. and Datta, K. (1998). Rhizobial lipochitooligosaccharide nodulation factors activate expression of the legume early nodulin gene ENOD12 in rice. Plant J. 14: 693-702.

Reeve, W.G., Tiwari, R.P., Wong, C.M., Dilworth, M.J. and Glenn, A.R. (1998). The transcriptional regulator gene *phr*R in *Sinorhizobium meliloti* WSM419 is regulated by low pH and other stresses. Microbiology 144: 3335-3342.

Ribbe, M., Gadkari, D. and Meyer, O. (1997). N_2-fixation by *Streptomyces thermoautotrophicus* involves a molybdenum-dinitrogenase and a manganese-superoxide oxidoreductase that couple N_2 reduction to the oxidation of superoxide produced from O_2 by a molybdenum-CO dehydrogenase. J. Biol. Chem. 272: 26627-26633.

Ribbe, M.W. and Burgess, B.K. (2001). The chaperone GroEL is required for the final assembly of the molybdenum-iron protein of nitrogenase. Proc. Natl. Acad. Sci. USA 98: 5521-5525.

Richardson, A.E., Viccars, L.A., Watson, J.M. and Gibson, A.H. (1995). Differentiation of *Rhizobium* strains using the polymerase chain reaction with random and directed primers. Soil Biol. Biochem. 27: 515-524.

Rickert, A.A., Soria, M.A. and Correa, O.S. (2000). The adaptive acid response in *Mesorhizobium* sp. World J. Microbiol. Biotechnol. 16: 475-480.

Rivas, R., Velazquez, E., Valverde, A., Mateos, P.F. and Martinez-Molina, E. (2001). A two primers random amplified polymorphic DNA procedure to obtain polymerase chain reaction fingerprints of bacterial species. Electrophoresis 22: 1086-1089.

Robleto, E.A., Kmiecik, K., Oplinger, E.S., Nienhuis, J. and Triplett, E.W. (1998). Trifolitoxin production increases nodulation competitiveness of *Rhizobium etli* CE3 under agricultural conditions. Appl. Environ. Microbiol. 64: 2630-2633.

Roche, P., Debellé, F., Maillet, F., Lerouge, P., Faucher, C., Truchet, G., Denarié, J. and Prome, J.C. (1991). Molecular basis of symbiotic host specificity in *Rhizobium meliloti* - *nod*H and *nod*PQ genes encode the sulfation of lipo-oligosaccharide signals. Cell 67: 1131-1143.

Roche, P., Maillet, F., Plazanet, C., Debellé, F., Ferro, M., Truchet, G., Promé, J.C. and Dénarié, J. (1996). The common *nod*ABC genes of *Rhizobium meliloti* are host-range determinants. Proc. Natl. Acad. Sci. USA 93: 15305-15310.

Rome, S., Fernandez, M.P., Brunel, B., Normand, P. and Cleyet-Marel, J.C. (1996). *Sinorhizobium medicae* sp. nov., isolated from annual *Medicago* spp. Int. J. Syst. Bacteriol. 46: 972-980.

Ronson, C.W., Bosworth, A.H., Genova, M., Gusbrandsen, S., Hankinson, T., Kwiatkowski, R., Robie, C., Sweeney, P., Szeto, W., Williams, M. and Zablotowicz, R. (1990). Field release of genetically-engineered *Rhizobium meliloti* and *Bradyrhizobium japonicum* strains. In: Nitrogen Fixation: Achivements and Perspectives (eds.) P.M. Gresshoff, L.E. Roth, G. Stacey, W.E. Newton, Chapman and Hall, New York, USA, pp. 397-403.

Rüberg, S., Tian, Z.-X., Krol, E., Linke, B., Meyer, F., Wang, Y., Pühler, A., Weidner, S. and Becker, A. (2003). Construction and validation of a *Sinorhizobium meliloti* whole genome DNA microarray: genome-wide profiling of osmoadaptive gene expression. J. Biotechnol. 106: 255-268.

Sadowsky, M.J., Cregan, P.B., Göttfert, M., Sharma, A., Gerhold, D., Rodriguez-Quinones, F., Keyser, H.H., Hennecke, H. and Stacey, G. (1991). The *Bradyrhizobium japonicum nol*A gene and its involvement in the genotype-specific nodulation of soybeans. Proc. Natl. Acad. Sci., USA 88: 637-641.

Sadowsky, M.J. and Graham, P.H. (2000). Root and stem nodule bacteria of legumes. In: The Prokaryotes: An Evolving Electronic Resource for the Microbiological Community (ed.) M. Dworkin, 3rd edition release 3.3 edn. Springer-Verlag, New York, USA.

Sanginga, N., Danso, S.K.A., Mulongoy, K. and Ojeifo, A.A. (1994). Persistence and recovery of introduced *Rhizobium* 10 years after inoculation on *Leucaena leucocephala* grown on an alfisol in southwestern Nigeria. Plant Soil 159: 199-204.

Sanjuan, J. and Olivares, J. (1989). Implication of *nif*A in regulation of genes located on a *Rhizobium meliloti* cryptic plasmid that affect nodulation efficiency. J. Bacteriol. 171: 4154- 4161.

Segovia, L., Young, J. and Martínez-Romero, E. (1993). Reclassification of American *Rhizobium leguminosarum* biovar *phaseoli* type I strains as *Rhizobium etli* sp. nov. Int. J. Syst. Bacteriol. 43: 374-377.

Selbitschka, W., Jording, D., Nieman, S., Schmidt, R., Pühler, A., Mendum, T. and Hirsch, P. (1995). Construction and characterization of a *Rhizobium leguminosarum* biovar *viciae* strain designed to assess horizontal gene transfer in the environment. FEMS Microbiol. Lett. 128: 255-263.

Selenska-Pobell, S., Gigova, L. and Petrova, N. (1995). Strain-specific fingerprints of *Rhizobium galegae* generated by PCR with arbitrary and repetitive primers. J. Appl. Bacteriol. 79: 425-431.

Selenska-Pobell, S., Evguenieva-Hackenberg, E., Radeva, G. and Squartini, A. (1996). Characterization of *Rhizobium 'hedysari'* by RFLP analysis of PCR amplified rDNA and by genomic PCR fingerprinting. J. Appl. Bacteriol. 80: 517-528.

Selvaraj, G. and Iyer, V.N. (1981). Genetic transformation of *Rhizobium meliloti* by plasmid DNA. Gene 15: 279-283.

Sharma, P.K., Kundu, B.S. and Dogra, R.C. (1993). Molecular mechanism of host-specificity in legume-*Rhizobium* symbiosis. Biotechnol. Adv. 11: 741-779.

Silva, C., Vinuesa, P., Eguiarte, L.E., Martínez-Romero, E. and Souza, V. (2003) *Rhizobium etli* and *Rhizobium gallicum* nodulate common bean (*Phaseolus vulgaris*) in a traditionally managed milpa plot in Mexico: Population genetics and biogeographic implications. Appl. Environ. Microbiol. 69: 884-893.

Simões-Araújo, J.L., Rodrigues, R.L., Gerhardt, L.B.D., Mondego, J.M.C., Alves-Ferreira, M., Rumjanek, N.G. and Margis-Pinheiro, M. (2002). Identification of differentially expressed genes by cDNA-AFLP technique during heat stress in cowpea nodules. FEBS Lett. 515: 44-50.

Singleton, P.W. and Tavares, J.W. (1986). Inoculation response of legumes in relation to the number and effectiveness of indigenous *Rhizobium* populations. Appl. Environ. Microbiol. 51: 1013-1018.

So, R.B., Ladha, J.K. and Young, J.P. (1994). Photosynthetic symbionts of *Aeschynomene* spp. form a cluster with bradyrhizobia on the basis of fatty acid and rRNA analyses. Int. J. Syst. Bacteriol. 44: 392-403.

Soberón-Chávez, G., Nájera, R., Espín, G. and Moreno, S. (1991). Formation of *Rhizobium phaseoli* symbiotic plasmids by genetic recombination. Mol. Microbiol. 5: 909-916.

Somasegaran, P. and Hoben, H.J. (1994). Handbook for Rhizobia. Springer-Verlag, New York.

Spaink, H.P. (1995). The Molecular basis of infection and nodulation by rhizobia: The ins and outs of sympathogenesis. Annu. Rev. Phytopathol. 33: 345-368.

Spector, M.P., Aliabadi, Z., Gonzalez, T. and Foster, J.W. (1986). Global control in *Salmonella typhimurium*–two-dimensional electrophoretic analysis of starvation-inducible, anaerobiosis-inducible, and heat shock-inducible proteins. J. Bacteriol. 168: 420-424.

Stackebrandt, E., Frederiksen, W., Garrity, G.M., Grimont, P.A.D., Kämpfer, P., Maiden, M.C.J., Nesme, X., Rosselló-Mora, R., Swings, J., Trüper, H.G., Vauterin, L., Ward, A.C. and Whitman, W.B. (2002). Report of the ad hoc committee for the re-evaluation of the species definition in bacteriology. Int. J. Syst. Evol. Microbiol. 52: 1043-1047.

Stephens, J.H.G. and Rask, H.M. (2000). Inoculant production and formulation. Field Crop. Res. 65: 249-258.

Strain, S.R., Whittam, T.S. and Bottomley, P.J. (1995). Analysis of genetic structure in soil populations of *Rhizobium leguminosarum* recovered from the USA and the UK. Mol. Ecol 4: 105-114.

Streit, W. R., Schmitz, R. A., Perret, X., Staehelin, C., Deakin, W. J., Raasch, C., Liesegang, H. and Broughton, W. J. (2004). An evolutionary hot spot: the pNGR234b replicon of *Rhizobium* sp. strain NGR234. J. Bacteriol. 186: 535-542.

Sullivan, J.T., Patrick, H.N., Lowther, W.L., Scott, D.B. and Ronson, C.W. (1995). Nodulating strains of *Rhizobium loti* arise through chromosomal symbiotic gene transfer in the environment. Proc. Natl. Acad Sci. USA 92: 8985-8989.

Sullivan, J.T. and Ronson, C.W. (1998). Evolution of rhizobia by acquisition of a 500-kb symbiosis island that integrates into a *phe*-tRNA gene. Proc. Natl. Acad. Sci., USA 95: 5145-5149.

Sullivan, J.T., Trzebiatowski, J.R., Cruickshank, R.W., Gouzy, J., Brown, S.D., Elliot, R.M., Fleetwood, D.J., McCallum, N.G., Rossbach, U., Stuart, G.S., Weaver, J.E., Webby, R.J., de Bruijn, F.J. and Ronson, C.W. (2002). Comparative sequence analysis of the symbiosis island of *Mesorhizobium loti* strain R7A. J. Bacteriol. 184: 3086-3095.

Sy, A., Giraud, E., Jourand, P., Garcia, N., Willems, A., de Lajudie, P., Prin, Y., Neyra, M., Gillis, M., Boivin-Masson, C. and Dreyfus, B. (2001). Methylotrophic *Methylobacterium* bacteria nodulate and fix nitrogen in symbiosis with legumes. J. Bacteriol. 183: 214-220.

Thurman, N.P., Lewis, D.M. and Jones, D.G. (1985). The relationship of plasmid number to growth, acid tolerance and symbiotic efficiency in isolates of *Rhizobium trifolii.* J. Appl. Bacteriol. 58: 1-6.

Tiwari, R.P., Reeve, W.G., Dilworth, M.J. and Glenn, A.R. (1996a). Acid tolerance in *Rhizobium meliloti* strain WSM419 involves a two-component sensor-regulator system. Microbiology 142: 1693-1704.

Tiwari, R.P., Reeve, W.G., Dilworth, M.J. and Glenn, A.R. (1996b). An essential role for *act*A in acid tolerance of *Rhizobium meliloti.* Microbiology UK 142: 601-610.

Turner, S.L. and Young, J.P. (2000). The glutamine synthetases of rhizobia: phylogenetics and evolutionary implications. Mol. Biol. Evol. 17: 309-319.

Valencia-Morales, E. and Romero, D. (2000). Recombination enhancement by replication (RER) in *Rhizobium etli.* Genetics 154: 971-983.

van Berkum, P., Beyene, D., Bao, G., Campbell, T.A. and Eardly, B.D. (1998). *Rhizobium mongolense* sp. nov. is one of three rhizobial genotypes identified which nodulate and form nitrogen-fixing symbioses with *Medicago ruthenica* [(L.) Ledebour]. Int. J. Syst. Bacteriol. 48 Pt 1: 13-22.

van Berkum, P., Terefework, Z., Paulin, L., Suomalainen, S., Lindström, K. and Eardly, B.D. (2003). Discordant phylogenies within the *rrn* loci of rhizobia. J. Bacteriol. 185: 2988-2998.

van Dillewijn, P., Soto, M.J., Villadas, P.J. and Toro, N. (2001). Construction and environmental release of a *Sinorhizobium meliloti* strain genetically modified to be more competitive for alfalfa nodulation. Appl. Environ. Microbiol. 67: 3860-3865.

van Dillewijn, P., Villadas, P.J. and Toro, N. (2002). Effect of a *Sinorhizobium meliloti* strain with a modified *put*A gene on the rhizosphere microbial community of alfalfa. Appl. Environ. Microbiol. 68: 4201-4208.

van Rhijn, P. and Vanderleyden, J. (1995) The *Rhizobium*-plant symbiosis. Microbiol. Rev. 59: 124-142.

van Rossum, D., Schuurmans, F.P., Gillis, M., Muyotcha, A., Vanverseveld, H.W., Stouthamer, A.H. and Boogerd, F.C. (1995). Genetic and phenetic analyses of *Bradyrhizobium* strains nodulating peanut (*Arachis hypogaea* L.) roots. Appl. Environ. Microbiol. 61: 1599-1609.

Vázquez, M., Santana, O. and Quinto, C. (1993). The NodI and NodJ proteins from *Rhizobium* and *Bradyrhizobium* strains are similar to capsular polysaccharide secretion proteins from Gram-negative Bacteria. Mol. Microbiol. 8: 369-377.

Velázquez, E., Cruz-Sanchez, J.M., Mateos, P.F. and Martinez-Molina, E. (1998). Analysis of stable low-molecular-weight RNA profiles of members of the family Rhizobiaceae. Appl. Environ. Microbiol. 64: 1555-1559.

Velculescu, V.E., Zhang, L., Vogelstein, B. and Kinzler, K.W. (1995). Serial analysis of gene-expression. Science 270: 484-487.

Vinuesa, P., Rademaker, J.L.W., de Bruijn, F.J. and Werner, D. (1998). Genotypic characterization of *Bradyrhizobium* strains nodulating endemic woody legumes of the Canary Islands by PCR-restriction fragment length polymorphism analysis of genes encoding 16S rRNA (16S rDNA) and 16S-23S rDNA intergenic spacers, repetitive extragenic palindromic PCR genomic fingerprinting, and partial 16S rDNA sequencing. Appl. Environ. Microbiol. 64: 2096-2104.

Wang, E.T., van Berkum, P., Sui, X.H., Beyene, D., Chen, W.X. and Martínez-Romero, E. (1999). Diversity of rhizobia associated with *Amorpha fruticosa* isolated from Chinese soils and description of *Mesorhizobium amorphae* sp. nov. Int. J. Syst. Bacteriol. 49: 51-65.

Welsh, J., Chada, K., Dalal, S.S., Cheng, R., Ralph, D. and McClelland, M. (1992). Arbitrarily primed PCR fingerprinting of RNA. Nucleic Acids Res. 20: 4965-4970.

Werner, D., Wilcockson, J., Stripf, R., Mörschel, E. and Papen, H. (1981). Limitations of symbiotic and associative nitrogen fixation by developmental stages in the systems *Rhizobium japonicum* with *Glycine max* and *Azospirillum brasilense* with grasses, e. g., *Triticum aestivum*. In: Biology of Inorganic Nitrogen and Sulfur (eds.) H. Bothe and A. Trebst, Springer-Verlag, Berlin, Germany, pp. 299-308.

Willems, A. and Collins, M.D. (1993). Phylogenetic analysis of rhizobia and agrobacteria based on 16S rRNA gene sequences. Int. J. Syst. Bacteriol. 43: 305-313.

Willems, A., Coopman, R. and Gillis, M. (2001a). Phylogenetic and DNA-DNA hybridization analyses of *Bradyrhizobium* species. Int. J. Syst. Evol. Microbiol. 51: 111-117.

Willems, A., Doignon-Bourcier, F., Goris, J., Coopman, R., de Lajudie, P., de Vos, P. and Gillis, M. (2001b). DNA-DNA hybridization study of *Bradyrhizobium* strains. Int. J. Syst. Evol. Microbiol. 51: 1315-1322.

Wood, D.W., Setubal, J.C., Kaul, R., Monks, D.E., Kitajima, J.P., Okura, V.K., Zhou, Y., Chen, L., Wood, G.E., Almeida, N.F., Woo, L., Chen, Y., Paulsen, I.T., Eisen, J.A., Karp, P.D., Bovee, D., Chapman, P., Clendenning, J., Deatherage, G., Gillet, W., Grant, C., Kutyavin, T., Levy, R., Li, M.J., McClelland, E., Palmieri, A., Raymond, C., Rouse, G., Saenphimmachak, C., Wu, Z., Romero, P., Gordon, D., Zhang, S., Yoo, H., Tao, Y., Biddle, P., Jung, M., Krespan, W., Perry, M., Gordon_Kamm, B., Liao, L., Kim, S., Hendrick, C., Zhao, Z.Y., Dolan, M., Chumley, F., Tingey, S.V., Tomb, J.F., Gordon, M.P., Olson, M.V. and Nester, E.W. (2001). The genome of the natural genetic engineer *Agrobacterium tumefaciens* C58. Science 294: 2317-2323.

Xu, L.M., Ge, C., Cui, Z., Li, J. and Fan, H. (1995). *Bradyrhizobium liaoningense* sp. nov, Isolated from the Root-Nodules of Soybeans. Int. J. Syst. Bacteriol. 45: 706-711.

Yanagi, M. and Yamasato, K. (1993). Phylogenetic analysis of the family Rhizobiaceae and related bacteria by sequencing of 16S rRNA gene using PCR and DNA sequencer. FEMS Microbiol. Lett. 107: 115-120.

Yeh, K.C., Peck, M.C. and Long, S.R. (2002). Luteolin and GroESL modulate in vitro activity of NodD. J. Bacteriol. 184: 525-530.

Young, J.P.W., Downer, H.L. and Eardly, B.D. (1991). Phylogeny of the phototrophic *Rhizobium* strain BTAi1 by polymerase chain reaction-based sequencing of a 16S rRNA gene segment. J. Bacteriol. 173: 2271-2277.

Young, J.P.W. (1992). Molecular Population Genetics and Evolution of Rhizobia. In: The N_2 Fixation and its Research in China (ed.) G. Hong, Springer-Verlag, Berlin, Germany, pp. 366-381.

Young, J.P.W. (1996). Phylogeny and taxonomy of rhizobia . Plant Soil 186: 45-52.

Young, J.P.W. and Haukka, K.E. (1996). Diversity and phylogeny of rhizobia. New Phytol. 133: 87-94.

Young, J.M., Kuykendall, L.D., Martínez-Romero, E., Kerr, A. and Sawada, H. (2001). A revision of *Rhizobium* Frank 1889, with an emended description of the genus, and the inclusion of all species of *Agrobacterium* Conn 1942 and *Allorhizobium undicola* de Lajudie *et al.* 1998 as new combinations: *Rhizobium radiobacter, R. rhizogenes, R. rubi, R. undicola* and *R. vitis*. Int. J. Syst. Evol. Microbiol. 51: 89-103.

Zahran, H.H. (1999). *Rhizobium*-legume symbiosis and nitrogen fixation under severe conditions and in an arid climate. Microbiol. Mol. Biol. Rev. 63: 968-989.

Zahran, H.H. (2005). Rhizobial nitrogen fixation in agriculture: biotechnological perspectives. In: Microbial Biotechnology in Agriculture and Aquaculture (ed.) R.C. Ray, Science Publishers, Enfield, New Hampshire, USA, pp. 71-100.

8

Influence of Microorganisms/ Microbial Products on Water and Sediment Quality in Aquaculture Ponds

Claude E. Boyd and Orawan Silapajarn*

INTRODUCTION

Considerable interest has been evinced in using microorganisms or their derivatives in aquaculture for controlling diseases, improving nutrition, and enhancing pond bottom soil and water quality (Verschuere et al., 2000). Some have applied the name probiotics to these products (Moriarty, 1999; Verschuere et al., 2000). However, this name is usually reserved for microorganisms in entirety, or their components that are beneficial to the health of the host organism (Irianto and Austin, 2002). As microorganisms or their derivatives are used to improve bottom soil and water quality, and are not probiotics according to Irianto and Austin's definition, the term microbial products will be used here to describe live microorganisms or their derivatives that have become items of commerce for use in aquaculture ponds.

Application of microbial products as sediment and water quality enhancers in aquaculture dates back to at least 25 years (Boyd, 1995a; Boyd and Gross, 1998). There is widespread use of these products in some sectors. For example, Gräslund et al., (2003) reported that 86% of shrimp farmers surveyed in Thailand applied these products to ponds or mixed them with shrimp feed. The products consisted of living bacteria, often

Department of Fisheries and Allied Aquacultures, Auburn University, Alabama 36849 USA
**Corresponding author: E-mail: ceboyd@acesag.auburn.edu*

Bacillus spp., or products prepared from cultures of microorganisms that contain organic acids, vitamins, micronutrients, and enzymes. Farmers claimed that microbial products increased decomposition of organic matter in sediment and that living microorganisms applied to ponds outcompeted pathogenic bacteria (Gräslund et al., 2003). Although the use of microbial products is most popular in Thailand and other Southeast Asian nations, they are promoted worldwide as manufacturers and vendors seek to expand markets. It is not known how many brands and varieties of microbial products are used in aquaculture, but one of the authors (CEB) collected advertisements for over 30 different brands during the past five years.

Microorganisms conduct many important functions in aquaculture ponds to include photosynthesis, decomposition of organic matter, nutrient recycling, transformation of substances between oxidized and reduced states, and detoxification of waste matter and pollutants. Interactions among microorganisms may be important in reducing the abundance of pathogenic species through competition. Microorganisms also serve as food for aquaculture species, particularly the larval forms of some species. Nevertheless, practical aquaculturists seldom have much knowledge about aquatic microbiology and how microorganisms influence pond ecosystems and aquaculture production. Thus, the purpose of this chapter is to: (1) describe in simple terms the influence of microorganisms on sediment and water quality in aquaculture ponds, (2) consider the status of research on microbial products for improving pond sediment and water quality, and, (3) provide an overview of the effectiveness of microbial products.

MICROORGANISMS, SEDIMENT, AND WATER QUALITY

Phytoplankton

Microscopic algae, known as phytoplankton, live suspended in water, and they are a dominant component of pond ecosystems. These organisms are ubiquitous and enter ponds via the water supply, from the air, and soil. They are autotrophic microorganisms containing chlorophyll and other light sensitive pigments. They require only light, water, warmth, carbon dioxide and other inorganic nutrients to conduct photosynthesis and grow. The well-known, general equation for photosynthesis is:

$$6\ CO_2 + 6\ H_2O \xrightarrow[\text{Nutrients}]{\text{Light}} C_6H_{12}O_6 + 6\ O_2$$

In photosynthesis, plants remove carbon dioxide from the water and reduce inorganic carbon in carbon dioxide to organic carbon in sugar.

Light energy is transformed to chemical energy in sugar. Molecular oxygen is released to the water as a by-product. Phytoplankton uses the sugar as an energy source, but they also synthesize an array of other compounds that comprise their bodies from the primary photosynthate (Boyd and Goodyear, 1971). Molecular oxygen from photosynthesis is the primary source of dissolved oxygen in most water bodies.

Phytoplankton responds to light and high nutrient concentrations in aquaculture ponds to form dense "blooms" which impart color and turbidity to the water. The life span of individual phytoplankton cells seldom exceeds two weeks, and they die and settle at the bottom continually (Boyd and Tucker, 1998). Photosynthesis by phytoplankton is a major source of organic matter in aquaculture ponds. Phytoplankton is the base of the food web in ponds (Fig. 8.1), but aquaculture production can be greatly enhanced by applying feed. Production of most species may be 10 to 20 times greater in ponds with feeding than in ponds where production is based entirely on natural food sources (Boyd and Tucker, 1998).

Nutrients from aquaculture feeds enter pond waters through uneaten feed, feces, and metabolic wastes, and stimulate phytoplankton growth. In ponds with high feed inputs, phytoplankton abundance may become

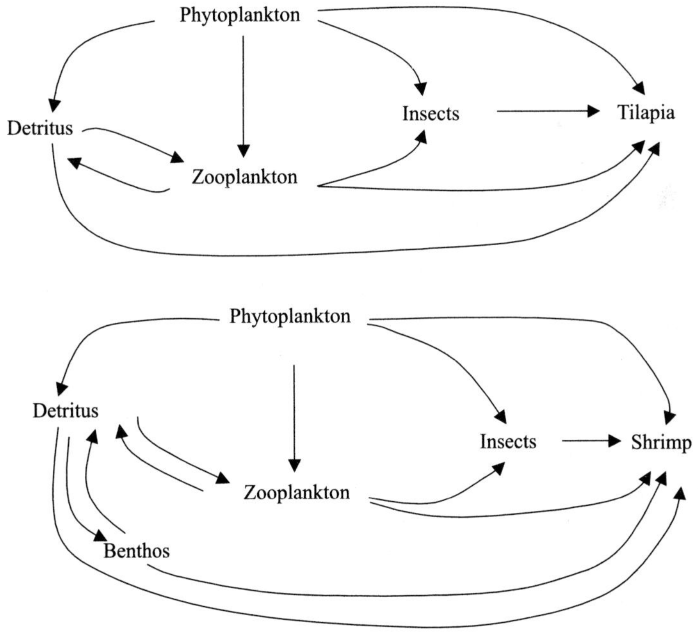

Fig. 8.1. The food web in a tilapia pond and in a shrimp pond

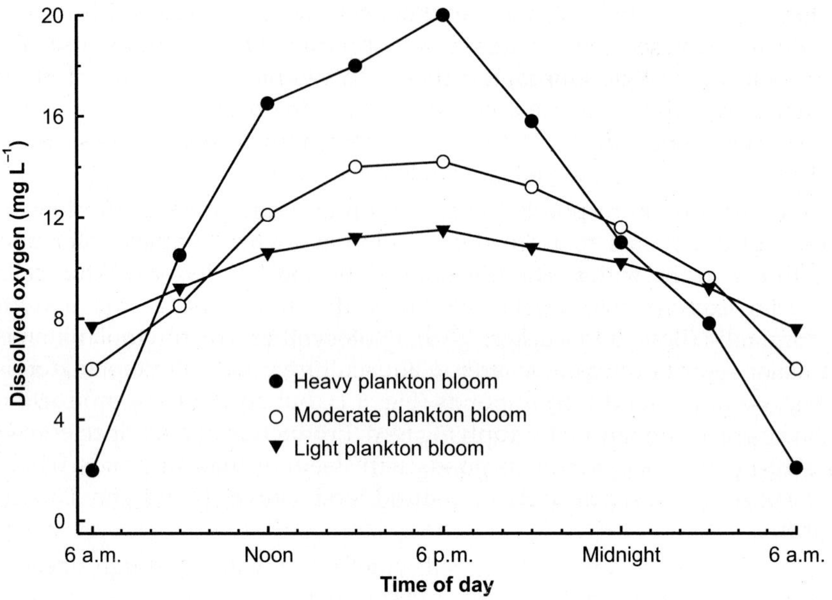

Fig. 8.2. Dissolved oxygen concentration over a 24-hour period in a pond with a dense phytoplankton bloom and a pond with a moderate phytoplankton bloom

excessive. Dense blooms of phytoplankton may cause low concentration of dissolved oxygen at night (Fig. 8.2). Also, high inputs of organic matter in pond bottoms can lead to anaerobic conditions at the soil-water interface. Blue-green algae, sometimes called cyanobacteria, are favored by high feed inputs, and compounds excreted by these species may be absorbed by fish and shrimp causing an unpleasant taste of the flesh – a phenomenon usually referred to as off-flavor (van der Ploeg et al., 1992, 1995; Boyd, 2003). Moreover, under certain conditions, some species of blue-green algae may be toxic to aquaculture species (Paerl and Tucker, 1995). Blue-green algae also form surface scums that can cause shallow thermal stratification. These scums sometimes die suddenly leading to depletion of dissolved oxygen (Boyd and Tucker, 1998). Thus, dense blooms of phytoplankton and the presence of certain species of blue-green algae may be highly detrimental.

A moderate density of phytoplankton obviously is desirable in aquaculture ponds, and certain forms such as species of green algae and diatoms are preferable to blue-green algae. Nevertheless, the authors are unaware of any attempts to alter phytoplankton dynamics in ponds by applying cultures of species of living phytoplankton considered desirable in aquaculture. Liming material and fertilizers are commonly applied to stimulate growth when phytoplankton abundance is low. Pond managers reduce feed and fertilizer inputs when phytoplankton is overly abundant, and some may flush nutrients and phytoplankton from ponds by water exchange. Copper sulfate and other algaecides are sometimes put into ponds in an attempt to control excessive phytoplankton (Boyd and Tucker, 1998).

A healthy phytoplankton bloom is especially important in ponds with feeding. Ammonia uptake by phytoplankton is a major factor preventing high concentrations of this potentially toxic substance in pond waters (Tucker and van der Ploeg, 1993).

We find it peculiar that enterprising individuals are not promoting and selling phytoplankton cultures to improve conditions for fish and shrimp culture in ponds. Maybe this is because there is a general realization by aquaculturists that water is colored by the presence of phytoplankton. Bacteria, fungi, and actinomycetes, on the other hand, are not visible to the naked eye even though there may be many millions of their cells in a liter of water.

Benthic Algae

Benthic algae require little sun light and most aquaculture ponds have benthic algae communities because light penetrates to their bottoms. However, in ponds with relatively clear water, dense mats of benthic algae commonly called lab-lab develop. Some types of fish culture depend upon lab-lab as a food source. A moderate abundance of benthic algae is desirable in shrimp ponds because benthos is a good food source for recently stocked postlarval shrimp (Boyd and Tucker, 1998).

Excessive growth of benthic algae can be harmful because highly anaerobic conditions develop beneath algal mats. Thus, most aquaculture production systems are managed to encourage phytoplankton that increase turbidity and limit light penetration to the bottom.

Saprophytic Microorganisms

Aerobic Decomposition

Saprophytic microorganisms consist of fungi, actinomycetes, and bacteria that decompose dead organic matter (substrate) originating in dead plankton, feces, and uneaten feed, and prevent it from accumulating to a high concentration in pond water and sediment. All three forms occur in ponds, but bacteria are by far the most numerous and important. In the presence of dissolved oxygen, organic carbon is decomposed by respiration of aerobic bacteria to carbon dioxide and water, as illustrated below:

$$C_6H_{12}O_6 + 6\,O_2 \rightarrow 6\,CO_2 + 6\,H_2O$$

Energy released by the oxidation reaction is captured and used by microbial cells. Nutrient recycling occurs because ammonia, phosphate, and other mineral compounds are released (mineralized) from organic combination by decomposition. Decomposition also consumes dissolved oxygen, and if large amounts of organic matter are present, the surrounding water may become depleted of dissolved oxygen. Photosynthesis and respiration are very different as biochemical reactions, but from an ecological standpoint, they are reverse reactions. In photosynthesis, carbon dioxide and other inorganic nutrients are removed from the water and oxygen is added, carbon dioxide is reduced, and energy is captured. In respiration, oxygen is removed from the water, carbon dioxide and other inorganic nutrients are returned, carbon is oxidized, and energy is released.

Bacteria live freely in water, but they are particularly abundant on organic particles suspended in the water or deposited at the bottom. Bacteria and organic particles covered by them are rich in protein and other nutrients, and all an excellent food source for filter-feeding fish (Schroeder, 1978), shrimp (McIntosh, 1999), and mollusks (Shumway et al., 2003). In fact, a new system for producing shrimp through a heterotrophic-based food web has been highly successful in Belize (McIntosh, 1999). A combination of mixed grain and shrimp feed are applied to ponds in amounts greater than shrimp can consume. The high organic matter input with continual mixing by intensive mechanical aeration (up to 30 horsepower ha^{-1}) results in an abundance of organic particles in the water. These particles become coated with bacteria and form a suspended floc in the water. The ponds basically function like a bioreactor, and the shrimp feed upon the floc. Shrimp excrete large amounts of ammonia and the bacteria remove this ammonia from the water and convert it to protein that can be used by the shrimp. Shrimp production exceeds 10 tons ha^{-1} per crop, and more than 20 tons ha^{-1} per crop have been achieved with this methodology (Boyd and Clay, 2002).

Anaerobic Decomposition

In the absence of dissolved oxygen, certain bacteria can decompose organic matter by fermentation. The end products of fermentation are carbon dioxide and carbon compounds such as alcohols, formate, propionate, and lactate (Boyd, 1995a). For example, fermentation of glucose gives ethanol:

$$C_6H_{12}O_6 \rightarrow 2\,CH_3CH_2OH + 2\,CO_2$$

Fermentation does not degrade organic matter as completely as aerobic decomposition, because a part of the organic carbon in substrate is not converted to carbon dioxide. In the example of ethanol production above, only one-third of the organic carbon in glucose was converted to carbon dioxide. The organic by-products accumulate in zones where fermentation is occurring, and these compounds can be toxic to aquatic organisms.

Certain bacteria decompose organic matter, including by-products of fermentation, to carbon dioxide, water, and other mineral compounds in the absence of molecular oxygen. These bacteria are called chemotrophic bacteria because they use oxidized, inorganic chemical compounds rather than molecular oxygen as hydrogen and electron acceptors in respiration (Boyd, 1995b). Denitrifying bacteria obtain oxygen from nitrate by reducing it to nitrite, ammonia, or nitrogen gas. A representative equation for denitrification is:

$$6\ NO_2^- + 5\ CH_3OH \rightarrow 5\ CO_2 + 3\ N_2 + 7\ H_2O + 6\ OH^-$$

Nitrogen gas (N_2) diffuses into the air and does not pose a toxicity threat to culture animals. However, ammonia and nitrite released by certain denitrifying bacteria can be toxic to fish, shrimp, and other aquaculture organisms.

Certain bacteria utilize iron and manganese oxides and hydroxides as sources of oxygen in respiration (Boyd, 2000). When organic carbon is mineralized to carbon dioxide in respiration, hydrogen ion is released:

$$CH_3COOH + 2\ H_2O \rightarrow 2\ CO_2 + 8\ H^+$$

These bacteria have a metabolic pathway in which hydrogen ions are combined with oxygen from iron or manganese compounds, as follows:

$$Fe\ (OH)_3 + 3\ H^+ = Fe^{2+} + 3\ H_2O$$

$$MnO_2 + 4\ H^+ = Mn^{2+} + 2\ H_2O$$

Thus, iron- and manganese-reducing bacteria mineralize organic carbon completely to carbon dioxide, water, and other mineral compounds. Ferrous ion (Fe^{2+}) and manganese ion (Mn^{2+}) released as metabolites accumulate in sediment pore water or in the water column where the process is occurring. Sediment containing ferrous iron will be darker than oxygenated sediment, and it may be jet black in color. Most of us, perhaps at sometime, may have walked on the bottom of a freshly drained pond where the sediment surface was brown or light gray, and having looked back, observed that our footprints were black or dark gray. The sediment on top was oxidized and contained ferric iron (Fe^{3+}), while the underlying sediment revealed by our footprints was reduced and contained ferrous iron. If a few hours later we returned to the pond, we also may have observed that our footprints had turned brown. This resulted from the oxidation of Fe^{2+} to Fe^{3+} by exposure to oxygen in the air.

Although reduced iron and manganese are not highly toxic, they oxidize and precipitate when mixed with oxygenated water. The precipitate can mechanically occlude the gills of fish and shrimp and cause stress or mortality. Iron and manganese precipitates also can cause discolored areas (mottling) on the surfaces of shrimp and other shellfish, making them less desirable in the market.

Sulfur-reducing bacteria oxidize organic carbon using sulfate as a source of molecular oxygen. The reaction occurs in roughly the same manner as those for iron- and manganese-reducing bacteria illustrated above. The hydrogen ion released when organic carbon fragments are oxidized is combined with sulfate to form sulfide and water as illustrated below:

$$SO_4^{2-} + 4\,H^+ \rightarrow S^{2-} + 4\,H_2O$$

Sulfides, and especially hydrogen sulfide, are highly toxic to fish, shrimp, and other aquatic animals. Sulfides have a strong odor (similar to rotten eggs) and can be detected easily by smelling the water or sediment. Many aquaculturists think sulfide causes the sediment to be black. Actually, the black color of sediment is from ferrous iron. Only dark-colored sediment will contain sulfide, but sulfide is not found in all dark-colored sediment unless the redox potential is quite low.

Methane bacteria use carbon dioxide as a source of molecular oxygen. A typical pathway for methanogenesis is:

$$CH_3COOH + 2\,H_2O \rightarrow 2\,CO_2 + 8\,H^+$$

$$8\,H^+ + CO_2 \rightarrow CH_4 + 2\,H_2O$$

Methane diffuses into the air and is rarely toxic to aquatic animals.

The saprophytic microorganisms in ponds perform the useful service of decomposing organic matter to prevent it from accumulating to high concentrations. The activity on anaerobic microorganisms is particularly important, for most organic matter in ponds accumulates in sediment where anaerobic conditions prevail below a depth of a few millimeters (Fig. 8.3). The oxidation-reduction (redox) potential depicted in Fig. 8.3 is an index of the degree of oxidation or reduction of sediment. Oxygenated water with more than 1 mg L^{-1} of dissolved oxygen will have a redox potential of about 0.56 millivolt (mV) (Boyd, 1995b). Nitrite appears at a redox value of about 0.3 or 0.4 mV and ferrous appears when the redox falls to about 0.2 mV. At a redox of 0.1 mV or less, hydrogen sulfide may be present.

In a single year, 2 or 4 tons of organic matter may reach pond bottoms in uneaten feed and feces of the culture species, and 10 to 20 tons more may be produced by photosynthesis. Without anaerobic microbial degradation,

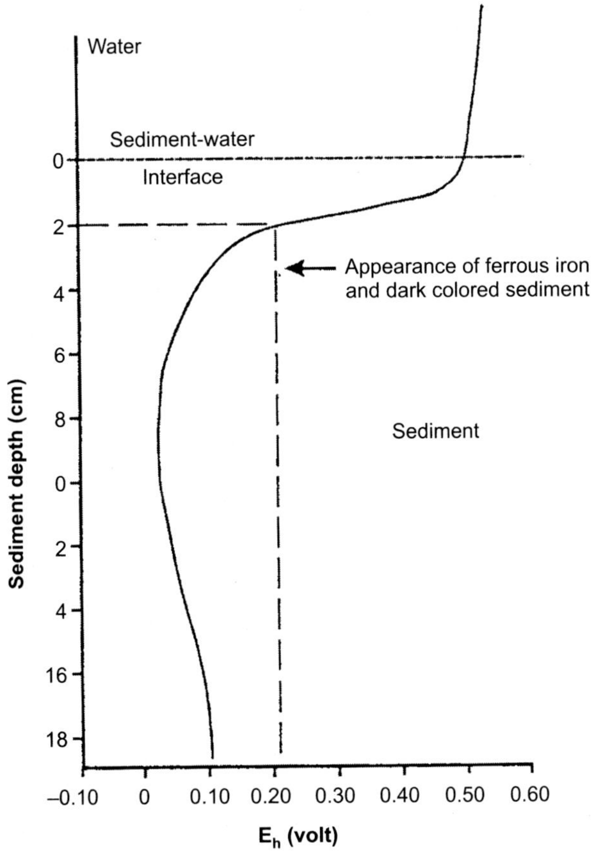

Fig. 8.3. Oxidation-reduction (redox) potential at different depths in pond water and sediment

the pond would soon have a deep layer of organic matter in the bottom. However, analyses of pond sediment revealed that concentrations of organic matter seldom exceed 4 to 6% (Boyd et al., 1994; Steeby et al., 2004). A study of tilapia ponds in Thailand (Thunjai et al., 2004) revealed that there was no strong relationship between pond age and organic carbon concentrations in pond bottoms (Table 8.1). The maximum concentration of organic carbon found in samples from 35 ponds was 3.39%. The practice of liming to prevent acidity and drying and tilling of bottoms between crops was sufficient to maintain healthy pond bottoms for at least 40 years.

In studies of pond soil organic matter, results may be reported as a percentage of organic carbon or as a percentage of organic matter. As a general rule, soil organic matter is about 58% carbon (Nelson and

Table 8.1. Average bottom soil organic carbon concentrations in tilapia ponds of different ages in Thailand

Pond age (years)	Number of samples	Mean	Standard deviation
0-10	14	1.90	0.49
11-20	13	1.66	0.47
21-30	5	2.19	0.32
31-40	3	2.49	0.79

Sommers, 1982). Thus, for practical purposes, a factor of two can be used to switch between organic carbon and organic matter.

In a pond, the water column normally is aerobic containing 4 or 5 mg L^{-1} dissolved oxygen or more. However, oxygen cannot enter sediment rapidly, and in aquaculture ponds, bacteria in sediment use oxygen faster than it can move downward by diffusion and by infiltration of the water in which it is dissolved. Thus, aerobic respiration only occurs up to a depth of a few millimeters. Fermentation can occur at any depth in the anaerobic sediment, but saprophytic, chemotrophic bacteria usually are stratified within the sediment with denitrifying organisms occurring at the bottom of the aerobic zone, followed in order by manganese- and iron-reducing bacteria, sulfate-reducing bacteria, and methane-produced bacteria (Fig. 8.4).

Toxic metabolites entering pond water from anaerobic sediment are oxidized to non-toxic form. However, if the rate at which they are released into the water exceeds the rate at which they are oxidized, an equilibrium concentration of one or more toxic metabolites will occur. This concentration may exceed the toxic threshold for aquaculture species. Thus, one of the objectives of pond management is to maintain an oxidized layer at the sediment-water interface to retard the movement of toxic metabolites from sediment into the water column (Fig. 8.5). This oxidized layer can be encouraged by drying pond bottoms between crops to hasten organic matter decomposition, using moderate stocking and feeding rates to prevent excessive organic matter input, and applying mechanical aeration to supplement the dissolved oxygen supply and encourage movement of oxygenated water over the pond bottom. Drying of pond bottoms also oxidizes reduced substances in the soil so that these compounds can be used as oxygen sources for chemotrophic, saprophytic bacteria during the next crop. In freshwater ponds, the pore water of wet sediment (bottom soil) only holds 8.2 mg L^{-1} dissolved oxygen at 25°C. When the soil dries, air containing 20.95% oxygen (209,500 mg L^{-1}) can move into the soil pores to accelerate aerobic decomposition and oxidation of ferrous iron, nitrite, sulfide, and other reduced substances. The soil also

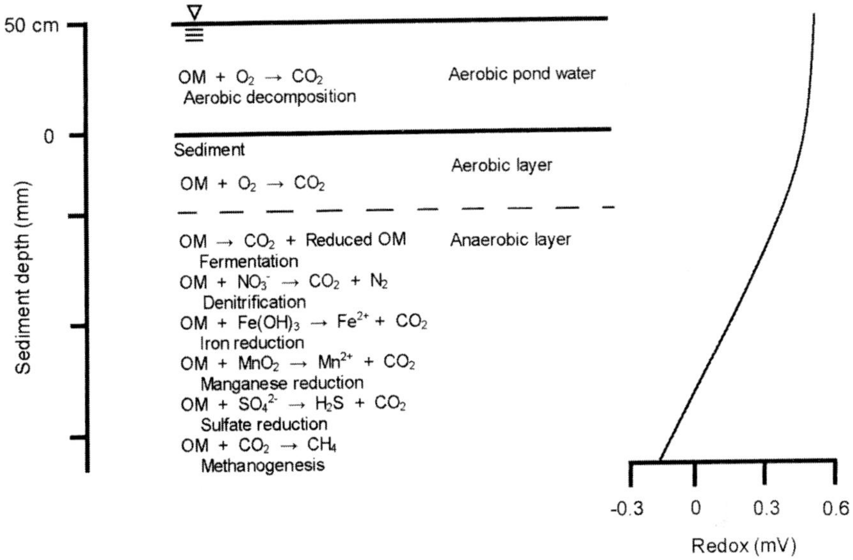

Fig. 8.4. Stratification of microbial processes by depth in pond water and sediment OM = Organic matter

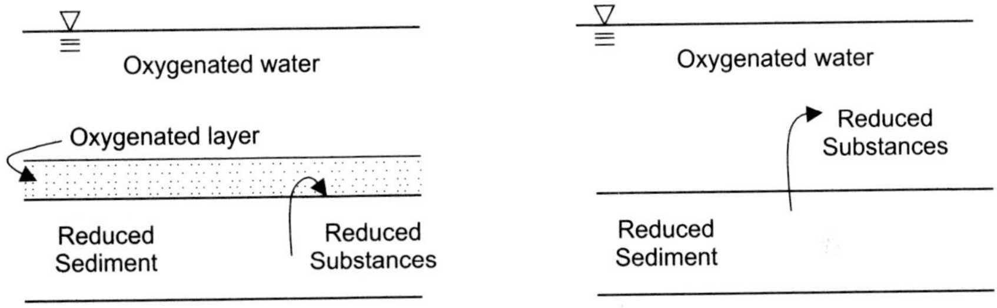

Fig. 8.5. Illustration of a "good" (left) versus "bad" (right) condition at the sediment-water interface

cracks as clay shrinks on drying (Fig. 8.6) to augment soil oxidation. Further enhancement of soil aeration can result from tilling the bottom to a depth of 10 cm with a disk harrow or other tilling implement.

In some ponds, water seeps into ponds in low areas or is retained in depressions to prevent drying. Sodium nitrate ($NaNO_3$) is a strong oxidant that can be applied over wet areas in pond bottoms to effect oxidation (Boyd, 1995c). The usual application rate is 20 to 40 g $NaNO_3$ m^{-2}. Sodium nitrate is also applied to ponds during the culture period to poise the redox potential at the soil-water interface at a level that prevents toxic metabolites from diffusing into the water column (Boyd, 1995c).

Fig. 8.6. A waterlogged pond bottom immediately after draining (upper) versus a dry pond bottom following exposure to air for two weeks (lower)

Chemoautotrophic Microorganisms

Metabolites from anaerobic micrcorganisms in the sediment diffuse into pond water. When this occurs, dissolved oxygen will oxidize reduced inorganic compounds, e.g. nitrite to nitrate, ferrous iron to ferric iron, and sulfide to sulfate. Bacteria that oxidize reduced inorganic compounds and use the energy released to transform carbon dioxide to organic carbon often accelerate these transformations. These bacteria are called chemoautotrophic microorganisms. Although the process by which chemoautotrophic microorganisms synthesize organic carbon is ecologically analogous to photosynthesis, biochemically, it is a very different pathway. Oxidation of reduced inorganic compounds allows them to be used again by chemotrophic, saprophytic bacteria. Also, the process transforms potentially- toxic compounds into non-toxic ones.

Ammonia is the major nitrogenous excretory product of aquaculture species. Microorganisms decomposing organic matter also release it. Thus, high ammonia concentration is a potentially harmful condition in aquaculture. Oxidation of ammonia to nitrate by nitrifying bacteria is the most important process acting to limit ammonia accumulation (Feliatra et al., 2004). A slightly alkaline pH (7.5 to 8.0) and plenty of dissolved oxygen is essential for effective nitrification. Thus, liming and mechanical aeration can be used to enhance this important process (Boyd and Tucker, 1998)

Nitrification, ferrous iron oxidation, and sulfide oxidation are illustrated by the equations below:

Nitrification

$$NH_4^+ + 1\frac{1}{2}\,O_2 \rightarrow NO_2^- + H_2O + 2\,H^+$$

$$NO_2^- + \frac{1}{2}\,O_2 \rightarrow NO_3^-$$

Iron oxidation

$$4\,FeCO_3 + O_2 + 6\,H_2O \rightarrow 4\,Fe(OH)_3 + 4\,CO_2$$

Sulfur oxidation

$$H_2S + 2\,O_2 = H_2SO_4$$

It should be emphasized that all of microbially-mediated oxidations require dissolved oxygen, and nitrification has an especially high demand for oxygen. Each milligram of ammonia nitrogen oxidized to nitrate nitrogen requires 4.57 mg L^{-1} dissolved oxygen. In aquaculture ponds, nitrification may use almost as much oxygen as organic matter decomposition in the water column (Boyd and Gross, 1999). Also, nitrification is a major source of acidity in aquaculture ponds (Hunt and Boyd, 1981).

Nitrification is a two-step process in which species of *Nitrosomonas* oxidize ammonia to nitrite and species of *Nitrobacter* oxidize nitrite to nitrate (Metting, 1992). Under certain conditions that are not completely understood, the oxidation of ammonia to nitrite proceeds faster than the conversion of nitrite to nitrate (Boyd and Tucker, 1998). This can cause the nitrite concentration to reach a toxic level. Elevated chloride concentration greatly decreases the uptake of nitrite by culture animals in freshwater. Thus, ponds may be treated with sodium chloride to provide 50 to 100 mg L^{-1} chloride and serve as a countermeasure against nitrite toxicity (Tavares and Boyd, 2003).

FACTORS CONTROLLING BACTERIA IN PONDS

Resting stages of all common species of bacteria occur in soil, water, and air. When a pond is filled with water at the beginning of a crop, bacterial cells and spores are present. They respond to substrate and a bacterial flora develops in proportion to the amount of organic matter present and the suitability of other environmental conditions for microbial growth. Chemoautrotrophic bacteria do not need organic matter, and respond to the amount of inorganic substrate present, e.g. concentrations of ammonia, ferrous iron, nitrite, sulfide, etc. The major factors affecting bacterial activity are temperature, dissolved oxygen concentration, amount of moisture, and quantity and quality of substrate. When conditions are favorable for growth, bacteria grow exponentially. Some species can double their numbers in less than an hour.

Bacteria respond to warmth and their growth rate and physiological processes increase roughly two-fold for each 10°C increase in temperature up to a temperature of about 40°C. Moisture is not a factor in bacterial growth during culture periods in ponds. However, when ponds are drained of water between crops, drying initially increases the rate of decomposition because oxygen from the air can enter the soil as described previously. However, when the pond bottom becomes too dry, microbial activity declines greatly (Fig. 8.7) due to lack of moisture (Boyd et al., 2002).

Bacteria, the dominant decomposers in pond sediment, decompose organic matter most rapidly under slightly alkaline conditions (pH 7.5-8.5) as illustrated in Fig. 8.8. Thus, application of agricultural limestone to acidic pond bottoms can greatly accelerate organic matter decomposition. Recommended application rates for agricultural limestone are provided in Table 8.2.

The carbon/nitrogen ratio in pond soils normally is between 6 and 12 (Boyd, 1995b). In some locations, ponds have been constructed on soils with high concentrations of native organic matter with a wide carbon/nitrogen ratio. For example, bottom soils of some brackishwater ponds in

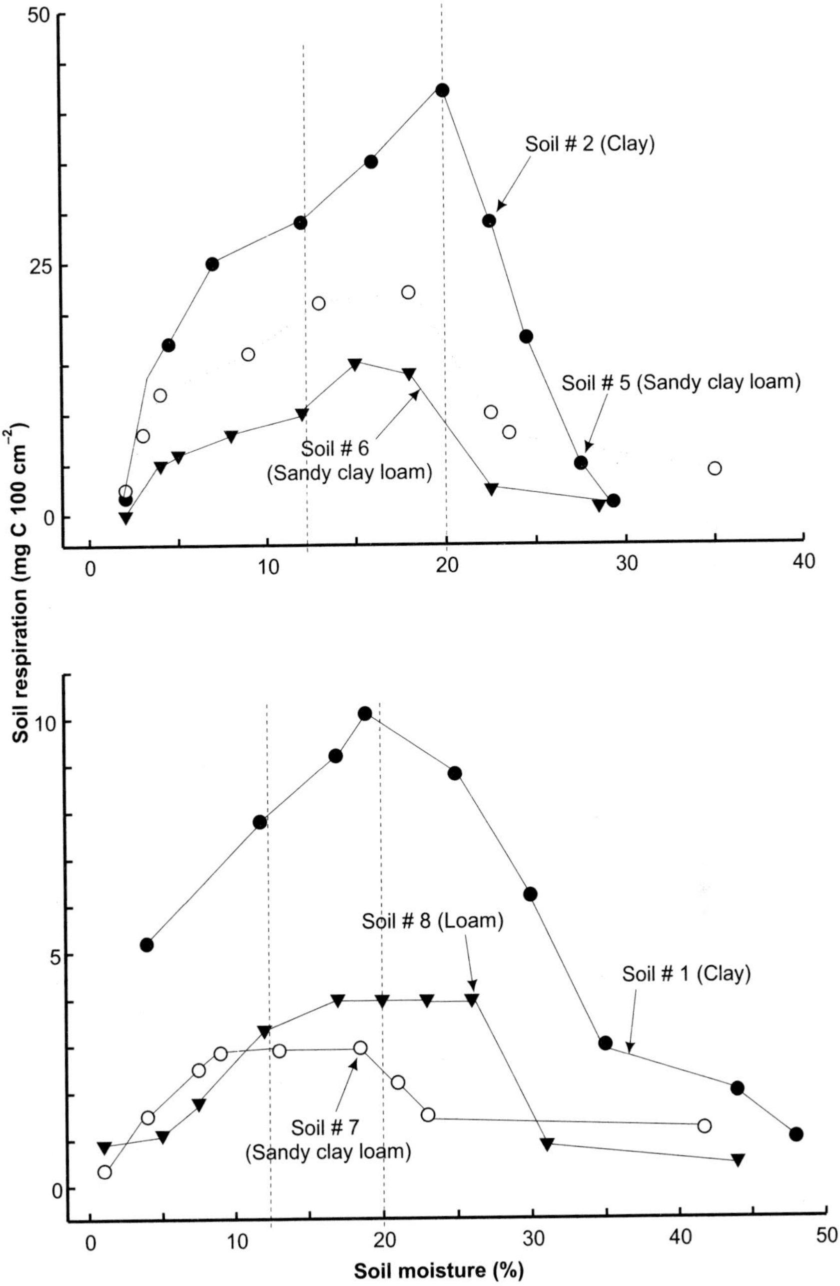

Fig. 8.7. Effect of soil moisture content on soil respiration in a pond bottom during draining

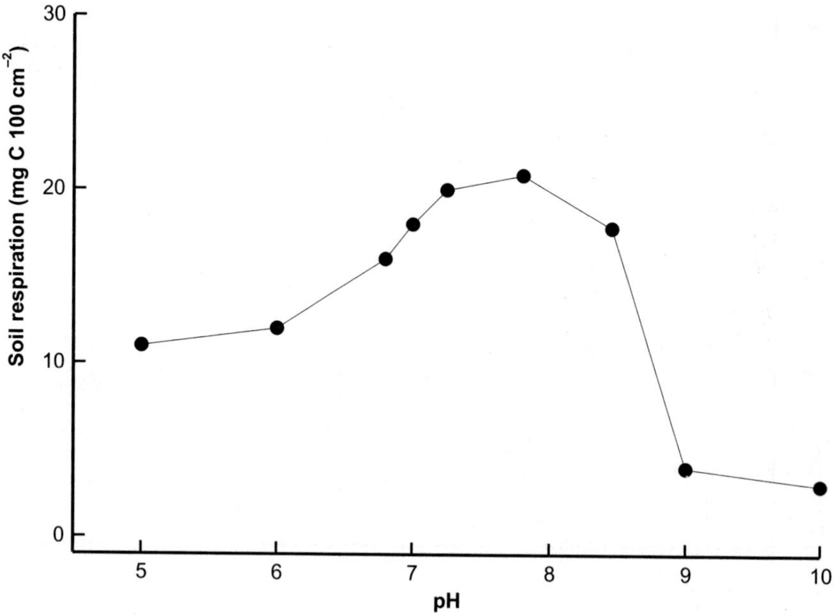

Fig. 8.8. Effect of soil pH on soil respiration at the bottom of an empty pond

Table 8.2. Suggested rates for application of agricultural limestone to pond bottom soils of different pH

Soil pH (Standard units)	Agricultural limestone (kg ha^{-1})
7.0 or above	0
7.0 - 6.5	500
6.5 - 6.0	1000
6.0 - 5.5	2000
5.5 or less	3000

Ecuador that were constructed in former mangrove areas where soils contained 8 to 14% organic carbon and had carbon/nitrogen ratios of 25 to 30 (Sonnenholzner and Boyd, 2000a). Such organic matter does not decompose readily, but it decomposes enough to cause poor soil condition during the culture period. Bacteria decompose organic matter most rapidly when the carbon/nitrogen ratio is 6 to 12. Application of nitrogen fertilizer at 200 to 400 kg ha^{-1} can accelerate the decomposition of resistant, native organic matter. Of course, organic soils tend to be acidic, so agricultural limestone also should be applied.

If limestone and urea or ammonium fertilizers are applied simultaneously, the bottom should be tilled to mix the fertilizer into the soil and prevent ammonia loss to the air by volatilization at elevated pH.

ENZYMES

Fresh organic residues usually consist of feed ingredients, feces, or dead plants and animals of relatively large particle size compared to bacterial cells. Bacteria cannot absorb the large molecules in these residues. Physical forces, feeding activity by animals, and other factors reduce the particle size of organic residues, and soluble compounds leach out. This physical process by which organic matter is broken down into smaller particles is called comminution (Boyd, 1995b).

Catabolism is the enzymatic degradation of substrate for smaller molecules by microorganisms. For example, hydrolytic bacteria convert cellulose, hemicellulose, and pectin to sugars, hydrolyze fat, split proteins, break phenolic rings, and cause many other alterations in organic compounds. The hydrolysis of cellulose to simple sugar molecules is illustrated below:

$$(C_6 10_{10} O_5)_n + n\ H_2O = n\ C_6H_{12}O_6$$

This reaction requires the enzyme cellulase, and only microorganisms that produce cellulase can hydrolyze cellulose readily.

An enzyme is a protein produced by living cells that functions as a catalyst in chemical reactions. Enzymes, like other chemical catalysts, lower the activation energy for reactions so that they occur more rapidly (Fig. 8.9). A catalyst (or enzyme) is not used up in the reaction that it catalyzes, and it is released back into the reaction medium to function again. Many different enzymes that function both inside and outside bacterial cells are necessary to fragment organic residues into carbon fragments small enough for use in microbial respiration. Extracellular enzymes produced and excreted into the environment by microbial cells catalyze initial fragmentation of substrate molecules to a size small enough for absorption. If bacteria are growing rapidly, they will excrete an abundance of extracellular enzymes. In a situation where there is a great abundance of substrate, enzyme blocking can reduce the activity of extracellular enzymes.

Enzymes function according to a template principle (Fig. 8.10). The enzyme surface fits into sites on one substrate surface to effect splitting of the substrate molecule. The enzyme and the substrate particles separate and the enzyme can function again (Fig. 8.10). However, if there is an abundance of substrate, the enzyme may attach to the wrong site on the substrate, and when this occurs, the enzyme does not split the substrate and it is not released from the substrate – hence, the term enzyme blocking. In such a case, additions of extracellular enzymes to the medium might encourage greater rates of decomposition or other microbial processes.

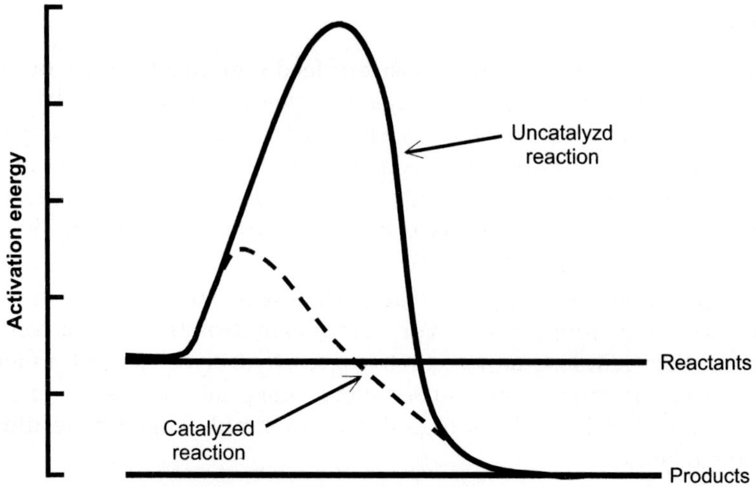

Fig. 8.9. Illustration of decrease in activation energy of reaction by a catalyst

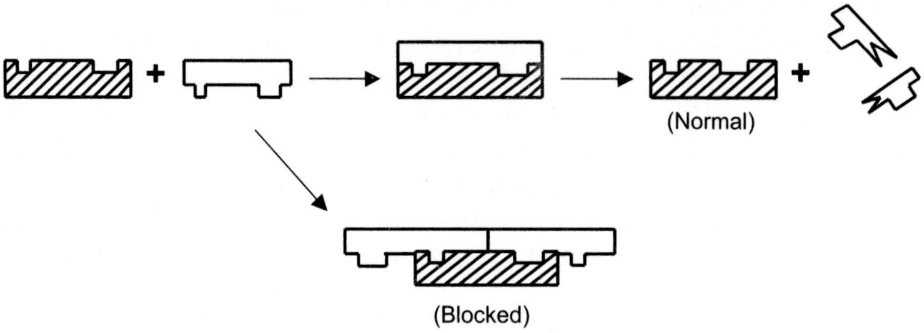

Fig. 8.10. Illustration of normal enzyme-substrate activity and of enzyme blocking

Enzymes inside microbial cells break absorbed substrate into even smaller pieces for use in respiration to release energy for cells.

Cultures of microorganisms and especially yeast cultures, are homogenized to rupture cells and release enzymes. Solutions or dried preparations from homogenized cultures of microorganisms are supplemented with living microorganisms, vitamins, trace elements, and other substances to prepare microbial products for aquaculture ponds.

THE ARGUMENT FOR MICROBIAL PRODUCTS

The discussion above clearly reveals that a healthy microbial flora is essential for maintaining good sediment and water quality in aquaculture ponds. When microbial activity does not progress fast enough, organic matter accumulates in sediment, there will be zones in the pond bottom with anaerobic conditions at the soil-water interface, the concentration of dissolved organic matter will increase in the water column, concentrations of potentially toxic metabolites such as carbon dioxide, ammonia, nitrite, and sulfide may exceed acceptable limits, and dissolved oxygen concentration will decline in the water column. The results of these changes on fish and shrimp populations include poor appetite and inefficient feed conversion, stress, greater susceptibility to disease, slow growth, increased mortality, and less efficient production.

Vendors claim that application of microbial products will enhance natural processes such as organic matter degradation, nitrification, denitrification, sulfide oxidation, and degradation of toxic pollutants. Although it is true that the microbial flora of ponds can carry out all of these processes and more, the important question is: *Will microbial products actually enhance sediment and water quality and improve environmental conditions for shrimp and fish culture?*

STUDIES OF MICROBIAL PRODUCTS

Ponds have a natural microbial flora capable of conducting the entire range of biogeochemical transformations necessary for healthy functioning of natural ecosystems. Microbial populations respond to the availability of substrate as illustrated in Fig. 8.11. If substrate is added to a resting culture, the bacteria will adjust to the new substrate and grow exponentially until the substrate is depleted or until metabolites accumulate to toxic concentrations. The abundance of bacteria in the culture then declines.

If pond water has ammonia, natural nitrifying bacteria will oxidize the ammonia to nitrate. If ammonia concentration increases, the population of nitrifying bacteria will respond to the increase by rapidly increasing in abundance and physiological activity. When ammonia concentrations decline, the abundance of nitrifying bacteria will decline. Nitrification rates are sensitive to ammonia concentration, but they are also regulated by environmental variables such as temperature, dissolved oxygen concentration, and pH. If nitrification does not increase in response to increasing ammonia concentration, it is because some environmental variable is not within the satisfactory range, and not because there is a shortage of nitrifying bacteria. The same principle applies to organic matter decomposition and other microbial processes. It does not seem likely that applications of bacterial inocula to most ponds would influence the rate of bacterially mediated processes.

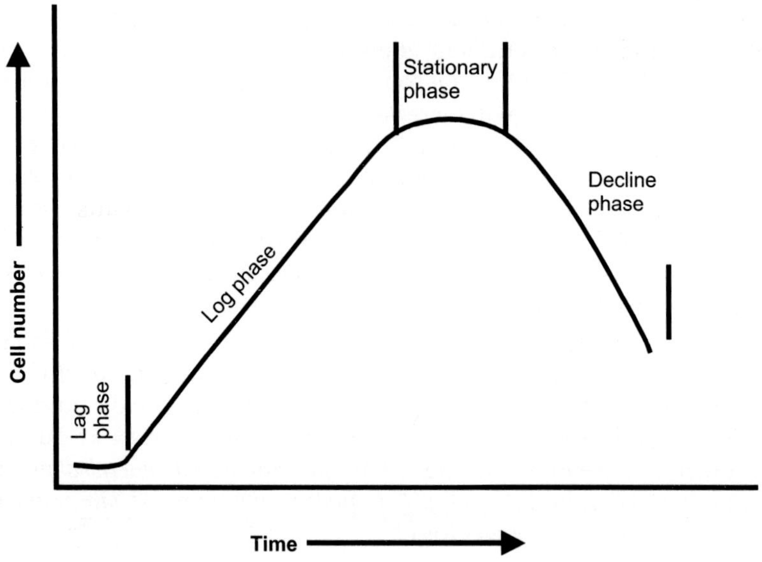

Fig. 8.11. Growth pattern in a microbial culture

Boyd and Gross (1998) summarized results of research on microbial products and concluded that published studies did not provide clear evidence of large benefits to sediment and water quality or production of culture species following treatment of ponds with these products. In fact, studies conducted on channel catfish culture (Boyd et al., 1984; Tucker and Lloyd, 1985; Chiayvareesajja and Boyd, 1993; Boyd and Pipoppinyo, 1994; Queiroz et al., 1998; Queiroz and Boyd, 1998) failed to demonstrate significant difference in concentrations of important water and sediment quality variables between ponds treated with microbial products and control ponds.

Since the review by Boyd and Gross (1998), Mischke (2003) studied two commercial enzyme preparations as water quality enhancers for channel catfish ponds. He concluded that neither product improved pond water quality at the application rates recommended by the vendors. He suggested that the most efficient way to keep the pond environment suitable for microbial processes is through normal pond management. Maintaining adequate dissolved oxygen concentrations through aeration and careful feeding practices is a more viable strategy for water quality maintenance than adding expensive commercial formulations of microbial products. It should also be noted that liming of ponds and dry-out of pond bottoms between culture cycles can be a useful tool in maintaining good sediment quality (Boyd et al., 2002).

Microbial products probably are used more widely in shrimp culture than in other types of aquaculture. Nevertheless, there is virtually no literature revealing that application of these products to shrimp ponds improved water and soil quality during culture periods. Sonnenholzner and Boyd (2000b) applied several microbial products to bottoms of Ecuadorian shrimp ponds during dry-out between crops. No effect on the rate of organic matter decomposition could be measured.

Most of the published studies on microbial products have been for channel catfish culture in the southeastern United States. These ponds had standing crops of 4,000 to 8,000 kg ha^{-1} and maximum feed inputs of 60 to 100 kg ha^{-1} per day. Water exchange was not used, but aeration was applied at night. Results from such culture systems should be applicable to tilapia, shrimp, and many other types of pond aquaculture.

In view of the lack of documentation of benefits of microbial products on pond sediment and water quality and the widespread use of these wares, Boyd and Gross (1998) recommended further research into the topic. It could be that microbial products only produce improvements when specific problems exist, and possibly, these conditions did not occur in pond and laboratory trials mentioned above. Also, the advertising literature is often based on laboratory studies where high concentrations of microbial products were used. In practice, it is too expensive to use such high concentrations, so the vendor recommends much lower doses for ponds. This is a common problem with the adoption of amendments in pond management. If there is a known effect of a certain agent that would be beneficial in ponds, the agent may be adopted for aquaculture even though the appropriate concentration is not known. Moreover, the agent is often used for prophylactic purposes rather than for treatment of a specific problem.

Independent parties not directly interested in sales of the products should conduct considerable research on microbial products in pond aquaculture. This research should elucidate the effects of microbial products on aquatic ecosystems, identify conditions under which these products can be beneficial, and develop appropriate doses and methods of application. In addition, the economic benefits that can accrue from microbial treatments should be determined. Until these findings are available, one should be cautious about microbial products. Although no damage to fish or shrimp crops or to the environment have been reported, Gräslund et al., (2003) mentioned that microbial products for aquaculture may not be produced under controlled conditions. It is possible that they may contain cells with acquired antibiotic resistance. Gräslund et al., (2003) further cautioned that such products could possibly introduce an antibiotic resistance into the microbial community of ponds that may potentially spread to shrimp pathogens. Moreover, one might spend considerable money on these products and receive little in return.

Some researchers have reported better survival of shrimp and fish in hatchery tanks and in ponds treated with microbial products (Garriquez and Arevalo, 1995; Queiroz and Boyd, 1998; Rengpipat et al., 1998; Irianto and Austin, 2002). This effect was attributed to competitive exclusion of pathogenic organisms by competition with the microorganisms added to the available space, nutrients, and dissolved oxygen. The topic of disease control through use of microbial products will not be reviewed here.

CONCLUSION

Microorganisms are extremely important in the ecology of aquaculture ponds. Photosynthetic forms produce organic matter and dissolved oxygen. Saprophytic species decompose organic matter (substrate) originating in uneaten feed, feces, and dead plankton under both aerobic and anaerobic conditions preventing them from accumulating to high concentrations. Chemotrophic bacteria oxidize ammonia, nitrite, ferrous iron, and sulfide. Microorganisms are ubiquitous and rapidly increase in abundance and physiological activity in response to added substrate. For example, when organic matter is applied to a pond, the number of decomposer microorganisms will increase exponentially. When the substrate is depleted, they will decline in number but remain as spores or resting cells. The main factors affecting their abundance and activity are warmth, adequate moisture, a suitable pH, the presence or absence of dissolved oxygen, and the quantity and quality of substrate. There is usually some imbalance in one or more of these factors if microorganisms are not functioning properly in a pond.

Many shrimp and fish farmers, and especially those in Asia, apply microbial inocula and enzyme products derived from microbial cultures to ponds to enhance microbial activity and improve environmental conditions for aquaculture. Unfortunately, this practise is not supported by biological principles. If microorganisms are not functioning properly in a pond, it is more likely because environmental conditions are not favorable or the input of organic matter is so great that it cannot be decomposed without depleting the dissolved oxygen concentration at the sediment-water interface. Most studies have not shown significant improvement in water or sediment quality following applications of microbial products. Nevertheless, because of the widespread use of microbial products, more research on this topic is urgently needed.

REFERENCES

Boyd, C.E. and Goodyear, G.P. (1971). The nutritive quality of food in ecological systems. Arch. Hydrobiol. 69: 256-270.

Boyd, C.E., Hollerman, W.D., Plumb, J.A. and Saeed, M. (1984). Effect of treatment with a commercial bacterial suspension on water quality in channel catfish ponds. Prog. Fish-Cult. 46: 36-40.

Boyd, C.E., Tanner, M., Madkour, M. and Masuda, K. (1994). Chemical characteristics of bottom soils from freshwater and brackishwater aquaculture ponds. J. World Aquac. Soc. 25: 517-534.

Boyd, C.E, and Pipoppinyo, S. (1994). Factors affecting respiration in dry pond bottom soils. Aquaculture 120: 282-294.

Boyd, C.E. (1995a). Chemistry and efficacy of amendments used to treat water and soil quality imbalances in shrimp ponds. In: Proceedings of the Special Session on Shrimp Farming (eds.) C.L. Browdy and J. S. Hopkins, World Aquaculture Society, Baton Rouge, Louisiana, USA, pp. 183-199.

Boyd, C.E. (1995b). Bottom Soils, Sediment, and Pond Aquaculture. Chapman and Hall, New York, USA, pp. 348.

Boyd, C.E. (1995c). Potential of sodium nitrate to improve environmental conditions in aquaculture ponds. World Aquac. 26(2): 38-40.

Boyd, C.E. and Gross, A. (1998). Use of probiotics for improving soil and water quality in aquaculture ponds. In: Advances in Shrimp Biotechnology (ed.) T.W. Flegel, National Center for Genetic Engineering and Biotechnology, Bangkok, Thailand, pp. 101-106.

Boyd, C.E. and Tucker, C.S. (1998). Pond Aquaculture Water Quality Management. Kluwer Academic Publishers, Boston, USA, pp. 700.

Boyd, C.E. and Gross, A. (1999). Biochemical oxygen demand in channel catfish *Ictalurus punctatus* pond waters. J. World Aquac. Soc. 30: 349-356.

Boyd, C.E. (2000). Water Quality, an Introduction. Kluwer Academic Publishers, Boston, USA, pp. 330.

Boyd, C.E. and Clay, J.W. (2002). Evaluation of Belize Aquaculture, Ltd: A superintensive shrimp aquaculture system. The World Bank, Network of Aquaculture Centres in Asia-Pacific, World Wildlife Fund, and Food and Agriculture Organization of the United Nations, Rome, Italy, pp. 17.

Boyd, C.E., Wood, C.W., and Thunjai, T. (2002). Aquaculture pond bottom soil quality management. Pond Dynamics/Aquaculture Collaborative Research Support Program, Oregon State University, Corvallis, Oregon, pp. 41.

Boyd, C.E. (2003). Off-flavor in pond-cultured marine shrimp. In: Off-Flavors in Aquaculture (eds.) A.M. Rimando and K.K. Schrader, American Chemical Society, Washington, USA, pp. 45-53.

Chiayvareesajja, S. and Boyd, C.E. (1993). Effects of zeolite, formalin, bacterial augmentation, and aeration on total ammonia nitrogen concentrations. Aquaculture 116: 33-45.

Feliatra, F., Nursyiwani, N., and Bianchi, M. (2004). Role of nitrifying bacteria in purification process of brackishwater ponds (tambak) in Riau Province, Indonesia. J. App. Aquac. 19: 51-62.

Garriquez, D. and Arevalo, G. (1995). An evaluation of the production and use of live bacterial isolate to manipulate the microbial flora in the commercial production of *Penaeus vannamei* postlarvae in Ecuador. In: Proceedings of the Special Session on Shrimp Farming (eds.) C.L. Browdy and J.S. Hopkins, World Aquaculture Society, Baton Rouge, Louisiana, USA, pp. 53-59.

Gräslund, S., Holmström, K., and Wahlström, A. (2003). A field survey of chemicals and biological products used in shrimp farming. Marine Pollution. Bull. 46: 81-90.

Hunt, D. and Boyd, C.E. (1981). Alkalinity losses resulting from ammonium fertilizers used in fish ponds. Trans. Amer. Fish. Soc. 110: 81-85.

Irianto, A. and Austin, B. (2002). Probiotics in aquaculture. J. Fish Diseases 25: 633-642.

McIntosh, R.P. (1999). Changing paradigms in shrimp farming, 1. General description. Global Aquac. Advocate 2(4/5): 40-47.

Metting, F.B, Jr. (1992). Soil Microbial Ecology. Marcel Dekker, New York, USA, pp. 646.

Mischke, C.C. (2003). Evaluation of two bio-stimulants for improving water quality in channel catfish, *Ictalurus punctatus*, production ponds. J. Appl. Aquac. 14: 163-169.

Moriarty, D.J.W. (1999). Diseases control in shrimp aquaculture with probiotic bacteria. In: Microbial Biosystems: New Frontiers (eds.) C.R. Bell, M. Brylinsky, and P. Johnson-Green, Proceedings of the 8th International Symposium on Microbial Ecology, Atlantic Canada Society for Microbial Ecology, Halifax, Canada.

Nelson, D.W., Sommers, L.E. (1982). Total carbon, organic carbon, and organic matter. In: Methods of Soil Analysis, Part 2 – Chemical and Microbiological Properties (eds.) A.L. Page, R.H. Miller and D.R. Keeney, American Society of Agronomy and Soil Science Society of America, Madison, Wisconsin, USA, pp. 539-579.

Paerl, H.W. and Tucker, C.S. (1995). Ecology of blue-green algae in aquaculture ponds. J. World Aquac. Soc. 26: 109-131.

Queiroz, J.F. and Boyd, C.E. (1998). Effects of a bacterial inoculum in channel catfish ponds. J. World Aquac. Soc. 29: 67-73.

Queiroz, J.F., Boyd, C.E. and Gross, A. (1998). Evaluation of bio-organic catalyst in channel catfish, *Ictalurus punctatus*, ponds. J. App. Aquac. 8: 49-61.

Rengpipat, S., Rukpratanporn, S., Piyatiratitivorakul, S. and Mensaveta, P. (1998). Probiotics in aquaculture: A case study of probiotics for larvae of the black tiger shrimp (*Penaeus monodon*). In: Advances in Shrimp Biotechnology (ed.) T.W. Flegel, The National Center for Genetic Engineering and Biotechnology, Bangkok, Thailand, pp. 117-182.

Schroeder, G.L. (1978). Autotrophic and heterotrophic production of microorganisms in intensively manured fish pond and related fish yields. Aquaculture 14: 303-325.

Shumway, S.E., Davis, C., Downey, R., Karney, R., Kraeuter, J., Parsons, J., Rheault, R., and Wikfors, G. (2003). Shellfish aquaculture – in praise of sustainable economics and environments. World Aquac. 34(4): 8-10.

Sonnenholzner, S. and Boyd, C.E. (2000a). Chemical and physical properties of shrimp pond sediments in Ecuador. J. World Aquac. Soc. 31: 358-375.

Sonnenholzner, S. and Boyd, C.E. (2000b). Managing the accumulation of organic matter deposited on the bottom of shrimp ponds. Do chemical and biological probiotics really work? World Aquac. 31(3): 24-28.

Steeby, J.A., Hargreaves, J.A., Tucker, C.S., and Kingsbury, S. (2004). Accumulation, organic carbon and dry matter concentration of sediment in commercial channel catfish ponds. Aquac. Eng. 30: 115-126.

Tavares, L.H.S. and Boyd, C.E. (2003). Possible effects of sodium chloride treatment on quality of effluents from Alabama channel catfish ponds. J. World Aquac. Soc. 34: 217-222.

Thunjai, T., Boyd, C.E., and Boonyaratpalin, M. (2004). Bottom soil quality in tilapia ponds of different age in Thailand. Aquac. Res. 35: 698-705.

Tucker, C.S. and Lloyd, S.W. (1985). Evaluation of a commercial bacterial amendment for improving water quality in channel catfish ponds. Mississippi Agriculture and Forestry Experiment Station, Mississippi State University, US, Research Report 10:1-4

Tucker, C.S, and van der Ploeg, M. (1993). Seasonal changes in water quality in commercial channel catfish culture ponds in Mississippi. J. World Aquac. Soc. 24: 473-481.

van der Ploeg, M., Tucker, C.S., and Boyd, C.E. (1992). Geosmin and 2-methylisoborneol production by cyanobacteria in fish ponds in the southeastern United States. Water Sci. Tech. 25: 283-290.

van der Ploeg, M., Dennis, M.E., and Deregt, M.Q. (1995). Biology of *Oscillatoria* cf. *chalybea*, a 2-methylisoborneol producing blue-green alga of Mississippi catfish ponds. Water Sci. Tech. 31: 173-180.

Verschuere, L., Rombaut, G., Sorgeloos, P. and Verstraete, W. (2000). Probiotic bacteria as biological control agents in aquaculture. Microbiol. Mol. Biol. Rev. 64: 655- 671.

9

Linking Ecotechnology and Biotechnology in Aquaculture

J. Oláh[1*], *Mike Poliouakis*[2] *and Claude E. Boyd*[2]

INTRODUCTION

Aquaculture is an important aspect of world fishery production. Data from the Food and Agriculture Organization (FAO) of the United Nations show that the global fishery capture increased until about 1990, but since then it has remained at around 90 million tons year^{-1}. Aquaculture production has been steadily increasing and reached 40 million tons or about 30% of the total world fishery production in 2002. The demand for fishery products will increase in the future in response to the growing world population (Brown et al., 1998). Thus, aquaculture production must increase or there will be a scarcity of fishery products.

Natural resources available for agriculture are continually diminishing or face increasing pressure from growing population; some agricultural techniques cause significant environmental problems; while other techniques are not sustainable (Brown et al., 1998). Techniques that rely on high inputs of petroleum products are likely to lead to problems in the future. We need to adopt agricultural techniques that conserve natural resources as much as possible and that are sustainable. Aquaculture must be considered a component of agriculture, and it competes with traditional agriculture for resources such as land, water, fertilizers, feedstuffs, etc. Environmentalists have become concerned about the efficiency of resource use in agriculture (Naylor et al., 2000) and question its overall contribution to the world's food supply.

[1]*Department of Environmental Management, Tessedik High School, Szarvas, Hungary*
[2]*Department of Fisheries and Allied Aquacultures, Auburn University, Alabama 36849 USA*
**Corresponding author: J. Olah E-mail: saker@szarvasnet.hu*

Expertise from traditional ecology can help make agriculture and aquaculture more sustainable. It can show us what we need to quantify, and it can give us a good catalog of tested relationships from natural systems by which to understand systems made by people. It can help us to understand the environmental and social effects of agricultural practises. It can help us to assess techniques, decide which techniques to apply, where and how to apply them, and the extent to which they can be applied. When used this way, it is called "ecotechnology".

The Hungarian wetlands may serve as an example with quantification. The ecological and economic value of wetlands was not fully appreciated until recently. Many wetlands were converted to traditional farmlands. Besides direct loss due to disappearance of the wetlands, this conversion was not always economically viable and often led to unanticipated ecological problems. Ironically, some of this farmland is being converted back to semi-wetlands by way of usage in aquaculture, an inadvertent reclamation of the wetlands in a different form. Aquaculture supplies require protein and dietary variety, and can be an excellent source of income. Aquaculture is one of the fastest growing fields of agriculture. The techniques of environmentally sound aquaculture are fairly well understood, and reasonable codes of practise have been developed (Boyd, 2003). The increasing interest in responsible aquaculture provides a good opportunity to combine ecotechnology, aquaculture, conservation, and modified reconversion of wetlands. This chapter uses nitrogen pathways, fluxes, and nutrient cycles to illustrate the ideas.

USING ECOTECHNOLOGY TO QUANTIFY FISH FARMING SYSTEMS

An ecotechnological survey to quantify integrated fish farming systems in the Mekong Delta has been outlined (Oláh and Pekár, 1997). This chapter describes ecotechnological principles and applications.

Sustainability has become fundamental. Nutritionally open and oil-based agriculture became the primary source of ozone depletion, acid rain and global warming. According to Odum (1971), a 10-fold increase in the application of agricultural chemicals and fossil energy is required to double agricultural crop yield. In Hungary, industrialization of agriculture doubled the crop at a cost of over a 100-fold increase in the quantity of the inorganic fertilizers and hauling (Oláh and Oláh, 1996).

In contrast, world aquaculture remained basically on sustainable principles of organic recycling and integration. Fish culture technologies in Asia, giving over 85% (by volume) of the global aquaculture production, are still largely characterized by simple, low input integrated farming methods. Moreover, in countries like Nepal, Laos, and Bhutan, where aquaculture was not traditionally practised, integrated fish culture was introduced and well established in the past few decades. Also, in some Asian countries, especially in Bangladesh. India and China, relatively small seasonal undrainable monsoon fed waterbodies, not originally meant for fish culture, are now integrated and utilized with low input aquaculture (Csavas, 1993), without endangering the multipurpose functions of these waterbodies (Oláh and Sinha, 1984; Oláh et al., 1990).

NEED TO QUANTIFY

The lack of quantification is a continuing weakness of aquaculture data on integrated small-scale farms. An unknown but significant portion of aquaculture production never reaches a market where representative sampling may be possible. A consequence is the underestimation of subsistence aquaculture (Csavas, 1993). Edwards's attempt (1993) to produce quantitative appraisal demonstrates the urgent need for quantification in integrated fish farming systems.

WHAT AND HOW TO QUANTIFY

Sustainability in renewable natural resource consumption is approached widely by holistic system analysis. We have randomly sampled 65 recently published books on ecology (Table 9.1). They demonstrate the increasing need for quantification of the ecosystem services for almost all human activities. The holistic approach has been started in aquaculture (Folke and Kautsky, 1992) but it has not received enough skill and attention.

TOWARDS A SELF-CONTAINING ECOLOGY

In defining ecology, priority has been given to the quantitative function of the ecosystem structure as the target of ecology rather than duplicating the efforts of other disciplines. The danger of deformation to global ecology requires a clearer distinction between ecology as a functional environmental science and the chemistry, microbiology, botany, and zoology, as structural environmental sciences. The real questions of ecology are: (1) how far the energy and the building blocks of organic and inorganic molecules will be able to reproduce the information in the physical, chemical, microbiological, botanical, animal and human structural units; and (2) how this ecosystem metabolism gets along with human technologies.

Table 9.1. Ecological disciplines

Habitat	*Function*
Terrestrial Ecology	Productivity of Biosphere
Soil Ecology	Aquatic Productivity
Forest Ecology	Functioning of Ecosystems
Grassland Ecology	Mineral Cycling
Aquatic Ecology	Food Chain Ecology
Marine Ecology	Ecosystem Energy Flow
Freshwater Ecology	
Plankton Ecology	*Management*
Benthic Ecology	Landscape Ecology
Limnology	Watershed Management
Hydrobiology	Ecology of Nature Protection
Running Water Ecology	Wildlife Management
Wetland Ecology	Management of Population
	Economy of Nature
Organism	Natural Resource Management
Microbial Ecology	Applied Ecology
Plant Ecology	Ecosystem Restoration
Animal Ecology	Water Quality Management
Human Ecology	Ecology of Sewage Purification
Insect Population	Ecology of Water Populations
	Invertebrate Pollution Ecology
Habitat+Organism	Environmental Toxicology
Marine Microbiology	Water Supply Ecology
Freshwater Microbiology	Ecology of Acid Precipitation
Ecology of Soil Microorganisms	Ecology of Global Warming
Ecology of Aquatic Bacteria	System Analysis and Simulation
	Algal Culture
Structure	Mass Cultivation of Invertebrates
Aquatic Chemistry	Ecology of Aquaculture
Ecological Biochemistry	Agroecology
Ecology of Distribution	Crop Ecology
Population Ecology	Ecology of Intercropping
Ecology of Communities	Integrated Animal Husbandry
Ecology of Diversity	
Thermal Ecology	

THE CONTENT OF ECOLOGY

Ecology may be defined as a scientific study of the unified gross metabolism of supraindividual structures. The principles, methods, techniques and procedures of particular disciplines, such as chemistry, are applied to quantify integrated functions in the gross metabolism of the supraindividual system. In practise, most of these methods were developed to measure the activity of single species. Ecology applies them to quantify the same activity in supraindividual units of the environment.

THE BOUNDARIES OF ECOLOGY

The only aspect that is not represented in the old traditional disciplines is the functional ecological approach that quantifies energy flow and mineral spiraling in a given unit of the biosphere, that is the gross metabolism at the level of higher organization. No other scientific discipline does this specific task. Higher organization units including even most of the economical entities, technologies and territorial objects of humans are also the proper objects of ecological research.

THE METABOLISM OF ECOSYSTEMS

Environmental metabolism incorporates all the catabolic, anabolic and intermediary pathways and transfer routes of the nutrients and energy between inorganic and organic compartments. It concentrates on the functional evolution of the highest organization levels, but integrates the lowest ones, starting from elementary particles and arriving at the noosphere, through the elements, molecules, organisms, populations, communities, ecosystems, landscapes, biomes, and biosphere. It applies the data, information, principles, concepts, and methods accumulated on structural and functional hierarchies in several scientific disciplines.

ECOSYSTEM METABOLIC FUNCTIONS

Metabolism on each organization level has three specific functions:

(1) Obtaining chemical energy from fuel molecules of the absorbed sunlight. This catabolic metabolism supplies the energy for mineral spiraling on each level of biological organization.
(2) Converting exogenous nutrients into building blocks.
(3) Coordinating catabolism and anabolism.

ENVIRONMENTAL EFFECTS

The environment influences ecosystem metabolism in three major ways:

(1) The environment is the ultimate supplier of the raw materials and energy.

(2) Every metabolic reaction, hence every organism as a whole, is affected greatly by the physicochemical environment. In this context, forces of global dimensions have a direct bearing on forces of molecular dimensions. Environment sets the stage for metabolism.

(3) Competitive biotic environment operates and regulates the environmental metabolism at each level of organization: competition for structural space, driving force of energy and building block nutrient molecules. Somewhat paradoxically, this competitive biotic environment is highly endangered by disappearing biodiversity and biodisparity. Under the impact of uncontrolled technological development, each organization level, and ultimately the biosphere, are rapidly loosing their competitive power.

THE CONCEPT OF ECOTECHNOLOGY

Human activity is the most effective environmental mediator in the exploitation of nature. Technology has become a fundamental component of the ecosystem. It has to be considered as an integral part of ecology. Ecotechnology is a basic discipline of the ecology dealing with the natural resource consumption: how the human activities explore, influence and modify the natural resources while producing commodities in an economic and socially satisfactory way.

THE DEFINITION OF ECOTECHNOLOGY

The main functions of ecotechnology are to develop and promote resource consumption that is ecologically sound, keeping in mind the key concept that humans are a part of nature rather than apart from nature (Mitsch and Jorgensen, 1989). All the definitions published so far express need for cooperation between humans and nature, the focus on designing human society to accord with its natural environment, and the human partnership with nature (Table 9.2). Definitions also emphasize environmental manipulation and the use of technology for ecosystem management, that is the operational aspects (Odum, 1962, Uhlman, 1983).

CONNECTION TO OTHER TECHNOLOGIES

It is useful to delineate the boundaries of ecotechnology and define how it is connected to related fields (Table 9.3) based on the ideas of Mitsch and

Table 9.2. Definitions of ecotechnology

Brief term	Author
Environmental manipulation by man	Odum, 1962
Partnership with nature	Odum, 1971
Use of technology for ecosystem management	Uhlman, 1983; Straskraba, 1984, 1985
Application of ecosystems principles	Ma, 1985
Human society with its natural environment	Mitsch, 1988
Cooperation between humans and nature	Mitsch and Jorgensen, 1989
Ecological resource consumption	Oláh and Pekár, 1997

Table 9.3. Contrasting ecotechnology with related technologies

Parameters	Biotechnology	Environmental Engineering	Ecological Engineering	Ecotechnology
Unit	Cell	Technologies	Operational	Noosystems
Disciplines	Genetics	Sanitary engineering	Technology	Ecology
Control	Genetic	Economy	Technical	Organisms CNP
Regulation	Genetic	Management	Mechanical	Synorganic
Designed	Biologists	Constructors	Engineers	Succession
Diversity	Changed	Established	Directed	Protected
Energy cost	High	High	Medium	Low
Energy	Fossil	Fossil	Fossil	Solar
Environment	In vitro	Unit facility	Constructed	Natural
Target	Conquer	Operate	Modify	Harmonize

Jorgensen (1989). Ecology belongs to biology. This is why in the broad sense of the word biotechnology also contains ecotechnology.

In practise, biotechnology and bioengineering include only the manipulation of genes and infraindividual organisms. Biomanipulation is genetically controlled and regulated, comes with a high cost of fossil energy, and done in vitro. Environmental engineering is the former sanitary engineering. It works with unit technologies in pollution control and waste disposal, and it is controlled by financial limits and regulated by management.

Ecological engineering applies ecotechnology. Directed diversity is modified in a constructed environment with only a moderate cost of fossil energy. Synorganically regulated noosystems ("mind systems") are the units of ecotechnology with as much natural environment as possible. Ecological succession is harmonized with protected organisms, done with a low cost of solar energy, other energies, and nutrient supplies.

QUANTIFICATION IN ECOTECHNOLOGY

Ecological quantification measures the structural components and the functional processes of ecosystems. Ecotechnological quantification does the same plus quantifies the technological parameters and processes in resource consumption practise. This is a promising approach to quantify the environmental sustainability of technologies.

QUANTIFICATION OF SUSTAINABILITY IN RESOURCE CONSUMPTION

Energy analysis in fish farming systems (Oláh and Sinha, 1986) and nitrogen metabolism on the landscape level (Oláh and Oláh, 1996) are good quantitative indicators of sustainability. The nitrogen atom participates in every kind of resource consumed by humans. It is relatively easy to quantify the nitrogen cost of products and services. Well-established conversion factors are already available to calculate how much nitrogen is released to the environment during production processes. By applying them the environmental cost of products or services may be computed.

Released nitrogen contributes to principal environmental problems such as ozone production, acid rain and global warming. Nitrate, being the most stable form of the nitrogen atom, moves down gradients and through the substrate of the watershed with the surface and subsurface waters. This character may be used for verification of the computed nitrogen metabolism if the nitrate pool is measured in these waters. The amount of the measured nitrate in groundwater or in river water has to correspond to what have been calculated as the total released component of nitrogen metabolism.

"Emergy" analysis differs from economic analysis because instead of using the dollar value of goods, services and resources, it calculates in terms of energy. This is stable and reliably comparable, regardless of political boundaries and historical perspective. Emergy can be conceptualized as energy memory, since it is a measure of all the energy previously required to produce a given product or process (Ulgiati, 1995). Emergy has been previously known as embodied energy or simply as the industrial energy cost of a product. Emergy is the energy of one type required in transformation to generate flow or storage (Huang et al., 1995). As a result of the first energy crisis, many papers were published on the cost of several products that are used during fish farming operations (Oláh and Sinha, 1986). A more recent study was an emergy analysis of channel catfish culture in the United States (Ortega et al., 1999)

ECOTECHNOLOGICAL SURVEY ON INTEGRATED AQUACULTURE

Quantification is based on three principles: (1) It has to be carried out, as far as possible, on the farm by personal interviews and by in situ field measurements. (2) To realize the first principle, the simplest field methods have to be adopted. (3) However, all those parameters that are necessary, i.e. the nitrogen spiraling and emergy flow have to be included.

ECOTECHNOLOGICAL ONFARM SURVEY

A typical survey starts with data collection on the structure of an integrated fish farm (Table 9.4) by personal interview (Table 9.5). Geographical location and farm size parameters help in understanding landscape integration. Detailed landuse pattern quantification collects area percentages of the different crops, including vegetables and fruit trees, as well as the number or weight of all livestock. This helps to calculate the energy resources and emergy flow together with nitrogen spiraling. Pond morphometry forms the basis of the same calculations for the pond metabolism and also helps to understand the state and variety of the physical environment and pond infrastructure. Fish stressors have to be noted and measured.

These items might be investigated: pond renovation activities, overflow, runoff, dike repair, desilting, distance from the public canal, flood occurrence, water limitation, supply timing and duration, tidal-gravity-pump, and inlet-outlet. A water budget is a measure of how much water is in what part of the system for how long, and, if possible, the condition of the water while it is in that part of the system. The quantification of crop yields and livestock production is required to calculate the whole farm metabolism.

Organic and inorganic nutrient applications are to be collected for the whole farm, and for each subunit: inorganic fertilization, manuring, composting, green plant application or feeding, or any other nutrient introduction (Table 9.6). Quantification of the biological energy resources and flow is illustrated in Table 9.7. Most of the parameters for nutrient budgeting are measured in the laboratory with routine methods (Table 9.8).

It is interesting to note that the ecotechnological survey of an aquaculture farm requires collection of almost exactly the same information needed to make an assessment of factors contributing to pollution loads and other negative environmental impacts (Boyd et al., 2000). Therefore, efforts to develop better management practises for aquaculture should be linked with application of ecotechnology.

Table 9.4. Farm, pond and fish stressor structure

Location and Size	Pond Morphometry	Fish Stressor Environment
Province, District	Pond size, m^2	pH
Village,	Actual water depth, cm	Dissolved oxygen at sunrise, mg dm^{-3}
Farm or farmer name	Maximum water depth, cm	Free ammonia-N, mg dm^{-3}
Landuse pattern, %	Minimum water depth, cm	Sediment gases, liter m^{-2}
Total farm size, ha	Soft sediment depth, cm	Desease events
Pond code	Hard sediment depth, cm	Fish kill occurences
Pond age, year	Total sediment depth, cm	
	Pond shape	
	Dyke height, cm	
	Dyke width, cm	
	Dyke slope	

Table 9.5. Management parameters

Pond Renovation	Pond Water Supply	Fish Stock Management
Dewatering	Distance from public canal, m	Stocking density, fish ha^{-1}
Drying	Flood occurrence	Stocking structure, species, fish ha^{-1}
Desilting	Water supply (date and duration)	Stocking size (individual weight, g)
Dike repair	Inlet-outlet (tidal, gravity, pump)	Daily growth rate, g
Deweeding	Water retention time, days	Survival rate, %
	Water exchange rate, days	Fish yield, kg ha^{-1} cycle
	Water budget	Production cycles in days
		Crop yields, kg ha^{-1} cycle
		Production cycles in days
		Livestock production, kg ha^{-1} $cycle^{-1}$
		Production cycles in days

Table 9.6. Nutrient application

Nutrient for Crops	Nutrient for Fishponds	Nutrient for Animals
Irrigation, m^3 ha^{-1} y^{-1}	Liming, kg ha^{-1} y^{-1}	Feeding, kg y^{-1}
Fertilization: type, kg ha^{-1} y^{-1}	Manure: type, kg ha^{-1} y^{-1}	Green, kg y^{-1}
Silting, kg ha^{-1} y^{-1}	Silt removal, kg ha^{-1} y^{-1}	Cereal, kg y^{-1}
Manuring: type, kg ha^{-1} y^{-1}	Urea, kg ha^{-1} y^{-1}	Grounded, kg y^{-1}
Composting, kg ha^{-1} y^{-1}	Superphosphate, kg ha^{-1} y^{-1}	Pelleted, kg y^{-1}
Feeding: type, kg ha^{-1} y^{-1}	Feeding: type, kg ha^{-1} y^{-1}	Grazing, kg y^{-1}
Fertilization frequency	Fertilization frequency	
Manuring frequency	Manuring frequency	
Feeding frequency	Feeding frequency	

Table 9.7. Biological energy resources and flow, farm and pond carbon flow

Natural Fish Food Resources	Biological Energy Resources	C flow, kg C $year^{-1}$
Bacterioplankton, cm^{-3}	Primary production, g C m^{-2} day^{-1}	Autochton input
Chlorophyll a, μg dm^{-3}	Macrophyte cover, %	Allochton input
Zooplankton, number dm^{-3}	Population associated with pond ha^{-1}	Output
Seston detritus, particles cm^{-3}	Livestock associated with pond ha^{-1}	Marketed
Sediment detritus, g dw m^{-2}		Released
Benthic animals m^{-2}		Respired
Zootecton m^{-2}		Retention

Table 9.8. Nutrient status and nitrogen spiral, farm and pond nitrogen budget

Water Nutrient Status	Sediment Nutrient Status	N budget kg N y^{-1}
Secchi-disc visibility, cm	Organic matter, mg g^{-1} dw	Input
Visual water color	Total nitrogen, mg g^{-1} dw	Output
Total organic carbon mg dm^{-3}	Total phosphorus, mg g^{-1} dw	Marketed
NH_4 -N, μg dm^{-3}	Adsorbed NH_4 -N, mg g^{-1} dw	Released
NO_3 -N, μg dm^{-3}		Retention
PO_4 -P, μg dm^{-3}		

NITROGEN CYCLING IN FISH-CUM-LIVESTOCK ECOSYSTEMS

The energy cost of one ton of fish produced in a recycling system was 367 GJ (Gega Joules), equivalent to 8.5 tons of crude oil. Most of this energy was burned to substitute for the natural cycling of nitrogen in order to maintain a healthy environment for proper fish growth. One ton of fish in a fish-cum-cattle undrainable pond in India had an energy cost of only 10 GJ (Oláh and Sinha, 1986). An energy intensive strategy was forced on developed nations during industrialization. As a result, 1 ton of wheat now consumes 50 times more energy than 50 years ago in traditional closed agriculture where the nitrogen was recycled (Oláh et al., 1990a). In modern agriculture, oil is burned to let the nitrogen run through crop fields directly to rivers and groundwaters.

Fortunately fishermen and fish growers have been more efficient. Unfortunately the quantitative aspect of the nitrogen balance, budget, cycle and metabolism of their activities is very poorly investigated. The only nitrogen laboratory which was operated and is still existing for simultaneous measurement of almost all nitrogen patways in fish-cum-livestock ponds was established in our institute, in Hungary. During the

Table 9.9. Ph.D. thesis works on nitrogen and fishpond ecosystem supervised by J. Oláh and available for reading at the Fisheries Research Institute, Szarvas, Hungary

Ph.D. students	Thesis subject
Toth, O.E. 1976	Amino acid and protein cycling
Kintzly, A. 1978	Nitrogen fertilization in polyculture
El Samra, M.I. 1980	N_2-fixation, NH_4 and NO_3 uptake
Abdelmoneim, M.A. 1982	Denitrification
Csengeri, I. 1982	Lipids in fish ponds
El Sarraf, W.M. 1983	Amino acid uptake and ammonification
Eross, I. 1983	Fish feeding and body composition
Papp, Z. 1983	Pond amino acid quantification
Janurik, E. 1984	Nitrogen quantification methodology
Pekar, F. 1984	Nitrification
Farkas, J. 1985	Bacterial populations
Talaat, K.M. 1986	Fish population growth
Kovacs, Gy. 1987	Domestic sewage fishponds
Ayyappan, S. 1988	Waterhyacinth processing fishponds
Szabo, P. in prep	Sediment N compartment, diffusion
Nezami, B.S. 1989	River nutrient load
Esteky, A.A. 1989	Sediment nutrient accumulation
Liptak, M. in prep	Bacterial production

last 20 years, 18 Ph.D. thesis were completed on different aspects of nitrogen and fishpond ecosystems (Table 9.9).

Our results on the nitrogen metabolism in fish-cum-duck and fish-cum-pig ecosystems are synthesized below. Hundreds of measurements were condensed to a single figure in order to quantify the unit transfer rate from one compartment to other: kg N ha^{-1} y^{-1}.

WATER AND SEDIMENT NITROGEN COMPARTMENT

In fish-cum-duck and fish-cum-pig ecosystems, 13 compartments in the water column and sediment have been monitored. A special cook-book for nitrogen compartment and trasfer rate measurement was published with detailed description of glassware, reagent preparation, procedure and calculation (Oláh and Janurik, 1985). The methods for all results in this chapter may be found there or in our other publications.

The ranges varied widely in both ecosystems (Table 9.10). The highest maxima were accumulated etheir during the peak summer such as for Particulate Amino Acid (PAA-N) in the body protein of planktonic organisms, or at the end of the growing season with decreasing sunshine as for NH_4-N. The NO_3-N maxima arrived at the very beginning of the

Table 9.10. Water and sediment nitrogen compartments in fish-cum-duck and fish-cum-pig ecosystems. Particulate amino acid (PAA), Dissolved combined amino acid (DCAA) Dissolved free amino acid (DFAA). Ranges and averages: µg N dm^{-3} except sediment absorbed NH_4-N and sediment PAA: µg g^{-1} dry sediment

Compartments	Fish-cum-duck	Fish-cum-pig
Water		
PAA-N	115-478	243-2813
DCAA-N	47-881	105-974
DFAA-N	0.6-11.3	2.3-8.9
NH_4 -N	69-893	97-4590
NO_2 -N	42-132	13-278
NO_3 -N	227-1530	94-1310
Sediment		
PAA-N	1976	1,750
DCAA-N	761	681
DFAA-N	23	103
NH_4-N	346-3483	1099-7499
Absorbed NH_4-N	58-181	11-122
NO_2-N	70	22-107
NO_3-N	30-117	64-166

culture period with filling river water. The most dominating nitrogen compartments occurred in the sediment of both the fish-cum-duck and fish-cum-pig ecosystems. Almost 2 mg PAA-N g^{-1} dry sediment was present in the second half of the growing season. This compartment represents the nitrogen content of living and dead organic particles. The average number of bacteria in the sediment of these heavily loaded ecosystems was 8.1×10^9 g^{-1} dry weight as measured with direct count of epifluorescent microscopy. Calculating with an average bacterial nitrogen content of 2.8 fg N $cell^{-1}$, it was estimated that the total nitrogen content of the sediment bacterial biomass was as small as 22.6 µg g^{-1} dry sediment, i.e. around only 1% of the total PAA-N. The nitrogen content of the benthic invertebrates was even less, only 2 µg g^{-1} dry sediment. Based on these calculations, it was concluded that 98.9% of the total PAA-N in the sediment is enclosed in dead particles. This compartment serves as an important nitrogen reserve decomposing and releasing dissolved free amino acids (DFAA-N) and further ammonia nitrogen. The bulk of this regenerated ammonia is absorbed and accumulated on fine clay particles or on organic complexes forming the second most important nitrogen compartment in these fish-cum-livestock ecosystems.

NITROGEN TRANSFER RATES

In fish-cum-livestock ecosystems, the nitrogen cycle operates primarily on nitrogen load, which is released directly by animals living in close contact with the ponds or introduced indirectly as collected manure if animals live far from the ponds. The organic nitrogen of dissolved and particulate proteins and peptides are cut into small molecules of free dissolved amino acids by extracellular enzymes of bacteria, and taken up by them or to a smaller extent taken up by certain physiological groups of algae, mostly blue-greens. Part of the assimilated amino acids is respired and released as NH_4-N. The amino acid uptake in fish-cum-livestock ecosystems is given in Table 9.11. On an average, glycine and glutamic acid uptake rates dominated among the individual acids. The methionine uptake rate was high in fish-cum-pig ecosystem but low in fish-cum-duck ponds. Upto 64% of the assimilated glycine and glutamic acid was respired, compared to 13% of methionine and valine in fish-cum-duck ponds (Toth et al., 1984).

Ammonia as the end product of amino acid respiration, or "ammonification", together with nitrate as its oxidized form, are the key inorganic nitrogen molecules for autochtonous protein production, utilizing the energy and reducing power of photosynthetically generated carbohydrates in the ponds. Their uptake rates were measured with Michaelis-Menten uptake kinetics. Ammonia was taken up preferably in both the fish-cum-duck and fish-cum-pig ecosystems (Table 9.12). Although the nitrate uptake rate was lower, especially in fish-cum-duck ponds, it is still an important nitrogen pathway, transporting daily a significant amount of inorganic nitrogen building material for protein synthesis.

Biological N_2-fixation, as an important nitrogen income in many natural waters has only secondary significance in these organic loaded ecosystems

Table 9.11. Amino acid uptake in fish-cum-livestock ecosystems, $\mu g\ dm^{-3}\ h^{-1}$ (El Sarraf, 1983)

Amino acid	Gross V_{max}	Respired %
Fish-cum-duck		
Glycine	6.0	45
Methionine	2.3	13
Lysine	2.4	17
Valine	1.5	13
Phenylalanine	2.0	36
Glutamic acid	6.7	64
Protein hydrolysate-N	9.3	22
Fish-cum-pig		
Glycine	8.5	43
Methionine	10.8	21

Table 9.12. NH_4-N and NO_3-N uptake in fish-cum-livestock ecosystems, μg N dm^{-3} h^{-1} (El Sarraf et al., 1982)

Ecosystem	NH_4 -N V_{max}	NO_3 -N V_{max}
Fish-cum-duck	21	14
Fish-cum-pig	24	19

Table 9.13. N_2-fixation, nitrification and denitrification in the water column and sediment of fish-cum-livestock ecosystems kg ha^{-1} 100 day^{-1} (Abdelmoneim et al., 1986, Oláh et al., 1987, Pekár et al., 1989)

Nitrogen transfer	Water column	Sediment	Total
Fish-cum-duck			
N_2-fixation	7.8	65	72.8
Denitrification	36	23	59
Nitrification	320	462	782
Fish-cum-pig			
N_2-fixation	5.7	6.1	11.8
Denitrification	30	42	72
Nitrification	642	361	1003

(Table 9.13). Filamentous blue-green algae, which are the dominant and more succesful N_2-fixing organisms in the water column, cannot develop significant biomass under the grazing and bioturbating pressure of dense fish populations. Moreover, the water column is never deficient of inorganic nitrogen which favors blue-green blooms. The rate of N_2-fixation measured with the acetylene reduction method was somewhat higher in the sediment especially in fish-cum-duck ponds. It was unclear why this level of N_2-fixation appeared in these sediments that were full of dissolved ammonium ions (as it was quantified with Reeburgh interstitial extracting procedure, Table 9.14). Combined inorganic nitrogen, and especially ammonium, are known to have a negative influence on N_2-fixation. An anaerobic nitrogen-fixer, *Clostridium pasteurianum*, was the dominant N_2-fixing organism in these sediments, reaching a population density of 10^6-10^7 cell g^{-1} dry wt. In this, ammonium represses the synthesis of nitrogenase but does not affect the activity of the enzyme. In aerobic N_2-fixers, ammonia represses the synthesis of nitrogenase and inhibits its activity as well. It was speculated that N_2-fixation might be a way to remove reduction equivalents from reduced sediments. One way to live without oxygen in such an environment is to use surplus electrons to utilize space and to transform nitrogen and energy. The electron reception function seems more pronounced ecologically than the commonly considered N_2-fixation aspect of this type of bacteria thriving in these waste processing ecosystems.

Table 9.14. Nitrogen budget of 20 years (1970-1990) on our commercial fish-cum-duck-alfalfa-rice farm of Horvathpuszta with around 300 ha, tons 20 y^{-1}. Average nitrogen contents used in analysis: manure: 0.5, fish feed: 1.6, duck feed: 1.9, alfalfa: 2.7, fish: 2.5, rice: 2.4, duck: 2.7%

Input		Ouput	
Inorganic fertilization	212	Alfalfa yield	46
Duck feed	156	Fish yield	86
Fish feed	17	Rice yield	64
Manure	14	Duck yield	84
Atmospheric deposition	90	Draining water	170
Filling water	255	Denitrification	236
Alfalfa N_2-fixation	210	Total output, $farm^{-1}$	686
Fish pond nitrogenfixation	292	Total output, ha^{-1} y^{-1}	0.114
Total input, $farm^{-1}$	1246	Total input-output, $farm^{-1}$	560
Total input, ha^{-1} y^{-1}	0.207	Total input-output, ha^{-1} y^{-1}	0.093

Another possibility of bacterial life in oxygen defficient sediment is denitrification. Denitrifiers perform an electron reception function reducing the nitrate into N_2O or N_2 gases, which then leave the system and release nitrogen into the atmosphere, at the same time decreasing the electron pool of the sediment. Although the measured rate of denitrification was significant in both ecosystems (Table 9.13), it was lower than in other aquatic systems. Nitrification was the most significant nitrogen transfer pathway in both ecosystems. In fish-cum-pig ponds the nitrogen cycled on this route surpassed 1 ton ha^{-1} y^{-1}.

NITROGEN CYCLING AND RETENTION

Management parameters with synchronized compartment and transfer rate measurement made the construction of a nitrogen cycle and budget possible. In these ecosystems the driving forces are the manure introduced by duck and pig populations, and the sun. River water also supplied much nitrate, as did atmospheric deposition. Biological N_2-fixation was significant only in the fish-cum-duck ponds, though not yet known why. One possible explanation is the higher specific phosphorus content of the duck manure. The cycling of nitrogen has a similar pattern in both ecosystems. However, the pig fed pond cycle operates on a more intensive level due to the higher nitrogen input introduced by manure. Pig ponds are more loaded than duck ponds. The influence of this higher load appears on transfer rates operating both in the sediment and in the water column.

The measured rate of amino acid respiration by bacterial ammonification was 329 kg ha^{-1} y^{-1} in fish-cum-pig and 226 kg ha^{-1} y^{-1} in fish-cum-duck pond. Earlier this was considered as the dominant pathway of ammonium production but now seems to form only the smaller part of total ammonia regeneration. The ammonia release was 774 + 193 kg ha^{-1} in fish-cum-pig and 630 + 163 kg ha^{-1} in fish-cum-duck ponds by invertebrates and fishes. Planktonic and benthic invertebrates grazing on algae as well as fish feeding on invertebrates, release around 75% of the assimilated nitrogen. Quantitative findings for natural waters support this conclusion (Liao and Lean, 1978; Gardner et al., 1983; Mitamura and Saijo, 1986; Tatrai, 1987).

A major pathway of nitrification was as a transfer route oxidizing ammonia to nitrate, up to 1 ton $ha^{-1}y^{-1}$ while producing an equivalent amount of bacterial organic carbon (C), 125 kg C ha^{-1} y^{-1} in a food chain. During this process, 3.4 tons of oxygen was combined and stored to cover the shortages of electron acceptor demand. Based on balance calculations, almost half of this combined oxygen is respired in the most reduced sediment layers in the nitrate respiration pathway.

To eliminate as many electrons as possible, special bacterial populations proved to reduce nitrate not only to nitrogen gas but also further to ammonia, displacing four more electrons per each nitrogen atom. This bacterial activity was described from very reduced ecosystems: ricefield sediment (MacRae et al., 1968), anaerobic soil (Buresh and Patrick, 1978), anaerobic costal marine sediment (Koike and Hattori, 1978; Sorenson, 1978), digested sludge (Kaspar et al., 1981), and bovine rumen (Kaspar and Tiedje, 1981).

Sediment harboring important nitrogen compartments and operating significant transfer routes functions as a nitrogen accumulator. Subtracting the total nitrogen output removed with harvested fish, water outflow and denitrification from the total nitrogen input, an accumulation values of 218 kg N ha^{-1} for fish-cum-duck and 294 kg N ha^{-1} for fish-cum-pig were received. These nitrogen amounts were trapped in the sediment during the growing season. However, it was not measured how much nitrogen was denitrified, volatilized or seeped out after the growing season when sediments become dried and exposed to sun and wind. There is significant nitrogen loss during the dry season of four to five months. To answer this question, the 20-year nitrogen budget in the commercial fish-cum-duck-alfalfa-rice farm of around 300 ha was analyzed. The 20-year average value for annual accumulation was only 93 kg ha^{-1} (Table 9.14) as compared to the above values accumulated in experimental ponds during the growing season. Seepage, denitrification and volatilization are important during the dry period emphasizing the technological importance of the management parameters. The nitrogen budgeting model of this commercial farm was verified with sediment nitrogen measurement. Fortunately, the total nitrogen content in the sediments of the commercial farm was studied 20 years ago. The measurements gave a farm average of 1.62 mg N g^{-1} dry wt (Fabry, 1975). The sediment nitrogen content of the same 19 ponds was reexamined. The farm average after 20 years increased to 5.10 mg N g^{-1} dry sediment. The annual rate of nitrogen accumulation calculated from these values was 174 μg N g^{-1} dry sediment. This rate of accumulation corresponds well with the budgeted accumulation values calculated for the farm. This analysis emphasizes the nutrient processing and accumulating nature of fish-cum-livestock ecosystems.

ENVIRONMENTAL COST OF AQUACULTURE ON HUNGARIAN LOWLAND

The Hungarian Lowland surrounded by the Carpathian Mountains was originally a huge inland delta for the upstream Danube river system, serving the functions of water retention and nutrient spiraling. The floodplain was 39,000 km^2. This extensive wetland system processed the upstream nutrient loads of the huge Danube watershed. The trapped nutrient produced a river valley with its extensive floodplain full of diverse fish and wildlife habitats. The spawning and nursing grounds were saturated with natural food and maintained very high fish yield and rich wildlife. This fertile wetland produced sufficient green plant biomass to feed around 6-8 million livestock, much of it for export. The fine network of the so-called "fok-györök" system operated a very productive floodplain husbandry with grassland, gallery forestry, plumb-apple-walnut plantation and rich aquaculture.

The greatest single engineering work in Europe changed drastically the landuse pattern on this riverine landscape in the second half of the 19th century. The original nutrient processing function and resource consumption activities were converted to an artificial and completely dry cropland. The drying is further accelerating today by greenhouse gas accumulation. Then came the rapid spread of the industrialized agriculture. The degradation has been quantified in a study of national nitrogen metabolism (Oláh and Oláh 1996).

As the result of industrialized agriculture, the following nitrogen load had challenged the river system in Hungary in 1989: (1) 617×10^3 N tons of inorganic fertilizers; (2) 201×10^3 N tons of organic manure; (3) 146×10^3 N tons of rain nitrogen; (4) 20×10^3 N tons of human population release; and (5) 10×10^3 N tons of direct industrial nitrogen–altogether 1094×10^3 tons of total nitrogen. Nitrate, as the final oxidized and thermodynamically most stable end product of the landscape nitrogen metabolism, has increased in the River Tisza from 1.5 μg ml^{-1} in 1968 to 15.6 μg ml^{-1} in 1984. It was around 0.1 μg ml^{-1} in the Danube at Budapest at the end of the 19th century. From these simple figures any ecologist understands the terrible environmental degradation.

POTENTIAL OF AUQACULTURE VERSUS AGRICULTURE

The national nitrogen budget and calculated landscape nitrogen metabolism was shown through the trends in riverine nitrate concentrations (Oláh and Oláh, 1996). The consumable farm nitrogen output for most agroecosystem types of agriculture and aquaculture was collected (Fig. 9.1). Detailed nitrogen cycling data were collected and evaluated by Frissel (1978) for 65 agricultural agroecosystems. The nitrogen output from each particular agroecosystem against protein production in the same (or similar) ecosystem was presented below. The wide range of intensity from shifting agriculture to intensive dairy farming required a logarithmic scale. A similar evaluation of the protein production potential has already been published (Sinha and Oláh, 1982). By plotting the 30 published production values against stocking figures in a wide range of fish rearing ecosystems, a near linear response was obtained. Taking into account the various latitudes where these experiments were conducted, all the fish production values were converted to 365 days. Fish production depended primarily on stocking density if environment and food were sufficient.

When the production of aquaculture systems was compared with the agricultural systems, it is surprising that natural waters may compete for productivity with the majority of agricultural systems. Properly managed natural waters have the potential to produce protein as efficiently as agriculture, but much more sustainably. Constructed fishpond ecosystems may produce animal protein, as much or even more than the peak yields of plant protein produced in any analyzed arable ecosystems. To appreciate this potential, it is emphasized that there is very significant loss when the produced plant protein is converted to animal protein in any of the terrestrial animal husbandries.

In contrast, industrial aquaculture is not as efficient. In industrialized fish factories, the production potential is almost unlimited, at least theoretically. In practise, however, there are questions: what is the energy cost and how much environment is to be consumed to produce one ton of protein?

Water is an important and scarce resource, and it should be used efficiently. Aquaculture appears to be a highly water-intensive endeavor. However, in reality, aquaculture does not consume large amounts of water

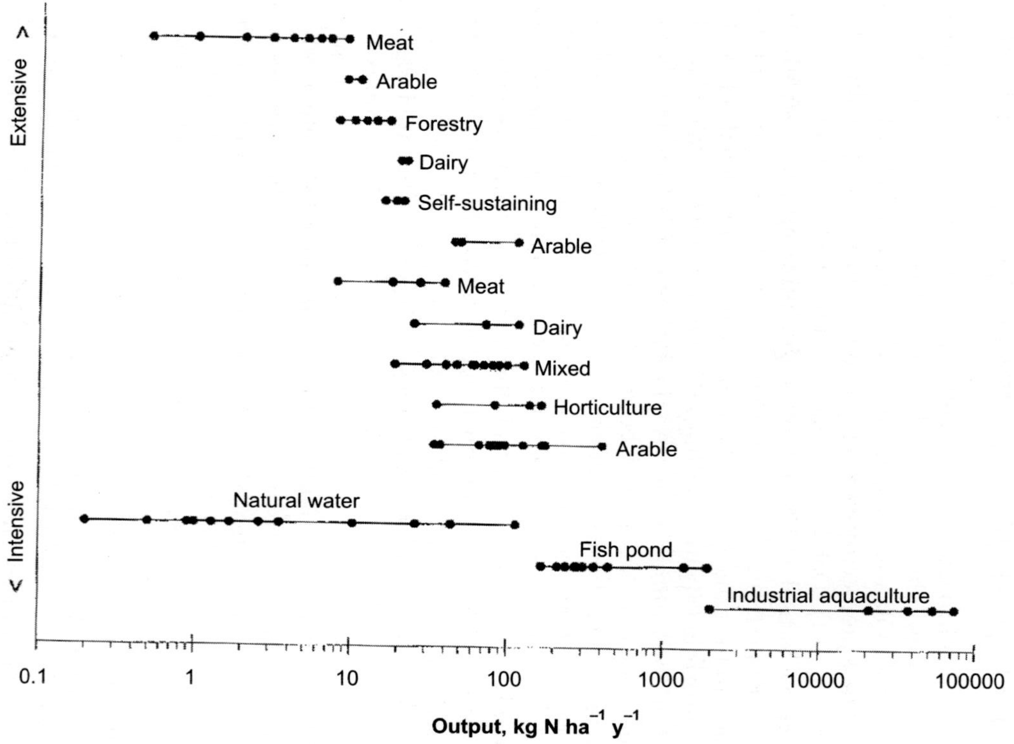

Fig. 9.1. Nitrogen output from agriculture and aquaculture agroecosystems based on Frissel (1978) for agriculture, and Sinha and Oláh (1982) for aquaculture

for most of the water passes through aquaculture systems and continues downstream where it can be used for other activities. According to Boyd (2005) the average, annual consumptive water use by typical agricultural crops in the southeastern United States ranged from 80 cm for soybeans to 123 cm for rice. Catfish ponds used between 130 and 220 cm of water depending on water source and method of production. However, the consumptive water index value for catfish ranged from US $0.39 to US $0.66/m^3, while values for agricultural crops ranged from US $0.07 to US $0.12 m^{-3}. Thus, from an economic standpoint, aquaculture obtains more value from water than does traditional agriculture.

FARM INPUT AND CONSUMABLE OUTPUT

The relationship between farm input and consumable farm output was quantified using Frissel's data for agricultural farms and our own data on five Hungarian aquacultural farming systems. The input and output of each food producing system were expressed in nitrogen atoms (Fig. 9.2). Solid line plots indicate the nitrogen efficiencies from 1 to 200%. When output was higher than input, N_2-fixation was not quantified properly or the soil or sediment organic nitrogen was consumed. As expected, farm output from arable systems is much higher than the output from livestock systems. All arable system output efficiencies fall between 30 and 100% of input. Up to a farm input of 150 kg N ha^{-1} y^{-1}, the output efficiencies are often close to the 66% mark. There are slightly lower efficiencies at higher farm inputs. In livestock systems, the range and scattering of data points is much wider. About half of the farms are within the efficiency range of 10 to 30%. As expected, the results in mixed systems are between those of arable and livestock systems.

Clearly different from the normal agricultural ecosystems analyzed by Frissel, in all five Hungarian aquacultural technologies only manure was applied as a source for the molecules needed for fish protein synthesis. There was no inorganic fertilization. This cheap and effective food producing biotechnology is realized through bacterial plus algal-based aquatic food chains. The industrialized aquaculture technologies were not plotted into this input-output analysis because of their environmentally unacceptable by-products. The fish-cum-goose ponds produced consumable farm output with the most efficient nitrogen transfer machinery. In this ecosystem, the significant streamflow and baseflow transported bulk of the nitrogen. The groundwater nitrate content ranged between 8-19 μg ml^{-1}. If the nitrogen input is not counted, the nitrogen transfer efficiency is around 100%. Another advantage of this animal protein producing system is its high nitrate removal capacity from the

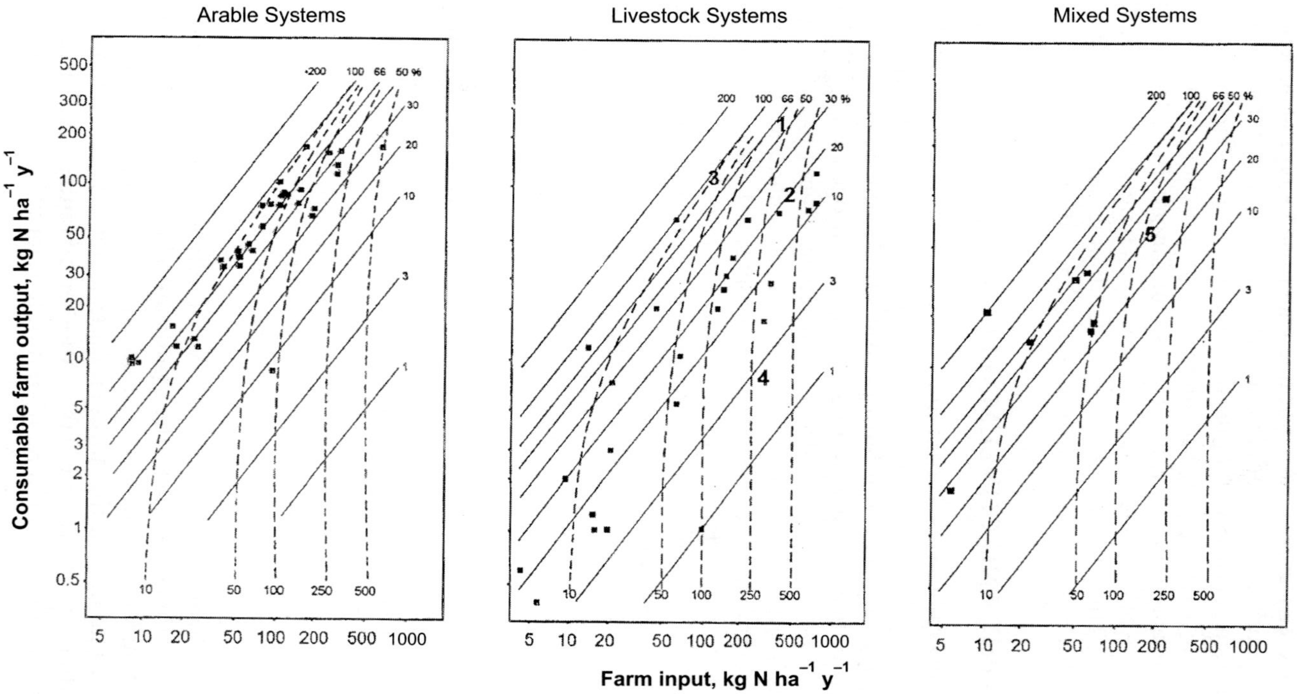

Fig. 9.2. Nitrogen efficiency (solid line %) and environmental load (dotted line kg N ha^{-1} y^{-1}) as functions of farm nitrogen input and consumable output: 1. Fish-cum-duck, 2. Fish-cum-pig (Oláh et al., 1994a), 3. Fish-cum-goose (Oláh et al., 1990), 4. Macrophyte choked carp pond (Oláh and Szabó, 1986), 5. Fish-cum-duck, alfalfa and rice rotation (Oláh et al., 1994b)

ground and stream water. Unfortunately, under the present western economic policy, this type of ecosystem service, i.e. the purification of nitrite polluted ground and stream water, is largely taken for granted.

The highest consumable farm output was produced in the fish-cum-duck integration with nitrogen transfer efficiency of above 50%. On an experimental scale, the fish and duck consumable output together has reached the very high value of 620 kg N ha^{-1} $year^{-1}$ (Pekár et al., 1993).

The fish-cum-pig ponds represent an indirect integration. The pigs were not kept directly in the ponds, only their manure was introduced daily into the ponds, so pig meat production was not considered here in the farm output calculations. However, even without this, the farm output was much higher than in any of the terrestrial animal husbandries except only one. An example of the mismanaged aquaculture in Hungary is the detailed study of the macrophyte choked carp pond ecosystem. In this shallow and large pond on the Hortobagy Puszta, the specialized transfer routes direct the nitrogen into the macrophyte biomass of *Potamogeton pectinatus*, instead of the fish protein synthesis (Oláh and Szabó, 1986). The efficiency is well below 3%. In the aquacultural rotation system of the fish-cum-duck-alfalfa-rice, an analysis of the mixed systems has been made. This complex agroecosystem utilizes input nitrogen with 30% efficiency. Its consumable output is the second highest among the quantified mixed systems.

ENVIRONMENTAL COSTS OF FARMING

Besides high consumable farm output and nitrogen efficiency, these aquacultural ecosystems are able to receive, process and retard waste. They function similar to true natural wetlands and so can substitute partly for the original nutrient processing activity of the wetland system once on the floodplain.

In Fig. 9.2, the dotted line indicates the equal absolute losses of the nitrogen from the food producing systems. This nutrient loss is greatest in open agricultural technologies. Absolute nitrogen loss in certain farming systems at the intensive end of the range may reach the extremely high value of 500 kg N ha^{-1} y^{-1}. In livestock rearing systems, the absolute nitrogen loss is usually higher than in arable systems. The integration of livestock production with the aquacultural wetlands could help decrease this loss.

The absolute total nitrogen loss is composed of several fluxing pathways. Plotting the nitrogen leached to surface and groundwater or lost to the atmosphere gives further details (Fig. 9.3). Surface and subsurface runoff may reach almost 100 kg N ha^{-1} y^{-1} at higher farm input,

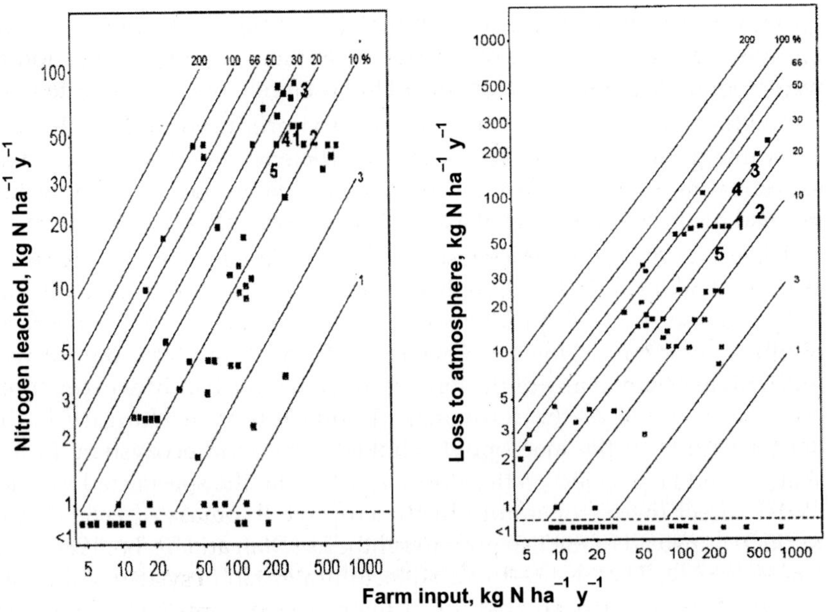

Fig. 9.3. Nitrogen leached and loss to atmosphere as function of farm nitrogen input in arable, livestock and mixed agroecosystems Figures 1-5 for aquaculture systems: 1. Fish-cum-duck, 2. Fish-cum-pig. (Oláh et al., 1994a), 3. Fish-cum-goose (Oláh et al., 1990), 4. Macrophyte choked carp pond (Oláh and Szabó, 1986), 5. Fish-cum-duck, alfalfa and rice rotation (Oláh et al., 1994b)

but is scattered on a wide range according to soil texture and gradient. In the five Hungarian aquacultural farming systems, the values actually represent the nitrogen content of the draining water. On solid clay, seepage is minimal. The amount of drained nitrogen ranged between 28 and 81 kg N ha^{-1} y^{-1}. The highest value was measured in the fish-cum-goose ponds due to the short water retention time. A project (Szabó, 1994) has clearly demonstrated that Hungarian fishponds received more nitrogen from the water supplying rivers than from water released back during the autumnal draining for harvest.

Nitrogen loss to the atmosphere usually surpasses leached or drained nitrogen. The highest value reaches the amount of 200 kg ha^{-1} y^{-1}. In the terrestrial agroecosystems, the loss to the atmosphere comes from denitrification and ammonia volatilization. The atmospheric loss from aquacultural ponds is usually the result of the denitrification. Volatilization may occur only under a very limited condition, at high pH values. Fortunately the high pH situation is usually accompanied by a very low ammonia concentration. As a result, volatilization is negligible in fish producing ecosystems.

Even on the dried floodplain area of the Carpathian Basin, aquaculture can retard the polluting nitrogen from industrialized terrestrial farming. It may come directly with filling river water, in which the nitrate concentration has reached the annual average value of 15.6 $\mu g\ ml^{-1}$ at the peak period of the agriculture industrialization, in 1984. The agricultural waste could be processed through integration with animal husbandry. Aquaculture receives agricultural waste as nitrogen rain. The annual nitrogen retention per hectare was measured and calculated: 218 kg in fish-cum-duck (Oláh et al., 1994a), 294 kg in fish-cum-pig, 12 kg in fish-cum-goose (Oláh et al., 1990b), 125 kg in macrophyte choked carp pond (Oláh and Szabó, 1986) and 93 kg in fish-cum-duck-alphalpha-rice rotation system (Oláh et al., 1994b).

ENERGY COST OF AQUACULTURE

The thin layer of air that covers our planet is being increasingly affected by greenhouse gases and by ozone consuming gases. The modern economy still takes for granted the relative stability of the Earth's climate.

The energy cost of human society is high. Unrenewable fossil fuels are consumed for dwellings, food and fiber production, transportation, and luxury. These carbon concentrates contain only a small amount of nitrogen, but in the course of energy utilization at high temperature, the dinitrogen molecules of the air are oxidized to combine nitrogen. This nitrogen then moves to the atmosphere and comes back with rain or with dry deposition (H.T. Odum, 1971).

Aquaculture has low energy expenditure for food production as compared to animal husbandry (Table 9.15). Only 0.5 ton of oil is burned to produce 1 ton of fish in an integrated fishpond ecosystem. The same figure is 7.68 tons in an industrialized recycling system and 14.37 tons in a beef-producing system. The nitrogen released to the atmosphere while burning oil to produce one ton of animal is only 4.8 kg in integrated fishponds, but as high as 126.3 kg in lamb-producing systems. Much less pollution, better waste utilization and highest protein yields per hectare come from fish as compared to crops and animals.

The use of energy in modern, industrialized aquaculture is much greater than that of integrated aquaculture systems. The main use of energy in industrialized agriculture is for pumping water into culture systems and mechanical aeration to improve dissolved oxygen concentrations (Boyd and Tucker, 1998). Energy use for pumping water and aeration in channel catfish culture in the United States ranges from 1,200 to 9,000 kilowatt hour ha^{-1} (Boyd and Tucker, 1995). The average is probably around 3,000 to 4,000 kilowatt hour ha^{-1}. Additional energy is

Table 9.15. Energy costs of animal protein production (Oláh and Sinha, 1986)

Agroecosystems	kJ for 1 g protein	Oil consumed, t for 1 t animal	N to atmosphere, kg from oil for 1 t animal
Animal husbandry			
Eggs	552	2.36	20.36
Broilers	624	2.67	23.02
Pork	779	3.34	28.74
Milk	1101	4.72	40.62
Beef	2869	12.31	106.25
Feed beef	3349	14.37	123.57
Lamb	3424	14.69	126.34
Aquaculture			
Integrated pond	132	0.56	4.87
Inorganic fertilized pond	275	1.18	10.10
Feed pond	418	1.79	15.40
Aerated pond	468	2.00	17.20
Cage culture	622	2.66	23.00
Recycling system	1791	7.68	66.20

used for applying feed from mechanical feeders and harvesting fish with seines drawn by tractors.

RECONSTRUCTION OF FLOODPLAIN AQUACULTURE

Aquaculture in the Carpathian Basin is the most sustainable consumption of the unrenewable energy and the wisest use of natural resources. It has several advantages: (1) high protein producing potential, (2) healthy protein from fish, (3) the energy is cost effective, and (4) a healthier natural environment. Priority has to be assigned for proper nutrient and organic waste recycling through fish farming on a larger scale.

Enlargement of fish and wildlife rearing areas, both with constructed fishponds and by restoring a reasonable part of the dried floodplain, would substitute for the former functions of the inland riverine delta. Nutrient processing, removing pollutants, or retarding natural resource depletion are more vital today than before the river regulation in the second half of the last century. The industrialization of agriculture resulted in an enormous increase in the use of nitrogen fertilizers. The national nitrogen budget has been calculated for the peak industrialization period and for 1993, and the nitrogen input sources have been reconstructed, all of which characterized the nitrogen load condition before river regulation that is on the intact riverine floodplain network (Table 9.16). There was almost a 10-fold increase of nitrogen input. This huge nitrogen load on the

Table 9.16. National nitrogen budget measured for the years of 1988, 1993 in Hungary and reconstructed for the intact floodplain of the 19th century, 10^3 t y^{-1}

Balance	Before regulation	1988	1993
Input			
Nitrogen rain	19	146	105
Inorganic fertilizer	0	617	124
Industry	0	10	8
Total input	115	495	425
	134	1268	662
Output			
Food export	-	60	45
Outflowing rivers	-	600	480
Total output	-	660	525
Accumulation, kg ha^{-1} y^{-1}	-	65	15

present agricultural floodplain is accompanied by a dramatic decrease of the nitrogen retarding wetland area from 39,000 km^2 to 67 km^2 . This deformed landscape metabolism results in a toxic condition on the Hungarian Lowland. The restoration of well-designed floodplain aquaculture is a good strategy that can be realized only together with nature protection, wildlife production and the development of ecotourism.

Floodplain aquaculture should yield cheaper and more nutritious foods within reach of the poor. Development of floodplain aquaculture in this sound direction will help achieve economic progress and generate substantial employment for the countryside. It has potential for exports, as fish consumption in developed countries is on the increase. However, with the reconstruction along the River Tisza and its tributaries, a more important target would be achieved, that is the restoration of a healthy environment.

Floodplain aquaculture can adversely affect local hydrology. Construction of aquaculture facilities on floodplains can increase flood levels by blocking flow along the floodplain. In the United States, the Natural Resources Conservation Service limits pond construction to 60% of the total floodplain area (Boyd et al., 2003).

CONCLUSION

The current emphasis on adoption of better management practises to make aquaculture more environmentally and socially responsible is directed mainly at industrial farms. These farms may be located in developed nations or in developing nations, but their products are sold exclusively in the developed world. Adoption of better practises can conserve resources, prevent negative impacts, resolve conflicts between aquaculture projects and neighbors, and increase the prospects for sustainability (Boyd and Clay, 1998).

The majority of world aquaculture is conducted in developing Asian nations to provide food for individual families or for sale in local markets. Therefore, application of better practises in small-scale Asian aquaculture could have much greater environmental and social benefits than application of better practises in industrial, export aquaculture. As such, programs to improve industrial aquaculture should be continued, but much greater emphasis should be directed to teaching small-scale farmers how to apply systems of linked ecotechnology and biotechnology.

REFERENCES

Abdelmoneim, M.A., Oláh, J. and Szabo, P. (1986). Denitrification in water bodies and sediments of Hungarian shallow waters. Aquacultura Hungarica (Szarvas) 5: 133-146.

Boyd, C.E. and Tucker, C.S. (1995). Sustainability of channel catfish farming. World Aquacul. 26(3): 45-53.

Boyd, C.E. and Clay, J.W. (1998). Shrimp aquaculture and the environment. Scientific American, June 1998, 278(6): 42-49.

Boyd, C.E. and Tucker, C.S. (1998). Pond Aquaculture Water Quality Management. Kluwer Academic Publishers, Boston, Massachusetts, USA, pp. 700.

Boyd, C.E., Queiroz, J., Lee, J., Rowan, M., Whitis, G.N., and Gross, A. (2000). Environmental assessment of channel catfish, *Ictalurus punctatus*, farming in Alabama. J. World Aquacul. Soc. 31(4): 511-544.

Boyd, C.E. (2003). The status of codes of practice in aquaculture. World Aquacul. 34(2): 63-66.

Boyd, C.E., Queiroz, J.F., Whitis, G.N. Hulcher, R., Oakes, P., Carlisle, J. Odom,D.Jr., Nelson, M.M., and Henstreet, W.G. (2003). Best management practices for channel catfish farming in Alabama. Special Report 1, Alabama Catfish Producers, Montgomery, Alabama, USA.

Boyd, C.E. (2005). Water use in aquaculture. World Aquaculture 35(3): 12-15 and 70.

Brown, L.R., Gardner, G. and Halweil, B. (1998). Beyond Malthus, sexteen dimensions of the population problem. Worldwatch Paper 143, Worldwatch Institute, Washington, DC, USA, pp. 89.

Buresh, R.J. and Patrick, W.H. Jr. (1978). Nitrate reduction to ammonium in anaerobic soil. Soil Sci. Soc. Amer. J. 42: 913-918.

Csavas, I. (1993). Aquaculture development and environmental issues in the developing countries of Asia. In: Environment and Aquaculture in Developing Countries (eds.) R.S.V. Pullin, H. Rosenthal and J.L. Maclean, ICLARM Conf. Proc. 31: pp. 74-101.

Edwards, P. (1993). Environmental issues in integrated agriculture-aquaculture and wastewater-fed fish culture systems. In: Environment and Aquaculture in Developing Countries (eds.) R.S.V. Pullin, H. Rosenthal and J.L. Maclean, ICLARM Conf. Proc. 31, pp. 139-170.

El Sarraf, W.M., Oláh, J., El Samra, M.J., Abdelmoneim, M.A. and Pekár, F. (1982). Inorganic and organic nitrogen uptake in the water and sediment of Hungarian shallow waters. Workshop on Eutrofication of shallow lakes: Modeling, monitoring and management. 29 Aug.3 Sept.1982, Veszprem, Hungary.

El Sarraf, W.M. (1983). Amino acids and microbial activity in Hungarian shallow waters. Ph.D. thesis, Budapest, Hungarian Academy of Sciences, Hungary.

Fabry, G. (1975). Sediment chemistry of Szarvas fish ponds with aquacultural relation. MEM Press, Budapest, Hungary, p. 92.

Folke, C. and Kautsky, N. (1992). Aquaculture with its environment: prospects for sustainability. Ocean and Coastal Management 17: 5-24.

Frissel, M.J. (1978). Cycling of mineral nutrients in agricultural ecosystems. Development in agricultural and managed-forest ecology. Elsevier, Amsterdam, The Netherlands.

Gardner, W.S., Nalepa, T.F., Slavens, D.R. and Laird, G.A. (1983). Patterns and rates of nitrogen release by benthic Chironomidae and Oligochaeta. Can. J. Fish. Aquat. Sci. 40: 259-266.

Huang, S.-L., Wu, S.-C. and Chen, W.-B. (1995). Ecosystem, environmental quality and ecotechnology in the Taipei metropolitan region. Environmental Engineering. 4: 233-248.

Kaspar, H.F. and Tiedje, J.M. (1981). Dissimilatory reduction of nitrate and nitrite in the bovine rumen: nitrous oxide production and effect of acetylene. Appl. Environ. Microbiol. 41: 705-709.

Kaspar, H.F., Tiedje, J.M. and Firestone, R.E. (1981). Denitrification and dissimilatory nitrate reduction to ammonium in digested sludge. Can. J. Microbiol. 27: 878-885.

Koike, I. and Hattori, A. (1978). Denitrification and ammonia formation in anaerobic coastal sediments. Appl. Environ. Microbiol. 35: 278-282.

Liao, C.F.-H. and Lean, D.R.S. (1978). Nitrogen transformation within a trophogenic zone of lakes. J. Fish. Res. Board Can. 35: 1102-1108.

MacRae, D.C., Ancajas, R.R. and Salandanan, S. (1968). The fate of nitrate nitrogen in soma tropical soils following submergence. Soil Sci. 105: 327-334.

Ma, S. (1985). Ecological engineering: Application of ecosystems principles. Environ. Conservation, 12: 331-335.

Mitamura, O. and Saija, Y. (1986). Urea metabolism and its significance in the nitrogen cycle in the euphotic layer of Lake Biwa. IV. Regeneration of urea and ammonia. Arch. Hydrobiol. 107: 425-440.

Mitsch, W.J. (1988). Ecological engineering and ecotechnology with wetlands: Applications of systems approaches. In: Advances in Environmental Modelling: Proceedings of Conference on State of the Art of Ecological Modelling, June, 1987, Venice, Italy. Elsevier, Amsterdam, The Netherlands.

Mitsch, J.W. and Jorgensen, S.E. (eds.) (1989). Ecological Engineering. An Introduction to Ecotechnology. John Wiley & Sons, New York, USA, pp. 472.

Naylor, R.L., Goldburg, R.J., Primavera, J.H., Kautsky, N., Beveridge, M.C.M., and five others. (2000). Effect of aquaculture on world fish supplies. Nature 405: 1017-1024.

Odum, E.P. (1971). Fundamentals of ecology, 3rd edition, Saunders, Philadelphia, USA.

Odum, H.T. (1962). Man in the ecosystem. In: Proceedings Lockwood Conference on the Suburban Forest and Ecology. Bull. Conn. Agr. Station 652. Storrs, CT, pp.57-75.

Odum, H.T. (1971). Environment, Power and Society. John Wiley & Sons, New York, USA, pp. 331.

Oláh, J. and Sinha, V.R.P. (1984). Principles and methods of monitoring of perennial undrainable pond ecosystems in tropical monsoon lands. Aquacultura Hungarica 4: 103-110.

Oláh, J. and Janurik, E. (1985). Methods to measure nitrogen cycling in shallow waters. MEM Press, Budapest, Hungary, p. 226.

Oláh, J. and Sinha, V.R.P. (1986). Energy cost of carp farming systems. Aquacultura Hungarica 5: 219-226.

Oláh, J. and Szabó, P. (1986). Nitrogen cycle in a macrophyte covered fish pond. Aquacultura Hungarica, 5: 165-178.

Oláh, J., Toth, L. and El Samra, M.I. (1987). Biological nitrogen fixation in shallow waters. Academic Press, Budapest, Hungary, p. 191.

Oláh, J., Ayyappan, S. and Purushotaman, C.S. (1990a). Structure, functioning and management in rural undrainable fishponds. Aquacultura Hungarica 6: 77-95.

Oláh, J., Vörös, G., Körmendi, S. and Pekár, F. (1990b). Groundwater nitrate biomanipulation in typha-covered ponds with goose-cum-fish cultura. EIFAC/FAO Symposium on Production Enhancement in Still-Water Pond Culture. Prague, Czechoslovakia, 15-18, May 1990. 1: 107-113.

Oláh, J., Pekár, F. and Szabó, P. (1994a). Nitrogen cycling and retention in fish-cum-livestock ponds. J. Appl. Ichthyol. 10: 341-348.

Oláh J., Szabó, P., Esteky, A. A. and Nezami, S.A. (1994b). Nitrogen processing and retention on a Hungarian carp farm. J. Appl. Ichtyol. 10: 335-340.

Oláh, J. and Oláh, M. (1996). Improving landscape nitrogen metabolism on the Hungarian lowland. Ambio 25(5): 331-335.

Oláh, J. and Pekár, F. (1997). Ecotechnological and socio-economic analysis of fish farming systems in the freshwater area of the Mekong Delta. Cantho University Press, Hungary, pp. 124.

Ortega, E., Queiroz, J.F., Boyd, C.E., and Ferraz, J.M.G. (1999). Emergy analysis of channel catfish farming in Alabama, USA. First Approach: A General Overview. In: First Biennial Energy Analyses Research Conference – Energy Quality and Transformities. University of Florida, Gainesville, Florida, USA.

Pekár, F., Oláh, J. and Toth, L. (1989). Nitrification in fish ponds and in natural waters. II.Nitrification in open waters and in sediment. Vizugyi Kozlemenyek 71: 295-310.

Pekár, F., Kiss, L., Szabó, P. and Oláh, J. (1993). Pond processing of high organic load in a fish-cum-duck culture system in Hungary. In: From Discovery to Commercialization. Abstracts of contributions presented at the International Conference Worlds Aquaculture 1993, Torremolinos, Spain, 26-28 May, 1993, p. 253.

Sinha, V.R.P. and Oláh, J. (1982). Potential of freshwater fish production in ecosystems with different management levels. Aquacultura Hungarica 3: 201-206.

Sorensen, J. (1978). Capacity for denitrification and reduction of nitrate to ammonia in a coastal marine sediment. Appl. Environ. Microbiol. 35: 301-305.

Straskraba, M. (1984). New ways of eutrophication abatement. In: Hydrobiology and Water Quality of Reservoirs (eds.) M. Straskraba, Z. Brandl, and P. Procalova, Acad. Sci., Cesce Budejovice, Czechoslovakia, pp. 37-45.

Straskraba, M. (1985). Simulation Models as Tools in Ecotechnology Systems. Analysis and Simulation, Vol. II. Academic Verlag, Berlin, Germany.

Szabó P. (1994). Quality of effluent water from earthen fish ponds in Hungary. J. Appl. Ichtyol. 10: 326-334.

Tatrai, I. (1987). The role of fish and benthos in the nitrogen budget of Lake Balaton, Hungary. Arch. Hydrobiol. 110: 291-302.

Toth, E.O., Oláh, J. and Papp, Z.G. (1984): Amino acid compartment in the water and sediment of Hungarian shallow waters. Aquacultura Hungarica 4: 111-134.

Uhlman, D. (1983). Entwicklungstendenzen der Ökotechnologie. Wiss. Z. Tech. Univ. Dresden 32: 109-116.

Ulgiati, S., Brown, M.T., Bastianoni, S. and Marchettini, N. (1995). Emergy-based indices and ratios to evaluate the sustainable use of resources. Ecological Eng. 5: 519-531.

10

Marine Microbial Biotechnology and Aquaculture – An Overview

P.K. Pandey and C.S. Purushothaman***

INTRODUCTION

The potential of the ocean as the basis for new biotechnology remains largely unexplored. The vast number of marine organisms, primarily microorganisms, is yet to be identified. Even for known organisms, there is insufficient knowledge to permit their intelligent management and applications. Oceanic organisms are of enormous scientific interest, as they constitute a major share of the Earth's biological resources (Ray and Edison, 2005). Also, marine organisms often possess unique structures, metabolic pathways, reproductive systems, and sensory and defence mechanisms because they have adapted to extreme environments ranging from the cold polar seas at –2°C to the great pressures of the ocean floor, where hydrothermal fluids spew forth. Most major classes of the Earth's organisms are primarily or exclusively marine, so the oceans represent a source of unique genetic information.

Interest in marine biotechnology has been growing in recent years and the latest developments in molecular biology promise fundamentally new approaches and opportunities for identifying, using and managing biological resources from the seas and other bodies of water (Akkermans et al., 1994). These resources have broad potential as technologies for application in many fields including health, agriculture, bioremediation and manufacturing. These technologies will also enhance understanding the basic biology of the seas and their pivotal role in preserving the global environment. With their vast genetic and physiological diversity, marine

Central Institute of Fisheries Education, Versova, Mumbai – 400061, India
*E-mail: *pkpandey_in@yahoo.co.uk **cspuru@indiatimes.com*

microorganisms will be the sources for new classes of biochemicals and processes, including pharmaceuticals, polymers, enzymes, vaccines, and diagnostic and analytical reagents.

Aquatic biotechnology may be defined as the application of science and engineering in the direct or indirect use of aquatic organisms or parts, or products of living aquatic organisms in their natural or modified forms. It includes several components: aquaculture biotechnology (fish health and broodstock optimization); aquatic bioprocessing (obtaining valuable compounds from marine organisms); and aquatic bioremediation (use of microorganisms to degrade toxic chemicals in the aquatic environment). The short-term opportunities are primarily in aquaculture biotechnology, while long-term opportunities are expected in the form of marine pharmaceuticals, aquatic bioremediation, bioinformatics and biosensors.

In recognition of the potential of the seas to contribute to economic growth and development of both industrialized and developing nations, marine biotechnology institutes have been established during the past decade in many countries, including the United States, Japan and Germany. Japan, in particular, has made major investments in marine biotechnology, with the government spending US$ 70 million to $100 million annually on R&D, and industry making an even larger investment of US$ 280 million to $ 400 million annually (Zilinskas et al., 1994; NACA/FAO, 2000). Contrary to this, the US Government invested less than US$ 44 million on marine biotechnology R&D in the fiscal year 1992 (BIC, 1995; NACA/FAO, 2000) mostly on sponsorship of academic research, while the US industry spent an estimated $25 million in 1992 (Zilinskas et al., 1994). Promising results are emerging from these activities. An overview of the recent microbial biotechnology applications in marine science and aquaculture has been discussed in this chapter. It identifies research priorities and specific opportunities that offer potential for significant developments.

MARINE MICROBIAL BIODIVERSITY

Biological diversity nowhere in the biosphere is as great as in the seas and the extent of this diversity becomes increasingly evident as scientists investigate new environments (Smith, 1999; Grommen and Verstraete, 2002). The bacterial group *Prochlorococcus*, discovered only a decade ago, may be the most abundant component of phytoplankton in the sea (Fuhrman and Campbell, 1998). These tiny organisms uniquely contain the photosynthetic pigment divinyl chlorophyll a and b, and are the major primary producers in tropical and subtropical waters (i.e. some 75% of the world's oceans). They contribute between 10 and 80% of the total local primary production (Goericke and Welschmeyer, 1993; Liu et al., 1997). Giant tubeworms (*Riftia* spp.) were recovered from areas adjacent to deep-ocean thermal vents, and novel mussels (*Bathymodiolus* spp.) that farm methanotrophic bacteria on their gill tissue were discovered around methane seeps in the Gulf of Mexico (Nix et al., 1995). Most newly described species have been and will continue to be microorganisms, although it is clear that new marine plants and animals also await discovery.

Less than 1% of the living bacteria, marine and terrestrial, have been isolated and described (Akkermans et al., 1994; BIC, 1995). The marine environment represents a particularly fertile source for new bacteria, as evidenced by the discovery of unusual "cold water" archaea 100-500 m deep in the oceans (De Long, 1992). These archaea comprise a high percentage of the total bacterial ribosomal RNA present in seawater samples (Pichard and Paul, 1991), yet they have not been isolated in pure culture and described. Pursuit of this research may provide new understanding of oceanic processes and once these archaea are cultivated, perhaps a novel source of products can be obtained. Countless other marine bacteria are yet to be cultured (Towner and Cockayne, 1993); advances in microbial culture technology will enable scientists to isolate, describe and possibly, exploit these organisms.

Cold habitats offer good sources of useful genes and novel natural products with activity at low temperatures (Cowan, 1997). At low temperatures, the rate of enzymatic reactions, the fluidity of cellular membranes, and the affinity of uptake and transport systems decrease (Phadtare et al., 2000). Therefore, biomolecules of organisms living in cold habitats may show distinctive physical properties. The Arctic is a representative cold habitat, which remains one of the least explored, studied and understood places on Earth. The studies on Arctic bacterial diversity have been restricted to the marine environment (Knoblauch et al., 1999; Sahm et al., 1999; Ravenschlag et al., 2001). The study on the bacterial diversity of the Arctic throws light into the biological mechanisms of adaptation to and tolerance of cold environments. Psychrophilic and psychrotolerant bacteria are defined by optimal growth at temperatures below 20°C and normal growth at temperatures as low as 0°C; psychrotolerant bacteria are distinguished by growth at temperatures above 20°C (Morita, 1975). Enzymes produced by psychrophilic and psychrotolerant bacteria are remarkably stable after long-term storage and occasional freeze-thaw cycles (Irwin et al., 2001). Therefore, psychrophilic and psychrotolerant bacteria have been focused for the isolation of cold-active enzymes (Davail et al., 1994; Huston et al., 2000; Secades et al., 2001; Nakagawa et al., 2003). Cold-active serine alkaline protease was isolated from a psychrophilic *Pseudomonas* strain (Zeng et al., 2003). Protease is an important enzyme of the medical, environmental, food, chemical and other industries (Storer, 1991; James and Simpson, 1996; Vermeij and Blok, 1996).

Other newly appreciated microorganisms of marine biological diversity are viruses. Electron microscopy and PCR (Polymerase Chain Reaction) technology have revealed abundant numbers of viral particles in seawater samples. Specific viruses that infect species of marine phytoplankton have been cultured from coastal as well as open ocean sites (Suttle et al., 1990). This is an exciting discovery for at least two reasons. Viral infection of higher forms of marine life almost certainly affects global ocean processes such as photosynthesis. Marine viruses will also provide new materials for the development of genetic and biotechnological tools that can be used to study and manipulate marine organisms. For example, marine viruses could be used to genetically engineer higher forms of marine life (Gildberg and Mikkelsen, 1998).

MARINE ECOLOGY

For developing new applications for marine products and processes, analyzing global climate change and improving fisheries management, scientists need to develop a fundamental understanding of marine organisms, and their specific adaptations to and interactions with their environmental ecosystem (Grommen and Verstraete, 2002). Developments in molecular techniques are rapidly expanding the capabilities for research on marine ecology. In his classic 'Paradox of the Plankton', Hutchinson asked how is it possible to maintain so many phytoplankton species in a homogeneous environment with only a few potentially limiting resources (Hutchinson, 1961). The newly recognized bacterial diversity makes matters even more complicated. Although there have been many possible explanations, including the suggestion that ubiquitous viruses may help to maintain diversity (Fuhrman and Suttle, 1993), the paradox still perplexes us today. It also reminds us how much we have yet to learn about biodiversity.

The new molecular tools should be applied to gain insight into the basic molecular and cellular processes by which marine organisms adapt to extreme environments. Examples of basic research that may lead to commercial applications are gene sequencing and the study of special glycoproteins that inhibit ice crystal growth in the tissues of Antarctic fish. These substances, along with gene sequence information for key marine species, may prove useful in industrial and medical preservation processes. Apart from that, a thorough understanding of the marine ecological systems must be developed in order to specify the baseline level of function, and to monitor and predict the potential changes and perturbations of systems due to physical, chemical or biological impacts. The development of predictive models for analyzing the potential global climate changes depends on the acquisition of fundamental information on molecular regulatory mechanisms of photosynthesis in the seas.

STATUS OF AQUACULTURE SECTOR

A definition of aquaculture, derived from the Food and Agriculture Organization (FAO, 2001) is "the culture of aquatic organisms including fish, molluscs, crustaceans and aquatic plants". Culture implies some form of intervention in the rearing process to enhance production, such as regular stocking, feeding, protection from predators, etc. Culture also implies individual or corporate ownership of the stock being cultivated. For finfish, aquaculture is analogous to the livestock and poultry industries - animals are raised in captivity and then, slaughtered and sold.

Population growth, urbanization and rising per capita incomes have led the world fish consumption to more than triple over the period 1961-2001, increasing from 28.0 to 96.3 million tonnes. Per capita consumption has been multiplied by a factor of 1.7 over the same period and in many countries, this trend is expected to continue in forthcoming decades. In the context of stagnant production or slow growth from the capture fisheries, only aquaculture expansion can meet this growing global demand. Acknowledging the challenges that this relatively new industry may face in coming years and the need to prepare for the sustainable growth of the sector, FAO carried out a study on global aquaculture production outlook to evaluate its potential to meet the projected demand for food fish in 2020 and beyond (http://www.globefish.com).

According to the FAO statistics, aquaculture's contribution to global supplies of fish, crustaceans and molluscs continues to grow, increasing from 3.9% of the total production by weight in 1970 to 27.3% in 2000. Aquaculture is growing more rapidly than all other animal food producing sectors. Worldwide, the sector has increased at an average compounded rate of 9.2% per year since 1970, compared with only 1.4% for capture fisheries and 2.8% for terrestrial farmed meat production systems. The growth of inland aquaculture production has been particularly strong in China, where it averaged 11.5% per year between 1970 and 2000 compared with 7.0% per year in the rest of the world over the same period. Mariculture production in China increased at an average annual rate of 14% compared with 5.4% in the rest of the world (http://www.aquanic.org/courses/marine/course%20materials/FAO%20World%20Aquaculture%20Production.htm).

The past 20 years have seen Asian aquaculture evolve from a traditional practice to a science-based activity and grow into a significant food production sector, contributing more to national economies and providing better livelihoods for rural and farming families (Roth et al., 2000). Aquaculture used to be regarded as an infant in comparison with crop and livestock husbandry, and capture fishery (Smith, 1999). It has since matured into a better organized economic sector characterized by more state patronage and stronger private sector participation in most parts of Asia. Asia dominates the world in aquaculture production and the sector is extremely diversified in species, technologies and farming systems employed. Many governments place priority on aquaculture development; however, there are various threats and constraints to its growth. Asia's contribution to the total world production in 1997, following the International Standard Statistical Classification for Aquatic Animals and Plants (ISSCAAP, 1999) grouping, was finfish – 89%; crustaceans (marine) 80%; freshwater crustaceans – 94%; molluscs – 88%; aquatic plants – 98%, and miscellaneous animals and products – 99%. The region provides 91% of the global aquaculture production. In 1997, the combined aquaculture production was 32.63 million tonnes valued at US$ 41.95 billion, an increase of 144 and 117% in weight and value, respectively, compared to those in 1988. Aquaculture production in the region has been growing at a rate nearly five times faster than landings from capture fisheries, and its share in total fisheries landings in the region increased from 32 to 50% between 1988 and 1997 (FAO, 2001).

The two major influences on Asian aquaculture development policy are the recent broadening from technical and economic objectives towards social objectives for aquaculture development that include poverty alleviation, livelihood development and food security, and the links made between sustainable aquaculture practices and trade (Edwards, 1998; DFID/FAD/NACA/GoB, 2001). An increasing recognition of the importance of small-scale, socially oriented aquaculture is taking place, and recent initiatives have been made at the regional level to focus governments and regional organizations on this issue (GESAMP, 1996). The role of the fish farmer is changing from merely raising fish to being a part of a chain for the production and delivery of safe, high quality products to the consumer. The link made to production practices and their impact on the environment on one hand and trade on the other hand has been reflected in recent regional and national programmes (Insull and Shehadeh, 1996; NACA/FAO, 2000).

According to FAO (2001) estimates, the global production of scampi in 1987-88 was about 27,000 tonnes, of which about 19,000 tonnes were from aquaculture. Aquaculture production increased to 23,000 tonnes in 1990, which then formed about 74% of the global production (31,000 tonnes). The steadily increasing production fell in 1992 due to the severe disease problem in Taiwan. However, this fall in production was partially offset by increased production in other countries like India and Vietnam. As much as 92% of *Macrobrachium rosenbergii* still originates in Asia. Major production in the rest of the world was contributed by South, Central and North America (2,200-2,400 tonnes year^{-1} or 7.7%). The contribution from Africa (0.2%), and the Pacific (< 0.1%) are relatively insignificant. Most of the African production is in one of its smallest countries, Mauritius.

PRESENT STATUS OF AQUATIC BIOTECHNOLOGY

The rising number of R&D initiatives around the world points to an increasingly competitive aquatic biotechnology sector. Countries such as Australia, France, Germany, Italy, Japan, Norway, Sweden, the United Kingdom and the United States have significant research and development programmes. Among the developing countries, China and India are making considerable investments in aquatic biotechnology. Aquatic biotechnology is now at the stage where practical applications are being identified and commercialized. Governments and the public are increasingly aware of the potential of aquatic biotechnology to generate new scientific knowledge, and produce useful products and processes. The public expects that government will ensure that biotechnology activities and products are safe for the marine and freshwater environment, living marine organisms and human health. They also see a role for government help in capitalizing on the potential of biotechnology to increase jobs and economic growth. This is especially true for coastal and rural public who see the potential for a competitive aquaculture sector to supplant the jobs lost due to productivity gains or reduced stock biomasses in the wild fisheries.

The USA was the first country to establish a specific programme in aquatic biotechnology (Centre for Marine Biotechnology, University of Maryland). It is the largest aquatic biotechnology institute in North America and the second largest in the world after Japan. The public sector in the USA spends over US$ 50 million annually for research and development in aquatic biotechnology. Japan spends US$ 1 billion (80% from industry and 20% from government) and expects to inject an additional $ 200 million over the coming years, reflecting its assessment of aquatic biotechnology as the greatest remaining technological and industrial frontier.

Efforts made by India are very small in comparison to the Japanese campaign. However, while Japan and the USA are leaders in bioprocessing aspects of aquatic biotechnology, and Canada is strong in aquaculture biotechnology, India is an emerging power in these fields. Due to the presence of well-established fisheries science expertise, the Indian knowledge base in aquatic biotechnology is quite strong and is viewed by many as a competitive strength. Nevertheless, aquatic biotechnology research in India continues to occur mostly in small teams with specialized expertise.

Aquaculture biotechnology industry in Canada, representing about 5% of the total bio-industry of Canada, has a disproportionately high percentage of the global market for their products and services besides the fact that Canada accounts for less than 0.3% of world aquaculture production. Canadian scientific strengths include diagnostics and vaccines for fish and shellfish diseases as well as genetic characterization to support broodstock optimization. An overwhelming percentage of the work to date is with natural biotechnology products as opposed to genetically modified biotechnology products. Canada has developed transgenic fish and is in the process of patent protection of this technology (Hew, 1989). However, nations like India and China are developing similar technology, and will soon offer products to world markets.

MARINE MICROBIAL BIOTECHNOLOGY

Apart from exploring the capabilities of the 30,000 known species of marine microorganisms to develop new classes of human vaccines, medicines and other medical products, chemical products, enzymes and industrial processes, marine biotechnology also provides new tools and is expected to gain significant advancements in such areas as biomaterials, health care diagnostics, aquaculture and seafood safety, bioremediation and biofilms and corrosion.

Non-conventional Products from the Seas

Aquatic bioprocessing spans the sectors of agriculture, environment and health. Tropical and sub-tropical marine environments harbour a wide diversity of animals and plants. Many marine organisms are sessile and must employ sophisticated methods to compete for a place to anchor. This characteristic is reflected in part by a metabolism that produces enormously diverse bioactive products – many with no terrestrial counterparts. Recent research has uncovered unicellular and multicellular microorganisms that are unique to the marine world. Marine bacteria are emerging as a significant chemical resource (Fenical, 1993). Before this vast biological resource can be developed and exploited, however, scientists must address the fundamental questions concerning the nutritional requirements of marine microorganisms (Fenical and Jensen, 1993), and the amounts and characteristics of biochemicals they produce.

Although it is vital to cultivate marine microorganisms that produce important products; the alternative approach of transferring genes of interest into non-marine microorganisms should also be examined. For example, the capability to produce a marine microbial polysaccharide (Cohen, 1999), a complex molecule that could be useful as a food additive or a water-resistant adhesive, could be transferred to an easily grown bacterium e.g. *Escherichia coli*, *Bacillus subtilis* or *Pseudomonas aeruginosa* (Davies et al., 1993; Hoyle et al., 1993). This approach might be more effective in some cases rather than cultivating the marine organism or recreating artificially the long and complicated production pathway for the polysaccharide.

Pharmaceuticals

The aquatic environment has not been explored to nearly the extent of the terrestrial environment. Therapeutic natural products found in terrestrial plants and microorganisms were the basis of early drug development (Tan et al., Chapter 10 in Volume I of this series–Microbial Biotechnology in Agriculture and Aquaculture) and continue to be explored and used as sources of new drugs, albeit with declining rates of success. Drugs and other agents developed from plants include antibiotics (including anti-tumour agents), cardiovascular drugs, anti-inflammatory agents, narcotics, analgesics and most recently, anticholestemics. By contrast, exploration and use of marine resources, especially microorganisms (bacteria, fungi, and microalgae), for drug development has barely begun. Screening programmes have discovered algae, corals, sponges and tunicates that produce compounds showing antibiotic, anti-tumour, anti-viral or anti-inflammatory properties (Lemos et al., 1985, 1991; Mearns-Spragg et al., 1997, 1998). Since the discovery of penicillin in 1928 to the Taq DNA polymerase obtained from *Thermus aquaticus* (Yellowstone hot spring) in 1989, nearly 50,000 natural products have been discovered from microorganisms (Datta and Licata, 2003). Over 10,000 of these are reported to have biological activity and over 100 microbial products are in use today as antibiotics, anti-tumour agents and agrochemicals (Carte, 1996). A microbial metabolite (from *Alteromonas* spp.) has been developed with anti-HIV potential as reverse transcriptase inhibitor from marine microbes isolated from the tissues of Bermudian marine sponge (Stierle, 2002). The antimicrobial drugs from marine actinomycetes have been widely studied and reviewed (Okazaki and Okami, 1972; Goodfellow and Haynes, 1984; Umezawa, 1998).

In spite of such successes in drug discovery from microorganisms, marine microorganisms have received very little attention. The difficulty in the search of metabolites from marine bacteria is mainly due to the non-culturability of the majority (over 99%) of them (Hugenholtz and Pace, 1996). The studies made by the scientists at the Scripps Institution of Oceanography show that marine bacteria are capable of producing unusual bioactive compounds that are not observed in terrestrial sources (Fenical, 1993). As work continues, it is expected that anti-parasitic, pesticidal, immune-enhancing, growth-promoting and wound-healing chemicals will be found in marine microorganisms. However, the real value of biotechnology is in the potential to clone the genetic material of these aquatic organisms to harness the unique compounds without depleting or harming the natural resources.

In health related research, seaweeds and seaweed-derived extracts have shown antifungal, antibacterial, anticoagulant, antithrombotic, antiviral, antihelminthic and antiarteriosclerotic properties in humans. Marine natural products are likely to provide major new products in the history of drugs (Fenical and Jensen, 1993). A number of bioactive substances from the marine environment have already been isolated and characterized with great promise for the treatment of human diseases (Boyd et al., 1999a). The compound manoalide from a Pacific sponge, for example, has spawned more than 300 chemical analogues, with a significant number of these going on to clinical trials as anti-inflammatory agents (Jha and Zi-rong, 2004).

Sea has got plenty of metabolites and other resources in living or dead form. Sponges (37%), coelenterates (21%) and microorganisms (18%) are the major sources of biomedical compounds followed by algae (9%), echinoderms (6%), tunicates (6%), molluscs (2%) bryozoans (1%), etc. (Blunt et al., 2004). Much more research efforts may be employed to determine the seasonal factors and life cycle or reproduction states, which are linked with natural production of an agent. Factors influencing production may include diet, physical and chemical conditions, distribution by phylogenetic affiliation, geographic location, water depth or associations with symbiotic microorganisms. The knowledge of these factors will be important in developing methods for producing selected metabolites, either from whole organisms, or in vitro from cell or tissue cultures of plants and animals. Many of these compounds are very large and complex molecules, requiring very elaborate biochemical processes; as a result, it can be difficult to synthesize them or clone all the genes for standard production through fermentation.

Nutraceuticals

Another area of growing interest is the discovery of marine nutraceuticals. These are natural products from marine organisms that possess health benefits without any adverse side effects. Nutraceuticals constitute a variety of substances used in the food supplement and "natural health" markets. The omega-3 polyunsaturated fatty acids (eicosapentaenoic acid and docosahexaenoic acid) obtained from fish rich in oil are examples of nutraceuticals (Hopwood and Sherman, 1990; Metz et al., 2001). These polyunsaturated fatty acids (PUFAs) are known to reduce the risk of heart disease, through lowering the levels of triglycerides in the body. Professor Wright worked with a team of scientists at Ocean Nutrition Canada Ltd. to develop a new phytosterol derivative of these fatty acids that simultaneously lowers cholesterol and triglyceride levels (Ewart et al., 2002).

The neutraceuticals are one of the major components included in the growing market for nutritional supplements, herbal or "natural" remedies and other forms of non-traditional medicine. This industry has grown dramatically in the last few years as a number of companies have developed and expanded the mass market for nutritional supplements and "natural" remedies. The alternative medicine market has grown to nearly US$ 30 billion in the US in 1999, with $14.7 billion spent on nutritional supplements alone (NBJ, 1999).

Enzymes

Many enzymes isolated from the marine environment may have unique, industrially beneficial characteristics due to the extreme environmental conditions faced by the organisms. Enzymes produced by marine bacteria are important in biotechnology due to their range of unusual properties. Some are salt-resistant, a characteristic that is often advantageous in industrial processes. The extracellular proteases are of particular importance, and can be used in detergents and industrial cleaning applications, such as in cleaning reverse-osmosis membranes. *Vibrio* spp. have been found to produce a variety of extracellular proteases. *Vibrio alginolyticus* produces six proteases, including an unusual detergent-resistant, alkaline serine exoprotease. This marine bacterium also produces collagenase, an enzyme with a variety of industrial and commercial applications, including the dispersion of cells in tissue culture studies (Toma and Honma, 1996). Similarly, *Vibrio costicola* isolated from marine environment from India, produced L-glutaminase under solid state fermentation (Prabhu and Chandrasekharan, 1996).

Japanese researchers have developed methods to induce a marine alga to produce large amounts of the enzyme superoxide dismutase, which is used in enormous quantities for a range of medical, cosmetic and food applications. Other research has demonstrated the presence of unique haloperoxidases (enzymes catalyzing the incorporation of halogen into metabolites) in algae. These enzymes could become valuable products, because halogenation is an important process in the chemical industry (Cohen, 1999).

An unusual group of marine microorganisms from which enzymes have been isolated, is the hyperthermophilic archaea (Woese et al., 1990), which can grow at temperatures over 100°C and, therefore, require enzyme systems that are stable at high temperatures. Archaea typically are found in extreme environments such as hot springs, animal guts, hydrothermal vents, sewage sludge digesters and hypersaline habitats, including the Great Salt Lake. The most able starch hydrolyzing enzymes from archaea and marine bacteria have been reported by several researchers (Brown et al., 1990; Brown and Kelly, 1993; Bibel et al., 1998; Bertoldo and Antranikian, 2002). Thermostable enzymes offer distinct advantages, many still to be discovered, in research and industrial processes (Tonkova, 2006). Thermostable DNA-modifying enzymes such as polymerases, ligases and restriction endonucleases already have important research and industrial applications (Ward and Ray, 2006). Hot springs in Yellowstone National Park provided the first archaeon (*Thermus aquaticus*) from which thermostable DNA polymerases were isolated. These novel enzymes (Taq® polymerases) became the basis for the polymerase chain reaction (PCR), a useful technique for studying the genetic material. Comparable enzymes continue to be discovered.

Most enzymes involved in the primary metabolic pathways of thermophilic bacteria and archaea are dramatically more thermostable than their counterparts living at moderate temperatures. Genes for amylolytic enzymes of the hyperthermophilic bacterium *Thermotoga maritime* have been analyzed (Bibel et al., 1998). Expanded study of enzymes from thermophilic marine microorganisms will contribute to the understanding of mechanisms of enzyme thermostability and should enable the identification of enzymes suitable for industrial applications as well as modification of enzymes to enhance thermostability.

Biomolecular Materials

Several researches have proved that marine biochemical processes can be exploited to produce new biomaterials. The mechanisms used by marine diatoms, coccolithophorids, molluscs and other marine invertebrates to generate elaborate mineralized structures on a nanometre scale (less than a billionth of a metre in size) are equally exciting. Nanometre-scale structures can have unusual and useful properties. Research that will enhance understanding and allow engineering of the processes for creating these bioceramics promises to revolutionize the manufacture of medical implants, automotive parts, electronic devices, protective coatings of pharmaceutical capsules and tablets and other novel products (Bunker et al., 1994).

Problems of biofouling (Armstrong et al., 1999; Boyd et al., 1999a, b) can be counteracted by biomaterials, which have long been recognized as an extensive and costly proposition (Voskerician et al., 2003). Bacterial biofilms form slime layers that increase drag on moving ships, interfere with transfer on heat exchangers, block pipelines and contribute to corrosion on metal surfaces (Clare, 1996). Bacterial and micro-algal colonization of surfaces is accompanied by settlement of invertebrate larvae and algal spores, eventually leading to "hard fouling" and the need for costly cleaning. The most effective anti-fouling coatings have utilized toxic chemicals, such as copper and organotins. There is an urgent need for non-toxic biofouling control strategies, due to the heightened recognition of the impact that toxic coatings can have on the environment (Keevil et al., 1999). Research is needed on the attachment mechanisms of marine organisms and the natural products they employ to prevent fouling of their own surfaces.

Molecular approaches to characterize biofilm structure and development offer considerable potential for finding novel biofouling prevention strategies (Clare, 1996). It is now possible to determine the genes and pathways involved in the regulation and synthesis of bacterial adhesive polymers. Considerable progress has been made in understanding the nature and expression of surface polymers produced by microorganisms such as the N_2-fixing *Rhizobium* spp. (Zahran, 2005, Chapter 3 in Volume I of this series) and the opportunistic pathogen *Pseudomonas aeruginosa* (Davies et al., 1993). Similar approaches can be applied to marine biofilm bacteria, to find the genetic determinants of adhesive production and the environmental factors that regulate synthesis (Armstrong et al., 2001).

Molecular biology techniques can also be applied to determine the basis of natural anti-fouling mechanisms. Many marine plants and animals remain free of attached bacteria, either because they produce repelling compounds or because their surface structure neutralizes bacterial adhesives. A product generated by the sea grass *Zostera marina* (eelgrass) is an effective agent for preventing fouling by bacteria, algal spores, and a variety of hard-fouling barnacles and tubeworms (Todd et al., 1993). Molecular characterization of natural fouling resistance could provide new strategies for fouling control. Potential applications include prevention of fouling in industrial pipelines or heat exchangers, improved design of trickling filters or aquaculture circulation systems, and control of biofilm infections of medical implants and prosthetic devices (Boyd and Turner, 1998).

Biomonitors

Biosensors, bio-indicators, and diagnostic devices for medicine, aquaculture and environmental monitoring can be obtained from marine organisms (Yoo et al., 2004). Natural biosensors such as the crab's sensing antennae monitor the water to detect dissolved substances ranging in chemical complexity from simple salts to pheromones. Marine biosensors, including monoclonal/polyclonal antibodies and DNA probes, can be used to detect and monitor substances and pathogens in the oceans. The development of biosensor kits for field use would give scientists the tool to more accurately determine the fate of chemicals and pathogens that have been released into the aquatic environment. A highly sensitive sodium (Na^+) transfer tithe biosensor was designed using frog bladder membrane to measure paralytic shellfish poisons in toxic marine plankton populations (Cheun et al., 1996; Yoo et al., 2004).

The enzymes responsible for bioluminescence are a type of biosensors. The *lux* genes, which encode these enzymes, have been cloned from marine bacteria such as *Vibrio fischeri* and transferred successfully to a variety of plants and other bacteria (Dunlap and Callahan, 1993). The *lux* genes typically are inserted into a gene sequence, or operon, that is functional only when stimulated by a defined environmental feature. The enzymes responsible for toluene degradation are synthesized only in the presence of toluene. When *lux* genes are inserted into a toluene operon, the engineered bacterium glows yellow-green in the presence of toluene. This genetically engineered system is useful in the biodegradation of toluene (Cohen, 1999).

Another type of biomonitor that holds great promise is the gene probe, which can be used to identify organisms that pose health hazards or may be useful in research (Amundson et al., 2001). Specific gene probes can be employed, for example, to detect human pathogens in seafood and recreational waters, fish pathogens in aquaculture systems, microorganisms capable of mediating desired chemical transformations (e.g. toxic chemical degradation, carbon dioxide assimilation and metal reduction), and specific fish stocks in fish migration and recruitment studies.

Biopesticides

Natural marine products have the potential to replace chemical pesticides and other agents used to maximize crop yields and growth. On a large scale, the use of biopesticides has remained insignificant representing only a small fraction of the global US$ 8 billion pesticide market. *Bacillus thuringiensis (Bt)* alone accounts for 90% of the US $160 million biopesticide market (Jarvis, 2001). A marine biopesticide, in use today, has been developed from a bait worm's toxin known to ancient Japanese fishermen. This natural pesticide has demonstrated activity against larvae of the rice stem borer, the rice plant skipper and the citrus leaf miner, among other pests. More recently, scientists discovered novel compounds in marine algae and marine sponges containing symbiotic microorganisms (Armstrong et al., 2001). These compounds were found to promote growth, stimulate germination and increase root lengths in test plants. Several sponge and nudibranch species produce terpenes, a broad class of aromatic compounds used in solvents and perfumes and known to deter feeding by fish. Extracts derived from these same sponge and nudibranch species also demonstrated powerful insecticidal activity against two species–grasshoppers and the tobacco hornworm (Cardellina, 1986).

Biofertilizers

Historically, macroalgae are used as soil fertilizers in coastal regions all over the world. The rational background for this interesting utilization of macroalgae or their extracted residues is the increase in water-binding capacity and mineral composition of the soil (Critchley and Ohno, 1998).

Biological N_2-fixation (BNF), the fixing of atmospheric nitrogen by microbes and making it available to plants, could be harnessed to improve the soil fertility and productivity of crops (Mekonnen et al., 2002). These microorganisms are often referred to as biofertilizers. However, biofertilizers also include microorganisms that solubilize and mineralize phosphorus (Garg et al., 2001: Garg and Bhatnagar, 2005 Chapter 9 in Volume I of this series) to make it available for plants and phytoplankton. Many microorganisms have the ability to fix nitrogen. These include *Azospirillum, Azotobacter, Rhizobium, Sesbania,* algae and mycorrhizae, while *Pseudomonas striata, Bacillus megaterium* and *Aspergillus* are among the other microorganisms that solubilize phosphorus. Biofertilizers have been used in aquaculture in Kenya, the United Republic of Tanzania, Zambia and Zimbabwe (Juma and Konde, 2002).

Biomass for Energy Production

Of all primary energy production, approximately 40% occur in the seas. In this process, oceanic plants (phytoplankton, seaweeds, seagrasses) take up CO_2, and with light energy from the sun, convert it to organic carbon (primarily sugars) and oxygen (Borowitzka, 1998a, b). The seas contain 50 times as much CO_2 as does the atmosphere, and it is estimated that primary production incorporates 35×10^{15} grams of carbon into marine biomass annually. This abundant source of fuel for energy production has not been tapped commercially because it is not competitive with other easily harvested traditional sources of biomass. However, very recently, high oil-producing algae are targeted for production of biodiesel, chemically modified natural oil that is emerging as an exciting new option for diesel engines (Peterson and Reece, 1994).

APPLICATION OF SEA-BASED NEW PROCESSES

Bioremediation

Aquatic bioremediation holds considerable potential for the environment-friendly clean up of polluted harbours and waterways, as well as decontaminating estuaries and other coastal communities using microbes and other filtering macroorganisms (Rao and Sudha, 1996). These problems include catastrophic spills of oil in harbours, shipping lanes and around oil platforms; movement of toxic chemicals from land through estuaries into the coastal oceans (Ray et al., 1992); disposal of sewage sludge, bilge waste and chemical process wastes; reclamation of minerals such as manganese; and management of aquaculture and seafood processing waste.

Without enhanced understanding of the unique conditions in marine environments, the full potential of marine organisms and processes to contribute new waste treatment and site remediation technologies, cannot be realized. For example, redox states can fluctuate in coastal and estuarine sediments. The impact of changing redox conditions on biodegradation of environmental contaminants must be understood before waste management and remediation strategies and predictive models can be developed for contaminated sediments. Since naturally thriving mercury-tolerant bacteria are rare and cannot be grown easily in culture, researchers at Cornell University, USA inserted the metallothionein gene into *Escherichia coli,* which grows well in culture. The genetically engineered bacteria are placed inside a bioreactor that efficiently removes mercury from water. The bacteria are later incinerated and the accumulated mercury is recovered (EC, 2002).

Bioprocessing

The emerging discipline of bioprocess engineering involves the application of biological science in manufacturing to produce products such as biopharmaceuticals and natural bioactive agents (Ward, 1991). Bioprocess engineering requires an understanding of the biological system employed (such as a marine organism), isolation and purification of a product, and translation of the product into a stable, efficacious and convenient form.

An emerging area of interest is the potential of marine algae, bacteria and fungi to produce unusual chemical structures with no parallels in terrestrial organisms (Fenical, 1993). Small-scale studies have indicated the richness of marine microorganisms as sources for novel lead structures (Fenical and Jensen, 1993). Especially, the phylogenetically archaic cyanobacteria produce numerous substances which exhibit antioxidative effects, PUFAs, heat-induced proteins and immunologically effective virostatic compounds. A market estimate of microalgal products is given in Table 10.1.

Table 10.1. Market estimations of microalgal products (Pulz and Gross, 2004, modified)

Product group	Product	Retail value (US$ × 10^6)	Development
Biomass	Health foods	1250-2500	Growing
	Functional foods	800	Growing
	Feed additives	800	Fast-growing
	Aquaculture	700	Fast-growing
	Soil conditioners	500	Promising
Colouring substances	Astaxanthin	<150	Starting
	Phycocyanin	>10	Stagnant
	Phycoerythrin	>2	Stagnant
Antioxidants	β-carotene	>280	Promising
	Tocopherol	150	Stagnant
	Antioxidant extracts	100-150	Growing
	PUFA extracts	10	Growing
Special products	Toxins	1-3	Growing
	Isotopes	>5	Growing

Biofilms secreted by bacteria isolated from the surface of a tunicate prevented the settlement of barnacles and tunicate larvae (Holmstrom et al., 1992). Moreover, some seaweeds require surface bacteria to develop properly, atypical morphology being exhibited when grown in axenic culture (Tatewaki et al., 1983). Thus, in terms of chemical ecology, it seems likely that there is a 'protective' role for some strains of epibiotic bacteria present on surfaces that release chemicals that prevent biofouling by other organisms. Chemical interactions between different species of bacteria can affect the production and secretion of antimicrobial secondary metabolites (Patterson and Bolis, 1997; Mearns-Spragg et al., 1998; Burgess et al., 1999). Thus, the study of the chemical ecology of living surfaces of marine organisms, and the symbiotic relationships between them and their microbial flora can also provide important biotechnological information.

Agriculture

Scientists are also investigating the viability of transferring characteristics inherent in marine organisms into terrestrial counterparts. For example, winter flounder contains an anti-freeze gene that allows it to survive sub-zero water temperatures. The flounder anti-freeze gene has been cloned and inserted into yeast and higher plants to determine if it will protect agricultural crops from sudden freezing. A number of strategies using recombinant DNA technology and genetic transformation have been utilized to enhance crop freezing tolerance in recent years (Cushman and Bohnert, 2000; Iba, 2002). These approaches include: (1) the over-expression of biosynthetic enzymes for osmoprotectants (such as mannitol, proline, treholose or glycine betaine), (2) the constitutive expression of stress-induced proteins (such as late embryogenesis abundant proteins or heat shock proteins, (3) altering the enzyme activity of antioxidants (such as superoxide dismutase or glutathione S-transferase) that are involved in the detoxification of active oxygen species accumulated during the stress environment, and (4) expression of transcriptional factors (such as dehydration-responsive element or CRT-binding factors) that bind to water-deficit and cold-responsive genes. Recently, it was reported that over-expression of plasma membrane-associated phospholipase could enhance freezing tolerance in *Arabidopsis* (Li et al., 2004). The freezing tolerance enhancement involves the constitutive expression of a protein kinase in an oxidative stress signalling pathway (Shou et al., 2004). Similar work is underway for salt tolerance in plants using genes from micro-algal species living in salt water.

Aquaculture

Aquaculture, which has long been practiced in Asia and is increasingly popular in the US, Europe and Latin America, will benefit tremendously from the use of new molecular tools and processes. With worldwide seafood demand projected to increase by 70% in the next 35 years and harvests from capture fisheries stable or declining, aquaculture will have to produce seven times as much seafood as it generates now to supply the global demand by 2025 (BIC, 1995). The use of modern biotechnology to intervene in the rearing process and enhance production of aquatic species holds great potential not only to meet this demand, but also to improve competitiveness in aquaculture (Liao and Chao, 1997). Biotechnology has the potential to help aquaculture enhance a cultured organism's growth rate, reproductive potential, disease resistance and the ability to endure adverse environmental conditions such as warm or cold water. Additionally, species- and serotype-specific vaccines are being developed using the techniques of modern biotechnology against the disease agents and parasites that commonly affect marine organisms.

Improving Broodstock Management Practices

Transgenic technologies can enhance growth rates and market size, feed conversion ratios, resistance to disease, sterility issues and tolerance of extreme environmental conditions. In the shrimp aquaculture sector, transgenic shrimp have been reported (Mialhe et al., 1995), but there has been no successful development for commercial culture (Bachere et al., 1995; Benzie, 1998). Through biotechnology, we can now regulate the sex of some species of fish for aquaculture (Exadactylos and Arvanitoyannis, Chapter 14 in this Volume). Researchers use pedigreed fish to produce strains, which have three sets of chromosomes, instead of the normal two, by subjecting eggs to pressure shock. This process, known as triploidy, results in non-reproductive (sterile) fish. The objective of induced triploidy in aquaculture is the inhibition of ovarian development in female triploids. Triploid female finfish are sterile and retain the market characteristics of immature fish throughout the year.

Since many strains of cultured fish have only recently removed from the wild, aquaculture broodstock selection programmes can produce significant genetic improvements (Humphries et al., 2000). Traditional selective breeding programmes have been used to develop broodstock strains with desirable characteristics such as faster growth and disease resistance. However, such stocks are time consuming and expensive to develop. The application of cost-effective DNA fingerprinting techniques facilitates broodstock management and selective breeding by linking genetic markers with quantitative traits to identify particular stocks of fish (Wang and Hill, 2000). There is increasing interest in the creation of selective breeding programmes for marine shrimp, tilapia, common carp, rohu and catla in Asia and Africa (Gupta and Acosta, 1996).

Genetic fingerprinting also offers rapid analysis techniques to minimize the inbreeding associated with selective breeding programmes in plants and animals. The isolation of DNA markers that are useful for analyzing genetic variation will become very important in ensuring genetic diversity within individual aquaculture broodstocks. Induced triploidy also offers a means of reducing or eliminating genetic interaction between farmed and wild fish. Although only a small percentage of aquaculture production currently comes from genetically improved species (Gjedrem, 1997), support for the promotion of genetic improvement programmes is well entrenched in development circles.

Improving Growth and Development

Recombinant DNA technology should be exploited to produce transgenic aquatic organisms with enhanced growth and food conversion characteristics (Miyamoto, 1994). In addition, biotechnology should be applied to improve the methods for administering hormones, nutrients and other growth-enhancing substances while enhancing understanding of the pharmacokinetics of uptake and residues of administered substances (Pullin, 1994). Since cultured fish take a long time to grow to market size relative to other sources of protein such as poultry, technologies that accelerate growth are important to aquaculture operators. Although not yet used commercially, growth enhancement is now possible using newer biotechnology techniques. One use of modern biotechnology, the microinjection of growth hormone genes into fertilized salmon eggs, has proved effective in accelerating growth (Chandler et al., 1990). Growth rates that are 30 to 60% higher than normal are achieved by using this technology.

Improving Fish Nutrition

Nutrition is an area of critical importance in aquaculture as feed represents the most significant input cost for production amounting to approximately half of the operating expense in intensive culture. Research is being carried out using biotechnology to develop vegetable oils that mimic the characteristics of fish meal in order to lessen the dependence on fish meal in fish diets and to lower production costs for fish farmers. Furthermore, cultured fish that can tolerate more plant protein means reduced use of wild fish as a protein source. Product quality improvements can be achieved through the application of biotechnology to ingredient production, e.g. development of nutritionally enhanced (for instance, omega-3-fatty acid content) fish may prove to have therapeutic benefits on human health (Metz et al., 2001). Another potential use of modern biotechnology is in the production of feeds with specific characteristics not only to enhance the rate of growth, but also to influence digestibility and other desirable characteristics such as flesh colour and texture.

Health Care

Profitability and sustainability of aquaculture depend on lowering the cost of production by learning how to identify and control bacterial and viral diseases, and through reducing feed costs by substituting plant protein for animal protein in formulated feeds.

The ability to prevent disease through vaccination or early diagnosis of diseases common in cultured fish is of great value to the aquaculturist. As aquaculture expands, the use of biotechnology to produce fish vaccines and diagnostic tests for infectious diseases is increasing. Biotechnology has contributed to the development of vaccines against enteric red mouth disease, vibriosis and furunculosis in finfish, and gaffkemia in lobsters. According to the World Health Organization (fact sheet 194 website), a lot needs to be done to reduce the overuse and inappropriate use of antimicrobials. The emphasis in disease management should be on prevention, which is likely to be more cost-effective than remedial action. This may lead to less reliance on the use of chemicals (antimicrobials, disinfectants and pesticides), which largely treat the symptoms of the problem and not the cause (Planas and Cunha, 1999).

Several alternative strategies to the use of antimicrobial in disease control have been proposed and have already been applied very successfully in aquaculture. The use of antimicrobial drugs in a major producing country such as Norway has dropped from approximately 50 metric tonnes per year in 1987 to 746.5 kg in 1997 measured as active components. During the same time, the production of farmed fish in Norway increased approximately from 5×10^4 to 3.5×10^5 metric tonnes. The dramatic decrease observed in the consumption of antimicrobial agents is mainly due to the development of effective vaccines (Subasinghe, 1997; Lunestad, 1998), which very well illustrates the potential effectiveness of the procedure. Enhancing the nonspecific defence mechanisms of the host by immunostimulants, alone or in combination with vaccines, is another very promising approach (Raa, 1996; Sakai, 1999). Yasuda and Taga (1980) had already anticipated that bacteria would be found to be useful both as food and biological control agents of fish disease and activators of nutrient regeneration in aquaculture. The addition of probiotics is now a common practice in commercial shrimp hatcheries in Mexico (Rico-Mora et al., 1998). According to Browdy (1998), one of the most significant technologies that have evolved in response to disease control problems is the use of probiotics.

Considering the recent successes of these alternative approaches, the Food and Agriculture Organization of the United Nations (Subasinghe, 1997) defined the development of affordable yet efficient vaccines, the use of immunostimulants and nonspecific immune enhancers, the use of probiotics and bioaugmentation for the improvement of aquatic environmental quality as major areas for further research in disease control in aquaculture. The use of probiotic i.e. lactic acid bacteria in biocontrol of fish diseases has been briefly discussed in Chapter 1 of Volume I of this series (Ray and Edison, 2005). The results of this research will undoubtedly help to reduce chemical and drug use in aquaculture, and will make aquaculture products more acceptable to consumers. The application of biotechnology to disease diagnostics has led to the development of very sensitive diagnostic procedures based on fluorescent antibodies, enzyme linked immunoassays and PCR. For example, the threshold sensitivity of traditional diagnostic technologies can be as much as 100,000 bacteria, whereas that for PCR tests is as low as 1-10 bacteria. Combining these newer tests with traditional diagnostic techniques provides powerful tools for controlling disease outbreaks.

Conserving Genetic Resources

There is a strong belief that biotechnology tools, such as genomics, will make a significant impact in the near future (Singer and Daar, 2002). The preservation and enhancement of biodiversity in natural systems are an important task. Therefore, the research programmes should be encouraged and supported to maintain and enhance biodiversity in aquatic systems through cultivation and stocking of aquatic species. Biotechnology can be employed in two ways to conserve genetic resources of aquatic species.

The tools of biotechnology can be used to identify and characterize important aquatic germplasms, including endangered species. Genomes of aquatic species can be analyzed and characterized, and quantitative trait loci identified (Vrijenhoek, 1998). Also, biotechnology can be applied to improve the understanding of the molecular basis of gene regulation and expression as well as sex determination, and thereby improve methods for defining species, stocks and populations. Approaches include developing marker-assisted selection technologies, improving precision and efficiency of transgenic techniques, and improving technologies for the cryo-preservation of gametes and embryos. Ultimately, stocking certain areas with selected, cultivated species and strains could help maintain biodiversity in natural aquatic ecosystems.

Enhancing Biomedical Models

Aquaculture has important purposes other than food production. As they often adapt to extreme environments, marine organisms can provide unique models for research on biological and physiological processes. Studies of the developmental, cellular and molecular aspects of marine organisms as model systems will provide insights into the basis of disease mechanisms and pathogenesis in humans. By contrast, the use of mammalian organisms as a basis for the development of some types of human disease models may be neither feasible nor cost effective.

The progress in this arena will require the sophisticated molecular biology technologies to be adapted to marine organisms in order to enhance the understanding of their biological processes. For example, approaches for gene transfer into eggs have been developed for many terrestrial organisms, though not for most marine species. This technology is needed for analyses of gene regulatory systems and gene expression. In addition, methods need to be developed for culturing tissues from marine organisms. Cultured cell lines will provide opportunities for gene transfer and gene expression studies, and enhance the usefulness of marine species as biomedical research models. This is an important research area deserving financial support.

IMPACT OF THE SEAS ON THE ENVIRONMENT

Powerful tools are being developed to elucidate the many biogeochemical cycles that determine the fate of all the life-supporting elements on Earth. Scientists are beginning to understand and manipulate the molecular genetics and biology of esoteric metabolic pathways associated with the carbon, sulphur, phosphorus, iron and other biogeochemical cycles (Boyd and Silapajarn, Chapter 8 in this Volume). For example, based on the hypothesis that iron controls photosynthesis in the oceans, immunological probes were used to show that addition of iron to open ocean water off the Galapagos Islands significantly increased energy production (Kolber et al., 1994).

Marine biotechnology will be useful in assessing the role of the oceans in affecting climate change and the global carbon cycle. Molecular techniques can facilitate and enhance the measurements of CO_2 concentrations and total CO_2 inventories are being developed for global ocean models of carbon cycling. There is compelling evidence that the exchange of dissolved and particulate materials between the continental shelf and its boundaries is a significant factor affecting the flux of CO_2 and biogenic elements within the global ocean.

CHALLENGES – MARINE BIOTECHNOLOGY

Challenges that are now being faced by marine biotechnology are parallel to those of every other development in biotechnology. The future trends in biotechnology are likely to be influenced by the advances in the other technological fields. Some of the technologies that will make great impressions on biotechnology research and development may include information and communication technologies, materials technology, cognitive technology and nanotechnology.

Genetic and Ecological Impacts

Many of the applications of marine biotechnology will have ecological benefits. For example, there will be decreased use of wild fish as a protein source in fish meal following the development of cultured fish capable of tolerating more plant protein, and the subsequent reduction in therapeutant use following the development of disease-resistant fish.

However, there are concerns and issues that must be resolved before the use of marine biotechnology is more widely accepted. Aquaculture biotechnologies, such as vaccination or monosex culture, perhaps affect only the target organisms and their ecological risk is low or at least understandable. However, transgenics, such as carp that have the rainbow trout growth hormone gene, are considered by some as new organisms for which there is little existing information relevant to their behaviour, interaction or performance in the wild nor is there any appropriate theoretical basis for prediction.

If a transgenic organism with modified traits were to interbreed with local populations of the same species, the genetic structure of the population could be affected. However, the genetic impact of escaped organisms on wild populations can be eliminated if the organisms are rendered sterile before there is any possibility of escape.

A transgenic organism could have an impact on members of the same species or of other species and could cause environmental disruption. The impacts can include competition for limited resources such as space or food, displacement from a niche, predation on various life-history stages and physical alteration of habitat. Depending on size and the frequency of escape, the impact could be local and reversible or could be widespread and environmentally damaging if it were repetitive. These impacts can occur even if the organisms are sterile.

Adequate methodology and criteria for environmental risk assessment need to be established. This information is important to allow objective evaluations and risk assessments to be performed. In the meantime, transgenics are considered new organisms for which there are no existing data on behaviour, interaction or performance in the wild.

Research and Development

The process of research priority setting should be seen to be legitimate and fair (Daniels and Sabin, 1997). Aquaculture is technology-intensive and necessitates considerable research information. Maintaining a competitive edge requires the access to practical, commercially oriented research that continues to broaden the technology and knowledge base of the sector. However, decisions regarding the goals and objectives of research initiatives must reflect the views of all stakeholders.

Marine biotechnology, while having some impressive isolated success stories, is still early in its development. The technology is at an earlier stage of development than those of the health or agriculture industries. Although much improved, there is a lack of detailed scientific knowledge about aquatic organisms and the pathogens affecting them. Our understanding of transgenesis and its consequences remains limited. We need to continue to build our basic knowledge of gene function in fish systems to allow the prediction of the consequences of novel gene constructs that will be developed in the future. Long-term fundamental research continues to need support until the sector is of sufficient size to be self-sustaining. Government-funded biotechnology research programmes promote research that supports industry competitiveness and the generation of information necessary for fulfilling government's stewardship responsibilities regarding health, safety and the environment. In addition, biotechnology being a knowledge-based activity, attracting and retaining highly skilled and highly educated staff continues to be a concern for both government and industry.

When considering biosafety, the aquatic environment is seen to have special characteristics, beyond those of terrestrial environments. Fish habitats are continuous and, therefore, lack barriers preventing the spread of introduced organisms. In addition, the continuity and movement of water favour dispersal of organisms. There is a need to develop regulations that will protect the environment, and at the same time, allow the use of biotechnology and its applications. To do this, further research to determine potential environmental impacts, researches on risk assessment, analysis and management methodologies, and research on physical and biological containment techniques are needed. Future cooperation and expertise in the environmental assessment of fish biologics (vaccines) need to be in place and readily available. Other applied research priorities include the substitution of fish meal by plant protein and better delivery of therapeutants (e.g. by microencapsulation). Possible avenues for exploration and development would be towards the promotion of in vitro potency assays for vaccines to replace the use of fish in vaccination-challenge assays and genetic improvement of broodstock.

Fishery officers in Canada are using DNA probe technology to identify the fish caught by poachers. The genetic identification of confiscated fish samples has already led to a guilty plea in the prosecution of halibut poachers. Following are the success stories:

- Development of a new, red seaweed-based sea vegetable, now being sold by a Canadian company to clients in Asia.
- Isolation and identification of marine toxins contaminating Atlantic Canadian shellfish, including work on paralytic shellfish toxins,

diarrhetic shellfish toxins, domoic acid and, most recently, spiralides.

- Development of new analytical methodologies for measuring toxins.
- Discovery of new drug leads in extracts of marine organisms.
- Development and production of a series of analytical chemistry standards and reference materials for marine toxins and some common pollutants. These are sold worldwide.

CONCLUSION

Research should focus on learning about natural functions, regulation and production of substances generated by marine microorganisms, in order to identify potentially important agents. Refined test systems should be developed in order to identify selective agents produced by marine organisms. Rapidly developing assay technology can facilitate exploration of the bioactivity of newly discovered compounds. These assay methods, which employ specific receptors for known physiological agents, require only minute amounts of a test substance and can be automated, while traditional chemical tests require considerable amounts of test material and laborious measurement processes.

The growth and international competitiveness of the aquaculture industry of any country will be determined by the size of the resource investment in research and technology development. This investment should be made through a partnership of government agencies and the private sector. The role of government is to provide leadership in supporting research to enhance knowledge in important research areas, and to facilitate the transfer of promising results and technologies to the private sector. The major research issues in aquaculture are similar to those for other agricultural sectors, but the knowledge base for aquaculture is comparatively meagre. Development of this knowledge is a particular challenge due to the diversity of cultured aquatic species and the systems for their production.

The application of biotechnology promises significant benefits to both producers and consumers of aquaculture products. The use of genetically enhanced organisms may improve production efficiency through improvements in growth rates, food conversion, disease resistance, and product quality and composition. The application of biotechnology to aquaculture may also help conserve wild species and genetic resources, and provide unique models for biomedical research.

REFERENCES

Akkermans, A.D.L., Mirza, M.S., Harmsen, H.J.M., Blok, H.J., Herron, P.R., Sessitsch, A. and Akkermans, W.M. (1994). Molecular ecology of microbes: a review of promises, pitfalls and the progress. FEMS Microbiol. Res. 15: 185-194.

Amundson, S.A., Bittner, M., Meltzer, P., Trent, J. and Fornace, A.J., Jr. (2001). Induction of gene expression as a monitor of exposure to ionizing radiation. Radiat. Res. 156: 657-661.

Armstrong, E., Mckenzie, J.D. and Goldsworthy, G.T. (1999). Aquacultures of sponges on scallops for natural products research and antifouling. J. Biotechnol. 70: 163-174.

Armstrong, E., Yan, L., Boyd, K.G., Wright, P.C. and Burgess, J.G. (2001). The symbiotic role of marine microbes on living surfaces. Hydrobiologia 461: 37-40.

Bachere, E., Mialhe, E., Noel, D., Boulo, V., Morvan, A. and Rodriguez, J. (1995). Knowledge and research prospects in marine mollusc and crustacean immunology. Aquaculture 132(1-2): 17-32.

Benzie, J.A.H. (1998). Penaeid genetics and biotechnology. Aquaculture 164: 23-47.

Bertoldo, C. and Antranikian, G. (2002). Starch hydrolyzing enzymes from thermophilic archaea and bacteria. Curr. Opinion Chem. Biol. 6: 151-160.

Bibel, M., Brettl, C., Gosslar, U., Kriegshauser, G. and Liebl, W. (1998). Isolation and analysis of genes for amylolytic enzymes of the hyperthermophilic bacterium *Thermotoga maritime*. FEMS Microbiol. Lett. 158: 9-15.

BIC. (1995). Biotechnology for the 21st Century: New Horizons. Biotechnology Information Center, U.S. Government Printing Office; Washington, DC, USA, pp. 125.

Blunt, J.W., Copp, B.R., Munro, M.H.G., Northcote, P.T. and Prinsep, M.R. (2004). Marine natural products. Nat. Prod. Rep. 21: 1-49.

Borowitzka, M.A. (1998a). Fats, oils and hydrocarbons. In: Micro-Algal Biotechnology (eds.) M.A. Borowitzka and L.J. Borowitzka, Cambridge University Press, Cambridge, UK, pp. 257- 287.

Borowitzka, M.A. (1998b). Company news. J. Appl. Phycol. 10: 417.

Boyd, C.E. and Turner, M. 1998. Coliform organisms in waters of channel catfish ponds. J. World Aquacul. Soc. 29(1): 74-78.

Boyd, K.G., Adams, D.R. and Burgess, J.G. (1999a). Antibacterial and repellent activity of marine bacteria associated with algal surfaces. Biofouling 14: 227-236.

Boyd, K.G., Mearns-Spragg, A. and Burgess, J.G. (1999b). Screening of marine bacteria for the production of microbial repellants using a spectrophotometric chemotaxis assay. Mar. Biotechnol. 1: 359-363.

Browdy, C. (1998). Recent developments in penaeid broodstock and seed production technologies: Improving the outlook for superior captive stocks. Aquaculture, 164: 3-21.

Brown, S.H., Costantino, H.R. and Kelly, R.M. (1990). Characterization of amylolytic enzyme activities associated with the hyperthermophilic archaebacterium *Pyrococcus furiosus*. Appl. Environ. Microbiol. 56: 1985-1991.

Brown, S.H. and Kelly, R.M. (1993). Characterization of amylolytic enzymes, having both (alpha)-1,4 and (alpha)-1,6 hydrolytic activity from the thermophilic archaea *Pyrococcus furiosus* and *Thermococccus litoralis*. Appl. Environ. Microbiol. 59: 2614-2621.

Bunker, B.C., Rieke, P.C., Tarasevich, B.J., Campbell, A.A., Fryxell, G.E., Graff, G.L., Song, L., Liu, J., Virden, J.W. and McVay, G.L. (1994). Ceramic thin-film formation on functionalized interfaces through biomimetic processing. Science 264: 48-55.

Burgess, J.G., Jordan, E.M., Bregu, M., Mearns-Spragg, A. and Boyd, K.G. (1999). Microbial antagonism: A neglected avenue of natural products research. J. Biotechnol. 70: 27-32.

Cardellina, J.H. (1986). Marine natural products as leads to new pharmaceutical and agrochemical agents. Pure Appl. Chem. 58: 365-374.

Carte, B.K. (1996). Biomedical potential of marine natural products. Bioscience 46: 271-286.

Chandler, D.P., Welt, M., Leung, F.C. and Richland, W.A. (1990). Development of a rapid and efficient microinjection technique for gene insertion into fertilized salmonid eggs. Technical Report of Batelle Pacific Northwest Laboratory. USA, pp.1-22.

Cheun, B.S., Endo, H., Hayashi, Y., Nagashima, Y. and Watanabe, E. (1996). Development of an ultra high sensitive tissue biosensor for determination of shellfish poisoning, tetrodotoxin. Biosen. Bioelec. 11: 1185-1191.

Clare, A.S. (1996). Marine natural product antifoulants: Status and potential. Biofouling 9: 211-229.

Cohen, Z. (1999). Chemicals from microalgae. Taylor & Francis. London, UK.

Cowan, D.A. (1997). The marine biosphere: A global resource for biotechnology. Trends Biotechnol. 15: 129-131.

Critchley, T. and Ohno, M. (1998). Seaweed resources of the world. JICA, Yokusuka, Japan.

Cushman, J.C. and Bohnert, H.J. (2000). Genomic approaches to plant stress tolerance. Curr. Opin. Plant Biol. 3: 117-124.

Daniels, N. and Sabin, J.E. (1997). Limits to health care: Fair procedures, democratic deliberation and the legitimacy problem for insurers. Philos. Public Aff. 26: 303-350.

Datta, K. and Licata, J.V. (2003). Thermodynamics of the binding of *Thermus aquaticus* DNA polymerase to primed-template DNA. Nucleic Acids Res. 31(19): 5590-5597.

Davail, S., Feller, G., Narinx, E., and Gerday, C. (1994). Cold adaptation of proteins. Purification, characterization, and sequence of the heat-labile subtilisin from the Antarctic psychrophile *Bacillus* TA41. J. Biol. Chem. 269: 17448-17453.

Davies, D.G., Chakrabarty, A.M. and Geesey, G.G. (1993). Exopolysaccharide production in biofilms: Substratum activation of alginate gene expression by *Pseudomonas aeruginosa*. Appl. Environ. Microbiol. 59: 1181-1186.

DeLong, E.F. (1992). Archaea in coastal marine environments. Proc. Natl. Acad. Sci. USA 89: 5685-5689.

DFID/FAO/NACA/GoB (Government of Bangladesh). (2001). Primary health care in small-scale, rural aquaculture. Proceedings of a Workshop held in Dhaka, Bangaladesh, 27th-30th September 2000, FAO, Rome, Italy.

Dunlap P.V. and Callahan, S.M. (1993). Characterization of a periplasmic 3':5'-cyclic nucleotide phosphodiesterase gene, *cpdP*, from the marine symbiotic bacterium *Vibrio fischeri*. J. Bacteriol. 175(15): 4615-4624.

EC (2002). Wonders of life. Stories for life sciences research (from the fourth and fifth framework programmes). Office of Official Publications of the European Communities, Luxembourg.

Edwards, P. (1998). A systems approach for the promotion of integrated aquaculture. Aquacult. Econ. Manage, 2: 1-12.

Ewart, H.S., Cole, L.K., Kralovec, J., Layton, H., Curtis, J.M., Wright, J.L.C. and Murphy, M.G. (2002) Fish oil containing phytosterol esters alters blood lipid profiles and left ventricle generation of thromboxane A2 in adult Guinea pigs. J. Nutr. 132(6): 1149-1152.

FAO. (2001). Report of the Nineteenth Session of the Coordinating Working Party on Fishery Statistics. Nouméa, New Caledonia, 10-13 July 2001. FAO Fisheries Report. No. 656. Food and Agriculture Organization of the United Nations, Rome, Italy, pp. 91.

Fenical, W. (1993). Chemical studies of marine bacteria: Developing a new resource. Chem. Rev. 93(5): 1678-1683.

Fenical, W. and Jensen, P.R. (1993). Marine microorganisms: a new biomedical resource. In: Marine Biotechnology: Pharmaceutical and Bioactive Natural Products (eds.) D.H. Attaway and O.R. Zabrosky, Plenum Press, New York, USA, pp. 419-457.

Fuhrman, J.A. and Suttle, C.A. (1993). Viruses in marine plankton. Oceanography 6: 51-63.

Fuhrman, J.A. and Campbell, L. (1998). Marine ecology: Microbial microdiversity. Nature 393: 410- 411.

Garg, S.K., Bhatnagar, A., Kalla, A. and Narula, N. (2001). In vitro nitrogen fixation, phosphate solubilization, survival and nutrient release by *Azotobacter* strains in an aquatic system. Bioresour. Technol. 80: 101- 109.

Garg, S.K. and Bhatnagar, A. (2005). Use of microbial biofertilizers for sustainable aquaculture/fish culture. In: Microbial Biotechnology in Agriculture and Aquaculture, Volume I (ed.) R.C. Ray, Science Publishers, New Hampshire, USA, pp. 261-297.

GESAMP. (1996). The contributions of science to integrated coastal management. Rep. Stud. GESAMP, 61, Joint Group of Experts on the Scientific Aspects of Marine Environmental Protection, pp. 66.

Gildberg, A. and Mikkelsen, H. (1998). Effects of supplementing the feed of Atlantic cod (*Gadus morhua*) fry with lactic acid bacteria and immunostimulating peptides during a challenge trial with *Vibrio anguilarum*. Aquaculture 167: 103-113.

Gjedrem, T. (1997). Selective breeding to improve aquaculture production. World Aquac. 28: 33- 45.

Goericke, R. and Welschmeyer, N.A. (1993). The chlorophyll-labeling method: Measuring specific rates of chlorophyll a synthesis in cultures and in the open ocean. Limnol. Oceanogr. 38: 80-95.

Goodfellow, M. and Haynes, J.A. (1984). Actinomycetes in marine sediments. In: Biological, Biochemical and Biomedical Aspects of Actinomycetes (eds.) L. Oritz-Oritz, L.F. Bojali and V. Yakoleff, Academic press, Orlando, Fl., USA, pp. 453-472.

Grommen, R. and Verstraete, W. (2002). Environmental biotechnology: The ongoing quest. J. Biotechnol. 98: 113-123.

Gupta, M.V. and Acosta, B.O. (1996). Proceedings of the Third INGA Steering Committee Meeting, Cairo, Egypt, 8-11 July 1996. International Network of Genetics in Aquaculture, Manila, The Philippines.

Hew, C.L. (1989). Transgenic fish: Present status and future directions. Fish Physiol. Biochem. 1989; 7: 409-413.

Holmstrom, C., Rittschof, D. and Kjelleberg, S. (1992). Inhibition of settlement by larvae of *Balanus amphitrite* and *Ciona intestinalis* by a surface-colonizing marine bacterium. Appl. Environ. Microbiol. 58: 2111-2115.

Hopwood, D.A. and Sherman, D.H (1990). Molecular genetics of polypeptides and its comparison to fatty acid biosynthesis. Annu. Rev. Genet. 24: 37-66.

Hoyle, B.D., Williams, L.J. and Costerton, J.W. (1993). Production of mucoid exopolysaccharide during development of *Pseudomonas aeruginosa* biofilms. Infect. Immun. 61: 777-780.

Hugenholtz, P. and Pace, N.R. (1996). Identifying microbial diversity in natural environment: A molecular phylogenetic approach. Trends Biotechnol. 14: 190-197.

Humphries, S., Ruxton, G.D. and Metcalfe, N.B. (2000). Group size and relative competitive ability: Geometric progressions as a conceptual tool. Behav. Ecol. Sociobiol. 47: 113-118.

Huston, A.L., Krieger-Brockett, B.B. and Deming, J.W. (2000). Remarkably low temperature optima for extracellular enzyme activity from Arctic bacteria and sea ice. Environ. Microbiol. 2: 383-388.

Hutchinson, G.E. (1961). The paradox of plankton. Am. Naturalist 95: 137-145.

Iba, K. (2002). Acclimative response to temperature stress in higher plants: Approaches of gene engineering for temperature tolerance. Annu. Rev. Plant Biol. 53: 225-245.

Insull, D. and Shehadeh, Z. (1996). Policy directions for sustainable aquaculture development. FAO Aquacult. Newsl. 13: 3-8.

Irwin, J.A., Alfredsson, G.A., Lanzetti, A.J., Gudmundsson, H.M. and Engel, P.C. (2001). Purification and characterisation of serine peptidase from the marine phychrophile strain PA-43. FEMS Microbiol. Lett. 201: 285-290.

ISSCAAP (International Standard Statistical Classfication of Aquatic Animals and Plants). (1999). Fisheries Information, Data and Statistical Service. Fisheries Department, FAO, Rome, Italy.

James, J. and Simpson, B.K. (1996). Application of enzymes in food processing. Crit. Rev. Food Sci. Nutr. 36: 437-463.

Jarvis, P. (2001). Biopesticides: Trends and Opportunities. Agrow Reports, PJB Publications, Surrey, UK, pp. 98.

Jha, R.K. and Zi-rong, X. (2004). Biomedical compounds from marine organisms. Mar. Drugs 2: 123-146.

Juma, C. and Konde, V. (2002). Industrial applications for biotechnology: Opportunities for developing countries. Environment 44: 23-35.

Keevil, C.W., Godfree, A., Holt, D. and Dow, C. (1999). Biofilms in the Aquatic Environment. The Royal Society of Chemistry Press, Cambridge, UK.

Knoblauch, C., Jorgensen, B.B. and Harder, J. (1999). Community size and metabolic rates of psychrophilic sulfate reducing bacteria in Arctic marine sediments. Appl. Environ. Microbiol. 65: 4230-4233.

Kolber, Z.S., Barber, R.T., Coale, K.H., Fitzwater, S.E., Greene, R.M., Johnson, K.S., Lindley, S. and Falkowski, P.G. (1994). Iron limitation of phytoplankton photosynthesis in the equatorial Pacific Ocean. Nature 371: 145-149.

Lemos, M.L., Toranzo, A.E. and Barja, J.L. (1985). Antibiotic activity of epiphytic bacteria isolated from intertidal seaweeds. Microbiol. Ecol. 11: 149-163.

Lemos, M.L., Dopazo, C.P., Toranzo, A.E. and Barja, J.L. (1991).Competitive dominance of antibiotic producing marine bacteria in mixed cultures. J. Appl. Bacteriol. 71: 228-232.

Li, W., Li, M., Zhang, W., Welti, R. and Wang, X. (2004). The plasma membrane-bound phospholipase Dä enhances freezing tolerance in *Arabidopsis thaliana*. Nature Biotechnol. 22: 427-433.

Liao, I.C. and Chao, N.H. (1997). Developments in aquaculture biotechnolgy in Taiwan. J. Mar. Biotechnol. 5: 16-23.

Liu, H., Nolla, H.A. and Campbell, L. (1997). *Prochlorococcus* growth rate and contribution to primary production in the equatorial and subtropical North Pacific Ocean. Aquat. Microb. Ecol. 12: 39-47.

Lunestad, B.T. (1998). Impact of farmed fish on the environment, presentation by representative of Norway. In: Notes of the Workshop on Aquaculture/Environment Interactions: Impacts on Microbial Ecology. Canadian Aquaculture Society, Halifax, Canada, p. 53.

Mearns-Spragg, A., Boyd, K.G., Hubble, M.O. and Burgess, J.G. (1997). Antibiotics from surface associated marine bacteria. In: Proceedings of the Fourth Underwater Science Symposium. The Society of Underwater Technology, London, UK, pp. 147-157.

Mearns-Spragg, A., Bregu, M., Boyd, K.G. and Burgess, J.G. (1998). Cross species induction and enhancement of antimicrobial activity produced by epibiotic bacteria from marine algae and invertebrates after exposure to terrestrial bacteria. Lett. Appl. Microbiol. 27: 142-146.

Mekonnen, A.E., Prasanna, R. and Kaushik, B.D. (2002). Cyanobacterial N_2 fixation in presence of nitrogen fertilizers. Indian J. Exp. Biol., 40: 854-857.

Metz, J.G., Roessler, P., Facciotti, D., Levering, C., Dittrich, F., Lassner, M., Valentine, R., Lardizabal, K., Domergue, F., Yamada, A., Yazawa, K., Knauf, V. and Browse, J. (2001). Polypetide synthases produce polyunsaturated fatty acids in both prokaryotes and eukaryotes. Science 293: 290-293.

Mialhe, E., Bachere, E., Boulo, V., Cadoret, J.P., Rousseau, C., Cedeño, V., Sorairo, E., Carrera, L., Calderon, J. and Colwell, R.R. (1995). Future of biotechnology-based control of disease in marine invertebrates. Mol. Biol. Biotech. 4: 275-283.

Miyamoto, K. (1994). In: Recombinant Microbes for Industrial and Agricultural Applications (eds.) Y. Murooka and T. Imanaka, Marcel Dekker, Inc., New York, USA, pp. 771-785.

Morita, R.Y. (1975). Psychrophilic bacteria. Bacteriol. Rev., 39: 144-167.

NACA/FAO. (2000). Report of the Conference on Aquaculture in the Third Millenium. Conference on Aquaculture in the Third Millenium, 20-25 February 2000, Bangkok, Thailand. Network of Aquaculture Centres in Asia-Pacific, Bangkok, and Food and Agriculture Organization of the United Nations, Rome, Italy, pp. 120.

Nakagawa, T., Fujimoto, Y., Uchino, M., Miyaji, T., Takano, K. and Tomizuka, N. (2003). Isolation and characterization of psychrophiles producing cold-active beta-galactosidase. Lett. Appl. Microbiol. 37: 154-157.

NBJ. (1999). Complementary and alternative medicine. Nutrition Business Journal, January 1999 (http://store.yahoo.com/nbj/jan19comalme.html).

Nix, E., Fisher, C., Vodenichar, J., and Scott, K. (1995). Physiological ecology of a mussel with methanotrophic endosymbionts at three hydrocarbon seep sites in the Gulf of Mexico. Mar. Biol. 122: 605-617.

Okazaki, T. and Okami, Y. (1972). Studies on marine microorganisms. II. Actinomycetes in Sagami Bay and their antibiotic substances. J. Antibiotics 25: 461-466.

Patterson, G.L. and Bolis, C.M. (1997). Fungal cell-wall polysaccharides elicit an antifungal secondary metabolite (phytoalexin) in the cyanobacterium *Scytonema ocellatum*. J. Phycol. 33: 54-60.

Peterson, C. and Reece, D. (1994). Toxicology, biodegradability and environmental benefits of biodiesel. In: Biodiesel' 94 (eds.) R. Nelson, D. Swanson and J. Farrell, Western Regional Biomass Energy Programme, Golden CO., USA.

Phadtare, S., Yamanaka, K. and Inouye, M. (2000). The cold shock response. In: Bacterial Stress Responses (eds.) G. Storz and R. Hengge-Aronis, American Society for Microbiology, Washington, DC, USA, pp. 33-45.

Pichard, S.L. and Paul, J.H. (1991). Detection of gene expression in genetically engineered microorganisms and natural phytoplankton productions in the marine environment by mRNA analysis. Appl. Environ. Microbiol. 57: 1721-1727.

Planas, M. and Cunha, I. (1999). Larviculture of marine fish: Problems and perspectives. Aquaculture 177: 171-190.

Prabhu, G.N. and Chandrasekharan, M. (1996). L- Glutaminase production by marine *Vibrio costicola* under solid- state fermentation using different substrates. J. Marine Biotechnol. 4: 176-179.

Pullin, R.S.V. (1994). Exotic species and genetically modified organisms in aquaculture and enhanced fisheries: ICLARM's position. Naga 17(4): 19-24.

Pulz, O. and Gross, W. (2004). Valuable products from biotechnology of microalgae. Appl. Microbiol. Biotechnol. 65: 635-648.

Raa, J. (1996). The use of immunostimulatory substances in fish and shellfish farming. Rev. Fish. Sci. 4: 229-288.

Rao, P.S. and Sudha, P.M. (1996). Emerging trends in shrimp farming. Fishing Chimes 16(3): 25-26.

Ravenschlag, K., Sahm, K. and Amann, R. (2001). Quantitative molecular analysis of the microbial community in marine Arctic sediments (Svalbard). Appl. Environ. Microbiol. 67: 387-395.

Ray, R.C., Padmaja, G., Potty, V.P. and Balagopalan, C. (1992). Microbial biotechnology for detoxification of cyanide and pesticide wastes. In: New Trends in Biotechnology (eds.) N.S. Subba Rao, C. Balagopalan and S.V. Ramakrishna, Oxford & IBH Publishing Co. Pvt. Ltd., New Delhi, India, pp. 219-248.

Ray, R.C. and Edison, S. (2005). Microbial biotechnology in agriculture and aquaculture - an overview. In: Microbial Biotechnology in Agriculture and Aquaculture, Volume I (ed.) R.C. Ray, Science Publishers, Enfield, New Hampshire, USA, pp. 1-30.

Rico-Mora, R., Voltolina, D. and Villaescusa-Celaya, J.A. (1998). Biological control of *Vibrio alginolyticus* in *Skeletonema costatum* (Bacillariophyceae) cultures. Aquacult Eng. 19: 1-6.

Roth, E., Rosenthal, H. and Burbridge, P. (2000). A discussion of the use of the responsible fisheries. Food and Agriculture Organization of the United Nations, Rome, Italy, pp. 48.

Sahm, K., Knoblauch, C. and Amann, R. (1999). Phylogenetic affiliation and quantification of psychrophilic sulfate reducing isolates in marine Arctic sediments. Appl. Environ. Microbiol. 65: 3976-3981.

Sakai, M. (1999). Current research status of fish immunostimulants. Aquaculture 172: 63-92.

Secades, P., Alvarez, B. and Guijarro, J.A. (2001). Purification and characterization of a psychrophilic, calcium-induced, growth-phase-dependant metalloprotease from the fish pathogen *Flavobacterium psychrophilum*. Appl. Environ. Microbiol. 67: 2436-2444.

Shou, H., Bordallo, P., Fan, J.-B., Yeakley, J.M., Bibikova, M., Sheen, J. and Wang, K. (2004). Expression of an active tobacco MAP kinase enhances freezing tolerance in transgenic maize. Proc. Natl. Acad. Sci. USA. 101: 3298-3303.

Singer, P. and Daar, A. (2002). Of genetic mice and men. The Globe and Mail, 16 December 2002 .

Smith, P. (1999). Report of the ACIAR-NACA-DOF Regional Workshop on Key Researchable Issues in Sustainable Shrimp Aquaculture, 28-31 October 1996, Hat Yai, Network of Aquaculture Centres in Asia-Pacific, Bangkok, Thailand.

Stierle, A. (2002). Montana invests – Building a better Montana, Montana Technology of the University of Montana 2002: 1-4 (http://www. mcu.Montana.edu/musdata/results).

Storer, A.C. (1991). Engineering of proteases and protease inhibition. Curr. Opin. Biotechnol. 2: 606-613.

Subasinghe, R. (1997). Fish health and quarantine. In: Review of the State of the World Aquaculture. FAO Fisheries Circular No. 886. Food and Agriculture Organization of the United Nations, Rome, Italy, pp. 45-49.

Suttle, C.A., Chan, A.M. and Cottrell, M.T. (1990). Infection of phytoplankton by viruses and reduction of primary productivity. Nature 347: 467-469.

Tan, J.D., Tomita, F. and Asano, K. (2005). Actinomycetes biotechnology in agriculture and aquaculture. In: Microbial Biotechnology in Agriculture and Aquaculture, Volume I (ed.) R.C. Ray, Science Publishers, Enfield, New Hampshire, USA, pp. 299-334.

Tatewaki, M., Provasoli, L. and Pinter, I.J. (1983). Morphogenesis of *Monostroma oxysperma* (Kutz.) Doty (Chlorophyceae) in axenic culture, especially in bialgal culture. J. Phycol. 19: 404-416.

Todd, J.T., Zimmerman, R.C., Crews, P. and Alberte, R.S. (1993). Isolation of *p*-sulfoxycinnamic acid from *Zostra marina* L. (eelgrass) and the antifouling activity of phenolic acid sulfate esters. Phytochemistry 34: 401-404.

Toma, C. and Honma, Y. (1996). Cloning and genetic analysis of the *Vibrio cholerae* aminopeptidase gene. Infect. Immunol. 64(11): 4495-4500.

Tonkova, A. (2006). Microbial starch converting enzymes of the alpha amylase family. In: Microbial Biotechnology in Horticulture, Volume 1 (eds.). R.C. Ray and O.P. Ward, Science Publishers, Enfield, New Hampshire, USA, pp. 421-472.

Towner, K.J. and Cockayne, A. (1993). Molecular Methods for Microbial Identification and Typing. Chapman & Hall, London, USA, pp. 202.

Umezawa, H. (1998). Low molecular weight enzyme inhibitors and immunomodifiers. In: Actinomycetes in Biotechnology (eds.) M. Goodfellow, S.T. Williams and M. Mordarski, Academic Press, London, UK, pp. 285-325.

Vermeij, P. and Blok, D. (1996). New peptide and protein drugs. Pharm. World Sci. 18: 87-93.

Voskerician, G., Matthew, S.S., Rebecca, S.S., von Recum, H., James, M.A., Michael, J.C. and Robert, L. (2003). Biocompatibility and biofouling of MEMS drug delivery devices. Biomaterials 24: 1959-1967.

Vrijenhoek, R.C. (1998). Conservation genetics of freshwater fish. J. Fish Biol. 53 (Suppl. A): 394-412.

Wang, J. and Hill, W.G. (2000). Marker-assisted selection to increase effective population size by reducing Mendelian Segregation Variance. Genetics 154: 475-489.

Ward, O.P. (1991). Bioprocessing. Open University Press. Milton Keynes, London, UK.

Ward, O.P. and Ray, R.C. (2006). Microbial biotechnology in horticulture - An overview. In: Microbial Biotechnology in Horticulture, Volume 1 (eds.) R.C. Ray and O.P. Ward, Science Publishers, Enfield, New Hampshire, USA, pp. 345-396.

Ward, O.P. and Singh, A. (2005). Microbial technology for bioethanol production from agriculture and forestry wastes. In: Microbial Biotechnology in Agriculture and Aquaculture, Volume I (ed.) R.C. Ray, Science Publishers, Enfield, New Hampshire, USA, pp. 449-479.

Woese, C.R., Kandler, O. and Wheelis, M.L. (1990). Towards a natural system of organisms: Proposal for the domains Archaea, Bacteria, and Eucarya. Proc. Natl. Acad. Sci., USA 87: 4576-4579.

Yasuda, K. and Taga, N. (1980). A mass culture method for *Artemia salina* using bacteria as food. Mer 18: 53-62.

Yoo, J.S., Cheun, B.S., Park, I.S., Song, Y.C., Seo, Y., Kim, N.G., Shin, H.W. and Lee, J.H. (2004). Use of sodium transfer tissue biosensor (STTB) for monitoring of marine toxic organisms. J. Environ. Biol. 25: 431-436.

Zahran, H.H. (2005). Rhizobial nitrogen fixation in agriculture: Biotechnological perspectives. In: Microbial Biotechnology in Agriculture and Aquaculture, Volume I (ed.) R.C. Ray, Science Publishers, Enfield, New Hampshire, USA, pp. 71-100.

Zeng, R., Zhang, R., Zhao, J. and Lin, N. (2003). Cold-active serine alkaline protease from the psychrophilic bacterium *Pseudomonas* strain DY-A: Enzyme purification and characterization. Extremophiles 7: 335-337.

Zilinskas, R.A., Colwell, R.R., Lipton, D.W. and Hill, R. (1994). The Global challenge of marine biotechnology: A status report on marine biotechnology in the United States, Japan and other countries. National Sea Grant College Program and Maryland Sea Grant College, Madison, USA.

11

Microbial Degradation of Pesticides: Atrazine as a Case Study

Moshe Herzberg[*1], Carlos G. Dosoretz and Michal Green*

INTRODUCTION

Pesticides are chemicals, mostly synthetic organic compounds, used to control weeds and unwanted or harmful pests. Pesticides can be divided into ten major categories according to the target organisms they are designed to control (Table 11.1), and these categories can be further subdivided into a total of 22 groups according to their ability to kill, repel or regulate growth. The groups span many chemical classes, and the chemical structures of the various pesticides differ within categories as well as between categories. Based on their chemical composition, over 1,400 compounds are included in the classified lists of pesticides of the Compendium of Pesticide Common Names, of which some display activity towards more than one target organism (Wood, 2003). The five largest groups of pesticides include insecticides (targeting harmful or destructive insects), herbicides, fungicides, rodenticides and acaricides (Rittmann and Mc Carty, 2001).

Despite their benefits, pesticides have a wide range of toxic side effects with potential hazard to the environment. Most pesticides are toxic to human beings, and the toxicity can vary widely within each group of pesticides. The WHO has classified their toxic effects from class I (extremely hazardous) to class III (slightly hazardous) and also "active ingredients unlikely to present acute hazard". Most class I technical grade

Faculty of Civil and Environmental Engineering, Technion – Israel Institute of Technology, Haifa 32000, Israel, [1]*Present address: Evironmental Engineering Program, Department of Chemical Enginnering, Yale Universiy, P.O. 208282, 9 Hillhouse Avenue, New Haven, CT 06520-8286, USA*

**Corresponding author: E-mail: moshe.herzberg@yale.edu*

Table 11.1. Classification of common pesticides according to the target organism

Pesticide	Target organism
Acaricides	Mite (acarose) killers
Algicides	Control of algae
Avicides	Control of birds
Bactericides	Destruction of harmful microorganisms (disinfectants)
Fungicides	Control of fungi
Herbicides	Plants or weeds killers
Insecticides	Insect killers
Molluscicides	Control of snails or slugs
Nematicides	Control of nematodes
Rodenticides	Rats, mice, and other rodents killers

pesticides are banned or strictly controlled in the regulated industrialized world (Eddleston et al., 2002). Biologically-based pesticides such as pheromones and microbial pesticides are becoming increasingly popular and often are safer than traditional chemical pesticides.

Environmental risk (ER) quantifies the fate of synthetic organic pesticides in soil and water ecosystems as a function of pesticide chemical properties and biodegradability. ER comprise the following risk factors: (1) *persistence* (P), indicating how long the pesticide remains active in the environment; (2) *mobility* (M) indicating how easily the pesticide can move from where it is applied; (3) *nontarget toxicity* (NTo), indicating how toxic the pesticide is to organisms other than the pest; and (4) *usage* (U), indicating the amount applied per cultivated area per year, mode and timing of application. A simple mathematical expression of the environmental risk ($ER = P \times M \times NTo \times U$) shows that it is minimized (e.g. reaches its lowest level) when any of the risk factors is close to zero.

After a pesticide is applied to a cultivated field, it meets a variety of fates. Some may be lost to the atmosphere through volatilization, others carried away to surface waters by runoff, or broken down in the sunlight by photolysis. Pesticides in soil may be taken up by plants, degraded into other chemical forms, or leached downward, possibly to groundwater. The remaining is retained in the soil and continues to be available for plant uptake, degradation, or leaching. How much each individual pesticide meets any of these fates depends on many factors, such as the properties of the pesticide, including biodegradability; the properties of the soil, including microorganisms; the site conditions, including climate; and management practises (Trautmann et al., 1998).

Pesticide use has increased dramatically over the last four decades to an estimated 5.5×10^5 tons of active ingredient used in the United States and

2.6×10^6 tons used globally during 1995 (Asplin, 1997). The global market of pesticides reached a maximum in 1996 doing a business worth US$ 30×10^9, but a continuous decline has been observed since then. In 2002 the global market performance for crop protection products (excluding sales of herbicide tolerant and insect resistant seed) reached US$ 25.15×10^9 distributed as follows: North America 31.0%; West Europe 21.9%; East Europe 4.7%; East Asia 21.9%; Latin America 15.9% and Rest of World 4.6%. Of these sales 49.7% were herbicides, 21.6% fungicides, 25.3% insecticides and 3.4% others (Crop Protection Association, 2003a, b; European Crop Protection Association, 2003). The decreased usage of pesticides is primarily due to increasing pest resistance that has resulted in lower yields and a resurgence of vector-borne diseases such as malaria. At the same time, the many related health and environmental costs of intensive pesticide use have become starkly apparent.

Herbicides

Herbicides are by far the most commonly used pesticides in the United States and they account for about 60% of all pesticide sales (Asplin and Grube, 1998; CropLife America, 2003). US farmers spent an estimated $5.61 billion annually on herbicides, plus another $1.1 billion in application costs in 1997 (Ware, 2000). The heavy use of herbicides is confined to North America, Western Europe, Japan, and Australia. Without the use of herbicides, it would have been impossible to mechanize fully the production of cotton, soybeans, sugar beets, all grains, potatoes, and corn.

Herbicides are chemicals that adversely affect the physiological activity and development of plants, and are used to control vegetation by causing death or suppressing growth. Herbicides can be grouped on the basis of their chemical structure and physiological action, and on the timing and method of their application. Chemicals with similar structure usually produce the same type of physiological reaction in plants, and control similar species. According to their *use*, herbicides are classified into:

- *Selective herbicide* – implies that certain weeds are killed but most desirable plants are not significantly injured.
- *Non-selective* – refers to chemicals that are generally toxic to plants.

According to their *mode of activity*, herbicides are classified into:

- *Contact herbicides – foliage applied*: control weeds by direct contact with plant parts. Only the plant part contacted is controlled and, therefore, good coverage is necessary.
- *Translocated herbicides – foliage applied*: products move through the entire plant system in both the water stream and the food stream and affect the active growth centers.

- *Root or emerging shoot absorbed herbicides – soil applied*: referred to as residual herbicides.

The length of time the soil remains relatively weed-free depends upon the chemical used, amount applied, rainfall, soil type, and the plant species invading the treated area. Soil residual herbicides generally have little effect on plants when sprayed on foliage. The main effect is when they are absorbed by the root system and translocated throughout the plant. All may be *selective* or *nonselective* (Brooks, 2000; http://www.cnr.uidaho.edu/extforest/TPC14.pdf).

Classification of herbicides according to their *chemical active group* includes 40 different classes such as amide, benzoic acid, benzoylcyclohexanedione, carbamate, dinitroaniline, dinitrophenol, diphenylether, halogenated aliphatic, imidazolinone, nitrile, organophosphorus, phenoxy, triazine, uracil and others (Wood, 2003).

Herbicide classification according to their *site of chemical action* includes 27 different classes (Mallory-Smith and Retzinger Jr., 2003). The *s*-triazine-ring herbicides, which constitute the major group of the inhibitors of photosystem II (group 5), are among the most environmentally prevalent in global use today.

S-TRIAZINES

S-triazine compounds are characterized by a symmetrical substituted hexameric ring consisting of alternating carbon and nitrogen atoms:

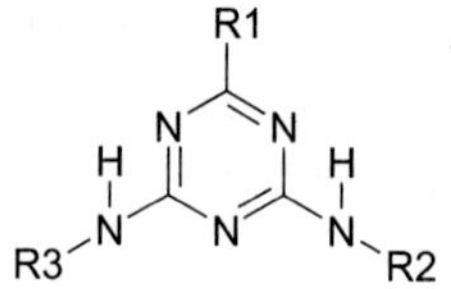

s-triazine

S-triazine herbicides are subdivided into three general types: chloro *s*-triazines, methylthio *s*-triazines, and methoxy *s*-triazines:

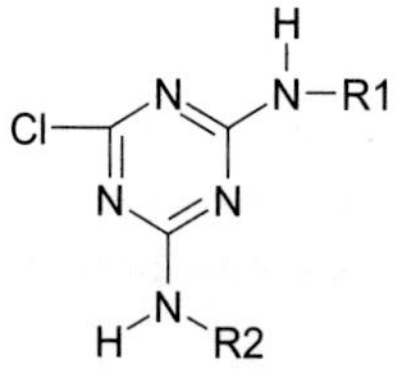

Chlorotriazine

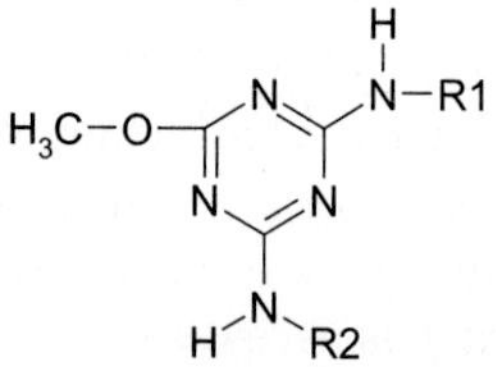

Methoxytriazine

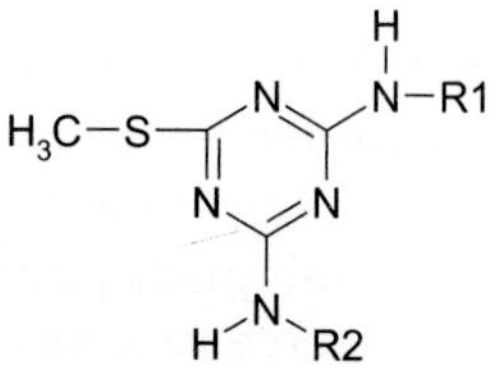

Methylthiotriazine

Table 11.2. Inhibitors of photosynthesis at photosystem II site A [*s*-triazine group noted in italics; Mallory-Smith and Retzinger (2003), Weed Technol. 17: 605–619]

Chemical Family		Common name	Trade name
Phenyl carbamate		Desmedipham	Betanex
		Phenmedipham	Spin-Aid
Pyridazinone		Pyrazon	Pyramin
		Ametryn	Evik
s-Triazine		*Atrazine*	Atranex, Various
		Cyanazine	Bladex
	Chlorotriazine	*Propazine*	Milo Pro
		Simazine	Princep, Various
		Terbuthylazine	#
		Trietazine	#
	Methoxytriazine	*Prometon*	Pramitol
		Terbumeton	#
		Desmetryn	#
	Methylthiotriazine	*Prometryn*	Caparol, Various
		Simetryn	#
Triazinone		Hexazinone	Velpar
		Metamitron	#
		Metribuzin	Sencor, Lexon
Triazolinone		Amicarbazone	BAY MKH 3586, BAY 314666
Uracil		Bromacil	Hyvar X

Not sold in the USA

S-triazines are primarily used as herbicides acting by binding to the quinone-binding protein in photosystem II and inhibiting photosynthetic electron transport. They constitute the major group of its class, as shown in Table 11.2 (Mallory-Smith and Retzinger Jr., 2003).

Two of the most commonly used *s*-triazines are atrazine (2-chloro-4-ethylamino-6-isopropylamino-*s*-triazine) and cyanazine (2-chloro-4-ethylamino-6-(1-cyano-1-methylethylamino)-*s*-triazine):

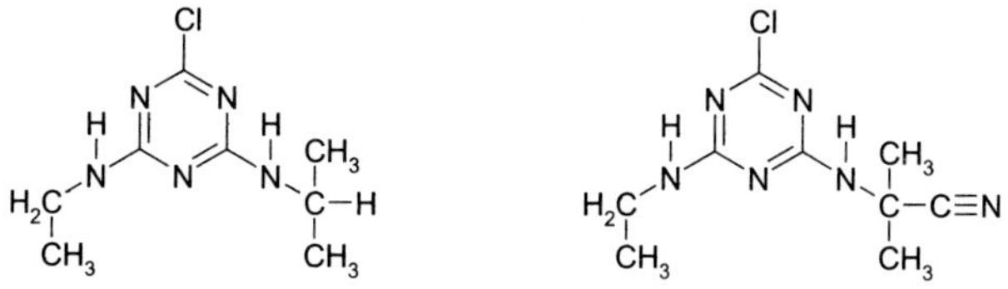

Atrazine   Cyanazine

Atrazine is a selective triazine herbicide used to control pre- and post-emergence broadleaved and grassy weeds in major crops such as corn, sorghum, sugarcane, pineapple, christmas trees, and other crops, and in conifer reforestation plantings. It is also used as a nonselective herbicide on non-cropped industrial lands and on fallow lands (Extonet, 1996; Ralebitso et al., 2002).

The fate of these compounds in the environment is directly correlated with the ability of microbes to metabolize them. *S*-triazines with $R_{1,2,3}$ groups of hydroxyl (cyanuric acid) or amino substitutes (melamine), are moderately biodegradable (Wackett and Hershberger, 2001).

Atrazine has been classified as a class C/possible human carcinogen (Loprieno et al., 1980). Confirmation for this classification was observed by Biradar and Rayburn (1995) who observed damage in ovary cells of hamsters exposed to atrazine. The EPA and European community standards of drinking water for this herbicide are 3 and 0.1 $\mu g\ L^{-1}$, respectively. Atrazine is moderately soluble in water (solubility of approximately 30 mg L^{-1}) and relatively persistent in soils (half-life in soil of between 15 and 100 days) (Protzman et al., 1999). Atrazine dealkylation metabolites, to be elaborated further in this chapter, are also regulated and their fate is also of grave concern (Brouwer et al., 1990; Kolpin et al., 1998). Chemical hydrolysis followed by biodegradation was considered to be the major mechanism for atrazine breakdown in soils as in aquatic environments. Hydrolysis of atrazine is rapid in acidic or basic environments, but is slower at neutral pH and addition of organic material increases the rate of hydrolysis. Atrazine is moderately to highly mobile in soils with low clay or organic matter content and does not strongly adsorb to sediments. As it does not adsorb strongly to soil particles and has a lengthy half-life, it has a high potential for groundwater contamination and is the second most common pesticide found in water wells. Bioconcentration and volatilization of atrazine are not environmentally important (Extonet, 1996; Ralebitso et al., 2002).

Due to the large amount of information published regarding the degradation of *s*-triazines in general and atrazine in particular, this chapter reviews in the first part only the latest publications regarding atrazine metabolism by different bacteria, the genetic diversity of atrazine degrading genes and the enzymes involved in atrazine mineralization. The second part of this chapter concentrates on the basic concepts underlying the microbial degradation of atrazine in bioreactors for its removal from polluted water.

BIODEGARADABILITY OF ATRAZINE

Even though a number of soil bacteria isolates found prior to the 1990s were shown to grow on *s*-triazines compounds (including cyanuric acid, melamine, cyclopropyl melamine and diethyl simazine) as the sole nitrogen source, none of the strains showed growth on atrazine (Cook et al., 1984; 1985). Detailed reviews and the theoretical basis for bacterial metabolism of *s*-triazine compounds is given by Cook (1987) and Erickson et al., (1989). Evidence for partial biodegradation of atrazine by oxidative dealkylation was shown by $^{14}CO_2$ evolution from several soil fungi incubated with ^{14}C-labeled ethyl or isopropyl atrazine (Kaufman and Blake, 1970). Chlorinated metabolites of atrazine were also identified in soil suggesting that the major atrazine biodegradation mechanism was biological oxidative dealkylation of the molecule (Li and Felbeck, 1972). Studies with *Rhodococcus* strains further supported the claim and showed that atrazine was partially biodegraded via oxidative dealkylation reactions. The oxidative dealkylation reactions were found to be intiated by cytochrome P-450 constituting atrazine monooxygenase and forming deisopropyl-atrazine (DIA) or deethyl-atrazine (DEA) (Behki and Khan, 1994; Nagy et al., 1995; Ellis et al., 2001). For the DEA pathway, the end product is probably 2-chloro-4-hydroxy-6-amino-1,3,5-triazine as no further mineralization was reported. For the DIA pathway, the degradation continues with either the above oxidation to the final product 2-chloro-4-hydroxy-6-amino-1,3,5-triazine, or hydrolytic reactions yielding the intermediary metabolite cyanuric acid. Dehalogenation of the dealkylated atrazine metabolites was found to occur by *s*-triazine hydrolase (TrzA) (Fig. 11.1) isolated and purified from *Rhodococcus corallinus* (Mulbry, 1994). While this enzyme enables the deamination of melamine, it failed to catalyze deamination or dechlorination of atrazine. Biodegradation of atrazine starting with oxidative dealkylation is shown in Fig. 11.1.

In the early 1990s, complete and fast biodegradation of atrazine and its dealkylated metabolites by a few bacterial strains, was reported. The isolated strains catabolize atrazine and its metabolites as the sole nitrogen source while releasing *s*-triazine ring carbon atoms as carbon dioxide (Mulbry, 1994; Mandelbaum et al., 1995; Radosevich et al., 1995; Struthers et al., 1998; Topp et al., 2000a, b). Most of the strains isolated transformed atrazine to hydroxyatrazine via hydrolytic dechlorination. Hydroxyatrazine is not a regulated chemical and its adsorption to soil is stronger than atrazine. It is also not known to be toxic to mammals (Clay and Koskinen, 1990). Therefore, technological applications of atrazine dechlorination have a major environmental importance.

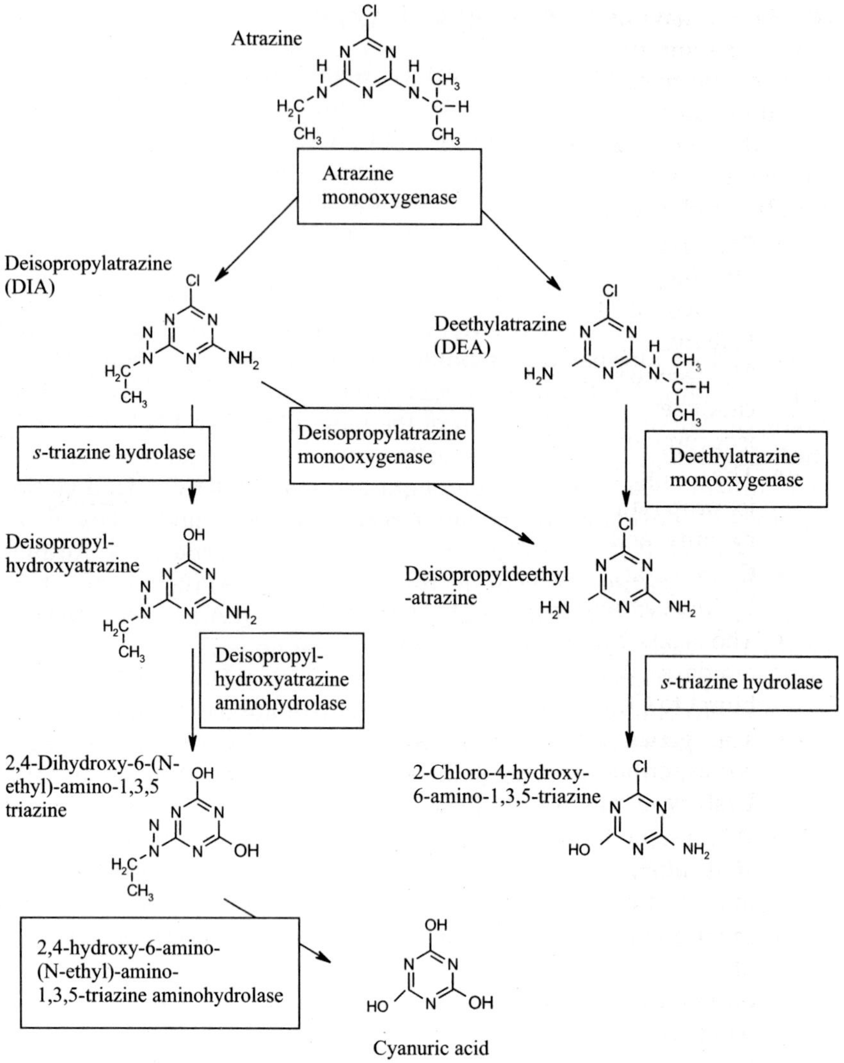

Fig. 11.1. Atrazine degradation by *Rhodococcus* spp. NI86/21, TE1; *Pseudomonas* spp. *192, 194; Streptomyces* sp. *PS1/5, Rhodococcus corallinus* NRRLB-15444R, and *Nocardia* sp. (http://umbbd.ahc.umn.edu/atr/atr_map.html, The University of Minnesota Biocatalysis/Biodegradation Database)

One of the first and fastest atrazine dechlorinating bacteria found was the *Pseudomonas* sp. strain ADP isolated and characterized by Mandelbaum et al., (1993, 1995). The atrazine degradation rate by the bacteria was several orders of magnitude higher than that reported by other researchers (Vink and van der Zee Sjoerd, 1997; Crawford et al., 1998), while leaving no toxic metabolites. *Pseudomonas* sp. strain ADP

(*P.* ADP), having a constitutive hydrolytic pathway for atrazine mineralization, utilizes the herbicide as its sole nitrogen source under aerobic conditions (citrate was used as the carbon and energy source). Fig. 11.2 illustrates constitutive degaradation of atrazine. The pathway is defined by six enzymatic steps with six different enzymes, which are encoded on a self-transmissible plasmid, *p*ADP-1 (de Souza et al., 1998a), a 108 kb complete sequenced plasmid (Martinez et al., 2001).

- The first enzyme, AtzA, catalyzes the hydrolytic dechlorination of atrazine, yielding hydroxyatrazine. The dechlorination of atrazine removes its herbicidal properties and reportedly decreases its toxicity.
- The second enzyme, AtzB (hydrohyatrazine ethylaminohydrolase), catalyzes hydroxyatrazine deamidation, yielding *N*-isopropylammelide.
- The third enzyme, AtzC (or *N*-isopropylammelide isopropylaminohydrolase), transforms *N*-isopropylammelide to cyanuric acid and isopropylamine.
- Cyanuric acid degradation occurs by three other enzymes encoded by three genes, *atz*D, *atz*E, *atz*F, located on the same plasmid.
- The AtzD hydrolyzes cyanuric acid to biuret and the next two amidases, AtzE and AtzF, are involved in the further catabolism of biuret to CO_2 and NH_3.
- The genes *atz*DEF were found by RT-PCR analysis to be cotranscribed on a single mRNA and are possibly regulated by LysR-type transcriptional regulator (Martinez et al., 2001).
- While *atz*DEF genes are organized in an operon-like structure, the *atz*A, *atz*B, *atz*C genes are dispersed on *p*ADP-1 and flanked by many transposable elements (Martinez et al., 2001).
- In addition, there is no evidence for regulatory elements upstream of the *atz*ABC genes, and transcription of *atz*A and *atz*B was observed by northern hybridization without atrazine as inducer (de Souza et al., 1998b).
- However, overexpression of *atz*ABC in *Pseudomonas* ADP and in *Chelatobacter heintzii* was observed recently only at very high concentration of atrazine of 55 mg L^{-1} in comparison to the absence of atrazine in the growth media (Devers et al., 2004).

DEGRADATION OF ATRAZINE BY DIFFERENT BACTERIA AND GENETIC DIVERSITY OF ATRAZINE DEGRADING GENES

The recent simultaneous reports of various atrazine-mineralizing pure cultures by different research groups (Yanze-Kontchou et al., 1994;

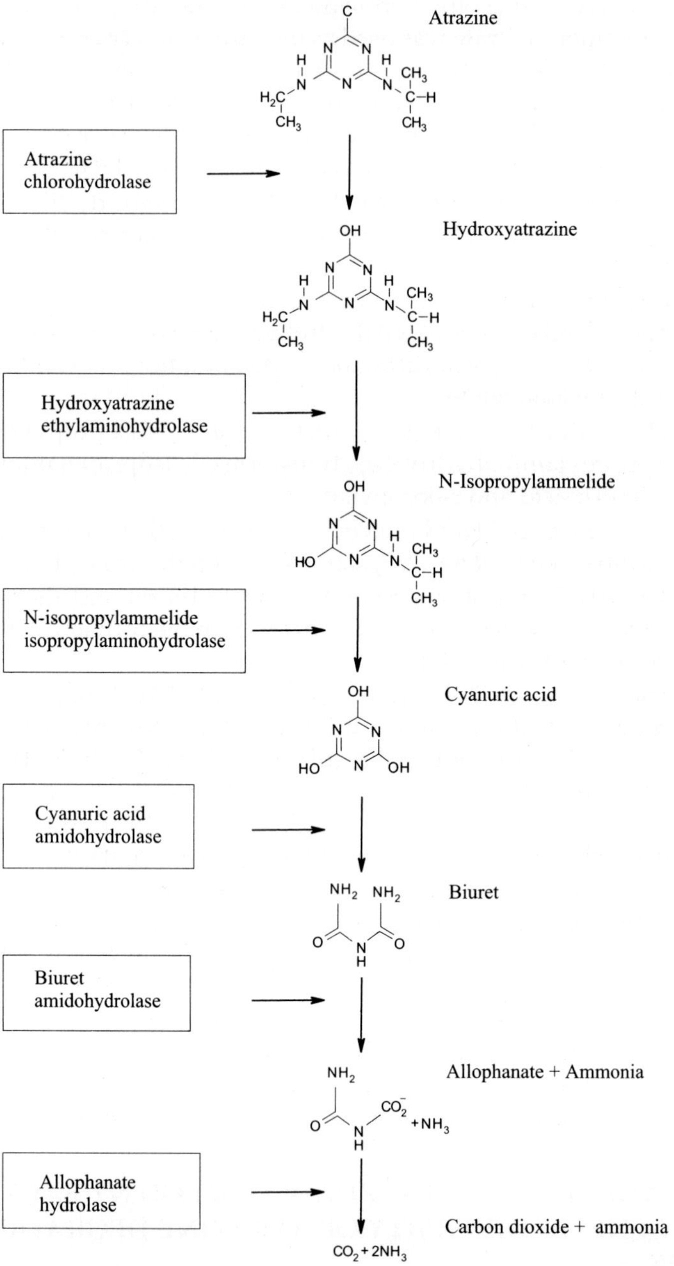

Fig. 11.2. Atrazine degradation by *Pseudomonas* sp. ADP, *Ralstonia* sp. M91-3, *Clavibacter* sp., *Agrobacterium* sp. J14a, and *Alcaligenes* sp. SG1 (http://umbbd.ahc.umn.edu/cya/cya_map.html, The University of Minnesota Biocatalysis/Biodegradation Database).

Mandelbaum et al., 1995; Radosevich et al., 1995; Bouquard et al.,1997) around the world after years of incomplete biodegradation of the atrazine molecule suggest a recent evolution in atrazine catabolism (Wackett et al., 2002). Support for this supposition is the high similarity of encoded enzymes which catabolize atrazine to cyanuric acid (de Souza et al., 1998b), a potential nitrogen source of many soil bacteria (Cook et al., 1985; Cook, 1987). A range of genera including *Clavibacter, Alcaligenes* SG1, *Agrobacterium* J14a, *Ralstonia* M91-3 isolated from different areas in the United States were found to express virtually identical homologs of the *atz*ABC genes.

de Souza et al., (1998a) showed that the average sequence identity of genes encoding enzymes for the catabolism of other commercially chlorinated compounds in different bacteria may range from 25 to 56% while in contrast, there is at most a 1% sequence difference within the *atz*ABC genes. The small degree of diveregence in the *atz*ABC genes versus other catabolic genes again suggests that the *atz*ABC genes were recently distributed globally from a single ancestor.

de Souza et al., (1998b) also showed that the *p*ADP-1 is a self-transmissible plasmid, an observation that suggests a potential molecular mechanism for the dispersion of the *atz*ABC genes to other soil bacteria. Analysis of plasmids in various atrazine-catabolizing bacteria show that different sized plasmids were found to harbor atrazine catabolism genes. This indicates that whole plasmid transfer is not the sole mechanism responsible for the spread of atrazine degradation genes (Wackett et al., 2002). The transposon sequences flanking the non-contiguous atrazine degradation genes (Martinez et al., 2001) also suggest that horizontal gene transfer and transposition may have contributed to the spread of atrazine degradation genes among bacteria (Seffernick et al., 2001a, b).

An example of the global spread of the *atz*A gene is illustrated by the atrazine-degrading bacterium identified as *Rhizobium* sp. strain PATR isolated from an agricultural soil treated with the herbicide (Bouquard et al., 1997). A comparison of amino acid sequences revealed that the hydrolase fragment of PATR enzyme has a 92% sequence identity with atrazine chlorohydrolase (AtzA) from *Pseudomonas* sp. strain ADP.

Another atrazine degrading gram-positive isolate was obtained from agricultural soils and identified on the basis of 16S ribosomal DNA sequencing as *Nocardioides* sp. strain (Topp et al., 2000a). The *Nocardioides* sp. strain converted atrazine through hydroxyatrazine to the end product *N*-ethylammelide by means of an atrazine chlorohydrolase and a hydroxyatrazine *N*-isopropylaminehydrolase. Although the *Nocardioides* atrazine degradation pathway appears to be identical to that recently reported for *Pseudomonas* ADP (de Souza et al., 1998a), the genes encoding the enzymes were not (*atz*ABC genes were undetectable by hybridization in *Nocardioides*). It was suggested that a heretofore undetected diversity in

Nocardioides genes is responsible for the same hydrolytic mechanism of *s*-triazine transformation found in *Pseudomonas* ADP.

In another study of Topp et al., (2000b), two of the soils which yielded *Nocardioides* sp. also yielded a *Pseudaminobacter* sp. that mineralized atrazine and contained hydrolases encoded by *atz*ABC genes. In this study, *Pseudaminobacter* sp. was reported to possess the ability to utilize atrazine as energy source (ethylamine and isopropylamine side chains) and without dechlorination of the molecule. In addition, the *Pseudaminobacter* were found to succesfully degrade many other *s*-triazines such as deethylatrazine, deisopropylatrazine, simazine, propazine, terbutylazine and premetryn.

The strain *Clavibacter michiganensis* ATZ1, part of an atrazine-mineralizing consortium isolated from agricultural soil (Alvey et al., 1995) is another example of a bacteria found to possess sequences homologous to the first three genes, *atz*ABC.

The existence of parts of *atz*ABC genes in different bacterial species enables the consecutive atrazine catabolism by a heterogeneous microbial consortium. A study carried by de Souza et al., (1998c) showed that two strains in bacterial consortium, *Clavibacter michiganese* ATZ1 and *Pseudomonas* sp. strain CN1 (mentioned earlier), collectively mineralized atrazine. *Clavibacter michiganese* ATZ1 contained homologs to the *atz*ABC genes and *Pseudomonas* sp. strain CN1 contained homologs to *atz*A and *atz*C, but not to *atz*B. *Clavibacter michiganese* ATZ1 did not metabolize cyanuric acid, which was metabolized by *Pseudomonas* sp. strain CN1. Atrazine degradation was faster by the bacterial consortium than by *Clavibacter michiganese* ATZ1 alone.

The recent evolution of atrazine chlorohydrolase (AtzA) was suggested in a research study of Seffernick and Wackett (2001a) and Seffernick et al., (2001b) by comparing the deamination of melamine (an industrial product used since the early 1900s) by melamine deaminase (TriA) for dechlorination of atrazine by atazine chlohydrolase (AtzA). Melamine was reclassified to biodegradable status in the 1960s after 30 years of being considered non-biodegradable. The two proteins showed an outstanding similarity, with 466 of 475 amino acids in identical sequence. Remarkably, the nine amino acid difference was exactly related to the nine nucleotide difference existing between *tri*A and *atz*A genes with no silent mutations in the two proteins. The results prompt the speculation that an intense selective pressure operating over a short time period could cause the evolution of melamine deaminase to atazine chlohydrolase.

ATRAZINE DEGRADING ENZYMES

The understanding of bacterial atrazine metabolism has been greatly advanced by isolating the steps of this process by cloning, sequencing and

Table 11.3 Atrazine-hydrolyzing microorganisms, recently isolated from geographically separated sites exposed to atrazine

Strain	Holomogs to *atz*ABC genes	Location where isolated	Reference
Pseudomonas ADP	+	Agricultural-chemical Dealership, Little Falls, MN	Mandelbaum et al., (1995)
Ralstonia M91-3	+	Agricultural soil, Ohio	Radosevich et al., (1995)
Clavibacter	+	Agricultural soil, Riverside, Calif	de Souza et al., (1998c)
Agrobacterium J14a	+	Agricultural soil, Nebraska	Struthers et al., (1998)
Alcaligenes SG1	+	Industrial settling pond, San Gabriel, La	de Souza et al., (1998b)
Rhizobium sp.	+	Doubs and Jura regions, France	Bouquard et al., (1997)
Rhodococcus corallinus NRRL B-15444R.	- (*trz*A)	Ontario and Quebec, Canada	Mulbry (1994)
Nocardioides sp.	–	Ontario and Quebec, Canada	Topp et al., (2000a)
Pseudaminobacter sp.	+	Outskirts of Paris, France	Topp et al., (2000b)
Clavibacter michiganese ATZ1	+	Handford, California	Alvey et al., (1995)
Pseudomonas sp. Strain CN1	+	Handford, California	Alvey et al., (1995)

expressing, in *Escherichia coli,* the genes involved in this pathway. The enzymes responsible for mineralization of atrazine through six degradation steps belong to the amidohydrolases superfamily of enzymes.

Amidohydrolase Superfamily

Amidohydrolases are distributed throughout the three domains of living organisms: Eubacteria, Archaea, and Eucarya. Members of the superfamily catalyze the hydrolysis of amides or the C-N bond of amines. The amidohydrolases are characterized by a conserved $(\beta\alpha)_8$ barrel structure (Holm and Sander, 1997; Thoden et al., 2001). Generally, members of this superfamily have conserved metal binding ligands and a common hydrolytic mechanism in which one or two metals are responsible for activating water for nucleophilic attack on the substrate, when the amino acids serving as metal ligands are maintained across the superfamily. Most of the enzymes in this family catalyze amide bond hydrolysis and hydrolytic displacement of amino groups from heterocyclic ring substrate. The majority of reactions catalyzed by the superfamily involve the hydrolytic removal of amino groups from purine and pyrimidine rings (adenosine deaminase and cytosine deaminase), or amide bond hydrolysis reactions (Seffernick and Wackett, 2001a, b; Seffernick et al., 2001b, 2002a, b). Sequence comparisons of atrazine degrading genes revealed that atrazine degrading enzymes belong to the amidohydrolase superfamily (Sadowsky et al., 1998).

Atrazine chlorohydrolase (AtzA)

Atrazine chlorohydrolase catalyzes the transformation of atrazine to hydroxyatrazine (de Souza et al., 1995):

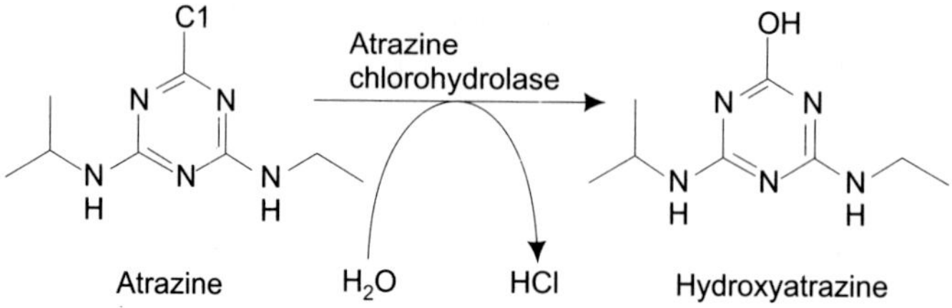

The AtzA enzyme from *Pseudomonas* strain ADP has been purified from cell-free extracts of *E. coli* (de Souza et al., 1996). The molecular weight of the AtzA holoenzyme is estimated to be 245,000. Sequence analysis showed that the protein is homologous to metalloenzymes within the amidohydrolase protein superfamily. The enzyme was suggested to have a mononuclear five-coordinate metal center that natively binds Fe(II) at the active site (Seffernick et al., 2002a, b).

The conversion of atrazine to hydroxyatrazine is a hydrolytic reaction as demonstrated by the incorporation of ^{18}O from ^{18}O-H_2O into the hydroxyl group of the product (de Souza et al., 1996). Substrate specificity studies show that only substrates containing a chlorine atom and an alkylamino side chain were hydrolyzed. Purified AtzA can catalyze hydrolysis of an atrazine analog substituted at the chlorine atom by fluorine, but not analogs containing the pseudohalide azido, methoxy, and cyano groups or the thiomethyl and amino groups. Moreover, atrazine analogs ranging in size from methyl to *t*-butyl with a chlorine substituent at C 2 and *N*-alkyl groups, underwent dechlorination by atrazine chlorohydrolase. Melamine was not degraded by the AtzA (Seffernick et al., 2000).

The K_m of AtzA was calculated to be 150 μM, and the V_{max} was 2.6 μmol of hydroxyatrazine per min per mg protein. Based on a holoenzyme molecular weight of 245,000, the equivalent K_{cat} is 11 s^{-1}. The K_{cat}/K_m value of *atzA* for atrazine is 7×10^4 $s^{-1}M^{-1}$ per unit indicating that the activity of the natural enzyme for the herbicide is reasonably high (de Souza et al., 1996).

Several other proteins show a low but significant amino acid identity with AtzA. All of these, urease-alpha subunit (urea amidohydrolase), cytosine deaminase, and imidazolone-5-propionate hydrolase, catalyze hydrolytic reactions with substrates involved in the metabolism of nitrogenous compounds (Sadowsky et al., 1998).

Hydroxyatrazine ethylaminohydrolase (AtzB)

Hydroxyatrazine ethylaminohydrolase catalyzes the transformation of hydroxyatrazine to N-isopropylammelide (Boundy-Mills et al., 1997):

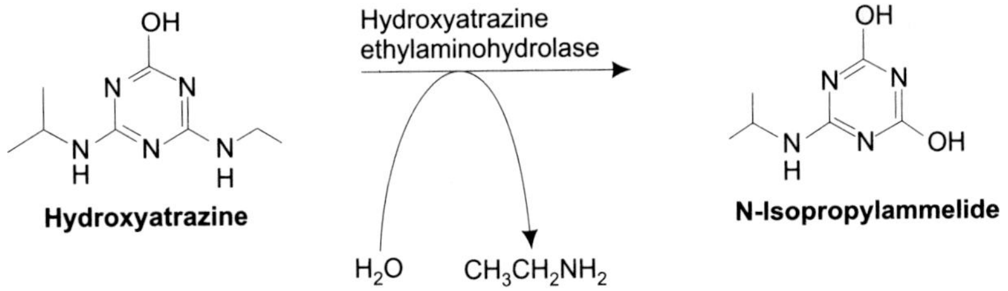

In experiments with crude extracts of *E. coli*, the observed activity of hydroxyatrazine ethylaminohydrolase demonstrated that hydroxyatrazine ethylaminohydrolase (the product of the *atz*B gene) can remove the N-ethylamino side chain of hydroxyatrazine but not that of atrazine (Boundy-Mills et al., 1997). In addition, AtzB was recently reported to dechlorinate 2-chloro-4-amino-6-hydroxy-*s*-triazine to yield ammelide (Seffernick et al., 2002a, b). AtzB was shown to have 25% identity to the TrzA protein, a protein involved in dechlorination and deamination of several *s*-triazines. The ethylamino side chain of hydroxyatrazine is completely removed by the enzyme in one step, in contrast to the mechanism of N-dealkylation using oxygen insertion which leaves the nitrogen attached to the ring and only the ethyl side chain is removed.

N-isopropylammelide N-isopropylaminohydrolase (AtzC)

N-isopropylammelide N-isopropylaminohydrolase metabolizes N-isopropylammelide (Sadowsky et al., 1998):

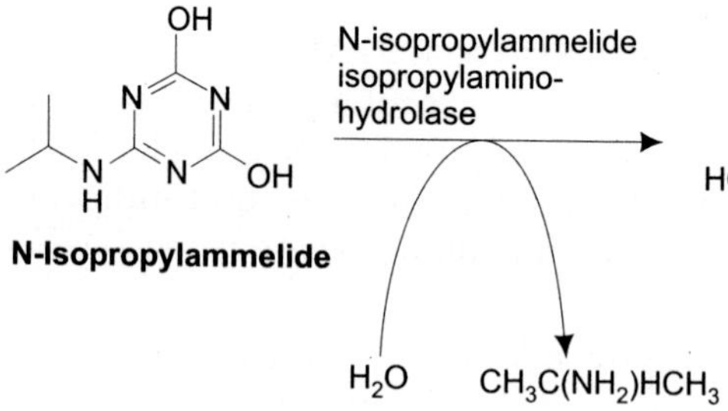

Sadowky et al., (1998) found in *Pseudomonas* sp. strain ADP the gene encoding the degradation of the AtzB product, N-isopropylammelide. The gene *atz*C was shown to encode the enzyme catalyzing the hydrolytic deamination of N-isopropylammelide to cyanuric acid and iso-propylamine. The enzyme, N-isopropylammelide N-isopropylaminohydrolase, was also found to be a new member of the amidohydrolase superfamily, based on its gene sequence. The *atz*C gene was cloned in *E. coli* and the enzymatic activity of a resting cell suspension using purified N-isopropylammelide showed that the product was cyanuric acid (Sadowsky et al., 1998). The size of the enzyme subunit is 44.9 kDa (dalton) and its holoenzyme molecular weight is 174,000 (Shapir et al., 2002).

Shapir et al., (2002) purified the AtzC enzyme and showed its substrate range. The K_m and K_{cat} values for the N-isopropylammelide reported are 406 μM and 13.3 s^{-1}, respectively. AtzC was also found to hydrolyze the herbicides simazine, terbuthylazine, and cyromazine contain N-ethyl, N-tertiary butyl, and N-cyclopropyl substituents, respectively. The enzyme is mostly effective in the deamination of N-ethylammelide (21-fold higher than for N-isopropylammelide). AtzC hydrolyzes several other *N*-substituted amino dihydroxy *s*-triazines and is only reactive with compounds that are metabolic intermediates of herbicides. The enzyme was found to have a catalytically essential five coordinate Zn(II) metal center in the active site (Shapir et al., 2002).

Cyanuric acid hydrolase (AtzD)

Many different soil bacteria can metabolize cyanuric acid, the product of the enzyme AtzC. Martinez et al., (2001) showed that *Pseudomonas* sp. strain ADP has the ability to hydrolyze cyanuric acid to biuret with the AtzD encoded by the *atz*D gene located on the *p*ADP-1 plasmid.

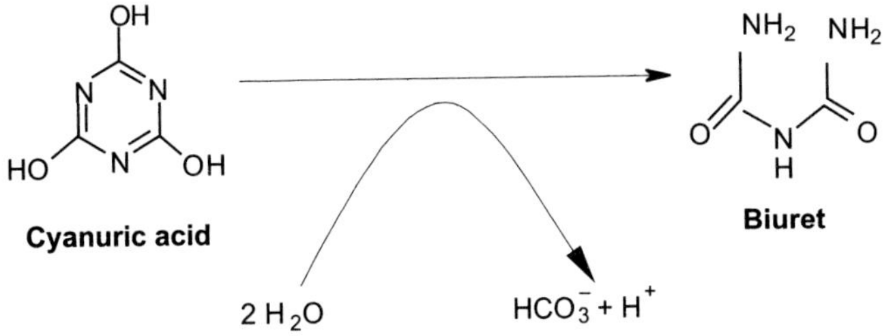

Fruchey et al., (2003) purified AtzD from *Pseudomonas* sp. strain ADP, and showed its substrate specificity and prevalence. The enzyme is one of two bacterial enzymes currently known to hydrolyze cyanuric acid to biuret. The other enzyme is the TrzD from *Pseudomonas* sp. strain NRRLB-12227. The two enzymes share 56% sequence identity (Fruchey et al., 2003). An overexpressed *atz*D gene cloned in *Escherichia coli* was used to purify the AtzD enzyme (Fruchey et al., 2003) and a molecular weight of 44 kDa of the AtzD was observed by SDS-PAGE (sodium dodecylsulfate-polyacrylamide gel electrophoresis). The k_{cat} and K_m values for cyanuric acid, in a steady state kinetic assays (under conditions of 25 mM Tris buffer at pH 8.2), were found to be 6.8 s^{-1} and 57 μM, respectivly. The enzyme was shown to have strict substrate specificity, when only cyanuric acid and N-methylisocyanuric acid were substrates (Fruchey et al., 2003). Another report of Martinez et al., (2001) also showed that the genes *atz*D, *atz*E, *atz*F are contiguous on the plasmid *p*ADP-1 and there is a possible coordinate control of these genes.

BIOREMEDIATION OF ATRAZINE FROM CONTAMINATED ENVIRONMENTS

The largest bioremediation project so far attempted of an atrazine contaminated site was carried out by Ciba-Geigy corporation at their atrazine production plant in St. Gabriel, Louisiana (Finklea and Fontenot, 1996). In this study, 190,000 m^3 of soil contaminated with atrazine was treated by a combination of land farming and bioaugmentation techniques. After six months of treatment, the total *s*-triazines range decreased from 100-300 mg L^{-1} to 2 mg L^{-1}. Another field case of bioremediation in South Dakota using chemically cross-linked and killed *E. coli* engineered to overproduce atrazine chlohydrolase (AtzA) was carried out by Strong et al., (2000) on 26 m^3 of soil. The combination of the non-viable bacteria encapsulating the AtzA enzyme and biostimulation (phosphate addition to the soil) showed a decrease in heavy contamination (up to 29,000 mg L^{-1}) to less than 100 mg L^{-1} of atrazine.

In the case of groundwater bioremediation, the development of a technology aimed at the simultaneous removal of atrazine and nitrate using a denitrifying biofilm reactor operating under non-sterile conditions, is described in the following section.

BIOREACTOR FOR THE SIMULTANEOUS REMOVAL OF ATRAZINE AND NITRATES FROM POLLUTED GROUNDWATER – CASE STUDY

The occurrence of high concentrations of both atrazine and nitrate is typical to groundwater in many intensive agricultural areas (Burkart et al., 1993; Giovanni, 1996). Studies of the metabolism of a limited number of recalcitrant chemicals by denitrifying bacteria illustrate the potential for the simultaneous removal of these two pollutants (Hutchins, 1991; Häggblom et al., 1993; Puhakka et al., 1995; Stolpe and Shea, 1995; Kuhlmann and Schoettler, 1996). Furthermore, the use of denitrifying bacteria in pollutant removal eliminates the need for aeration and the energy cost involved in the aerobic process, while their growth rate is much higher than that of other anaerobic bacteria. However, the large difference between nitrate and atrazine concentrations in groundwater (about five orders of magnitude) and the availability of only the side chain carbon atoms as an energy source, makes atrazine a poor electron donor for nitrate reduction (Cook, 1987). Moreover, the utilization of atrazine as a nitrogen source is not advantageous under anoxic conditions, since nitrate is usually more easily metabolized. Therefore, it can be assumed that simultaneous removal of both nitrate and atrazine from contaminated groundwater can only be accomplished by bacteria having a constitutive metabolic pathway for atrazine mineralization.

Based on considerations including atrazine degradation rate, a constitutive metabolic pathway for atrazine mineralization, the ability to use nitrate as electron acceptor, and the ability to form biofilm, the aforementioned *Pseudomonas* sp. strain ADP (*P.*ADP) was selected for the process aimed at the simultaneous removal of atrazine and nitrate. Bichat et al., (1999) studied the utilization of atrazine by *P.*ADP in the presence of exogenous nitrogen (NH_4, NO_3, urea and glycine) and determined that atrazine degradation was not markedly affected. Atrazine mineralization and the disappearance of nitrate in aquifer sediments under anoxic conditions were also reported by Shapir et al., (1998). Batch experiments using suspended cells of *P.*ADP indicated that *P.*ADP can simultaneously degrade atrazine and use nitrate as the electron acceptor (Katz et al., 2000a).

The simultaneous removal of atrazine and nitrate by *P.*ADP under anoxic fluidized bed reactor (FBR) was studied initially using sintered glass particles as the carrier material, while in a later experimental stage the sintered glass was replaced by GAC (granulated activated caron) as the carrier for the biofilm.

P.ADP growth and atrazine degradation in batch Experiments

Results obtained from batch experiments demonstrated that *P.*ADP kept the ability to degrade and mineralize atrazine in the presence of nitrate under both aerobic and anoxic conditions. In addition, the results confirmed the ability of *P.*ADP to grow under anoxic conditions. Results are given in Fig. 11.3.

The presence of nitrate only minimally affected *P.*ADP growth and its ability to degrade atrazine. Nitrate reduction and atrazine degradation were found to follow Monod kinetics with Monod's constants for

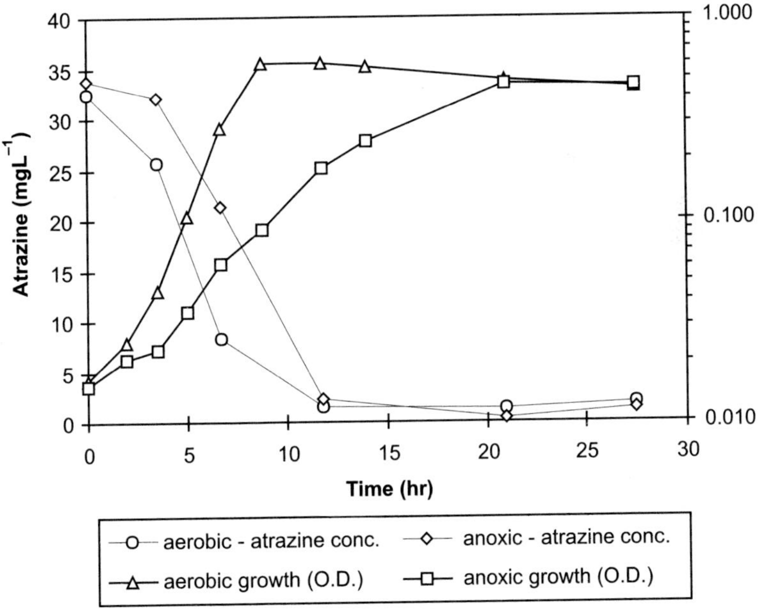

Fig. 11.3. *P.* ADP growth and atrazine degradation at aerobic and anoxic conditions batch experiments. [Katz et al., 2000b; Water Sci. Technol 41(4-5): 49-56]

denitrification: $K_S = 1.88$ mg L^{-1} NO_3^-, and $q_{max} = 4.66$ g NO_3^- g^{-1} VSS day^{-1} and for atrazine degradation: $K_s = 10.16 \pm 3.78$ mg L^{-1} and $q_{max} = 1.70 \pm 0.72$ g atrazine g^{-1} VSS day^{-1} (Katz, 1998).

Hydroxyatrazine ($C_8H_{14}ON_5$) was detected as the first metabolite in the mineralization of atrazine under anoxic conditions, which is in agreement with the reported metabolic pathway of atrazine degradation using *P.*ADP under aerobic conditions (de Souza et al., 1995, 1996). Moreover, short term experiments with atrazine, labeled with ^{14}C in the triazine ring, demonstrated an identical mineralization mechanism under both anoxic and aerobic conditions (Katz et al., 2000a).

Fluidized bed reactors (FBRs) with sintered glass as the carrier for the biofilm

Laboratory FBRs, each with 1 to 2.5 litre volume column, were filled with 300 - 700 ml sinter-glass particles (2 mm in diameter) used as a carrier for the bacteria (a carrier characterized by high surface area especially designed for biofilm). The carrier was fluidized by recirculating the fluid in the reactor with a recirculation ratio of 1:40. During start-up, the reactors were inoculated by *P.*ADP and operated under selective enrichment conditions consisting of atrazine as a sole nitrogen source and citrate as the electron donor. After the establishment of an active biofilm (about a week), the oxygen was replaced by nitrate as the electron acceptor. The nitrate daily load was 4.5 g L^{-1} (as NO_3^-) and the citrate load was according to a 5.1:1 weight ratio of citrate to N-NO_3 (Katz et al., 2000b). Phosphate was added as KH_2PO_4 and Na_2HPO_4 at 0.05 g L^{-1} each. The atrazine daily loading was about 0.1 g (L reactor)$^{-1}$ and two different influent atrazine concentrations were studied: 10-15 mg L^{-1} and 0.1 mg L^{-1}. The hydraulic retention time varied between 12 and 0.75 hours. A constant biomass concentration of about 5 g VSS L^{-1} was maintained by daily removal of excess biomass. The reactors operated under non-sterile conditions.

Nitrate removal

Under all reactor operational conditions, the denitrification efficiency of the FBR was always higher than 95%, with nitrite effluent concentrations close to zero. Results of a typical FBR run, operated at a 3 hour retention time, are given in Fig. 11.4.

Atrazine degradation in the FBRs under anoxic conditions

In all the FBRs studied and under all operational conditions, two different phases could be identified. In the first phase, atrazine degradation was observed to be close to 100%, and lasted for a period of 20 to 40 days. The second phase was characterized by a gradual decrease in atrazine degradation efficiency, reaching values of between 10 to 25% degradation after 85 to 120 days. In addition, a corresponding decrease in

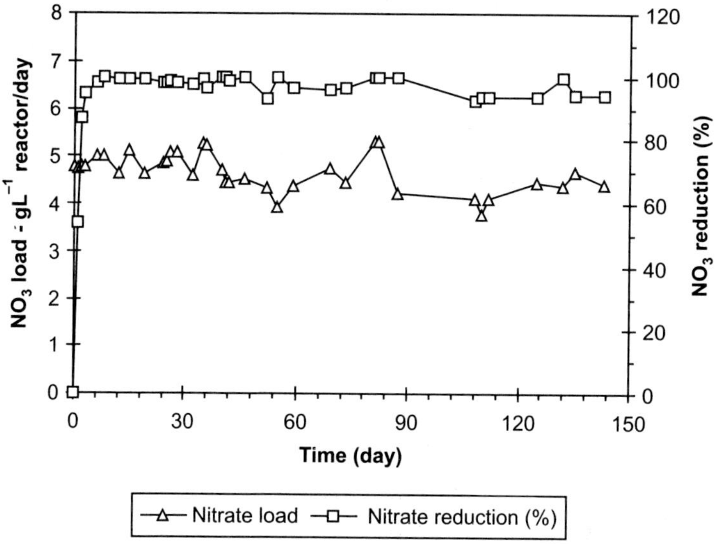

Fig. 11.4. Nitrate reduction in a FBR with sintered glass beads as the *P*.ADP biofilm carrier. [Katz et al., 2000b; Water Sci. Technol. 41(4-5): 49-56]

chlorohydrolase concentrations (the first enzyme in atrazine hydrolytic degradation pathway and is used as a measure for atrazine degradation) was observed.

To understand the reasons for the deterioration in the reactor performance, the growth of *P*.ADP in sterile and non-sterile chemostats was studied. The results showed that in the sterile chemostat, an almost constant chlorohydrolase concentration accompanied by an almost complete atrazine degradation efficiency, were maintained. In contrast, after 11 days the non-sterile chemostats showed a sharp decrease in atrazine degradation and chlorohydrolase concentration. The chemostat results indicated that the main reason for the decrease in atrazine degradation in the continuous reactors was due to contamination by foreign denitrifying bacteria that lacked the ability to degrade atrazine (non-sterile conditions were maintained) (Katz et al., 2001). Nevertheless, loss of atrazine degradation ability of *P*.ADP could also play an important role in the biofilm reactor deterioration (de Souza et al., 1998a, Herzberg, 2003).

FBRs with GAC as the carrier for the biofilm

To overcome the problem of reactor deterioration under non-sterile denitrification conditions, the use of GAC as a carrier for the biofilm of *P*.ADP in a FBR was studied (Herzberg et al., 2004, 2005). GAC typically has a very irregular surface with many holes, ridges and crevices, making it a good candidate for biofilm attachment and growth. In addition, due to its characteristic adsorption-desorption properties, GAC could potentially provide a favorable microenvironment for the bacteria. Atrazine adsorption was found to occur, even under conditions of partial penetration in the biofilm, provided that the biofilm has a patch-like structure, i.e. partial biofilm coverage of the particles (Herzberg et. al., 2003). The regular removal of excess biomass during FBR operation to ensure steady state conditions and to prevent bioparticle washout promotes this type of partial biofilm coverage.

GAC F400 (Calgon-Chemviron) was compared to non-adsorbing carbon ("Baker product", Calgon-Chemviron) particles taken from the GAC manufacturing process before the activation stage. Baker product particles were assumed to have the same physical properties as the GAC. The reactors were operated under the same conditions described above (refer section – FBRs with sintered glass as the carrier for the biofilm) with the following few exceptions: the atrazine daily loading was about 0.1 g L^{-1} reactor in the non-adsorbing reactor, and between 0.1 g L^{-1} reactor to 2.4 g L^{-1} reactor in the GAC reactor. Two different influent atrazine concentrations were studied: 5-25 mg L^{-1} and 0.8-0.08 mg L^{-1}. The hydraulic retention time varied between two and four hours. A constant biomass concentration of about 2 mg protein g^{-1} carrier was maintained on the Baker product and on the GAC by daily removal of excess biomass. All the reactors were operated under non-sterile conditions.

Atrazine degradation in the GAC reactor was determined based on influent and effluent concentrations as well as on GAC adsorption. During the start-up stage, the reactors were operated aerobically with atrazine as the sole source of nitrogen and the GAC reactor was preloaded with atrazine (47 mg atrazine g^{-1} GAC). At the end of the start-up period, the rate of atrazine degradation was similar in the two reactors and the oxygen was replaced by nitrate as the electron acceptor. Nitrate reduction responded linearly to the increase in nitrate loads (up to 4.6 g $liter^{-1}$ reactor day^{-1} as N) with nitrate reduction always exceeding 90% and effluent nitrate and nitrite concentration close to zero.

Atrazine removal in the "Baker product" FB reactor (non-adsorbing carrier)

The results of the reactor using Baker product as a carrier exhibited a similar pattern of deterioration of the atrazine degradation rate as in the sintered glass reactor with degradation efficiency decreasing by 80% after 30 days (Fig. 11.5). A significant reduction in the chlorohydrolase fraction was also observed after 30 days (Fig. 11.6).

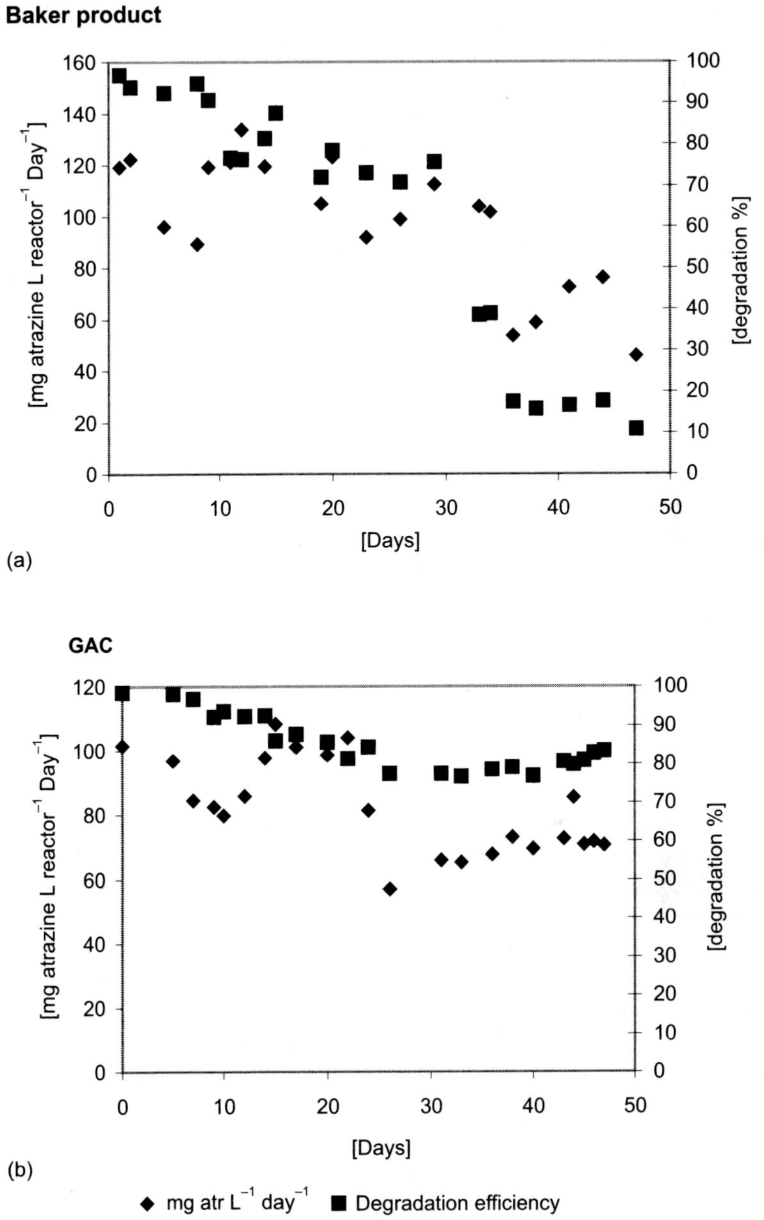

Fig. 11.5. Reactors' atrazine degradation rate and degradation efficiency. [Herzberg et al., (2004); J. Chem. Technol. Biotechnol. 79: 626-631]

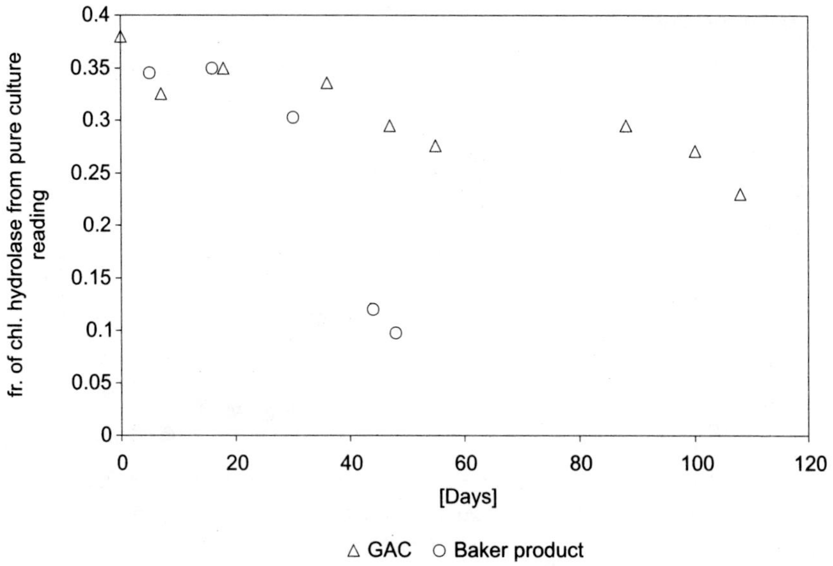

Fig. 6. Chlorohydrolase fraction in the reactors relative to its fraction in pure culture of *P.*ADP [Herzberg et al., (2004); J. Chem. Technol. Biotechnol. 79: 626-631]

Atrazine removal in the BGAC (biological granulated activated carbon) FB reactor

In contrast to the non-adsorbing carrier reactor, the atrazine degradation efficiency of the BGAC reactor remained fairly high during the entire experimental period of four months, with only 20% deterioration after 45 days of operation (Fig. 11.5). During the first 45 days of high atrazine influent concentration, the concentration of atrazine adsorbed to the GAC was almost constant (47 ± 2.2 mg atrazine g^{-1} GAC) and, therefore, atrazine removal was only due to biodegradation. In addition, the chlorohydrolase fraction also remained almost constant during this period (between 0.3 and 0.4) (Fig. 11.6).

The operation of the GAC reactor continued for another three months in order to study atrazine degradation stability and to determine the maximal volumetric and specific atrazine degradation rate. GAC particles adsorbed with atrazine (50 mg atrazine g^{-1} GAC) were added to the 2.5 liter reactor to a final weight of 700 g. The influent atrazine load was increased gradually from 100 to 2,500 mg atrazine L^{-1} reactor day^{-1}. A maximal atrazine volumetric and specific degradation rate of 0.820 ± 0.052 g atrazine L^{-1} reactor day^{-1} and 1.7 ± 0.4 g atrazine g^{-1} protein day^{-1} respectively (Fig. 11.7), were observed.

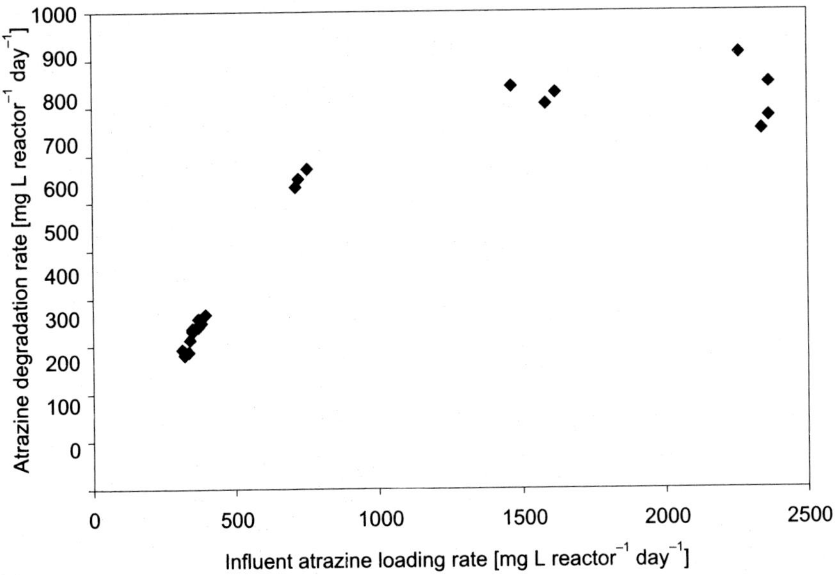

(a)

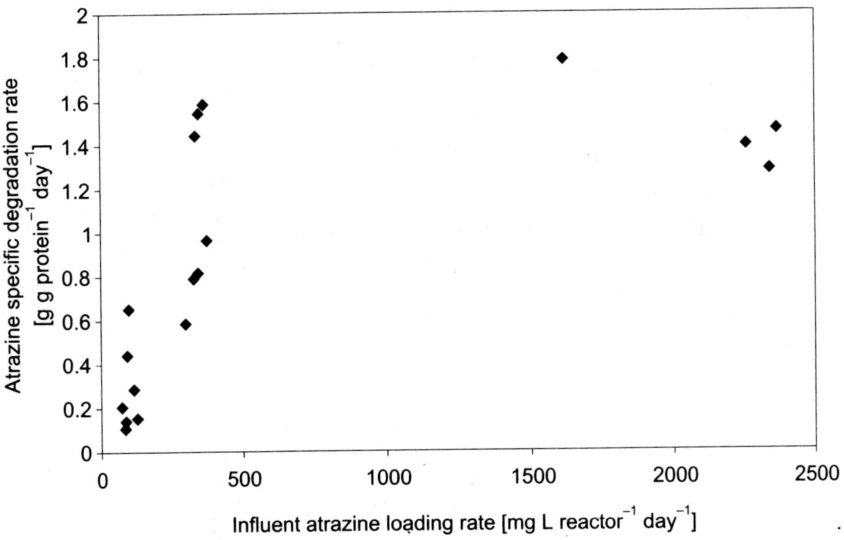

(b)

Fig. 11.7. Atrazine volumetric (a) and specific (b) degradation rates in the GAC reactor versus influent atrazine loading rate. [Herzber et al., (2004); J. Chem. Technol. Biotechnol. 2004, 79: 626-631]

A second BGAC reactor was used to investigate the performance of the reactor under conditions of low atrazine concentrations typical to contaminated groundwater. During this stage, the atrazine influent concentration was gradually dereased from 1 mg L^{-1} to 0.08 mg L^{-1} and the atrazine loading rate was decreased to 1 mg atrazine L^{-1} reactor day^{-1}. The retention time was also decreased from four to two hours. The results showed (Fig. 11.8) a continuous degradation of atrazine with a removal efficiency higher than 95%. At the lowest atrazine influent concentration studied, 0.080 mg L^{-1}, the effluent concentration was lower than the EPA standards for atrazine in drinking water (0.003 mg L^{-1}).

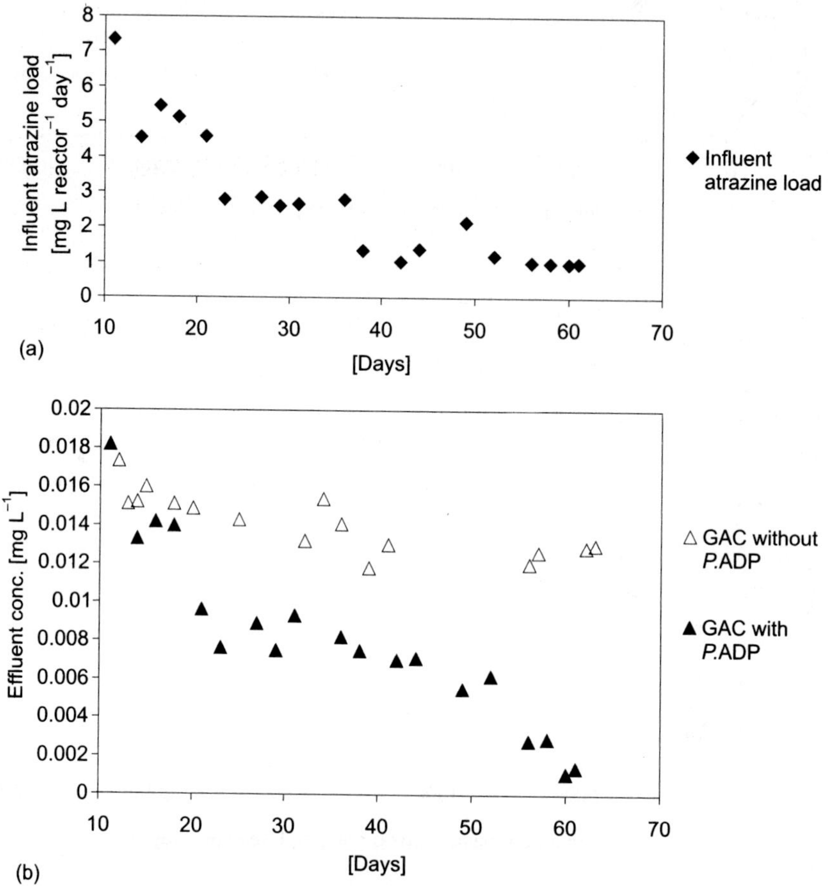

Fig. 11.8. Influent atrazine loading rate and effluent concentrations in the GAC reactor with and without *P.*ADP at low atrazine influent concentration. [Herzberg et al., (2004); J. Chem. Technol. Biotechnol. 2004, 79: 626-631]

In order to differentiate between biological degradation and GAC adsorption, another GAC reactor without bacteria was operated under the same conditions. The GAC particles in the non-biological reactor were pre-adsorbed with atrazine at a concentration similar to that of the biological reactor. A mass balance, including changes in the amount of atrazine adsorbed on the GAC, indicated that the biological degradation rate was between 10 to 30 times higher than the influent loading rate of atrazine (Fig. 11.9). The results also showed that effluent atrazine concentrations in the reactor without the bacteria were always higher (Fig. 11.8).

The stable atrazine degradation in the BGAC reactor, as opposed to the non-adsorbing carrier reactors, can probably be explained by the adsorption-desorption mechanism characterizing the BGAC carrier. In the non-adsorbing carrier reactors under conditions of partial substrate penetration, the deep biofilm layers starve and die, while faster growing foreign denitrifying bacteria penetrate and populate the carrier. In contrast, the adsorption-desorption mechanism in the BGAC reactor provides a unique microenvironment that preserves the viability of the deep biofilm layers consisting mainly of the original atrazine degrading *P.*ADP from the start-up period. In addition, the chances for penetration and growth of non-atrazine degrading bacteria (including *P.*ADP lacking the plasmid) in the deeper layers of the biofilm are lower due to selective conditions favoring atrazine degrading bacteria. The release of atrazine from the adsorbing carrier provides a nitrogen source and even a small energy source to these bacteria.

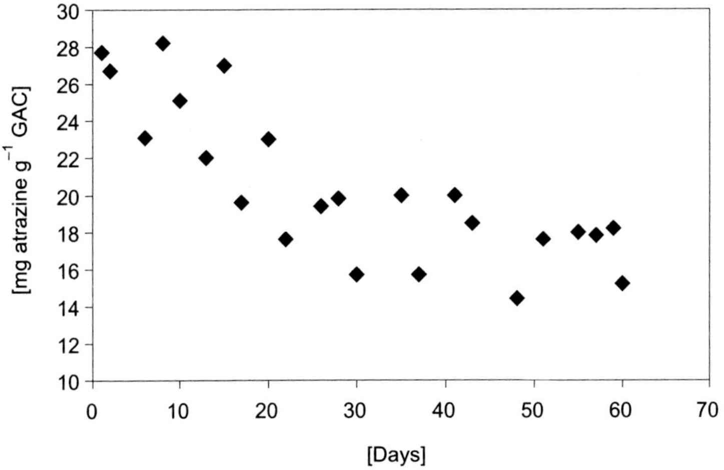

Fig. 11.9. Adsorbed atrazine concentration in the BGAC reactor at low atrazine influent concentration. [Herzberg et al., (2004); J. Chem. Technol. Biotechnol. 79: 626-631]

CONCLUSION

The wide use of pesticides is an integral part of modern agricultural production practises throughout the world. Despite their benefits, pesticides have a wide range of toxic side effects with potential hazards to the environment. The increase in the use of pesticides has resulted in massive accumulation in the environment, with persistence depending on the properties of each pesticide, site conditions, and management practises.

Herbicides are by far the most commonly used pesticides around the world and atrazine is one of the most widely used herbicide for broadleaved and grassy weeds in major crops such as corn, sorghum, sugarcane, different plantations and other crops. As it does not adsorb strongly to soil particles and has a lengthy half-life, it has a high potential for groundwater contamination and is the second most common pesticide found in water wells.

After several years when only incomplete and slow biodegradation of the atrazine molecule was reported, complete and fast biodegradation of atrazine and its dealkylated metabolites was reported for a few bacterial strains in the early 1990s. The isolated strains catabolize atrazine and its metabolites as the sole nitrogen source while releasing *s*-triazine ring carbon atoms as carbon dioxide. The recent concurrent reports of different pure bacterial strains mineralizing atrazine by various research groups around the world suggest ongoing evolution in atrazine metabolism.

In addition, the high sequence identity (more than 90%) ofthe enzymes that catabolize atrazine to cyanuric acid, a potential nitrogen source of many soil bacteria, reported for genera, suggests that *atz*ABC genes were recently distributed globally from a single ancestor.

Based on the intensive research of atrazine biodegradation over the past decade, new technologies for industrial scale atrazine bioremediation of contaminated sites have been developed and show promise. Further research and scaling up of new processes is still necessary for successful and cost effective bioremediation of atrazine polluted environments.

REFERENCES

Alvey, S. and Crowley, D.E. (1995). Influence of organic amendments on biodegradation of atrazine as a nitrogen source. J. Environ. Qual. 24: 1156-1162.

Asplin, A.L. (1997). Pesticides industry sales and usage, 1994 and 1995 market estimates. US EPA Report No. 733-K-94-001.

Asplin, A.L. and Grube, A.G.H. (1998). Pesticide industry sales and usage: 1996 and 1997 Market estimates. Office of prevention, Pesticides and Toxic Substances, US Environmental Protection Agency. 73-R-98-0001. Washington, DC 20460. USA, pp. 37.

Behki, R.M. and Khan, S.U. (1994). Degradation of atrazine, propazine and simazine by *Rhodococcus* strain B-30. J. Agric. Food Chem. 42: 1237-1241.

Bichat, F., Sims, G.K. and Mulvaney, R.L. (1999). Microbial utilization of heterocyclic nitrogen from atrazine. Soil Sci. Soc. Am. J. 63: 100-110.

Biradar, D.P. and Rayburn, A.L. (1995). Chromosomal damage induced by herbicide contamination at concentrations observed in public water supplies. J. Environ. Qual. 24: 1222-1225.

Boundy-Mills, K.L., de Souza, M.L., Mandelbaum, R.T., Wackett L.P. and Sadowsky, M.J. (1997). The *atz*B gene of *Pseudomonas* sp. strain ADP encodes the second enzyme of a novel atrazine degradation pathway. Appl. Environ. Microbiol. 63: 916-923.

Bouquard, C., Ouazzani, J., Prome, J.C., Michel-Briard Y., and Plesiant P. (1997). Dechlorination of atrazine by *Rhizobium* sp. isolate. Appl. Environ. Microbiol. 63: 862-866.

Brooks, R. (2000). The basics of weed control, UI Extension Forestry Information Series, Tree Planting and Care No. 14, University of Idaho, URL: http://www.cnr.uidaho.edu/extforest/TPC14.pdf.

Brouwer, W.W.M., Boesten, J.J.T.I. and Siegers. W.G. (1990). Adsorption of transformation products of atrazine by soil. Weed Res. 30: 123-128.

Burkart, M.R. and Kolpin, D.W. (1993). Hydrologic and land-use factors associated with herbicides and nitrate in near-surface aquifers. J. Environ. Qual. 22: 646-656.

Clay, S.A. and Koskinen, W.C. (1990). Adsorption and desorption of atrazine, hydroxyatrazine and *s*-glutathione atrazine on two soils. Weed Sci. 38: 262-266.

Cook, A.M. and Hutter, R. (1984). De-ethylsimazine: Bacterial dechlorination, deamination and complete degradation. J. Agric. Food Chem. 32: 581-585.

Cook, A.M., Beilstein, P., Grossenbacher, H. and Hutter, R. (1985). Ring cleavage and degradative pathway of cyanuric acid in bacteria. Biochem. J. 231: 25-30.

Cook, A.M. (1987). Biodegradation of *s*-triazine xenobiotics. FEMS Microbiol. Rev 46: 93-116.

Crawford, J.J., Sims, G.K., Mulvaney, R.L. and Radosevich, M. (1998) Biodegradation of atrazine under denitrifying conditions. Appl. Microbiol. Biotechnol. 49: 618-623.

CropLife America. (2003). The value of herbicides in U.S. crop production. NCFAP report, National Center for Food & Agricultural Policy. URL: http://www.croplifeamerica.org/public/pubs/benefits/FullText.pdf.

Crop Protection Association. (2003a). Every drop counts: Keeping water clean. URL: http://www.cropprotection.org.uk/content/resources/5_Pubs.asp.

Crop Protection Association. (2003b). 2002-Global sales of crop protection products. URL: http://www.cropprotection.org.uk/content/news/4_PR_archive.asp.

de Souza, M.L., Wackett, L.P., Boundy-Mills, K.L., Mandelbaum R.T. and Sadowsky M.J. (1995). Cloning, characterization and expression of the gene region from *Pseudomonas* sp. strain ADP involved in the dechlorination of atrazine. Appl. Environ. Microbiol. 61: 3373-3378.

de Souza, M.L., Sadowsky, M.J. and Wackett, L.P. (1996). Atrazine chlorohydrolase from *Pseudomonas* sp. strain ADP: Gene sequence, enzyme purification and protein characterization. J. Bacteriol. 178: 4894-4900.

de Souza, M.L., Wackett, L.P and Sadowsky, M.J. (1998a). The *atz*ABC genes encoding atrazine catabolism are located on self-transmissible plasmid in *Pseudomonas* sp. strain ADP. Appl. Environ. Microbiol. 64: 2323-2326.

de Souza, M.L., Seffernick, J., Martinez, B., Sadowsky M.J. and Wackett L.P. (1998b) The atrazine catabolism genes *atz*ABC are widespeard and highly conserved. J. Bacteriol. 180: 1951-1954.

de Souza, M.L., Newcombe, D., Alvey S. Crowley, D.E., Hay A., Sadowsky, M.J. and Wackett, L.P. (1998c). Molecular basis of a bacterial consortium: Interspecies catabolism of atrazine. Appl. Environ. Microbiol. 64: 178-184.

Devers M., Soulas, G. and Martin-Laurent, F. (2004). Real-time reverse transcription PCR analysis of expression of atrazine catabolism genes in two bacterial strains isolated from soil. J. Microbiol. Methods 56: 3-15.

Eddleston, M., Karalliedde, L., Buckley N., Fernando R., Hutchinson G., Isbister G., Konradsen F., Murray D., Piola J.C., Senanayake N., Sheriff R., Singh S., Siwach S.B. and Smit L. (2002). Pesticide poisoning in the developing world–a minimum pesticides list. The Lancet 360: 1163-1167.

Ellis, L.B.M., Hershberger, C.D., Bryan, E.M. and Wackett, L.P. (2001). The University of Minnesota Biocatalysis/Biodegradation database: emphasizing enzymes. Nucl. Acids Res. 29: 340-343.

Erickson, L.E. and Lee, H.K. (1989). Degradation of atrazine and related *s*-triazines. Crit. Rev. Environ. Cont. 19: 1-14.

European Crop Protection Association. (2003). ECPA Review 2002 > 2003. URL: http://www.ecpa.be/library/AR_brochures/annual_report_2003.pdf.

Extonet. (1996). Extension toxicology network, Pesticide Information Profiles, Atrazine. Oregon State University, URL: http://ace.orst.edu/cgi-bin/mfs/01/pips/atrazine.htm.

Finklea, H.C. and Fontenot, M.F. (1996). Accelerated bioremediation of triazine contaminated soils: a practical case study. Ciba-Geigy Corporation, St. Gabriel, LA, USA. In: Bioremediation (eds) H.D. Skipper and R.F. Turco, Madison, WI: Soil Sci. Soc. Amer. SSSA Special Publication 43: 221-235.

Fruchey, I., Shapir, N., Sadowsky M.J. and Wackett, L.P. (2003). On the origins of cyanuric acid hydrolase: Purification, substrates, and prevalence of *Atz*D from *Pseudomonas* sp. strain ADP. Appl. Environ. Microbiol. 69: 3653-3657.

Giovanni, R. (1996). Water quality decreasing in Brittany streams: nitrates and example of two pesticides, atrazine and lindane. Cybium 20: 143-162.

Häggblom, M.M., Rivera, M.D. and Young, L.Y. (1993). Influence of alternative electron acceptors on the anaerobic biodegradation of chlorinated phenols and benzoic acids. Appl. Environ. Microbiol. 59: 1162-1167.

Herzberg, M. (2003). Biological degradation of atrazine in FB denitrification reactor with activated carbon as the carrier for the biofilm. Ph.D. Thesis, Technion - Israel Institute of Technology, Israel.

Herzberg, M., Dosoretz, C.G., Tarre, S. and Green, M. (2003). Patchy biofilm coverage can explain the potential advantage of BGAC reactors. Environ. Sci. Technol. 37: 4274-4280.

Herzberg, M., Dosoretz, C.G., Tarre, S., Beliavsky M., Minz D. and Green M. (2004). Simultaneous removal of atrazine and nitrate using a biological granulated activated carbon (BGAC) reactor. J. Chem. Technol. Biotechnol. 79: 626-631.

Herzberg, M., Dosoretz, C.G. and Green, M. (2005). Increased biofilm activity in BGAC reactors. AIChe Journal 52: 1042-1047.

Holm, L. and Sander, C. (1997). An evolutionary treasure: Unification of a broad set of amidohydrolases related to urease. Proteins 28: 72-82.

Hutchins, S.R. (1991). Optimization BTEX biodegradation under denitrifying conditions. Environ. Toxicol. Chem. 10: 1437-1448.

Katz, I. (1998). Combined process for the removal of nitrates and herbicides from groundwater. Master thesis, Technion – Israel Institute of Technology, Israel.

Katz, I., Green, M., Ruskol, Y. and Dosoretz, C.G. (2000a). Characterization of atrazine degradation and nitrate reduction by *Pseudomonas* sp. strain ADP. Adv. Environ. Res. 4: 219-224.

Katz, I., Dosoretz, C., Ruskol, Y. and Green, M. (2000b). Simultaneous removal of nitrate and atrazine from groundwater. Water Sci. Technol. 41(4-5): 49-56.

Katz, I., Dosoretz, C.G., Mandelbaum, R.T. and Green, M. (2001). Atrazine degradation under denitrifying conditions in continuous culture of *Pseudomonas* ADP. Water Res. 35: 3272-3275.

Kaufman, D.D. and Blake, J. (1970). Degradation of atrazine by soil fungi. Soil Biol. Biochem. 2: 73-80.

Kolpin, D.W., Thurman, E.M. and Linhart, S.M. (1998). The environmental occurrence of herbicides: The importance of degradates in groundwater. Arch. Environ. Contam. Toxicol. 35: 385-390.

Kuhlmann, B. and Schoettler, U. (1996). Influence of different redox conditions on the biodegradation of the pesticide metabolites phenol and clorophenols. Int. J. Environ. Anal. Chem. 65: 289-295.

Li, G.C. and Felbeck, G.T. (1972). Atrazine hydrolysis as catalyzed by humic acids. Soil Sci. 114: 201-209.

Loprieno, N., Barale, R., Mariani, L., Presciuttini S., Rossi A.M., Sbrana I., Zaccaro L., Abbondandolo A. and Bonatti S. (1980). Results of mutagenicity tests on the herbicide atrazine. Mutat. Res. 74: 250.

Mallory-Smith, C.A. and Retzinger, J.E. Jr. (2003). Revised classification of herbicides by site of action for weed resistance management strategies. Weed Technol. 17: 605–619.

Mandelbaum, R.T., Wackett, L.P. and Allan D.L. (1993). Mineralization of the *s*-triazine ring of atrazine by stable bacterial mixed cultures. Appl. Environ. Microbiol. 59: 1695-1701.

Mandelbaum, R.T., Allan, D.L. and Wackett, L.P. (1995). Isolation and characterization of a *Pseudomonas* sp. that mineralizes the *s*-triazine herbicide atrazine. Appl. Environ. Microbiol. 61: 1451-1457.

Martinez, B., Tomkins, J., Wackett, L.P., Wing R. and Sadowsky M.J. (2001) Complete nucleotide sequence and organization of the atrazine catabolic plasmid *p*ADP-1 from *Pseudomonas* sp. strain ADP. J. Bacteriol. 183: 5684-5697.

Mulbry, W. (1994). Purification and characterization of an inducible *s*-triazine hydrolase from *Rhodococcus corallinus* NRRL B-15444R. Appl. Environ. Microbiol. 60 : 613-618.

Nagy, I., Compernolle, F., Ghys, K., Vanderleyden, J. and Demot, R. (1995). A single cytochrome P-450 system is involved in degradation of the herbicides EPTC (s-ethyl dipropylthiocarbamate) and atrazine by *Rhodococcous* sp. strain NI86/21. Appl. Environ. Microbiol. 61: 2056-2060.

Protzman, R.S., Lee, P.H., Ong, S.K. and Moorman, T.B. (1999). Treatment of formulated atrazine rinsate by *Agrobacterium radiobacter* strain J14A in a sequencing batch biofilm reactor. Water Res. 33: 1399-1404.

Puhakka, J.A., Herwing, R.P., Koro, P.M., Wolfe, G.V. and Ferguson, J.F. (1995). Biodegradation of chlorophenols by mixed and pure cultures from a fluidized-bed reactor. Appl. Microbiol. Biotechnol. 42: 951-957.

Radosevich, M., Traina, S.J., Hao, Y.L. and Tuovinen, O.H. (1995). Degradation and mineralization of atrazine by soil bacterial isolate. Appl. Environ. Microbiol. 61: 297-302.

Ralebitso, K.T., Senior, E. and van Verseveld, H.W. (2002). Microbial aspects of atrazine degradation in natural environments. Biodegradation 13: 11–19.

Rittmann, B.E. and McCarty, P.L. (2001). Environmental Biotechnology: Principles and Applications. McGraw-Hill International Edition, NY, USA.

Sadowsky, M.J., Tong, Z., de Souza, M.L. and Wackett, L.P. (1998). AtzC is a new member of the amidohydrolase protein superfamily and is homologous to other atrazine-metabolizing enzymes. J. Bacteriol. 180: 152-158.

Seffernick, J.L. Johnson, G., Sadowsky, M.J. and Wackett, L.P. (2000). Substrate specificity of atrazine chlorohydrolase and atrazine-catabolizing bacteria. Appl. Environ. Microbiol. 66: 4247-4252.

Seffernick, J.L. and Wackett, L.P. (2001a). Rapid evolution of bacterial enzymes: a case study with atrazine chlorohydrolase. Biochemistry 40: 12747-12753.

Seffernick, J.L., de Souza, M.L., Sadowsky, M.J. and Wackett, L.P. (2001b). Melamine deaminase and atrazine chlorohydrolase: 98 percent identical but fuctionally different. J. Bacteriol. 183: 2405-2410.

Seffernick, J.L., McTavish, H., Osborne, J.P, de Souza, M.L., Sadowsky, M.J. and Wackett, L.P. (2002a). Atrazine chlorohydrolase from *Pseudomonas* sp. strain ADP is a mettalloenzyme. Biochemistry 41: 14430-14437.

Seffernick, J.L., Shapir, N., Schoeb, M., Gilbert J., Sadowsy, M.J. and Wackett, L.P. (2002b). Enzymatic degradation of chlorodiamino-*s*-triazine. Appl. Environ. Microbiol. 68 : 4672-4675.

Shapir, N., Mandelbaum, R.T. and Jacobsen, C.A. (1998). Rapid atrazine mineralization under denitrifying conditions by *Pseudomonas* sp. strain ADP in aquifer sediments. Environ. Sci. Technol. 32: 3789-3792.

Shapir, N., Jeffrey, P.O., Gilbert, J., Sadowsky, M.J. and Wackett, L.P. (2002). Purification, substrate range and metal center of AtzC: the *N*-Isopropylammelide aminohydrolase involved in bacterial atrazine metabolism. J. Bacteriol. 184: 5376-5384.

Stolpe, N.B. and Shea, P.J. (1995). Alachlor and atrazine degradation in a Nebraska soil and underlying sediments. Soil Sci. 160: 359-370.

Strong, L.C., McTavish, H., Sadowsky, M.J. and Wackett, L.P. (2000). Field-scale remediation of atrazine-contaminated soil using recombinant *Escherichia coli* expressing atrazine chlorohydrolase. Environ. Microbiol. 2: 91-98.

Struthers, J.K., Jayachandran, K. and Moorman, T.B. (1998). Biodegradation of atrazine by *Agrobacterium radiobacter* J14a and use of this strain in bioremediation of contaminated soil. Appl. Environ. Microbiol. 2: 91-98.

Thoden, J.B., Phillips, G.N. Jr., Neal, T.M., Raushel, F.M. and Holden, H.M. (2001). Molecular structure of dihydroorotase: a paradigm for catalysis through the use of a binuclear metal center. Biochemistry 40: 6989-6997.

Topp, E., Mulbry, W.M., Zhu, H., Nour, S.M. and Cuppels, D. (2000a) Characterization of *s*-triazine herbicide metabolism by a *Nocardioides* sp. isolated from agricultural soils. Appl. Environ. Microbiol. 66: 3134-3141.

Topp, E., Hong, Z., Nour, S.M., Houot S., Lewis, M. and Cuppels, D. (2000b) Characterization of an atrazine-Degrading *Pseudaminobacter* sp. isolated from Canadian and French agricultural soils. Appl. Environ. Microbiol. 66: 2773-2782.

Trautmann, N.M., Porter, K.S. and Wagenet, R.J. (1998). Pesticides and Groundwater: A Guide for the Pesticide User. URL: http://pmep.cce.cornell.edu/facts-slides-self/facts/pest-gr-gud-grw89.html.

Vink, J.P.M. and van der Zee Sjoerd, E.A.T.M. (1997). Effect of oxygen status on pesticide transformation and sorption in undisturbed soil and lake sediment. Environ. Toxicol. Chem. 16: 608-616.

Wackett, L.P. and Hershberger, C.D. (2001). Biocatalysis and biodegradation: the microbial transformation of organic compounds. ASM Press, Washington, DC, USA.

Wackett, L.P., Sadowsky, M.J., Martinez, B. and Shapir, N. (2002). Biodegradation of atrazine and related *s*-triazine compounds: From enzymes to field studies. Appl. Microbiol. Biotechnol. 58: 39-45.

Ware, G.W. (2000). An introduction to herbicides. In: E.B. Radcliffe and W.D. Hutchison [eds.], Radcliffe's IPM World Textbook, URL: http://ipmworld.umn.edu , University of Minnesota, St. Paul, MN.

Wood, A. (2003). Compendium of Pesticide Common Names. Classified Lists of Pesticides. URL: http://www.alanwood.net/pesticides/class_pesticides.html.

Yanze-Kontchou, C. and Gschwind, N. (1994). Mineralization of the herbicide atrazine as a carbon source by a *Pseudomonas* strain. Appl. Environ. Microbiol. 60: 4297-4302.

12

Microbiological Technology for Extraction of Jute and Allied Fibres

M.K. Basak

GENERAL INTRODUCTION

A large number of plants are reported to yield fibres, but only a few are of commercial importance. Most of the fibre yielding plants belong to the families Malvaceae, Tiliaceae and Sterculiaceae in the order Malvales. Besides these, some other families are also fibre yielding, e.g. Papilionaceae, Linaceae, Urticaceae, Amaryllidaceae and so on. On the basis of origin, the fibres may be grouped into three categories (Maiti, 1980):

(a) Bast fibres – obtained from the bast or the bark of stem, e.g. jute, mesta, roselle, sun hemp, ramie and flax.

(b) Leaf fibres – fibres obtained from mesophyll of leaves, e.g. sisal, pineapple, manila, etc.

(c) Fruit (coir) and seed fibres (cotton).

Stem or Bast Fibres

The order Malvales is mainly producers of bast fibres, and Urticaceae and Linaceae yield stem fibre. The fibre strands of most bast fibres, unlike seed cotton, are composite structures composed of ultimate cells united end to end. The most important plants belonging to this group are jute (*Corchorus capsularis* L. and *C. olitorius* L.), flax (*Linum usitatissimum* L.), sun hemp

---

National Institute of Research on Jute and Allied Fibre Technology, 12, Regent Park, Kolkata – 700040, India

E- Mail: mkbasak2002@yahoo.com

(*Crotalaria juncea* L.), mesta (*Hibiscus cannabinus* L.), ramie (*Boehmeria nivea* L.), roselle (*Hibiscus sabdariffa* L.), Congo jute (*Urena lobata* L. and *U. sinuate* L.) and China jute (*Abutilion*).

Leaf Fibres

The fibre strands are composed of compact sclerenchyamtous fibres present in the ground tissue of monocot leaves, which offer mechanical strength. The most important leaf fibres are sisal (*Agava sisalana*), manila (*Musa textiles*) and pineapple (*Ananas comosus*).

Fruit Fibres

These are also sclerenchyamtous fibres associated with vascular bundles present and developed in the pericarp of the fruit. The most important fibre belonging to this group is the coir fibre, used mostly for making ropes, brushes and mattresses.

In this chapter, the microbial technology used for extraction of some of these commercially important fibres from crops such as jute, flax, sun hemp, etc. are discussed.

JUTE

Introduction

Jute has been cultivated in India for centuries. In Europe, towards the end of 18th century, East India Company of England introduced jute fibre in their search for a substitute of flax in packaging application. It was 1793 when the first dispatch of about 100 tonnes of raw jute was exported from India. During the 40 years from 1828 to 1868, the export of raw jute from erstwhile Bengal (comprising the present West Bengal province of India and Bangladesh) reached from 1,200 to 262,800 tonnes (NIRJAFT, 1998).

Jute and kenaf are primarily cultivated in developing countries like India, Bangladesh, China, Thailand, Nepal, etc. Table 12.1 presents the area, production and productivity of raw jute and allied fibres of some of the jute producing countries in the world. Before 1947, India exercised a monopoly in raw jute export. After India partition in 1947, the centre of jute production remained in West Bengal (India) while the jute fields were clustered in Bangladesh (then East Pakistan). At the time of partition of India in 1947, the area under jute plantation was only 260,000 ha with a production of about 1670,000 bales (1 bale = 180 kg). Table 12.2 outlines the history of raw jute production since 1947. The average acreage and production of jute increased slowly but steadily during different plan periods. In the first year of the Ninth Plan (1997-1998), coverage under jute increased to 920,000 hectares with fibre production of 9,725,000 bales in the

Table 12.1. Area, production and yield of jute, kenaf and allied fibres (2002 - 2003)

Country	Bangladesh	China	India	Myanmar	Nepal	Thailand
Area (100,000 ha)	4.36	0.56	10.25	0.59	0.12	0.33
Yield (q ha^{-1})	18.20	27.70	20.10	7.10	14.50	16.60
Production (100,000 bales)	44.20	8.60	114.40	2.32	0.94	3.10

Source: FAO Statistics, December 2004

Table 12.2. Trend in area, production and productivity of jute in India at 10 year intervals since 1947

Year	Area (100,000 ha)	Production (100,000 bales)	Productivity (q ha^{-1})
1947-1948	2.64	16.70	11.38
1957-1958	7.05	40.14	10.24
1967-1968	8.80	63.20	12.92
1977-1978	7.97	53.61	12.11
1987-1988	6.98	58.00	14.96
1997-1998	9.20	97.25	19.60
2002-2003	10.25	103.25	20.10

Source: Pathak (2001)

Table 12.3. World exports of products of jute, kenaf and allied fibres (1,000 tonnes)

Year	1998	1999	2000	2001	2002	2003
World	747.2	685.9	643.8	642.5	676.0	716.4
Bangladesh	395.7	433.7	378.0	409.2	400.6	391.9
China	7.5	6.6	6.5	5.3	9.1	14.4
India	242.5	161.0	177.8	151.0	189.9	243.3
Nepal	10.0	10.0	10.0	10.0	10.0	10.0
Thailand	10.4	8.0	7.2	6.3	5.2	4.1

Source: FAO Statistics, December 2004

country (Pathak, 2001). The world export of fibre products in major producing countries is shown in Table 12.3. In India, two popular species of jute are cultivated which belong to the family Tiliaceae viz. *Corchorus capsularis* (white jute) and *C. olitorius* ("tossa" jute).

Microbial Retting and Extraction of Fibre

Definition of Retting

Retting is defined as the process of separating the embedded fibre from the stem through partial rotting by immersion in water; this rotting is brought about by a complex enzymatic action of microbes naturally present in the retting water. Retting ranks as the single most important factor governing the quality of fibre (Ghosh, 1983; German, 1985). The important conditions for ideal retting are:

- The water should be non-saline and clear.
- The volume of water in the retting tank or ditch should be sufficient to allow jute bundles to float.
- Bundles, when immersed, should not touch the bottom.
- The same retting tank or ditch should not be used when water becomes dirty.

Harvest of Jute Plants

Harvesting at the proper age is emphasized so as to avail opportune retting time and to balance yield with quality. Theoretically, this vegetative fibre crop can be harvested at any stage of its growth. In early stage of growth upto 90 days, there is a mixture of immature and mature fibres, which produce good quality but lesser fibre. After 120 days of crop growth, the fibre is less flexible in nature but yields more. Maximum elongation of ultimate fibre cells takes place between 90-120 days, beyond this age there is a reduction in fibre cells and greater accumulation of lignin takes place. To obtain optimal ultimate fibre cell production and to fit in multiple cropping sequence it is better to harvest the crop between 100-120 days, thus maintaining a balance between yield and quality (Plate 12.1). Harvesting is done manually by sickle and left in the field for leaf shedding.

Preparation of 'Jak'

After harvest, the defoliated jute bundles are brought to the site where they are to be retted and placed side by side in the retting water with the bottom facing in one direction. This floating mass of bodies is known as 'jak' (Plate 12.2). A few stems of sun hemp (*Crotalaria juncea*) or "dhanicha" (*Sesbania aculeate*) are inserted into the bundles for better multiplication of retting microbes.

Covering and Steeping

The 'jak' is then covered with water hyacinth, paddy straw or grass. The floating 'jak' is weighted down 10-15 cm below the water surface with dry wooden logs, bricks, stones, etc. Mud clogs, banana plants or freshly cut logs of mango trees should not be used as weighing materials. The float should not touch the clay at the bottom of the ditch or tank. Submerging is important as any portion of the 'jak' or bundle sticking out above the water surface will not ret properly, or may not ret at all and create a defect in the fibre.

Plate 12.1. Harvesting of jute plants

Plate 12.2. A 'jak' (floating mass) of jute plants

Process of Retting

Retting is usually carried out by immersing the jute stems in water. At the appropriate temperature bacteria present on stems as well as in water attack the jute tissues softening them so that the residue/debris can be washed away, leaving the fibres intact. Although the process is basically a simple one, timing is important in obtaining good quality fibre. If the stem is under-retted, it will not be sufficiently softened and removing the bark will be difficult. Over-retting is equally harmful since the bacteria will not only break down the plant tissues but will also attack the fibre bundles, thus weakening the fibre.

The process of retting can be divided into two distinct phases: (a) the physical phase, and (b) the biochemical phase. The physical phase begins when the harvested plants (reeds) are steeped in water. The tissues absorb water and swell, releasing soluble components into the retting water. The released substances are carbohydrates, nitrogenous compounds and salts of different kinds, which facilitate the growth and multiplication of the microbes in the retting water prior to their entry into the reeds (Mandal et al., 1987; Mandal and Saha, 1997). These organisms develop and multiply by utilizing free sugars, pectins, hemicellulose and proteins of the plants as nutrients. Of the fermentable substances, free sugars are decomposed in the early stage of retting.

Extraction of Fibre

When retting is complete, bundles are taken out of water and fibre is extracted manually one by one i.e. single plant extraction (Plate 12.3) or by beat-break-jerk method i.e. multiple plant extraction (Plate 12.4). Single plant extraction is carried out on the embankment of a pond or tank. A retted bundle from the 'jak' is brought to the extraction site by placing the butt-end on the embankment, while the major portion of the bundle remains in the water. A portion of 8 or 10 cm from the base is extracted reed by reed; fibre from 5-6 such reeds are collected and held together by the hand to pull and extract the entire length slowly. In the beat-break-jerk process (Plate 12.4), 7-8 reeds are taken out from the retted bundle and malletted at the base by a wooden mallet. The reeds are then broken with the help of the knee at about one-third height from base, the broken sticks are eliminated manually by gripping the exposed fibre, and the remaining sticks are ejected out with successive jerks. For this process the labourer has to stand waist deep in the water to do the work. The single reed method gives untangled, straight fibre free from broken sticks. But in the beat-break-jerk process, the fibres may get somewhat entangled and retain broken sticks.

Plate 12.3. Single plant extraction of jute plants

Plate 12.4. Beat-break-jerk process of extraction of the jute plants

Washing and Drying

After extraction, the fibres are washed thoroughly in clean water to remove gum, dirt, sticks and decomposed plant remains adhering to the fibres. The fibres are then squeezed properly to remove water. Drying under mild sun is best (Plate 12.5). The wet fibre should be spread over a bamboo frame for two to three days. After complete drying, the fibres are tied in bundles for marketing.

Factors Which Affect Retting

The efficiency of retting depends on a number of factors as detailed below:

Plant age: As the jute plant grows older, the tissues become more and more matured. As a result, more and more lignin deposition takes place and retting takes a longer time. A compromise between fibre quality and yield is generally achieved if plants are harvested at small pod stage.

Retting water: It is known that retting is a biochemical process in which various decomposition products are formed. If the process is carried out in stagnant water, accumulation of these products causes hindrance to the growth and activity of the causative organisms. Very fast moving water, on the other hand, removes these toxic substances quickly but it also

Plate 12.5. Drying of jute fibres in the sun

carries away the microbial population along with it resulting in non-uniform retting. As such, retting is best carried out in slow flowing water (Kundu, 1961, 1964).

pH and temperature: Natural water from different sources having a pH between 6.0-8.0 is conveniently used for retting. The optimum temperature for jute retting was found to be around 34-36°C at which the process is completed in the minimum time due to the accelerated activity of the retting microflora. Generally, in the harvesting season the temperature varies from 30-36°C. With the gradual fall of temperature, the time taken for retting is also increased. Capsularis jute rets earlier than Olitorius (Roy and Mandal, 1967).

Depth of water: It was observed that microbial action in retting water is maximum at a depth of 15 cm from the surface of water and retting is quicker and better at this depth. Some microbial action is evidenced even up to a depth of 35 cm, but below this practically no effect has been observed (Basak et al., 1988).

Activators: Retting is accelerated in the presence of several activators. Natural activators like "dhanicha" (*Sesbania aculeate*) and sun hemp (*Crotalaria juncea*) plants are generally introduced into the jute stem bundles before they are put in water for retting. These leguminous plants being rich in nitrogen content, help the growth and activity of retting microbes by supplying additional nutrients to them.

Addition of cultures: A method for production of good quality jute by inoculating green jute plants with effective mixed bacterial cultures has been developed at NIRJAFT, Kolkata, (India). The method has been tried at the farmers' level and found to be very effective (Basak and Paul, 1988). Results in these trials have shown that fibre samples were improved by one to two grades over the corresponding untreated samples, accompanied by reduction in the retting period by 3-10 days due to effectiveness of retting cultures. The specific retting organisms added as inoculum are highly pectinolytic and non-cellulolytic and multiply very rapidly under normal conditions of retting, using gums and pectins as a source of carbon and energy. This is achieved by suppressing the growth and multiplication of other fibre degrading microflora present in situ. As a result, retting of jute is completed in a shorter period with removal of barks and with retention of fibre strength along with other properties of the fibre.

Quality of Fibre

The quality of bast fibres is a complex function which involves different quality parameters like strength, fineness, meshiness, surface structure of fibre strands, colour, lustre, etc. Good lustre and colour are associated with stronger fibre and all these are ultimately dependent on retting of the fibre. Meshiness, specks, roots and knotty structure cause hindrance in carding and subsequent spinning process, thereby deteriorating the quality of the yarn (Chakravarty et al., 1977). The chemical composition of fibres from capsularis and olitorius varieties of jute after retting is given in Table 12.4.

Mechanization in Extraction

Reducing the retting water requirement to the minimum and relieving the farmer of the drudgery of manual extraction in foul retting water are priorities. With this background in view, an alternative method has been evolved such as ribboning of green stems by mechanical devices and retting the ribbons in a small volume of water instead of stem retting. In ribbon retting, the volume of biomass to be retted comes down to about 40% only and ribbons are stripped out mechanically from the stem of mature jute plants, coiled and allowed to ret under water. Ribbon retting reduces the time of normal retting by four to five days. Use of efficient pectinolytic microbial inoculum improves the quality of fibre, reduces the time of retting and the environmental pollution. A mixed bacterial retting culture developed at NIRJAFT, was inoculated during ribbon retting of jute. The ribbon retted jute fibres were absolutely free from bark and of high grade. Moreover, the fibre filaments were stronger, with improved colour and finer texture compared to the conventional stem retted jute fibre (Banik et al., 2003; Munder et al., 2004).

Advantages of Ribboning

(i) Less water required for retting: Where there are shortages of water for retting, by using ribbons, twice as much fibre can be produced in the same volume of water.

(ii) Valuable plant nutrient returned to the soil: The green leaves which have a high content of nitrogen, phosphorus, potassium and calcium are left in the field and help to maintain soil fertility.

(iii) Less weight to transport: The green ribbons represent only about 40% of the total weight of the plant. By ribboning in the field, the farmer has much less material to transport to the retting area.

Table 12.4. Chemical composition (%) of jute (*Corchorus* spp.) fibre after retting

Constituent	*C. capsularis*	*C. olitorius*
α-cellulose	60.7	61.0
Pentosan	15.6	15.9
Lignin	12.5	13.2
Polyuronide	4.8	5.2
Acetyl value	3.5	2.9
Fat and wax	1.0	0.9
Nitrogenous matter	1.87	1.56
Ash	0.78	0.50

(iv) Retting may be carried out at any time: Water is scarce at harvest time in many jute growing areas and retting in insufficient water results in poor fibre. Ribbons can be readily dried and stored in bales until there is adequate water.

(v) Fibre washing is easier: When stems are retted, the farmer usually has to stand waist deep in water while stripping and washing the fibre. When ribbons are retted, the washing can be carried out in a tank on dry land.

Disadvantages of Ribboning

(i) Additional labour required for green stem stripping by hand: Work carried out at BJRI (Bangladesh Jute Research Institute), Bangladesh showed that to obtain 1,000 kg of dry fibre by ribboning, 72 man days would be required whereas to obtain the same quantity of fibre by whole plant, retting would require 52 man days.

(ii) High capital outlay: A single ribboner with maximum capacity can only deal with not more than one-half hectare in an eight hour working day. Therefore, when a large area has to be harvested, many ribboning machines need to be available and this means a large capital expenditure.

Various Types of Ribboner

Hand operated ribboners: In some countries, labour saving devices for stripping ribbons from the stem have long been in use. They are of very simple construction and can thus be constructed easily by the farmers.

Bamboo hook ribboners: A piece of bamboo about 2 m long is fixed firmly in the ground. Using a sharp knife, a 'V' or hook is cut in the upper end. The butt-end of the jute stem is softened with a hammer or mallet. The loosened raw butt is grasped by the operator, half is pulled round one part of the hook and the other half round the other part. The two parts of the raw bast are given a strong pull with both hands. The two ribbons remain in the hands while the stick is ejected to the front.

Bicycle hub ribboner: A horizontal board fitted with a hub from a bicycle wheel is used to reduce the physical effort needed to ribbon. The ribboning capacity varies depending on the skill and operational practice by the worker. The major advantage of manual ribboning is that the inner sticks are not broken and the cultivators can use the full sticks for fencing, cultivation of betel leaves, fuel in cooking, etc. at rural level.

Motor powered ribboning machine: Over the past two decades, the work on mechanizing the extraction of green barks from harvested plants has been carried out in various countries. The various types of ribboning machines so far developed remove the ribbons in different ways. Most of the machines operate as decorticators, which strike the stems by rotating blades, and break the sticks. The ribboner developed by NIRJAFT, however, peels off the green barks in the form of ribbons by compression in the pair of rollers (Paul et al., 2002). The best performance in most of the machines is achieved with a plant which is harvested when it is not over matured or not at an early stage. The critical parameter at the time of ribboning is, of course, the moisture content in stems. Jute plant ribboner comprises three pairs of rollers having a serrated surface at feed end, fluted surface in the middle and spined surface at the delivery end. The NIRJAFT ribboner (Plate 12.6) was demonstrated to jute cultivators at jute field site during harvest season. Although the capacity of the present model was estimated to be about 200 kg stems hr^{-1} when run by 1 HP (Horse Power) motor, the productivity can well be raised to the desired capacity by scaling up the design to suit the appropriate situation where the economic framework and systems exist for ribbon retting.

Retting and Environment

Retting of jute stems or ribbons give rise to various microbes, organic acids and gases in the retting water which have an unpleasant smell. Extraction of the fibres is done by standing directly in the retting water with heavy organic load or sitting on the bank of the retting ditch or pond. A lot of broken woody cores (jute sticks) are left in the retting water or around the retting site spoiling the aesthetics and contributing to siltation. The

Plate 12.6. Motor powered ribboning machine developed at NIRJAFT, Kolkata, India

extracted fibres are washed in the retting water. The washing further adds to the partially decomposed organic matter. During and immediately after retting, the dissolved oxygen levels in the water are too low for most of the biota except for the air-breathing fish to survive. With increased intensity and frequency of retting the situation becomes more acute. The water turns darker, frothy, more acidic, foul smelling and becomes a mosquito breeding centre. However, it is not reported to be a major issue in any of the jute producing countries (Alam, 1998).

Upgradation of Low Quality Barky Jute at Industrial Level

Due to bad growing conditions such as waterlogging, drought, attack of pests and incomplete retting, a significant percentage of jute in India contains fragments of bark mainly at the bottom portion of jute. Jute fibres with such bark fragments are considered to be of inferior quality from the spinning point of view. Jute obtained in the market sometimes contains lot of such barky portions. These barky portions cause processing difficulties and are either cut off before processing, or the jute is used for inferior quality products. The cut off portions are called root cuttings and are generally considered as waste material. These cuttings or barky jute which can be softened by a specific microbial culture. Fungi are generally superior in this respect. Several fungi were screened and one culture, later identified as *Penicillium corylophilum* Dierckx, was found suitable for softening the material at industrial level as these fungi grew profusely within a short time (48 hours) even with a very low moisture content. It was found that application of the fungal culture during processing of jute improved the processibility of the fibre at industrial level. The improvement in processibility of the fibre is attributed to upgradation of fibre quality due to pectinolytic activity of *P. corylophilum* and moisture retention of the fibre at optimum level during processing. The fungus was grown on rice husk, a cheap lignocellulosic agro-residue by solid state fermentation process. The enzymes i.e. hemicellulases, cellulases and pectinases, liberated in the medium were assayed (Sarkar et al., 2001). Pectinase activity in the medium was found high while the cellulase activity was very low. The high pectinolytic activity and low cellulolytic activity of the fungus help in improving the fibre quality by removing pectin without affecting the cellulose. It was also observed that the cellulose and lignin content remain almost unaffected and it is the pectin portion of the bark which is decomposed considerably by the treatment (Table 12.5). A number of mill trials based on this method showed that the treated barky jute can even be utilized in batches in admixture with other qualities of fibre without any deterioration in the product quality and without any processing difficulties. Table 12.6 describes the properties of the yarn manufactured in one of the jute mills with and without treatment

Table 12.5. Chemical constituents (dry weight basis) of untreated and fungal treated barky jute

Constituent %	Untreated Barky Jute	Fungal Treated Barky Jute
Ash	2.6	1.4
Fat and wax	3.2	1.6
Lignin	18.9	17.9
Pectin	3.1	1.4
Pentosan	13.9	16.0
Holocellulose	80.9	81.5
α-cellulose	59.8	60.3
Degree of polymerization	1340	1390

Table 12.6. Effect of fungal culture treatment on yarn properties

Sl. No.	Parameters	Fungal Culture Method	Conventional Method
1	Grist (lbs)	11.00	10.43
2	Yarn strength (lbs)	11.72	9.89
3	Strength CV%	15.80	20.66
4	Quality ratio	104.64	92.12
5	End breakage 100 sp hr^{-1}	170.40	180.60

of the fungus, *P. corylophilum*. The properties of fungal treated yarn were superior in all respects than that of conventional yarn. Chemical and microscopic studies clearly indicate that the fungus grows on the barky jute utilizing the pectinous portions as a carbon source and helps in softening the bark by removal of the pectin which acts as a binding material between the fibre strands (Basak et al., 1987; 1991). The fungus can be grown in large-scale utilizing rice husk – an agro waste by solid state fermentation method and applied on low quality jute fibres in the jute mills during processing (Basak et al., 2004).

Upgradation of Low Quality Jute by Fungal Culture at Rural Level

Due to scarcity of water, a great majority of jute crop remains barky. This barky jute fetches a lower price to the cultivators. A specific fungus, *Aspergillus flavus*, proved quite efficient in removing the bark in the cultivators' field itself without any detrimental effect on fibre strength and fineness. A large number of trials in the farmers' field also established the efficiency of the culture. A novel method of preservation of the culture was also developed by which the culture in a solid form (in admixture with kaolin powder) can retain its viability and activity for about three months.

The solid culture can be put into packets in a convenient manner in small polyethylene bags and transported. By using the culture, jute can be upgraded by 2-3 grades (Bhattacharyya and Basu, 1981; 1982).

OTHER ALLIED FIBRES

1. MESTA

Introduction

Two species of *Hibiscus,* i.e. *H. cannabinus* L. and *H. sabdariffa* L. var. *altissima,* are commercially known as mesta or kenaf and belong to the family Malvaceae. Both the species are important jute supplements and show wide adaptability unlike jute. *H. cannabinus* has its possible origin in Africa where it is found in the wild and also in subtropical regions of Asia, but controversy arises as to the origin of *H. sabdariffa*, which is attributed to both Africa and India.

Mesta (*H. cannabinus*)

The quality of mesta (*H. cannabinus*) depends on the conditions of retting on one hand, and on the age of the crop on the other. Compared to jute, mesta is of inferior quality with reference to different parameters like fineness, lustre, colour and amount of defects. Mesta varieties show poor performance in spinning, because the fibre is coarse, stiff, brittle and irregular in cross-section. Large amounts of specks, roots and barky matter at later ages of the crop is another drawback in mesta. Mesta cannot spin alone in jute machines unless it is mixed with jute in some proportion. It has been observed that time required for retting differs at various stages of harvesting and depending on the prevailing temperature of the season. As the age increases and the stems become mature, and if the temperature decreases, the time required for retting increases correspondingly.

Roselle (*H. sabdariffa*)

Roselle, like mesta, is a useful substitute for jute but in quality it is somewhat inferior to mesta. They are used in the manufacture of coarse sackings. The *H. sabdariffa* (HS) is more tolerant to drought than *H. cannabinus* (HC) and may be grown in low rainfall areas. It requires a long vegetative period of 180 to 237 days when sown in mid-March to mid-May and do not mature before the second week of November, irrespective of the sowing date (Chakravarty et al., 1977). *H. cannabinus* exhibits considerable variation in maturity and takes around 135-165 days to attain the maximum vegetative growth.

Harvest and Extraction

The crop is harvested at flowering or small pod stage. The kenaf (HC) could be harvested at 5-$5^1/_2$ month stages by the end of September or early October, while rosselle (HS) is harvested by mid-November – mid-December at 150 to 175 days for optimum fibre yields. After maturity, the mesta plants are either cut at the bottom or the whole plant is uprooted from the soil and put into the tank for retting. The crop is cut near the base and tied in bundles of 25-30 cm diameter and then removed for retting from the field. Retting practice followed for mesta is quite similar to that of jute. With good management practices, yield between 25-30 q ha^{-1} is obtained.

2. SUN HEMP

Introduction

Sun hemp (*Crotalaria juncea* L.) is an important multipurpose leguminous crop grown in India, Bangladesh, Brazil and African countries mainly for fibre, green manure as well as fodder. The crop belongs to the family Papillionaceae. In India till the 1960s it was a premier crop and used to occupy more than 200,000 hectares (ha) with 0.75 million tonnes of fibre annually and earned substantial foreign exchange, but subsequently the area was reduced and production declined.

The plants are quick growing, a 3-4 month crop–an erect annual with cylindrical and straight stem. Well drained alluvial soil with a loamy texture is well suited. In spite of having important natural properties such as green manuring and fodder, the fibre yield (10 q ha^{-1}) of the crop is much below the other bast fibre crops like jute and mesta (20 q ha^{-1}). This is mainly attributed to low fibre content (2-3%) of the crop in comparison to jute and mesta (5-6%). The fibre is of better quality having good lustre, higher tensile strength and more durable to exposure. The fibre is used in cordage making, valuable paper manufacture and cigarette paper. Its demand is increasing in the manufacture of specialized tissue paper and currency notes. Other diversified uses of sun hemp fibre as well as dry stick are in the preparation of carpets and mats, toys, cushions, boards, insulation products and sun hemp composites.

Sun hemp grows best in tropical climates of 23-30°C temperature and high relative humidity (85-90%) with 170-200 mm rainfall. Light soil with neutral pH between 6.0 to 7.5 is ideal for sun hemp cultivation. It can tolerate pH up to 9.0 but root nodule formation is hampered. Light soil facilitates nodulation. In low lying soil it grows vigorously but produces less quantity of coarse fibre.

Harvest and Extraction

Sun hemp matures in 100-120 days. For fibre it is harvested after 80-90 days at 50% flowering stage. March-April sown crop is harvested by early July. Mid-May sown crops are harvested in mid-September which produce highest fibre yield and quality.

Harvesting process is conventional for jute and kenaf. After harvest, 30 cm is cut from the plant top and removed for fodder. Plants are tied in 25 cm diameter bundles and kept exposed to sunlight for drying and defoliation. After two days, the bundles are taken to retting tank and kept standing in water to soften the hard base for 1-2 days and tied side by side in a platform or 'jak' kept across and pressed down with stone, bamboo, palm or jamun log, which do not stain and deteriorate fibre. Fibre is extracted from dried bundles by hand stripping.

3. RAMIE

Introduction

Ramie fibre obtained from the plant *Boehmeria nivea* L Gaud, a member of the Urticaceae, is one of the most important natural textile fibre. Besides high strength, its desirable length, durability, absorbency, lustre, resistance to rot and good spinning quality, make the fibre very useful for the manufacture of a wide variety of textile and cordage products. Ramie is a semi-perennial crop and it usually thrives for four to five years economically. The crop gives maximum yield in the second or third year. Soils rich in calcium and organic matter content are best suited for ramie cultivation. The propagation of ramie is generally done vegetatively by rhizome, stem cutting and layering. The rainy season i.e. May to September, is the optimum time for planting as the sprouts of rhizome get the opportunity to establish properly.

Harvest and Extraction

Ramie fibre is extracted from the freshly harvested green stalk (after defoliation) by a mechanical process called decortication. By decortication, the outer bark, central wood core and some portions of the gums and waxes of the bark are mechanically removed. The product obtained through decortication crude fibres contains 25-30% gum. By washing some percentage of gums and other substances are removed and the colour of the fibre improves. The gum content present in the crude decorticated fibre makes it weak, brittle and unspinnable due to inter-fibre adhesion. By reducing the gum content to 2-6%, the fibre becomes soft, fine and well separated for spinning in textile applications. A chemical degumming technique has been developed in which the duration of degumming and the concentration, volume and temperature of alkali solution have been optimized for maximum efficiency to reduce the gum content to around 2% level (Bhaduri and Ganguly, 2002). Although degumming with chemicals is effective, its use involves recurring cost of chemicals, power consumption and enormous quantities of water, besides environmental hazards.

Microbial treatment of decorticated fibre for five days with a mixed bacterial culture isolated from the rhizoshpere soil of ramie plant followed by chemical treatment with 1% alkali solution for one hour reduces the gum content to 6.1%. The degummed fibre shows increased fineness and retains good strength. Microbial pretreatment thus reduces the energy input by saving the time of chemical treatment by one hour. The combined microbial and chemical treatment is more economical and less drastic with acceptable fibre properties for textile processing (Basak et al., 2004).

4. FLAX

Introduction

Flax grown both for fibre and linseed belongs to the family Linaceae. While flax grown for the former is preferably of an unbranched type, flax for the latter is branched variety. It is mainly grown in Germany, Netherlands, Belgium, France and also in the United States. Linen is the fibre obtained from the flax plant. The plant grown for fibre purpose usually attains a height of 3 to 4 ft. The stems branch out only at the top. The fibre extends to the entire length of the stem and is situated between the epidermis and a layer of coarse woody fibre which adheres to the innermost core of the plant. Besides being cultivated for its fibres, the flax plant is also grown for its seed, which yields the valuable oil known as linseed. High yielding variety viz. FT895 developed by Central Research Institute for Jute and Allied Fibres (CRIJAF), Barrackpore, India gives a yield of 800 kg ha^{-1}

Harvest and Extraction

Two primary methods of retting, viz. dew retting and water retting, have been used traditionally to extract fibres for textile and commercial applications (Eassan and Molloy, 1996). In dew retting, the stalks are lightly spread over a grassy piece of land with care being taken to ensure that all the root ends are even. As a result of climatic conditions, the bacteria and moulds release or produce enzymes which convert the gummy substances into soluble material that may be partly washed away by the dew and rain. Despite the fact that highest quality flax fibre is produced by water retting, this practice has been largely discontinued in

Western Europe due to high costs, and the pollution and stench arising from fermentation of plant materials. Dew retting is now the most common practice for separating flax fibres even though some water retted fibre is still marketed. (Akin et al., 2004).

Enzymes have been considered for sometime as a potential replacement for dew retting of flax. The water retted linen is quite pronounced and almost silky in appearance. Flax that is over-retted is dull in appearance. The lustre plays a most important part in the preparation and spinning of flax fibre. As flax has the desirable qualities of lustre, heat transmission, flexibility, tensile strength and the ability to be dyed, it is used widely in the manufacture of table linen, summer suitings, dress goods and many other items, besides its use for high grade cordage and string. The general method for extraction of flax in Indian climate is half-retting in water. When the plants are half-retted, they are taken up and dried in the sun and staged. The fibres are then extracted in a scutching machine by passing the half-retted dried flax stalks through, the machine, and as a result, the fibres are separated from the woody stalks. After scutching, the fibres are thoroughly combed to separate the dry woody stalks or barky matter.

5. PINEAPPLE LEAF FIBRE

Introduction

Pineapple leaf fibre (*Ananas comosus* L.) is a multicellular lignocellulosic fibre obtained from the leaves of the plant. The pineapple was indigenous to Mexico and Brazil. Today, the principal source of supply is from Pacific islands, Hawaii and the Philippines. Soon after its introduction into the Philippine islands, the natives began using the fibre for the production of a fine, silky cloth known as "pina". The fibre is obtained from the leaves after the fruit has been harvested. The fibre is extracted from the leaves either by mechanical means or by retting the leaves in water. The pineapple leaf scratcher machine is used for scratching the leaves before retting. Since the leaves are collected after the harvest of the fruit, the fibre production would be more or less a by-product. The leaves of the plant produce strong, white, fine, silky fibres. The yield of the fibre is generally 2-3% of the weight of green leaves. The physical characteristics of pineapple leaf fibre clearly show that it is as good as the finer quality of jute, although about 10 times as coarse as cotton. Unlike jute, its structure is without mesh and the filaments are well separated.

Harvest and Extraction

By the usual retting process, fibres cannot be extracted from the leaves as the leaves contain an impervious layer of wax, and microbes cannot enter the leaves to cause extraction. If some minor mechanical scratching is imparted to the leaves, retting becomes very rapid and fibres can be extracted within four days. A combination of both the mechanical and microbial method has been developed and the technology has been demonstrated on a large scale. The handy scratcher designed and fabricated for this purpose was found to be quite suitable for use by small farmers.

6. SISAL FIBRE

Introduction

Sisal is a leaf fibre derived from the leaves of the plant *Agave sisalana* belonging to the family Amaryllidaceae. Sisal is grown in large plantations in Java, East and West Africa, and India. The fibre is used by the Mexican Indians not only for cordage but also for the manufacture of coarse cloth. The leaves are beaten by hand and the fibres removed, after which they are washed, bleached and dried in the sun. The long fibres are then either hand-spun or the individual fibres are woven directly into cloth.

The fibre's elasticity is good and its extension is considerably high. The principal use for sisal fibre is in the manufacture of commercial tying twines, ropes and cords. The fibre may also be called 'The Green Gold' for its unique properties like being strongest fibre, and tolerance to marine water. By using this fibre, there is a possibility of the agro-industry for manufacturing diversified products.

Harvest and Extraction

The leaves grow from the base of the plant in the form of a rosette and each leaf is cut by hand close to the ground. These leaves are hauled to the central plant for decortication – a process done by large machines which scrape the epidermis and pulp off and away from the fibres, which are then washed to get the fibres.

7. MANILA OR ABACA

Introduction

Abaca is a native plant of the Philippine islands and has long been utilized by the natives for the manufacture of what is known as sinamay cloth. The plant is perennial, which grows to a height of 9 to 18 ft. The first stalks are ready for harvesting about 20 months to three years after planting, depending on location and variety. After the first harvest, the plants are usually cut every six to eight months. Harvesting is done by hand with a sharp knife. The yield varies greatly but 500 kg of fibre per acre is considered a good crop. Single plant yields about half kg of fibre. The fibre is white and lustrous in appearance, light and stiff, easily separated, strong, and possesses great durability.

Extraction

To extract the fibre from leaves, the natives first make a slight incision just beneath the fibre at the end and giving a sharp pull, brings away a strip or ribbon of the outside skin containing the fibre. When the sufficient numbers of ribbons are thus obtained, they are scraped with a long knife blade. This method of decortication or fibre production from the plant is universal in the Philippines. After the above hand operation, the fibres are cleansed by washing and dried in the sun.

MICROBIOLOGY AND BIOCHEMISTRY OF RETTING

Retting occurs from the combined action of three kinds of microorganisms i.e. aerobic and anaerobic bacteria, and fungi. This is done by the action of microbial enzymes – pectinase, hemicellulases and proteinases. The microbial population present in retting water varies widely and normally as a source of water. The underground water has a low microbial population and may require inoculation before retting can be undertaken. Several aerobic bacteria of the genus *Bacillus* i.e. *B. subtilis, B. marcerans, B. corchorus, B. polymixa,* and anaerobic bacteria of the genus *Clostridium* i.e. *C. tertium, C. felsinum, C. aurantibutyricum and C. pectinovorum*, have been isolated from retting waters. Among the fungus, *Aspergillus niger, Macrophomina phaseoli, Mucor abundans, Chaetomium* sp. have been found to be good retting agents (Kundu, 1961). The aerobic and anaerobic bacteria are present from the first to the last day of retting, but their numbers in the reeds vary with the progress of retting. From one study, it has been observed that the bacterial count in retting water increased upto the seventh day when it reached the maximum and then there was a sharp fall in the number (Basak et al., 1998). Aerobic bacteria multiply quickly exhausting the dissolved oxygen available. Under depleted oxygen, the anaerobic microbes multiply and begin decomposing the material that binds the fibres.

The biochemical phase starts as soon as the bacteria enter the reeds. It has been suggested that retting bacteria entered the submerged jute plants through the cut ends (Ghosh et al., 1974) and also through the ruptures created on the surface of the plants due to swelling for absorption of water. During this phase, a series of biochemical reactions take place as a result of which the chemical composition, pH, Eh (redox potential), BOD (biological oxygen demand) and COD (chemical oxygen demand) of retting water change continuously. Pectin, hemicelluloses, reducing sugars, tannins and proteins are the principal substances removed during retting of jute (Kundu and Mukherjee, 1961). Of these, hemicelluloses and pectin form the major constituents, the amount of hemicelluloses being the highest. The loss of pectin is found to be almost same in both capsularis and olitorius varieties of jute, while the loss of hemicelluloses in the former species is almost double that in the latter, the loss of reducing sugars is more in capsularis (Roy and Kundu, 1965). It was observed that release of total sugars increased gradually during retting process up to the 13th day and beyond this period the increase was significant indicating a condition of over-retting, when decomposition of cellulosic components takes place. Release of galacturonic acid, which is a degraded product of pectin – the main binding material between the jute fibres, reaches the peak on the 13th day, when retting is complete. The maximum galacturonic acid release may be regarded as an indication of completion of retting and over-retting starts beyond this period, when decomposition of cellulosic fibre occurs as indicated by excess release of soluble sugars in the retting liquor. Due to these decompositions, several organic acids such as acetic, butyric, lactic, etc. are also produced. Specific enzymes secreted by the organisms cause degradation of the complex organic materials into simpler compounds which are then metabolized for their life process. However, if retting is continued beyond optimum period, microorganisms begin to disintegrate the hard lingo-cellulose of the cell fibre. Such a condition, known as over-retting, produces a weaker fibre that usually fetches a low price. During the process of retting accumulation of organic acids takes place. During retting acids are produced rapidly at higher temperature. Accumulation of decomposed products in the water hinders further activity of the microbes, thus delaying the process and making it imperfect. That is why if retting is carried out in slow flowing water the fibre quality will be the best because in such a case the accumulated products are also removed from the retting site by the slow flowing water. When retting is carried out in stagnant or controlled conditions it is advisable (i) to impart a rocking action or trampling of the material to eliminate the accumulated matters, or (ii) to replace all or part of the water periodically to reduce the level of acids in retting water.

CONCLUSION

From the above discussion it is clear that the extraction of fibres from the plants involves the action of enzymes from various microorganisms which degrade mainly pectins and other gummy substances to separate the fibres. The mediation of the chemical reactions associated with retting by catalytic proteins (enzymes) is a central feature of living systems. Living cells may be engineered to produce more enzymes or to produce different types of enzymes (protein engineering) with improved characteristics, specificity, stability and performance. Help of biotechnology may be sought in this regard. The fungi and the bacteria which are presently being used can be improved in respect of specific enzyme liberation and efficiency of the liberated enzymes to degrade pectin by genetic engineering. The strain with very high pectinase activity but without any or very low cellullase activity will be the ideal one for retting purpose.

Another novel application of enzyme in textile is in biopolishing. Here, cellulase enzyme is applied. In case of cotton to remove the fuzzy hairs from the surface of cotton yarn, treatment with cellulase enzyme is followed. Jute yarn also has many hairs on its surface which can effectively be removed by use of cellulase enzyme and make the yarn hair free. Biobleaching is another field where enzymes can be utilized to replace costly and environment hazard chemicals. Microbiology and biotechnology can thus play a vital role for overall improvement in the extraction and processing of jute and allied natural fibres.

REFERENCES

Akin, D.E., Henriksson, G., Evans, J.D., Peter, A., Adamsen, S., Foulk, J.A. and Roy, B.D. (2004). Progress in enzyme retting of flax. J. Natrural Fibres. 1(1): 21-47.

Alam, A. (1998). Retting and extraction of jute – Problems and perspectives. Proc. Int. Seminar Jute and Allied Fibres - Changing Global Scenario, 5-6 February 1998, Kolkata, India, p. 1.

Banik, S., Basak, M.K., Paul, D., Nayak, P., Sardar, D., Sil, S.C., Sanpui, B.C. and Ghosh, A. (2003). Ribbon retting of jute – a prospective and eco-friendly method for improvement of fibre quality. Industrial Crops Products 17: 183-190.

Basak, M.K., Sao, K.P. and Bhaduri, S.K. (1987). Scanning electron microscopic studies on microbial softening of jute. Indian J. Text. Res. 12: 154-157.

Basak, M.K. and Paul, N.B. (1988). An improved method of retting of green jute plants at farmers' level. Indian Agric. 32(2): 87-91.

Basak, M.K., Roy, A.R. and Sasmal, B.C. (1988). A study on the variation in quality of jute fibre due to retting factors. Jute Dev. J. 8(1): 1-3.

Basak, M.K., Bhaduri, S.K. and Sao, K.P. (1991) Physico-chemical nature of barky jute treated with *Penicillium corylophilum* Dierckx. Indian J. Fibre Text. Res. 16: 1752-1777.

Basak, M.K., Bhaduri, S.K., Banik, S., Kundu, S.K. and Sardar, D. (1998). Some aspects of biochemical changes associated with retting of green jute plants. Proc. Int. Seminar Jute and Allied Fibres - Changing Global Scenario, 5-6 February 1998, Kolkata, India, pp. 80-83.

Basak, M.K., Ganguly, P.K., Bhaduri, S.K. and Sarkar, S. (2004). Industrial production of *Penicillium corylophilum* on rice husk and its application in jute processing. Indian J. Biotechnol. 3: 70-73.

Basak, M.K., Majumder, P. and Bhaduri, S.K. (2004). Chemi-microbial method for degumming of ramie fibres. J. Textile Assoc. 64(5): 249-252.

Bhaduri, S.K. and Ganguly, P.K. (2002). Degumming of ramie fibre with recovery of degraded gum. Patent Application No. 17/DEL/2002 dated 11.01.2002.

Bhattacharyya, S.K. and Basu, M.K. (1981). Upgrading of low quality of jute by fungal culture. Food Farmg. Agric. 13(9/10): 170-172.

Bhattacharyya, S.K. and Basu, M.K. (1982). Kaolin powder as a fungal carrier. Appl. Environ. Microbiol. 44(3): 751-753.

Chakravarty, A.C., Roy, P.K., Bose, S.K. and Bandyopadhyay, S.B. (1977). Note on assessment of yield and quality of jute fibre from living jute plants. Indian J. Agric Res. 11(3): 191-192.

Eassan, D.L. and Molloy, R. (1996). Retting – a key process in the production of high value fibre from flax. Outlook Agric. 25: 235-242.

FAO (Food and Agricultural Organization). (2004). FAO statistical data, December 2004, Rome, Italy.

German, C.G. (1985). The retting of jute. FAO Agro Service Bull., No. 60, Rome, Italy.

Ghosh, T., Roy, A.B., Mandal, A.K. and Dasgupta, B. (1974). Path of entry of retting bacteria in jute and importance of aerobic phase of retting. Jute Chron. 9: 113-114.

Ghosh, T. (1983). Handbook on jute. FAO Plant Produc. Protec. Paper No. 51, Rome, Italy.

Kundu, A.K. (1961). Studies on the microbiological retting of jute. D. Phil. thesis, University of Calcutta, India.

Kundu, A.K. and Mukherjee, M.K. (1961). Nitrogen and jute retting II. Changes in amino acid concentration during retting. Proc. Nat. Acad. Sci. (India) 31: 207-210.

Kundu, A.K. (1964). Factors influencing retting of jute. Jute Bull. 27: 1-6

Maiti, R.K. (1980). Plant Fibres, 1st. Edn, Bishen Singh and Mahendra Pal Singh Publishers, Dehradun, India.

Mandal, T.C., Dutta, R.K. and Ojha, T.P. (1987). Principles of mechanical extraction of bast fibre. J. Agric. Eng. 24(2): 197-205.

Mandal, T.C. and Saha, M.N. (1997). Jute retting methods and mechanization. Cen. Res. Inst. Jute Allied Fibres (CRIJAF), Barrackpore, West Bengal, India.

Munder, F., Furr, C. and Hempel, H. (2004). Advanced decortication technology for untreated bast fibres. J. Natural Fibres. 1(1): 49-65.

National Research Institute for Jute and Allied Fibre Technology (NIRJAFT) (1998). Publication - Sixty Years of NIRJAFT, Jute Technological Research Laboratory (JTRL) to NIRJAFT, Kolkata, India.

Pathak, S. (2001). Technology for increasing jute production in India. CRIJAF, Barrackpore, West Bengal, India.

Paul, D., Banik, S., Das, B.K., Datta, A.K., Sil, S.C. and Sanpui, B.C. (2002). Study on extraction of green bark from jute plant by manual and power ribboner. Annual Report, NIRJAFT, India, 2002-03, pp. 7-9.

Roy, A.B. and Kundu, A.K. (1965). Investigation on the substances removed during retting of jute. Jute Bull. 28(6): 134-137.

Roy, A.B. and Mandal, A.K. (1967). Retting and quality of jute. Jute Bull. 30(4): 131-139.

Sarkar, S., Basak, M.K., Bhaduri, S.K., Ganguly, P.K. and Sardar, D. (2001). Studies on the enzymatic activity of *Penicillium corylophilum* produced by solid state fermentation on rice husk for improved processibility of jute fibre. Trends Carbohydrate Chem. 7: 85-89.

13

Microbial Bioconversions of Agri-Horticultural Produces into Alcoholic Beverages - The Global Scene

Nduka Okafor

INTRODUCTION

Man has consumed alcoholic beverages from time immemorial. Like food, the alcoholic beverages indigenous to any part of the world depend on: (1) the crops, which can grow in the climate pertinent to that region, and (2) the culture of the people.

In general, alcoholic beverages produced from cereals are known as beers, while those produced from other raw materials are known as wines. This distinction is not always strictly followed. For instance, Japanese sake produced from the cereal rice, is known as a wine, rather than as a beer. Spirits are those alcoholic beverages which consist predominantly of distillates.

TYPES OF ALCOHOLIC BEVERAGES

Beers

The commonest beers are those made from barley. Beers are also produced from other cereal like rice, sorghum, maize, etc.

Barley Beers

The word beer is derived from the Latin word *bibere,* meaning to drink. The process of producing beer is known as brewing. Beer brewing from barley

Department of Applied Microbiology and Brewing, Nnamdi Azikiwe University, Awka, Nigeria
E-mail: ndukaokafor@cs.com; or ndukaokafor1@yahoo.com

was practiced by the ancient Egyptians as far back as 4,000 years ago, but investigations suggest that Egyptians learnt this art from the people of the Tigris and Euphrates, where man's civilization is said to have originated. The use of hops is, however, much more recent and can be traced back to a few hundred years ago.

Types of Barley Beers

Barley beers can be divided into two broad groups: top-fermented beers and bottom-fermented beers. This distinction is based on whether the yeast remains at the top of brew (top-fermented beers) or sediments to the bottom (bottom-fermented beers) at the end of the fermentation.

(a) Bottom-fermented Beers

These are also known as lager beers because they were stored or "lagered" (from German *lagern* = to store) in cold cellars after fermentation for clarification and maturation. Yeasts used in bottom-fermented beers are strains of *Saccharomyces uvarum* (formerly *Saccharomyces carlsbergensis).* Several types of lager beers are known. They are Pilsener, Dortmund and Munich, named after Pilsen (Czech), and Dortmund and Munich (Germany), the cities where they originated. Most of the lager (70-80%) beers drunk in the world are of the Pilsener type. Bottom fermentation was a closely guarded secret in the Bavarian region of Germany, which has Munich as its capital. Legend has it that in 1842 a monk passed on the technique and the yeast to Pilsen. Three years later they found their way to Copenhagen, Denmark. Shortly after, German immigrants transported bottom brewing to the United States.

Pilsener Beer: This is a pale beer with a medium hop taste. Its alcohol content is 3.0-3.8% by weight. Classically it is lagered for two to three months, but modern breweries in Europe, etc. have substantially reduced the lagering time to about two weeks. The water for Pilsener brew is soft, containing comparatively little calcium and magnesium ions.

Dortmund Beer: This is a pale beer, but it contains fewer hops (and, therefore, is less bitter) than Pilsener. However, it has more body (i.e. it is thicker) and aroma. The alcohol content is also 3.0-3.8%, and is classically lagered for a slightly longer period of three to four months. The brewing water is hard, containing large amounts of carbonates, sulphates and chlorides.

Munich: This is a dark, aromatic and full-bodied beer with a slightly sweet taste, because it is only slightly hopped. The alcohol content could be quite high, varying from 2-5%. The brewing water is high in carbonates but low in other ions.

Weiss: *Weiss* beer of Germany and *steam* beer of California, US, both made from wheat, are bottom-fermented beers which are characterized by being highly effervescent.

(b) Top-Fermented Beers

Top-fermented beers are brewed with strains of *Saccharomyces cerevisiae.*

Ale: Whereas lager beer can be said to be of German or continental European origin, ale (Pale ale) is England's own beer. Unless the term 'lager' is specifically used, beer always refers to *Ale* in England. It is a pale, highly hopped beer with an average alcohol content of 4-5% (w/v) but sometimes as high as 10%. Hops are added during and sometimes after fermentation. It is, therefore, very bitter and has a sharp acid taste and an aroma of wine because of its high ester content. Mild ale is sweeter because it is less strongly hopped than the standard Pale ale. In Burton-on-Trent (England) where the best ales are made, the water is rich in gypsum (calcium sulphate). When ale is produced in places with less suitable water, such water may be 'burtonized' by the addition of calcium sulphate.

Porter: This is a dark-brown, heavy bodied, strongly foaming beer produced from dark malts. It contains fewer hops than ale and consequently is sweeter. It has an alcohol content of about 5%.

Stout: Stout is a very dark heavy bodied and highly hopped beer with a strong malt aroma. It is produced from dark or caramelized malt; sometimes caramel may be added. It has a comparatively high alcohol content, 5.0-6.5% (w/v) and is stored for up to six months, fermentation sometimes proceeding in the bottle. Some stouts are sweet, being less hopped than usual.

Raw Materials for Brewing

The raw materials used in brewing are: barley malt, adjuncts, yeasts, hops and water.

(a) Barley Malt

As a brewing cereal, barley has the following advantages. Its husk is thick, difficult to crush and adheres to the kernel. This makes malting as well as filtration after mashing, much easier than with other cereals, such as wheat. The second advantage is that the thick husk is a protection against fungal attack during storage. Thirdly, the gelatinization temperature (the temperature at which the starch is converted into a water-soluble gel) is 52-59°C, which is much lower than the optimum temperature of alpha (α)-amylase (70°C) as well as of beta (β) –amylase (65°C) of barley malt. The effect is to bring the starch into solution and to hydrolyze it in one operation. Finally, the barley grains even before malting contain very high amounts of β-amylase unlike wheat, rice and sorghum. Alpha-amylase is produced only in the germinated barley seed. Two distinct barley types are known. One with six rows of fertile kernel *(Hordeum vulgare)* and the other with two rows of fertile kernels *(Hordeum distichon)*. The six-row variety is used extensively in the US, whereas the two-row variety is used in Europe as well as in parts of the US. The six-row variety is richer in protein and enzyme content than the two-row variety. This high enzymatic content is one of the reasons why adjuncts are so widely used in breweries in the US. Adjuncts dilute out the protein i.e. increase the carbohydrate/protein ratio. If an all-malt beer is brewed from malts rich in protein, as in the six-row variety, this protein will find its way into the beer and give rise to hazes.

(b) Adjuncts

These are starchy materials, which were originally introduced because six-row barley varieties grown in the US produced malt that had more diastatic power (i.e. amylases) than was required to hydrolyze the starch in the malt. The term has since come to include materials other than would be hydrolyzed by amylase. For example, the term now includes sugars (e.g. sucrose) added to increase the alcoholic content of the beer. Starchy adjuncts, which usually contain little protein, contribute after their hydrolysis to fermentable sugars, which in turn increase the alcoholic content of the beverage. Adjuncts thus help bring down the cost of brewing because they are much cheaper than malt. They do not play much part in imparting aroma, colour or taste. Starch sources such as sorghum, maize, rice, unmalted barley, cassava, potatoes can or have been used, depending on the price. Corn grits (defatted and ground), corn syrup and rice are mostly widely used in the US.

When corn (maize) is used, it is milled so as to remove as much as possible of the germ and the husk which contain most of the oil of maize, which could form 7% of the maize grain. The oil may become rancid in the beer and thus adversely affect the flavour of the beverage if it is not removed. The de-fatted ground maize is known as corn grits. Corn syrups produced by enzymatic or acid hydrolysis are also used in brewing. Since adjuncts contain little nitrogen, all the needs for the growth of the yeast must come from the malt. The malt/adjunct ratio hardly exceeds 60/40. Soybean powder, preferably defatted, may be added to brews to help nourish the yeast. It is rich in nitrogen and vitamin B.

(c) Hops

Hops are the dried cone-shaped female flower of hop-plant *Humulus lupulus* (synomyn: *H. americanus, H. heomexicams, H. cordifolius*). It is a temperate climate crop and grows wild in northern parts of Europe, Asia and North America. It is botanically related to the genus *Cannabis,* whose only representative is *Cannabis sativa* (Indian hemp, marijuana or hashish). Nowadays hop extracts are preferred in place of the dried hops. The importance of hops in brewing lies in its resins, which provide the precursors of the bitter principles in beer, and the essential (volatile) oils that provide the hop aroma. Both the resin and the essential oils are lodged in lupulin glands borne on the flower. The addition of hops has several effects:

(i) Originally it was to replace the flat taste of unhopped beer with the characteristic bitterness and pleasant aroma of hops.

(ii) Hops have some anti-microbial effects especially against beer *sarcina (Pediococus damnosus)* and other beer spoiling bacteria.

(iii) Because of the colloidal nature of the bitter substances, they contribute to the body, colloidal stability and foam head retention of beer.

(iv) The tannins in the hops help precipitate proteins during the boiling of the wort; these proteins if not removed, cause a haze (chill haze) in the beer at low temperature.

(d) Water

The mineral and ionic content and the pH of the water have profound effects on the type of beer produced. Some ions are undesirable in brewing water: nitrates slow down fermentation, while iron destroys the colloidal stability of the beer. In general, calcium ions lead to a better flavour than magnesium and sodium ions. The pH of the water and that of malt extract produced with it control the various enzyme systems in malt, the degree of extraction of soluble materials from the malt, the solution of tannins and other colouring components, isomerization rate of hop humulone and the stability of the beer itself and the foam on it. Calcium and bicarbonate ions are most important because of their effect on pH.

Water is so important that the natural water available in great brewing centres of the world lent special character to beers peculiar to these centres. Water with a large content of calcium and bicarbonate ions, as is the case with Munich (Germany), Copenhagen (Denmark), Dublin (Ireland) and Burton-on-Trent (England), is suitable for the production of the darker and sweeter beers. The reason for this is not clear but carbonates in particular tend to increase the pH, a condition that appears to enhance the extraction of dark coloured components of the malt. Water of a composition ideal for brewing may not always be naturally available. If the production is of a pale beer without too heavy a taste of hops, and the water is rich in carbonates then it is treated in one of the following ways:

(i) The water may be burtonized by the addition of calcium sulphate (gypsum). Addition of gypsum neutralizes the alkalinity of the carbonates.

(ii) An acid may be added such as lactic, phosphoric, sulphuric or hydrochloric may be added. Carbon dioxide is released, but there is an undesirable chance that the resulting salt may remain.

(iii) Boiling may decarbonate the water.

(iv) Ion exchange resins may be used.

(e) Brewers Yeasts

Not all yeasts are suitable for brewing. Brewers' yeasts, in addition to producing the adequate quantity of alcohol, must also produce the various compounds that give beer its flavour. Of the two brewing yeasts, *Saccharomyces cerevisiae* sporulates more readily than *S. uvarum*. But perhaps the most diagnostic distinction between them is that *S. uvarum* is able to ferment the trisaccharides, raffinose made up of galactose, glucose and fructose. *S. cerevisiae* is capable of fermenting only the fructose moiety; in other words, it lacks the enzyme system needed to ferment meibiose, which is formed from galactose and glucose. Finally, *S. cerevisiae* strains have a stronger respiratory system than *S. uvarum* and this is reflected in the different cytochrome spectra of the two groups.

Brewery Processes

The processes involved in the conversion of barley malt to beer may be divided into the following stages:

1. Malting
2. Cleaning and milling of the malt
3. Mashing
4. Mash operation
5. Wort boiling treatment
6. Fermentation
7. Storage or lagering
8. Packaging

Of the above processes, malting is specialized and is not carried out in the brew-house. Rather, breweries purchase their malt from specialized malsters (or malt producers).

1. *Malting*

The purpose of malting is to develop amylases and proteases in the grain. These enzymes are produced by the germinated barley to enable it to break down the carbohydrates and proteins present in the grain to nourish it before its photosynthetic systems are developed adequately to support the plant. However, as soon as the enzymes are formed and before the young seedling has made any appreciable in-road into the nutrient reserve of the grain, the development of the seedling is halted by drying, but at temperatures which will not completely inactivate the enzymes in the grain. These enzymes are re-activated during mashing and are used to hydrolyze starch and proteins and release nutrients for the nourishment of the yeasts.

Modification or production of enzymes is complete in four to five days of the growth of the seedling; the extent being tested roughly by the sweet taste developed in the grain. Further reactions in the grain are halted by *kilning,* which consists of heating the 'green' malt in an oven, first at a relatively mild temperature until the moisture content is reduced from about 40% to 6%. Subsequently, the temperature of heating depends on the type of beer to be produced. At the end of malting, some changes occur in the gross composition of the barley grain as seen in Table 13.1. The barley malt, with its rich enzyme content, resembles swollen grains of unthreshed rice and can be stored for considerable periods before being used.

2. *Milling of Malt*

The purpose of milling is to expose particles of the malt to the hydrolytic effects of malt enzymes during the mashing process. The finer the particle,

Table 13.1. Composition of barley grain before and after malting

Fraction	Proportion (% dry weight)	
	Barley	Malt
Starch	63.0-65.0	58.0-60.0
Sucrose	1.0-2.0	3.0-5.0
Reducing sugars	0.1-0.2	3.0-4.0
Other sugars	1.0	2.0
Soluble gums	1.0-1.5	2.0-4.0
Hemicelluloses	8.0-10.0	6.0-8.0
Cellulose	4.0-5.0	5.0
Lipids	2.0-3.0	2.0-3.0
Crude protein (N x 6.25)	8.0-11.0	8.0-11.0
Albumin	0.5	5.0
Globulin	3.0	-
Hordein-protein	3.0-4.0	2.0
Glutelin-protein	3.0-4.0	3.0-4.0
Amino acids and peptides	0.5	1.0-2.0
Nucleic acids	0.2-0.3	0.2-0.3
Minerals	2.0	3.0
Others	5.0-6.0	6.0-7.0

therefore, the greater is the extract from the malt. However, very fine particles hinder filtration and prolong it unduly. The brewer has, therefore, to find a compromise particle size which will give him maximum extraction, and yet permit reasonably rapid filtration rate. No matter what is chosen, the crushing is so done as to preserve the husks, which contribute to filtration, while reducing the endosperm to fine grits.

3. *Mashing*

Mashing is the central part of brewing. It determines the nature of the wort, hence the nature of the nutrients available to the yeasts and, therefore, the type of beer produced. The purpose of mashing is to extract the maximum soluble portion of the malt and to enzymatically hydrolyze insoluble portions of the malt and adjuncts. In essence, mashing consists of mixing the ground malt and adjuncts at temperatures optimal for amylases and proteases derived from the malt. The aqueous solution resulting from mashing is known as wort. The two largest components of barley malt, in terms of dry weight, are starch (55%) and protein (10-12%). The controlled breakdown of these two components has a tremendous influence on beer character and is detailed below.

Starch Breakdown During Mashing

The key enzymes in the breakdown of malt starch are the α-and β-amylase. The temperature of optimal activity and destruction of these enzymes as well as their optimum pH are given in Table 13.2.

Mashing is affected by a combination of temperature, pH, time and concentration of the wort. When the temperature is held at 60-65°C for long periods, wort rich in maltose occurs because β-amylase activity is at its optimum and this enzyme yields mainly maltose. On the other hand, when a higher temperature around 70°C is employed dextrins predominate. Dextrins contribute to the body of the beer but are not utilized by yeast. Mash exposed to too high temperature will, therefore, be low in alcohol due to insufficient maltose production. The concentration of the mash is important. The thinner the mash the higher the extract (i.e. the materials dissolved from the malt) and the maltose content.

Table 13.2. Temperature optima of α- and β-amylases

Enzyme	Optimum temperature (°C)	Temperature of destruction (°C)	Optimal pH
Alpha-amylase	70.0	80.0	5.8
Beta-amylase	60.0-65.0	75.0	5.4

Protein Breakdown During Mashing

The breakdown of the malt proteins, albumins, globulins, hordeins and gluteins starts during malting and continues during mashing by proteases which breakdown proteins through peptones to polypeptides and polypetidases, which breakdown the polypeptides to amino acids. Protein breakdown has no pronounced optimum temperature, but during mashing it occurs evenly up to 60°C, beyond this proteases and polypetidases are greatly retarded. Proteoloytic activity in wort is, however, dependent on pH and for this reason wort pH is maintained at 5.2-5.5 with lactic acid, mineral acids or calcium sulphate. The optimum pH for β-amylase activity is about the same as that of proteolysis and as can be seen in Table 13.2, a fortunate coincidence for the maximum production of maltose and the breakdown of protein.

4. *Mashing Methods*

There are three broad mashing methods:

(a) Decoction method–where part of the mash is transferred from the mash tun to the mash kettle where it is boiled.

(b) Infusion method–where the mash is never boiled, but the temperature is gradually raised.

(c) Double mash method–in which the starchy adjuncts are boiled and added to the malt.

Decoction method: In this method the mash is mixed at an initial temperature of 35-37°C and the temperature is raised in steps to about 75°C. About one-third of the initial mash is withdrawn, transferred to the mash kettle, and heated slowly to boil, and returned to the mash tun, the temperature of the mash is gradually raised in the process. The enzymes in the heated portion are destroyed but the starch grains are cooked, gelatinized and exposed. Another portion may be removed, boiled and returned. In this way the process may be a one-, two- or three-mash process. In a three-mash process (Fig. 13.1) the initial temperature of 35-40°C favours proteolysis; the mash is held for about half an hour at 50°C for full proteolysis, then for about one hour at 60-65°C for saccharification and production of maltose, and at 70-75°C for two to three hours for dextrin production.

Infusion method: Infusion method is typically used to produce top-fermenting beers. The method involves grinding malt and a smaller amount of unmalted cereal, which may sometimes be precooked. The ground material (grist) is mixed thoroughly with hot water (2:1 by w/w) to produce a thick porridge-like mash and the temperature is carefully raised to about 65°C. It is then held at this temperature for a period varying from

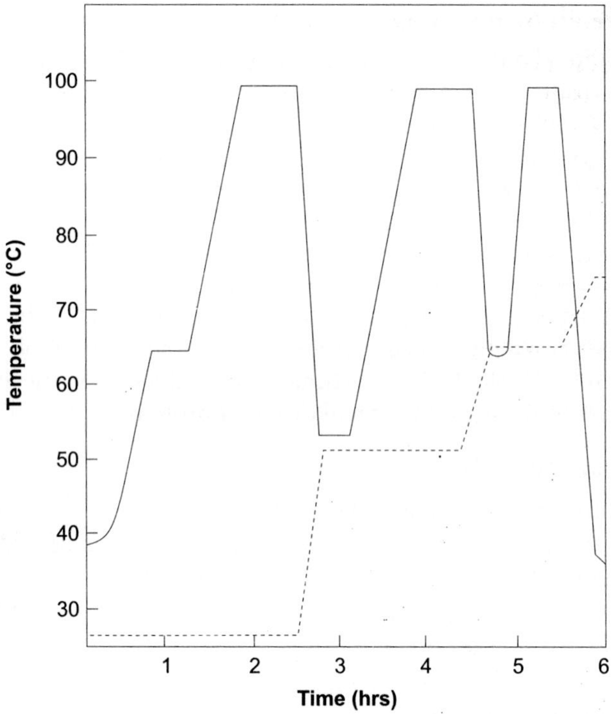

Fig. 13.1. Three stage decoction method. Broken lines indicate temperature of main mash. Unbroken lines indicate temperature of added portion of mash.

30 minutes to several hours. The enzyme acts principally on the starch and its degradation products in both the malted and unmalted cereal, and only a little protein breakdown occurs. It is more easily automated, but malt in which the proteins are already well degraded must be used since the high temperature of mashing rapidly destroys the proteolytic enzymes.

Double-mash method (also called the cooker method): This method was developed in the US because of its use of adjuncts. It has features in common with the infusion and the decoction method. In a typical American double mash method, ground malt is mashed with water at a temperature of 35°C. It is then held for an hour during the "protein rest" for proteolysis. Adjuncts are then cooked in an adjunct cooker for 60-90 minutes. Sometimes about 10% malt is added during the cooking. Hot cooked adjunct is then added to the mash of ground malt to raise the temperature to 65-68°C for starch hydrolysis and maintained at this level for about half an hour. The temperature is then increased to 75-80°C after which the mashing is terminated.

Mash Separation and Boiling

The conventional method of separating the husks and other solids from the mash is to strain the mash in a lauter (German word for clarifying) tub, which is a vessel with a perforated false bottom about 10 mm above the real bottom on which the husks themselves form a bed through which the filtration takes place. In recent times in large breweries, especially in the US, the Nooter strain master has come into use. Like the lauter tub, filtration is through a bed formed by the husks, but instead of a false bottom, straining is through a series of triangular perforated pipes placed at different heights of the bed. The strain master itself is rectangular with a conical bottom, whereas the lauter tub is cylindrical. Its advantage over others is that it can handle larger quantities than the lauter tub. Besides the lauter tub and the strain master, cloth filters located in plate filters and screening centrifuges are also used.

The wort is boiled for 1-1½ hours in a brew kettle (copper) which used to be made of copper (hence the name) but which, in many modern breweries, is now made of stainless steel. When corn syrup or sucrose is used as an adjunct, it is added as boiling commences. Hops are also added, some before, and some at the end of the boiling. The purpose of boiling is as follows:

(a) To *concentrate* the wort, which loses 5-8% of its volume by evaporation during the boiling.

(b) To *sterilize* the wort to reduce its microbial load before its introduction into the fermentor.

(c) To *inactivate* any enzymes so that no change occurs in the composition of the wort.

(d) To *extract* soluble materials from the hops, which not only aid in protein removal, but also in introducing the bitterness of hops.

(e) To *precipitate* protein, which forms large flocs because of heat denaturation and complexing with tannins extracted from the hops and malt husks. Un-precipitated proteins form hazes in the beer, but too little protein leads to poor foam head formation.

(f) To *develop* colour in the beer; some of the colour in beer comes from malting but the bulk develops during wort boiling. Colour is formed by several chemical reactions including caramelization of sugars, oxidation of phenolic compounds, and reactions between amino acids and reducing sugars.

(g) *Removal* of volatile compounds: Volatile compounds such as fatty acids, which could lead to rancidity in the beer, are removed. During the boiling, agitation and circulation of the wort help increase the amount of precipitation and floc formation.

5. *Pre-fermentation Treatment of Wort*

The hot wort is not sent directly to the fermentation tanks. If dried hops are used, then they are usually removed in a hop strainer. During boiling, proteins and tannins are precipitated while the liquid is still warm. Some more precipitation takes place when it is cooled to about 50°C. The "warm" precipitate is known as 'trub' and consists of 50-60% protein, 16-20% hop resins, 20-90% polyphenols and about 3% ash. Trub is removed either with a centrifuge, or a whirlpool separator, which is now common. When the temperature has fallen to about 50°C, further sludge known as 'cold break' begins to settle, but it cannot be separated in a centrifuge because it is too fine. In many breweries the wort is filtered at this stage with kieselguhr, a white diatomaceous Earth.

6. *Fermentation*

The cooled wort is now ready for fermentation. It contains no enzymes but it is a rich medium for fermentation. It has, therefore, to be protected from contamination. During the transfer to the fermentor, the wort is oxygenated at about 8 mg l^{-1} of wort in order to provide the yeasts with the necessary oxygen for initial growth. It is pumped by gravity into fermentation tanks and yeast is inoculated or 'pitched' in.

Top fermentation: This is used in England for the production of stout and ale, using strains of *S. cerevisiae*. Traditionally an open fermentor is used. A fishtail spray introduces wort so that it becomes aerated to the tune of 5-10 ml litre^{-1} of oxygen for the initial growth of the yeasts. Yeast is pitched in at the temperature of 15-16°C. The temperature is allowed to rise gradually to 20°C over a period of about three days. At this point it is cooled to a constant temperature. The entire primary fermentation takes about six days. Yeasts float to the top during this period, they are scooped off and used for future pitching. The transfer helps aerate the system and also enables the discarding of the cold break sediments. Sometimes aeration is also achieved by circulation with paddles and by the means of pumps. Nowadays, cylindrical vertical closed tanks are replacing the traditional open tanks.

Bottom fermentation: Wort is inoculated to the concentration of 7-15 x 10^6 yeast cells ml^{-1} of wort. The yeast cells then increase four to five times in number over three to four days. Yeast is pitched in at 6-10°C and is allowed to rise to 10-12°C, which takes some three to four days before it is cooled and maintained at 6-8°C for the next three to four days; it is cooled to about 5°C at the end of the fermentation. Carbon dioxide is released and creates a head called *Krausen* (actively fermenting wort), which begins to collapse after four to five days as the yeasts begin to settle. The total fermentation period may last from 7-12 days.

Following fermentation progress: The progress of fermentation is followed by wort specific gravity. During fermentation the gravity of the wort gradually decreases because yeasts are using up the extract. However alcohol is also being formed. As alcohol has a lower gravity than wort, the reading of the special hydrometer (known as a saccharometer) is even lower. The saccharometer reading does not, therefore, reflect the real extract, but an apparent extract, which is always lower than the real extract because of the presence of alcohol. In England and some other countries the extract is measured as the direct specific gravity at 60 F (15.5 C) x 1,000. Hence, with wort with sp. gr. of 1.053 the extract would be 1,053°. Outside England, the extract is measured in °Balling or °Plato. Both systems measure the percentage of sucrose required to give solutions of the same specific gravity. °Brix is used in the sugar industry, whereas °Balling (United States) and °Plato (Continental Europe) are used in the brewing industry.

7. Lagering (Bottom-fermented Beers) and Treatment (Top-fermented Beers)

(a) Lagering: At the end of the primary fermentation above, the beer, known as 'green beer', is harsh and bitter. It has a yeasty taste arising probably from higher alcohols and aldehydes. The green beer is stored in closed vats at a low temperature (around 0°C), for periods which used to be as long as six months in some cases, to mature and make it ready for drinking. During lagering secondary fermentation occurs. Yeasts are sometimes added to induce this secondary fermentation, utilizing some sugars in the green beer. The secondary fermentation saturates the beer with CO_2. Sometimes actively fermenting wort (*Kraeusen*) may be added. At other times CO_2 may be added artificially into the lagering beer. Materials which might undesirably affect flavour and which are present in green beer e.g. diacetyl, hydrogen sulphide, mercaptans and acetaldehyde are decreased by evaporation during secondary fermentation. An increase occurs in the desirable components of the beer such as esters. Any tannins, proteins and hop resins still left are precipitated during the lagering period.

Lagering used to take up to nine months in some cases. The time is now considerably shorter and in some countries the turnover time from brewing, lagering and consumption could be as short as three weeks. This reduction has been achieved by artificial carbonation and by the manipulation of the beer due to greater understanding of the lagering processes. Thus, in one method used to reduce lagering time, beer is stored at high temperature (14°C) to drive off volatile compounds e.g. H_2S, and acetaldehyde. The beer is then chilled at -2°C to remove chill haze materials, and thereafter it is carbonated. In this way lagering could be reduced from two months to 10 days.

Lagering gives the beer its final desirable organoleptic qualities, but it is hazy due to protein-tannin complexes and yeast cells. The beer is filtered through kieselguhr or through membrane filters to remove these. Some properties of lager beer are given in Table 13.3.

(b) Beer treatment (for top-fermented beers): Top-fermented beers do not undergo the extensive lagering of bottom-fermented beers. They are treated in casks or bottles in various ways. In some processes the beer is transferred to casks at the end of the fermentation with a load of 0.2-4.0 million yeast cells ml^{-1}. It is 'primed' to improve its taste and appearance by the addition of a small amount of sugar mixed with caramel. The yeasts grow in the sugar and carbonate the beer. Hops are also sometimes added at this stage. It is stored for seven days or less at a temperature of about 15°C. After priming, the beer is 'fined' by the addition of isinglass. Isinglass, a gelatinous material from the swim bladder of fish, precipitates yeasts cells, tannins and protein-tannin complexes. The beer is thereafter pasteurized and distributed.

8. *Packaging*

The beer is transferred to pressure tanks from where it is distributed to cans, bottles and other containers. The beer is not allowed to come in contact with oxygen during this operation; it is also not allowed to lose

Table 13.3. Some properties of lager beer

Property	Pilsener	US Lager Beer	Danish Pilsener	English Ale	English Stout	Munich Lowenbrau	Dortmund
Original extract content °pc	12.1	11.50-12.00	10.60	15.00	21.10	13.30	13.60
Real extract content °pc	5.30	5.50	3.10	5.00	8.70	6.40	5.50
Alcoholic content, wt %	3.50	3.40-3.80	3.90	5.20	6.70	3.60	4.20
Protein content, wt %	–	0.28-0.35	0.30	0.60	0.60	0.50	0.80
CO_2 content %	0.53	0.50	0.40	0.41	–	0.42	–
Colour, EBC	10.00	2.70	5.00	–	–	40.00	8.00
Air in bottle, ml	–	1.50	2.00	8.00	10.00	–	6.00
pH	–	4.20-4.50	4.00	–	–	–	–
Real degree of attenuation, %	–	60.00-75.00	69.00	66.00	59.00	48.00	60.00

CO_2 or to become contaminated with microorganisms. To achieve these objectives, the beer is added to the tanks under a CO_2 atmosphere, bottled under a counter pressure of CO_2, and all equipments are cleaned and disinfected regularly.

Bottles are thoroughly washed with hot water and sodium hydroxide before being filled. The filled and crowned bottles are passed through a pasteurizer, set to heat the bottles at 60°C for half an hour. The bottles take about half an hour to attain the pasteurizing temperature, remain in the pasteurizer for half an hour and take another half an hour to cool down. This method of pasteurization sometimes causes hazes and some of the larger breweries now carry out bulk pasteurization and fill containers aseptically.

Use of External Enzymes

In countries such as Nigeria external enzymes are widely used with malted or unmalted sorghum, or maize grits, on account of the high expense of barley malt which has to be imported with foreign exchange. Due to the absence of the husks of barley malt, the lauter tun has been replaced with cloth strainers. The beer quality remains as high as when barley was used. Consumers from Nigeria and outside do not notice any difference. Several different enzymes are used, beginning with thermostable α-amylase, then β-amylase, and a protease.

Sorghum (and Maize) Beers

Barley is a temperate crop. In many parts of tropical Africa, beer has been brewed for generations with locally available cereals. The commonest cereal used is *Sorghum bicolor* (*Sorghum vulgare)* known in the US as "milo", in South Africa as "kaffir" corn, and in some parts of West Africa as Guinea corn. The cereal, which is indigenous to Africa, is highly resistant to drought. Sorghum is often mixed with maize *(Zea mays)* or millets, *(Pennisetum* spp.). In some cases such as in Central Africa e.g. Zimbabwe, maize may form the major cereal. Outside Africa sorghum is not used normally for brewing except in the US, where it is occasionally used as an adjunct. The method of processing and the nature of the sorghum beers is remarkably similar throughout Africa. The main features are:

(i) They are usually pinkish in colour, sour in taste, and of fairly heavy consistency imposed partly by starch particles.
(ii) They are consumed without the removal of the organisms.
(iii) They are not aged or clarified.
(iv) They include lactic acid fermentation.

The tropical beers of Africa are known by different names in various parts of the world: "buru-kutu", "otika", and "pito"' in Nigeria, "maujek" among the Nandis in Kenya, "mowa" in Malawi, "kaffir beer" in South Africa, "merisa" in Sudan, "bouza" in Ethiopia and "pombe" in many parts of East Africa.

Sorghum beers are usually produced on small scales. It is only in South Africa that production has been undertaken in large breweries; elsewhere although considerable quantities are produced, this is done by small holders to satisfy small local clienteles. In South Africa, in fact, it is reported that three or four times more sorghum beer is produced and consumed than is the case with barley beer. The processes of producing the beer include malting, mashing and fermentation.

Malting

For malting sorghum beers, grains are steeped in water for periods varying from 16-46 hours. They are then drained and allowed to germinate for five to seven days, water being sprinkled on the spread out grains. At the end of this period, the grains are usually dried, often in the sun or in the South African system in an oven at 50-60°C. Kilning is however not done. In some parts, the dried malt may be stored and used over several months.

Contrary to opinions previously held by many, sorghum malt is rich in amylases, particularly α-amylase, although the ungerminated grain does not contain β-amylase, as is the case with barley. Sorghum has not received much attention as a brewing material except occasionally as an adjunct in the US.

It has been suggested that the saccharification of sorghum starch is brought about partly by the fungi which grow on the grains during their germination as well as by the germinated sprout. Some workers, however, have vehemently disputed this. The fungi so implicated are *Rhizopus oryzae, Botryodiplodia theobromae, Aspergillus flavus, Penicillium funiculosum* and *P. citrinum.*

Mashing

The malt is coarsely ground and mixed in a rough proportion (6:1, v/v) with water and boiled for about two hours. During boiling, starchy adjuncts in the form of dried powder of plantains, cassava flour or 'gari' a fermented food made from cassava (*Manihot esculenia* Crantz) or unmalted cereal may be added so that an approximate 1:2:6 proportion of the adjunct, malt and water is attained. It is filtered and is then ready for fermentation. In South African sorghum beer breweries, the adjunct consisting of boiled sorghum or maize grits is added after the initial souring of the mash.

Fermentation

Two types of fermentation take place during sorghum beer production: lactic acid fermentation, and alcoholic fermentation. In traditional fermentation, the dregs of a previous fermentation are inoculated into the boiled, filtered and cooled wort. This inoculum consists of a mixture of yeasts, lactic acid and acetic acid bacteria. The first phase of the fermentation is brought about by lactic bacteria mainly *Lactobacillus mesenteroides,* and *Lb. plantarum.* In sorghum beer breweries in South Africa, the temperature of the mash is held initially at 48-50°C to encourage the growth of thermophilic lactic acid bacteria that occur naturally on the grain, for 16-24 hours. The pH then drops to about 3.0-4.0. The sour malt is added to the previously cooked adjunct of un-malted sorghum or maize, and sometimes some more malt may be added. It is then cooled to 38°C and pitched with the top fermenting yeasts.

In the traditional method, yeasts and lactic acid bacteria are present in the dregs. The yeasts which have been identified in Nigerian sorghum beer fermentation are *Candida* spp., *S. cerevisiae,* and *S. chevalieri.* Fermentation is for about 48 hours during which lactic acid bacteria proliferate.

Thereafter, it is ready for distribution and consumption. No secondary fermentation of the kind is seen in lager beer, lagering or clarification is done. The live yeasts and the lactic acid bacteria are consumed in much the same ways as they are done in palm wine. In some localities the fermentation lasts a little longer and the flavour is influenced by a slight vinegary taste introduced by the action of acetic acid bacteria.

Sorghum beers usually contain large amounts of solids (Table 13.4) mainly starch, apart from the microorganisms. It is for this reason that some authors have considered these alcoholic beverages equivalent to foods. Sorghum beers from various parts of Africa are briefly described.

"Burukutu" and "Pito"

"Burukutu" and "Pito" are very popular in the more northern (grassland) parts of Nigeria and West Africa and are made from sorghum. It appears that the two names are synonyms for the same brew. The only difference between them is, perhaps, that adjuncts are added during the production of "burukutu", whereas they are not added in the production of "pito". The adjuncts used are cassava flour or 'gari'-the fermented food made from cassava with 0.3-1.5% lactic acid content. The beverages are usually produced from sorghum; however, maize alone or a mixture of maize and sorghum may be used. The stages in "burukutu" and "pito" preparation are malting, mashing, fermentation and maturation (Fig. 13.2).

Sorghum grains are steeped in water overnight. The grains are then drained of water and spread on banana leaves up to a depth of about 10 cm and covered with more leaves. The grains are watered on alternate days and occasionally turned over. Germination lasts for four to five days. The malt is sun-dried for a day or two before grinding. During "burukutu"

Table 13.4. Properties of South African sorghum beer

Properties	Small-scale	Factory
pH	3.50	3.40
Alcohol	0.10	3.00
Solids (% w/v)		
-Total	4.90	5.40
-Insoluble	2.30	3.70
Nitrogen (% w/v)		
-Total	0.084	0.093
-Soluble	-	0.014

production, adjunct in the form of 'gari' is added with water to a 'gari'–malt–water ratio of 1:2:6 (v/v) and left to ferment for two days. It is then boiled for four hours, cooled, and left to mature for two days.

The microflora on the sorghum malt includes many moulds (Table 13.5). It will appear that enzymes produced by these moulds along with the enzyme of the malt help to saccharify the starch in the grains. The moulds include species of *Mucor, Aspergillus* and *Penicillium.* Many types of yeasts are also present on the malt. During the fermentation of the 'gari'-sorghum mash, these yeasts and lactic acid bacteria flourish whilst moulds disappear. During the 48-hour fermentation period the pH drops from about 6.5 to 4.0, no doubt as a result of lactic activity. During maturation (48 hours), acetic acid bacteria form the bulk of the organisms resulting in further souring of the beverage. The persisting sweetish taste is due to several sugars identified in the beer (Table 13.6).

"Bouza"

It is an alcoholic beverage produced in Egypt. Formerly preferred by all classes, it is now consumed mainly among the lower income groups.

Table 13.5. Microorganisms involved in "Pito" production

	Organisms found on sorghum malt after 3 days of malting	Organisms found during fermentation	Organisms found during maturation
Mucor spp.	+	-	-
Aspergillus spp.	+	-	-
Penicillium spp.	+	-	-
Saccharomyces spp.	+	+	+
Candida spp.	+	+	+
Lactobacillus spp.	+	+	+
Streptococcus spp.	+	+	+
Acetobacter spp.	-	-	+

Grain (maize and/or sorghum)

↓

Soak for 2 days

↓

Malt (germinate) for 5 days

↓

Grind (or sun-dry and hold until used)

↓

Adjunct ('gari') added in "burukutu" production
(no adjunct in "pito")

↓

Mix mash with cold water

↓

Boil mash for 6 to 12 hours

↓

Filter through a fine mash

↓

Cool filtrate (Residue used as animal feed)

↓

Ferment overnight (mixed natural inoculum)

↓

Boil for 12 hrs

↓

Cool concentrate and add starter
(sediment from earlier brew)

↓

Incubate for 12 t/o 24 hours

↓

"Pito"

Fig. 13.2. Flow sheet for the traditional production of "pito" and "burukutu"

Egyptian "Bouza" is prepared either from wheat or maize, but the one prepared from wheat is more popular. To prepare "bouza", coarsely ground wheat (about 20% of the total) is placed in a large wooden basin and kneaded with water to form dough which is cut into thick loaves and lightly baked. The rest of the wheat (about 80% of the total) is germinated for three to five days, sun-dried and coarsely ground. The malted wheat and the crumbled bread are mixed in water in a wooden barrel. "Bouza" from a previous fermentation is added and the whole mixture is fermented for 24 hours at room temperature. The mash is sieved to remove large

Table 13.6. Sugars in "burukutu" beer

Sugar	Concentration (% wt)
Fructose	0.20
Mannose	0.65
Glucose	0.60
Galactose	0.36
Galacturonic acid	0.48
Sucrose	0.20
Maltose	0.42
Isomaltose	0.38
Raffinose	0.60
Maltotriose	0.56
Stachyose	0.43
Maltotrtrose	0.16

Table 13.7. Changes in pH and alcoholic content of market and laboratory "bouza" with increasing fermentation

Sample	24 hours	48 hours	72 hours
pH			
- Laboratory 'bouza'	4.05	3.65	3.50
- Market 'bouza'	3.90	3.80	3.68
Ethanol content (% w/w)			
- Laboratory 'bouza'	4.20	5.06	5.39
- Market 'bouza'	3.91	4.12	4.45

particles and the beverage is ready for drinking. The beverage has a pH of about 3.6 and an alcohol content of about 4%, but the pH drops while alcohol increases with further fermentation (Table 13.7).

"Talla" (tella)

It is an Ethiopian small producer beer with a smoky flavour derived from inverting the fermentation containers and "talla" collection pots over smouldering olivewood. "Talla" also acquires some smoke flavour from the toasted, milled and boiled cereal grains. During the toasting the grains are roasted until they begin to smoke slightly. The flow sheet for "talla" production is given in Fig. 13.3. In the production of "talla", powdered hop leaves and water are put in a fermentation vessel and allowed to stand for about three days. Ground barley or wheat malt and pieces of flat bread which is baked to burning point on the outside, are added. On the fifth day, hop stems are added in addition to cereal flour made by milling sorghum which is first boiled and then toasted. Water is added and it is left

Smoke-season an earthenware pot or barrel

↓

Add water and hops

↓

Let stand for 3 days

↓

Add powdered barley or wheat malt

↓

Add broken pieces of freshly baked flat bread made
from cereal flour and water

↓

Cover and let stand for 2 days

↓

Add pulverized hop stems and cereal flour slurry
(made by toasting, milling and boiling one or more kinds of grain)

↓

Let stand for 1 day

↓

Add more water, cover and ferment for 5 to 7 days

↓

Filtrate into a clean smoke-season pot

↓

"Talla"

Fig. 13.3. Flow sheet for the production of "talla"

to ferment for two days (i.e. to the seventh day). It is filtered and ready to drink.

"Busaa"

"Busaa" is an acidic alcoholic beverage consumed by the Luo, Abuluhya and Maragoli ethnic groups of Kenya. It is porridge-like and light brown in colour, and is warmed to 35-40°C before being consumed. A stiff dough made from maize flour and water is incubated at room temperature for three to four days. The fermented dough is pounded and then heated on a metal plate till it turns brown.

Malt is made from millet, allowed to germinate for three to four days, sun-dried, and ground to powder. A slurry is made in water of the millet malt powder and roasted maize flour dough and left to ferment for two to four days. It is filtered through cloth and the filtrate is "busaa".

The organisms responsible for fermenting the uncooked maize dough include *Candida krusei, S. cerevisiae, Lactobacillus helveticus, Lb. salivarus, Lb. brevis, Lb. viridescens, Lb. plantarurn, Pediococcus damnosus,* and *P. panulus.* The final product is the result of alcohol produced mainly by *S. cerevisiae.* The lactic acid in the beverage is produced by several lactobacilli.

"Merissa"

"Merissa" is a sour Sudanese alcoholic (up to 6%) beverage made from sorghum. It has a pH of about 4.0 and a lactic acid content of about 2.5%. Sorghum grains are malted, dried, and ground into a coarse powder. Unmalted sorghum is milled into a fine powder. One third of this powder is mixed with a little water and allowed to ferment at room temperature for 24-36 hours. The resulting fermented sour dough is heated without further water addition until the product caramelizes to give rise to "soorji." The cooled "soorji" is allowed to ferment for 4-5 hours in a mixture of malt, to which previous "merissa" is added as inoculum. The result is a vigorously fermenting product known as "dehoba." The remaining two-thirds of the sorghum flour is cooked in two equal amounts: one to a greyish brown paste, and the other cooked further to a brown paste. The two are mixed together and allowed to cool, resulting in a product called "futtura". Portions of the "futtura" are mixed with malt flour (about 5%) and added to the "dehoba" from time to time. Fermentation lasts from 8-10 hours, after which it is filtered to produce the resultant drink, "merissa".

Other Alcoholic Beverages

"Tej"

"Tej" is a mead (i.e. a wine made by fermenting honey) of Ethiopian origin. It is yellow, sweet, effervescent, and cloudy due to its yeast content. As it is expensive, it is beyond the reach of most Ethiopians and used only on special occasions. The wine may be flavoured with spices or hops and also by passing smoke into the fermentation pot before it is used. To prepare "Tej", the honey is diluted with water in the ratio of 1:2 to 1:5 i.e. to a liquid of 13-27% sugar (since honey contains about 80% sugar). Yeasts of the genus *Saccharomyces* spontaneously ferment the brew in about five days to give a yellow wine of 7-14% alcohol (v/v)

"Agadagidi"

Wines from bananas and plantains have the opaque, effervescent sweet-sour nature typical of African traditional alcoholic beverages. In Nigeria the best known is "agadagidi." This beverage is found in the cocoa growing areas of southwestern Nigeria where plantains provide shade for the young cocoa trees. The ripe fruits are peeled and soaked in water where the sugars dissolving from the preparation permit the development of yeasts and lactic acid, and result in a typical opaque effervescent wine. The alcohol content is about 1%.

"Mbege"

It is consumed in Tanzania mainly by those living near Mount Kilimanjaro. It is not a wholly banana/plantain wine as is the case with Nigerian "agadagidi." Rather, it is produced from a mixture of malted millet and fermented banana juice. The juice is produced by boiling the ripe banana followed by decantation. The banana infusion is mixed with cooked and cooled millet malt and allowed to ferment for four to five days.

Palm wines

Palm wines are produced from saps obtained from palms. Palms themselves are monocotyledonous plants of the family Palmae, tribe Cocoineae and are found in all parts of the tropical and sub-tropical world. The type of wine depends on the nature of the palm from which the sap is obtained. In India and Sri Lanka for example it is obtained from the coconut *Cocos nutifera,* in the Middle East and North Africa from the date palm, *Phoenix daetyhfera,* and in Brazil from several palms including *Morenia montana* and *Sheela a princeps.*

All palm wines are milky white effervescent liquids which have a sweetish taste and a variable alcoholic content ranging from 0.5 to about 7.0% (v/v), depending on the stage of fermentation at which they are consumed. In Africa the major palms from which palm wine is obtained are the oil palm *Elaeis guineensis* and the raphia palms *(Raphia hookeri* and *R. vinifera).* Raphia palms grow best in swampy land.

Obtaining the Sap

Sap is obtained from *Elaeis guineensis* in one of the following four ways:

(1) From a slit, usually triangular, made at the bottom of the immature male inflorescence. Not all such slits result in the yield of the sap. Where sap successfully oozes out, the surface of the triangular incision has to be cleared twice daily to prevent slimy materials from the plant blocking the sieve tubes. Sap obtained in this way produces wine which is particularly popular in West Africa including Nigeria and Ghana.

(2) The stem of the palm may be tapped just below the apical growth point.

(3) The stem may be tapped above the root system.

(4) The terminal bud of the felled tree may be cut and the sap collected.

This last method leads to the depletion of the tree population and is not favoured in Nigeria. In Ghana where it is popular, a government requirement is that a new palm is planted to replace the felled tree.

Raphia palms die once they have flowered and fruited. Sap is obtained from cutting the terminal bud when the flowers appear and sap issuing from it is collected. The method is used both in Nigeria and in Zaire. Raphia palm wine has not been studied in as much detail as that of the oil palm, probably due to the easier accessibility of the latter which grows on dry land rather than in swamps. The rest of the discussion will, therefore, be mainly on wine from the oil palm, *Elaeis* sp.

Composition of Palm Sap

The sap of the oil palm tree as it drips from the tree is a clear, sweet and syrupy liquid. It has a pH of about 7.0 and contains sucrose ranging from about 9-14%, as well as small amounts of glucose, fructose, raffinose and vitamins.

Microorganisms in Palm Wine

The microorganisms in palm wine are yeasts and bacteria. The majority of the yeasts are *Saccharomyces* spp. Other species which have been encountered include *Pichia, Schizosaccharomyces, Endomycopsis* and *Candida.* The predominant bacteria in palm wine are lactic acid bacteria. In early fermentation, *Enterobacteriaceae* such as *Serratia* may occur but they soon disappear. If the wine fermentation proceeds beyond about 12 hours, acetic acid bacteria begin to flourish. The distribution of the individual species of yeast and bacteria in palm wine is highly variable. The milky appearance and slight thickness of palm wine appears to be conferred on it by activities of gum producing *Leuconostoc* spp. When microorganisms are inoculated into palm sap or sugar solution, the typical palm wine milky white colour was produced by *Leuconostoc* while the yeast produced the typical odour.

The microorganisms enter the sap as it drips from the tree fermenting the sugary liquid and converting it to wine. The yeasts and bacteria enter the sap through a variety of routes: the activities of insects, the tapping utensils (i.e. tapping knife, funnel and collecting vessel) and the surface of the palm tree and the air.

Biochemistry of the Conversion of Palm Sap to Palm Wine

The utensil in which the sap dripping from the tree is collected is an unsterile one, which is used solely for this purpose. As a consequence, it already contains an inoculum of the microorganisms which convert the sap to wine.

The result of microbial activities is an effervescent beverage which is sweetish within the first 12 hours or so of the fermentation. Beyond that period the sucrose content drops rapidly and within 36 hours it is almost completely depleted as a result of concomitant production of lactic and acetic acids. Malic, pyruvic, succinic, and tartaric acids also contribute to the titrable acidity of the wine. All the acids, except tartaric, are absent from the sap. The wine is no longer acceptable to most people after about 36 hours.

The alcohol content of palm wine is variable, usually ranging from 1.5% to about 4% at the time it is consumed. When it is fermented for longer than, say, four days, the alcohol content could be up to 8-22% although it is then used more for distillation of the palm wine gin – "ogogoro" – rather than for drinking.

The Future of Traditional African Alcoholic Beverages

Most of the African traditional beverages are produced on a small scale. However, as urbanization and industrialization increase in the producer countries, the trend is likely to be towards the large-scale industrial production of the beverages. It is expected that the intrinsic nature of these beverages (opaqueness, sour-sweet taste, effervescence, etc.) will be retained, as any drastic changes will most probably result in new beverages that may be rejected.

A number of scientific stages will, however, be passed before industrialized large-scale production. The stages begin with isolating and identifying the microorganisms involved in the fermentation. Studying their physiology and improving them genetically, though not strictly necessary, may perhaps be undertaken before the beverages are produced experimentally on a small scale. Process improvement may take the form of using alternative substrates with more assured composition such as a nutrient sucrose solution in place of palm sap in palm wine before pilot, and later, a full industrial scale production is undertaken. Table 13.8 summarizes information on the beverages and shows that South Africa sorghum beer and palm wine are perhaps the most advanced, being produced in industrial and semi-industrial scales, respectively.

One major area of prior study necessary for large-scale production is the establishment of the appropriate 'palm wine' yeast or "Bouza" yeast, in the same way as a grape wine yeast is known. Currently several different yeasts have been identified in palm wine (Table 13.9), sometimes up to four different yeasts being found simultaneously. In a sample of "burukutu" beer, for instance, up to six different yeasts were found: *Saccharomyces cerevisiae* (12.5%), as percentage of total organisms including bacteria, *S. chevalieri* (1.3%), *Candida tropicalis* (2.3%), *C. guilliermondi* (0.61%), *Candida mycoderma* (0.05%), *Hansenula anomala* (1.4%).

Table 13.8. Traditional beverages of Africa and stages attained towards industrialization of each

Sl. No.	Beverage	Country/Region of Production	Substrate	Description	Stage of Industrialization
		Rain Forest Zone			
1	Palm wines	Southern Nigeria, Southern Ghana, Cameroon, Zaire	Sap from various palms	Milky white effervescent sour-sweet liquid	1, 2a, 3a, 3b, 4, 5
		Grassland Zone			
2	South African Sorghum Beer	Southern Africa, South Africa, Malawi, Zimbabwe, Zambia	Sorghum	Thin pinkish-brown sour drink	1, 2a, 3a, 3b, 4, 5, 6
3	Burukutu/ Pito	Nigeria	Sorghum and/or maize	Light brown sweetish liquid	1, 2a
4	Bouza	Egypt	Wheat	Pale yellow thick sour drink	1, 2a
5	Tella	Ethiopia	Barley, wheat, hops	Smokey flavour, tan to dark brown	1, 2a
6	Bussa	Kenya	Maize/millet	Porridge-like light brown	1, 2a
7	Merissa	Sudan	Sorghum	Opaque	1, 2a
8	Tej	Ethiopia	Honey	Yellow, about 12% alcohol	1
9	Agadagidi	Nigeria	Plantain	Thick effervescent	1
10	Mbege	Tanzania	Banana/ millet	Opaque	1

Key:

1. Organisms involved identified.
2a. Biochemical processes studied.
2b. Genetic improvement of organisms.
3a. Experimental production using pure cultures.
3b. Process improvement including use of alternate substrates.
4. Pilot production.
5. Industrial production.

In the process of industrializing these beverages, new and modern ways must be found to eliminate the tedium of traditional processes by adopting modifications based on better knowledge of the beverages. Thus, for example, in those beers where a smoky flavour is required, it may be more convenient to identify the predominant compounds in the smoke and add them or their analogues. Similarly, since the art of climbing the palm tree to

Table 13.9. Yeasts reported in different palm wines

Type of Palm Wine	Yeasts Encountered
Oil palm	*Sacharomyces pastorianus, S. ellipsoids, S. cerevisiae, S. chevalieri, S. vafe, S. marki, S. ludwigi, Pichia* sp., *Schizosaccharomyces pombe, Endomycopsis, Kloeckera apiculata.*
Raphia palm	*S. rosei, Candida* sp.
Palmyra palm	*S. pombe, Schevaliari, S. ludwigi*
Arenga palm	*S. cerevisiae*

obtain a sap is gradually dying, perhaps a completely synthetic palm wine consisting of a sucrose or molasses solution (suitably pasteurized) can replace palm wine. In summary, the scientific prospects for modernizing and industrializing African traditional alcoholic beverages are encouraging; however, they are still at the initial stage.

Grape Wines

Wine is by common usage defined as a product of the "normal alcoholic fermentation of the juice of sound ripe grapes". Nevertheless, any fruit with a good proportion of sugar may be used for wine production. Thus citrus, bananas, apples, and pineapples may be used to produce wine. Such wines are always qualified as fruit wines. If the term is not qualified, then it is regarded as being derived from grapes, *Vitis vinifera.* The production of wine is simpler than that of beer, in that there is no need for malting since sugars are already present in the fruit juice being used. This, however, exposes wine making to a greater contamination hazard.

Wine is today principally produced in countries or regions with mild winters, cool summers, and an even distribution of rainfall throughout the year. In North America, the US is the leading producer, most of the wine coming from the State of California and some from New York. In Europe the principal producers are Italy, Spain, France, and Russia. In South America, Argentina, Chile and Brazil are the major producers and in Africa, the producers are Algeria, Morocco and South Africa. Other producers are Turkey, Syria, Iran and Australia.

Processes in Wine Making

Crushing of Grapes

Selected ripe grapes of 21-23° Balling (refer to section on beer) are crushed to release the juice which is known as 'must', after the stalks which support the fruits have been removed. These stalks contain tannins that would give the wine a harsh taste if left in the must. The skin contains much of the materials which give wine its aroma and colour. For the production of red wines, the skins of black grapes are added to impart the color. Grapes for sweet wines must have a sugar content of 24- 28° Balling so that a residual sugar content is maintained after fermentation. The chief sugars in grapes are glucose and fructose; in ripe fruits they occur in about the same proportion. Grape juice has an acidity of 0.60-0.65% and a pH of 3.0-4.0 mainly due to malic and tartaric acids with a little citric acid. During ripening both levulose content and tartaric acid contents rise. Nitrogen is present in the form of amino acids, peptides, purines and small amounts of ammonium compounds and nitrates.

Fermentation

Yeast Used

The grapes themselves harbour a natural flora of microorganisms (the bloom) which in previous times brought about the fermentation and contributed to the special characters of various wines. Nowadays, the must is partially 'sterilized' by the use of sulphur dioxide, a bisulphite or a metabisulphite which eliminates most microorganisms in the must leaving wine yeasts. Yeasts are then inoculated into the must. The yeast which is used is *Saccharomyces cerevzstae* var. *ellipsoideus* (synonyms: *S. cerevisiae, S. ellipsoideus, S. vini.).* Other yeasts which have been used for special wines are *S. fermentati.* and *S. bayanus*. Wine yeasts have the following characteristics: (a) growth at the relatively high acidity (low pH), (b) resistance to high alcohol content (higher than 10%), and (c) resistance to sulphite.

Control of Fermentation

(a) Temperature: Heat is released, as is the case in fermentation, and the fermentation must be cooled.

(b) Yeast nutrition: Yeasts normally ferment the glucose preferentially although some yeasts, e.g. *S. elegans* prefer fructose. To produce sweet wine, glucose-fermenting yeasts are used leaving the fructose, which is much sweeter than glucose. Most nutrients including macro- and micronutrients are usually abundant in must; occasionally, however, nitrogenous compounds are limiting. They are then made adequate with small amounts of $(NH_4)_2\ SO_4$ or $(NH_4)_2HPO_4$.

(c) Oxygen: As with beer, oxygen is required in the earlier stage of fermentation when yeast multiplication is occurring. In the second stage when alcohol is produced, the growth is anaerobic and this forces the yeasts to utilize intermediate products such as acetaldehyde as hydrogen acceptors and hence encourage alcohol production.

Flavour Development

Although some flavour materials come from the grapes, most of them come from yeast action. The flavour of wine has been elucidated with gas chromatography and has been shown to be due to alcohol, esters, fatty acids and carbonyl compounds, the esters being the most important. Diacetyl, acetoin, fusel oils, volatile esters and hydrogen sulphide have received special attention. Autolysates from yeasts also have a special influence on flavour.

Ageing and Storage

The fermentation is usually over in three to five days. At this time 'pomace' formed from grape skins (in red wines) will have risen to the top of the brew. As has been indicated earlier for white wine, the skin is not included in the fermentation. Hence, at the end of this fermentation the wine is allowed to flow through a perforated bottom to remove "pomace", if any. When the pomace has been separated from the wine and the fermentation is complete or stopped, the next stage is 'racking'. The wine is allowed to stand until a major portion of the yeast cells and other fine suspended materials have collected at the bottom of the container as sediment or 'lees'. It is then 'racked', during which the clear wine is carefully pumped or siphoned off without disturbing the lees. The wine is then transferred to wooden casks (100-1000 gallons), barrels (about 50 gallons) or tanks (several thousand gallons). The wood allows the wine only slow access to oxygen. Water and ethanol evaporate slowly leading to air pockets that permit the growth of aerobic wine spoilers e.g. acetic acid bacteria and some types of yeasts. The casks are, therefore, regularly topped up to prevent the pockets. In modern tanks made of stainless steel the problem of air pockets is tackled by filling the airspace with an inert gas such as carbon dioxide or nitrogen.

During ageing desirable changes occur in the wine. These changes are due to a number of factors:

(a) Slow oxidation, since oxygen can diffuse slowly through the wood. Small amounts of oxygen also enter during the filling up. Alcohol reacts with acids to form esters and tannins are oxidized.

(b) Wood extractives also affect ageing by affecting the flavour.

(c) In some wines, microbial malolactic fermentation occurs. In this fermentation, malic acid is first converted to pyruvic acid and then to lactic acid. The reaction is responsible for the rich flavour developed during the ageing of some wines e.g. Bordeaux. Cultures, which have been implicated in this fermentation, are *Lactobacillus* sp. and *Leuconostoc* sp. A temperature of 11-16°C is best for ageing wines. High temperature probably functions by accelerating oxidation.

Clarification

The wine is allowed to age for a period ranging from two to five years, depending on the type of wine. At the end of the period some wines will have clarified naturally. For others, artificial clarification may be necessary. The addition of a fining agent is often practiced to help clarification. Fining agents react with the tannin, acid, and protein or with some added substance to give a heavy quick-settling coagulum. In the process of settling various suspended materials are adsorbed. The usual fining agents for wine are gelatin, casein, tannin, isinglass, egg albumin, and bentonite. In some countries the removal of metal ions is accomplished with potassium ferrocyanide known as 'blue fining'; it removes excess ions of copper, iron, manganese and zinc from wines.

Packaging

Before packing in bottles, the wine from various sources is sometimes blended and then pasteurized. In some wineries, the wine is not pasteurized, rather it is sterilized by filtration. In many countries the wine is packaged and distributed in casks.

Wine Defects

The most important cause of wine spoilage is due to microbial metabolism; less important defects are acidity and cloudiness. Factors which influence spoilage by bacteria and yeasts, include the following: (a) wine composition, specifically the sugar, alcohol and sulphur dioxide content; (b) storage conditions e.g. high temperature and the amount of air space in the container; and (c) the extent of the initial contamination by microorganisms during the bottling process. When proper hygiene is practiced, bacterial spoilage is rare. When spoilage occurs the microorganisms concerned are acetic acid bacteria which cause sourness in the wine. Lactic acid bacteria, especially *Leuconostoc* and sometimes *Lactobacillus*, also spoil wines. Various spoilage yeasts may also grow in wine. The most prevalent is *Brettanomyces*, a slow growing yeasts that grow in wine causing turbidities and off-flavours. Other wine spoilage yeasts are *Saccharomyces oviformis*, which may use up residual sugars in a sweet wine and *Saccharomyces bayanus*, which may cause turbidity and sedimentation in dry wines with some residual sugar. *Pichia membranefaciens* is an aerobic yeast which grows especially in young wines with sufficient oxygen. Other defects of wine include cloudiness and acidity.

Wine Preservation

Wine is preserved either by chemicals or by some physical means. The chemicals which have been used, include bisulphites, diethyl pyrocarbonate and sorbic acid. Physical means include pasteurization and sterile filtration. Pasteurization is avoided when possible because of its deleterious effect on wine flavour.

Classification of Wines

Grape wines may be classified in several ways. Some of the criteria include place of origin, colour, alcohol content and sweetness. The system to be described classifies wines into two groups: natural wines and fortified wines

The Natural Wines

Natural wines result from complete natural fermentation. Further fermentation is prevented because the sugar is to a large extent exhausted. Spoilage organisms such as acetic acid bacteria do not grow if air is excluded. Owing to the natural limit of sugar in grapes, the alcohol content does not usually exceed 12%. Natural wines are subdivided into still (without added CO_2), and sparkling (with added CO_2). Colour and sweetness also subdivide the wines. The colour pink or red is derived from the colour of the grape; white wine comes from a grade whose skin is light green, but whose juice is clear. White wine can also result from black (or deep red) grapes if the skin is removed immediately after pressing before fermentation. Red wine results when the skin of black grapes is allowed in the fermenting must. Pink wine results when the skin is left just long enough for some material to extract some skin colouring.

In general, the natural wines are usually consumed at one sitting once they are opened. It is for this reason that they are called 'table' wines and are intended to be part of the meal. They are usually served in generous amounts, partly because they contain less alcohol than the desserts and appetizers and partly because they do not have a high keeping quality once opened, compared with appetizers and dessert wines.

Desert and Appetizer Wines

The second broad groups of wines are desert or appetizer wines. As can be seen from their names, they are served at the beginning (appetizers) or at the end (dessert) of meals. They contain extra alcohol from distilled grape wine.

Broad Classification of Grape Wines

A. Natural Wines: 9-14% alcohol; nature and keeping quality mostly dependent on 'complete' yeast fermentation and protection from air.

1. Still wines (known as 'table wines' intended as part of meal); no carbon dioxide added
   (a) *Dry table wines:* (no noticeable sweetness) – (i) White, (ii) Rose (pink), and (iii) Red wine
   (b) *Sweet table wines:* (i) White, and (ii) Rose

Further naming of above depends on grape type, or region of origin.

2. Sparkling wines (appreciable CO_2 under pressure)
   (a) White (Champagne),
   (b) Rose (Pink Champagne) and
   (c) Red (Sparkling burgundy; cold duck)

B. Fortified (Desert and Appetizer) Wines: Contain 15-21% alcohol; nature and keeping quality depends heavily on addition of alcohol distilled from grape wine

1. *Sweet wines*
   (a) White (Muscatel, White Port, Angelica)
   (b) Pink (California Tokay, Tawny Port)
   (c) Red (Port, Black Muscat)
2. *Sherries* (White sweet or dry wines with oxidized flavour)
   (a) Aged types
   (b) Flor types
   (c) Baked types
3. *Flavoured specialty wines* (usually white Port base)
   (a) Vermouth (pale dry, French; Italian sweet types)
   (b) Proprietary brands led wines, partly to make them more potent, but also to preserve them from yeast spoilage. These are divided into three categories: (i) Sweet e.g. port, (ii) Sherries – sweet or dry, they originated from Portugal and are characterized by flavours induced by various degrees of oxidation; and (c) Flavoured wines e.g. vermouth; these are flavoured with herbs and other components, which are secrets of the producing firms.

Sparkling wines (especially champagne), sherry, and flavoured wines are explained briefly.

Sparkling Wines

Sparkling wines contain CO_2 under pressure before they are opened. They are called 'sparkling' because the gentle release of carbon dioxide from the wine after the bottle is opened gives the wine a sparkle. The best of the sparkling wines is produced in Champagne region in northern France, which has given its name to the wine. Champagne is produced either in the bottle or in bulk.

(a) Bottle Champagne: Champagne is usually a clear sparkling wine made from white wine. Pink or red champagne made from wines of the same colour are preferred by some. Champagne is produced by a second fermentation in the bottle of an already good wine. Not only does its production take a long time, but it also requires a complicated method which is difficult to automate and hence must be handled manually. For these reasons the drink is expensive. The usual process is described below. The parlance of champagne manufacture is understandably French. The method *Champenoise* to be described is the one used for making the best sparkling wines. After the must is fully fermented, the wines to be used for champagne are racked, clarified, stabilized and fined. A blend is then made from several different wines to give the desired aroma. This blend is known as 'cuvee'. The 'cuvee' should have an alcohol content of 9.5 to 11.5%, adequate acidity (0.7 to 0.9% titrable as tartaric acid) and light straw or light yellow colour. The SO_2 content should be low, otherwise SO_2 odour would show when the wine is poured, or worse still, the yeasts might convert SO_2 to forms of hydrogen sulphide, which would give the rotten egg odour.

The 'cuvee' is placed into thick walled bottles able to withstand the high pressure of CO_2 to be built up later in the bottle. For the secondary fermentation in the bottle, more yeast, more fermentable sugar, usually sucrose and nowadays, sometimes a small amount (0.05 to 0.1%) of ammonium phosphate is added. The yeast usually used is *Saccharomyces bayanus*. It is selected because it meets the following requirements encountered in secondary fermentation for Champagne production: it must grow at a fairly high alcohol content (10-12%), at high pressures and relatively low temperatures; the yeast must die or become inactive within a short time in order to prevent further fermentation after sugar (known as the 'dosage') is added once again before the final corking. Finally, the yeast must be able to form a compact granulated sediment after fermentation.

The amount of sugar added depends, on the expected CO_2 pressure of the sparkling wine. Sparkling wines have a pressure of about six atmospheres but not less than four. As a rough rule, 4 g of sugar l^{-1} will produce one atmosphere of CO_2 pressure. Therefore, the sugar added is about 24 g l^{-1}. Account is taken of any sugar present in the wine. Although the bottles are thick, if the fermentation is too rapid, temperature too high or sugar too high, the bottle may burst. Champagne bottles are, therefore, not re-used since a scratched bottle may burst.

The bottle with its mixture of wine, sugar and yeast is placed horizontally and allowed to ferment at a temperature of about 15-16°C;

this secondary fermentation takes two to three months. The secondarily fermented wine, now known as 'tirage,' is then stored still horizontally at a temperature of 10°C and remains undisturbed in that position for at least a year and sometimes up to five years. Much of the aroma of well-aged champagne appears to come from the complete reactions involving materials released from yeast autolysis.

The next stage is getting the sediment from the side to the neck of the bottle. Various methods are used by individual handlers. This transfer or *remuage* is achieved by first placing the bottleneck downwards at an angle in a rack with varying degrees of jolting. The bottles are next rotated clockwise and anti-clockwise on alternate days during which the bottle is gradually straightened to a perpendicular position. The process may take anything from two to six weeks, at the end of which the sediment finds it's way to the neck of the bottle. To remove the sediment of yeast cells, the neck of the bottle is frozen to 1-15°C; an ice plug which includes the sediment forms therein. The bottle is turned to an angle of about 45° C and opened. The ice plug is forced out by the pressure in the bottle. During this process of *disgorging* some CO_2 is lost, but sufficient is left to give the usual pop which launches a bride, a ship or a graduating student into the future! The lost wine is replaced from another bottle and the *dosage* is added. The *dosage* is a syrup consisting of about 60 g of sugar in 100 ml of well-aged wine. All champagnes, even those labelled dry, contain dosage; otherwise it would taste sour. Sweet champagne contains up to 10% sugar.

Bulk Production of Champagne: Champagne is sometimes produced in bulk in a large tank rather than in individual bottles. Bulk production is known as the Charmat process named after its inventor. When this is so produced, it is usually declared on the label to save the more labour-intensive bottle-made versions from unfair competition. The tank, which has a lining of an inert material such as glass, holds 500 to 25,000 gallons and is built to withstand 10-20 atmospheres as a safety measure. Valves control the pressure and cooling coils the temperature. Since the tanks are aerated, a rapid turnover is possible and 6-12 fermentations per year are made. Another reason for the rapid turnover is that a heavy yeast growth occurs which could lead to the production of off flavours, especially, if H_2S is allowed prolonged contact with the wine.

Tank fermented champagne is usually given a cold-stabilizing treatment to remove excess tartarates. It is filtered still cold and under pressure to remove yeast cells. The wine is then filled into bottles with *dosage* of the right kind. It is usual to introduce some sulphur dioxide into it just in case some yeasts were not removed by filtration. Sulphur dioxide also helps to prevent darkening of the oxidation which may occur as the wine takes up oxygen during the transfer process. The sulphur dioxide odour is usually noticeable when Charmat-prepared champagne bottles are opened. Furthermore, they lack the aroma conferred on well-aged bottle brands by the autolysis of yeast. Bulk champagnes are amenable to bulk production because they are more difficult to ferment owing to their higher tannin contents; they may also require higher fermentation temperatures.

THE DISTILLED ALCOHOLIC (OR SPIRIT) BEVERAGES

The distilled alcoholic or spirit beverages are those potable products whose alcohol contents are increased by distillation. In the process of distillation, volatile materials, which emanate directly from the fermented substrate or after microbial (especially yeast) metabolism, introduce substances which have a great influence on the nature of beverage. The character of the beverage is also influenced by such post-distillation processes as ageing, blending, etc. The components of spirit beverages, which confer specific aromas on them, are known as congeners.

Measurement of the Alcoholic Strength of Distilled

"Proof": In English-speaking countries, such as the US, Canada, England and Australia, the alcoholic content of spirit beverages (and also of non-potable alcohol) is given by the term "proof". The reason for the term is historical. Before the advent of the use of the hydrometer in alcoholic measurements, an estimate of the alcoholic content of a spirit beverage was obtained by mixing it with gunpowder. If the gunpowder still ignited, the spirit was satisfactory because it contained more than 50% (v/v) of alcohol. If it did not ignite, it was 'under proof' because it contained more than 50% water. Proof has nowadays become more clearly defined, although slight differences occur among countries. In the US proof spirit (i.e. 100° "proof", written 1000) shall be held to be that alcoholic liquor which contains one half its volume of alcohol of a specific gravity of 0.7939 at 15.6°C. In other words, the proof is always exactly twice the alcoholic content by volume. Thus, 100° proof spirit contains 50% alcohol. In the British system, proof spirit contains 57.1% by volume and 49.28% by weight of alcohol at 15.6°C.

It is customary to quantify large amounts of distilled alcoholic beverages or alcohol in 'proof gallons' for tax and other purposes. This term simply specifies the amount of alcohol in a gallon of spirit. Thus, a US proof gallon contains 50% ethyl alcohol by volume; a gallon of liquor at 120° proof is 1.2 proof gallon, and a gallon at 750 proof is 0.75 proof gallon. However, the ordinary US gallon is 3.785 litres, whereas the England (Imperial) gallon is larger, 4.546 litres. The England proof gallon is multiplied by 1.37 to convert it to a US gallon:

The "proof" is read off a special hydrometer.

Percentage by Weight: This is used in Germany for the ethanol content of a beverage or other liquid. The hydrometers are graduated at 15°C and the reference is with (Mendeleafs) table of density. This results in a scale independent of the ambient temperature.

Percentage by Volume: Many countries especially in Europe use the percentage by volume system. For most of them the hydrometer is calibrated at 15°C. France, Belgium, Spain, Sweden, Norway, Finland and Switzerland (which also used weight %), Brazil, Egypt and Russia use 20°C, Denmark and Italy 15.6°C, whereas many South American countries use 12.5°C. In many of these countries the specific gravity of alcohol used as reference differ slightly.

GENERAL PRINCIPLES IN THE PRODUCTION OF SPIRIT BEVERAGES

In general, the following steps are involved in the preparation of the above beverages. The details differ according to beverage.

Preparation of the Medium

In the grain beverages (whisky, vodka, gin) the grain starch is hydrolyzed to sugars with the enzymes of barley malt. In all the others no hydrolysis is necessary because sugars are present, as in brandy (grape sugar) and rum (cane sugar).

Propagation of Yeast Inoculum

Large distilleries produce hundreds of litres of spirits daily for which fermentation broth of many more times in volume are required. These broths are inoculated with up to 5-10% (v/v) of thick yeast broth. Although yeast is reused, there is still a need for regular inocula. In general, the inocula are made of selected alcohol-tolerant yeast strains usually *S. cerevisiae* grown aerobically with agitation and in a molasses base. Progressively larger volumes of culture may be developed before the desired volume is attained.

Fermentation

When the nitrogen content of the medium is insufficient, nitrogen is added usually in the form of an ammonium salt. As in all alcohol fermentation, the heat released from must be reduced by cooling and temperatures are generally not permitted to exceed 35-37°C. The pH is usually in the range of 4.5-4.7, when the buffering capacity of the medium is high. Higher pH values tend to lead to higher glycerol formation. When the buffering capacity is lower, the initial pH is 5.5 but this usually falls to about 3.5 during the fermentation. Contamination can have serious effects on the process: sugars are used up leading to reduced yields; metabolic products from the contaminants may not only alter the flavour of the finished product, but metabolites such as acids affect the function of the yeast. The most important contaminants in distilling industries are lactic acid bacteria notably *Lactobacillus.* Acetic acid bacteria may also produce acetic acid, which affects the flavour of the product.

Distillation

Distillation is the separation of more volatile materials from less volatile ones by a process of vaporization and condensation. Three systems used in spirit distillation are detailed.

Rectifying Stills

If the condensate is repeatedly distilled, the successive distillates will contain components, which are more and more volatile. The process of repeated distillation is known as rectification. Rectification is done in columns, towers or stills containing a series of plates at which contact occurs i.e. returned to the system. The alcohol-water mixture flows downwards and is stripped of alcohol by steam which is introduced from the bottom and flows upwards. Alcohol-rich distillate is withdrawn at the top of the column. Fusel oils separate out just above the point of entry of the mixture and are drawn off to another column. Volatile fractions are composed of esters and aldehydes. Whisky and brandy may be distilled successfully in a two-column still, but for high-strength distillates, at least three 'columns' and possibly four or five may be required. The above description is of the modern still, a modification of which is also used for producing industrial alcohol. Much older versions of stills continue to be used in some parts of the world.

Pot Still

These are traditional stills, usually made of copper. They are spherical at the lower-portion which is connected to a cooling coil. They are operated batch-wise. The first portion or 'heads' and the latter portion or 'tails', of the 'low wines' are usually discarded and only the middle portion is collected. Malt whisky, rum and brandy are made in the pot still. Its advantage is that much of the lesser aroma conferring compounds are collected in the beverage, thereby imparting a rich aroma to it.

Coffey (patent) Still

The Coffey still was patented in 1830 and the various modifications since then have not added much to the original design of the still. Its main feature is that it has a rectifying column besides the wash column in which the beer is first distilled.

Maturation

Some of the distilled alcoholic beverages are aged for a few years, often prescribed by legislation.

Blending

Before packaging, samples of various batches of different types of a given beverage are blended together to develop a particular aroma.

TYPES OF SPIRIT BEVERAGES

The beverages described below are whisky, brandy, rum, vodka, kai-kai (or Akpeteshi), schnapps and cordials.

Whisky

Whisky is the alcoholic beverage derived from the distillation of fermented cereal. Various types of whiskies are produced; they differ principally in the cereal used. Although many countries including Japan and Australia now produce whisky for export, the countries best associated with whisky are first and foremost, Scotland followed by Ireland, the US and Canada.

In all whisky producing countries the alcoholic content, materials and method of preparation are controlled by government regulations, which vary from country to country. In Scottish Malt Whisky, the barley is malted just as in beer making, but during the kilning smoke from peat is allowed to permeate the green (fresh) malt, so that the whisky made from the malt has a strong aroma of peat smoke derived mainly from phenol. In the US the principal types of whisky are rye and bourbon whiskies. Rye whisky is prepared from rye and rye malt, or barley and barley malt. Bourbon whisky is prepared from maize, preferably yellow maize, barley malt or wheat malt. A typical mash, which must contain 51% corn, may have a composition of this type: 70% corn, 15% rye and 15% barley malt.

The un-malted rye or corn is cooked to gelatinize it and hence to facilitate saccharification (or conversion to sugar) by the enzymes in the malt. The solids are not removed from the mash and the inoculating yeast sometimes contains *Lactobacillus,* whose lactic acid is said to improve the flavour of the whisky. Fermentation is usually carried out in a two-column Coffey type still.

All whisky is matured in wooden casks for a number of years, usually three or more. They may then be blended with various types (usually controlled by law) before bottling.

Brandy

Brandy is a distillate of fermented fruit juice. Thus brandy can be produced from any fruit, i.e. strawberries, paw-paw or cashew. However, when it is unqualified, the word brandy refers to the distillate from fermented grape juice. It is subjected to a distillation limitation of 170° proof (85%). The fermented liquor is double distilled, without previous storage, in pot stills. A minimum of two years maturation in oak casks is required for maturation. Brandies produced in other parts of France are merely *eau de vie* (water of life) and are not called brandy. Certain parts of Europe (e.g. Spain, where brandy is distilled from Jerez sherry and South America as well as the USA produce special brandies.

Rum

Rum is produced from cane or sugar by-products especially molasses or cane juice. Rum production is associated with the Caribbean, especially Jamaica, Cuba and Puerto Rico. It is also produced in the eastern US. Rum with a heavy body is produced from molasses, while light rum is produced from cane syrup using continuous distillation. During fermentation, the molasses is clarified to remove colloidal material which could block the still by the addition of sulphuric acid. The pH is adjusted to about 5.5 and a nitrogen source ammonium sulphate or urea may be added. For the heavier rums *Schizosaccharomyces pombe* is used, while *S. cerevisiae* is used for the lighter types. For maturation rum is stored in oak casks for 2-15 years.

Gin, Vodka and Schnapps

These beverages differ from whisky, rum and brandy in the following ways:

(a) Brandy, rum and whisky vary from pale-yellow straw coloured to deep brown, by extractives from wooden casks in which they are aged and which have sometimes been used to store molasses or sherry. To obtain consistent colour, caramel is sometimes added. Gin, vodka and schnapps are water-clear.

(b) The flavour of brandy and whisky is due to congenerics present in the fermented mash or must. For gin, vodka, and schnapps the congenerics derived from fermentation are removed and flavouring is provided (except in vodka) with plant parts.

(c) The raw material for their production is usually a cereal, but potatoes or molasses may be used. For gin maize is used, while for vodka rye is used. The cereals are gelatinized by cooking and mashed with malted barley. In recent times, amylases produced by fungi or bacilli have been used since the flavour of malt is not necessary in the beverage. Congeners are removed by continuous distillation in multi-column stills.

In gin production, the grain-spirits (i.e. without the congeners) are distilled over juniper berries, *Juniperus communis,* dried angelica roots, *Angelica officinalis* and others including citrus peels, cinnamon, nutmeg, etc. Russian vodka is produced from rye spirit, which is passed over specially activated wood charcoal. In other countries it is sometimes produced from potatoes or molasses. Schnapps are gin flavoured with herbs.

Cordials (Liqueurs)

Cordials are the American name for what are known as liqueurs in Europe. They are obtained by soaking herbs and other plants in grain spirits, brandy or gin or by distilling these beverages over the plant parts mentioned above. They are usually very sweet, being required to contain 10% sugar. Some well-known brand names of cordials are Drambui, Creme de menthe, Triple Sac, Benedictine and Anisete.

Kai-kai, Akpeteshi, or Ogogoro

Kai-kai is an alcoholic beverage widely consumed in West Africa. It is a gin produced by distilling fermented palm wine. At present it is produced on a small scale. Its development appears to have been hampered by being declared illicit in the days when these countries were colonies. Workers in Nigeria and Ghana have carried out studies, which have set the stage for the commercialization of kai-kai, or beverages derived from this.

CONCLUSION

To sum up, alcoholic beverages from traditional sources like cereals (barley, sorghum, etc.) and grapes have been practiced since time immemorial. However, these beverages are associated with the agro-climate, culture and civilization of the region where they are produced. In this chapter, the major alcoholic beverages produced and consumed in different parts of the world have been discussed. Still, there are some traditional (e.g. "toddy", date palm beer from India), non-traditional [e.g. wine from non-grape minor fruits like cashew (*Anacardium occidentale* L.) apple, sapota (*Manikara achras* L.), jamun (*Syzygium cumini* L.), etc.] and novel alcoholic beverages [e.g. sweet potato (*Ipomoea batatas* L.) wine and beer] which are to be brought into limelight. These aspects would be a fruitful area for future research.

REFERENCES

Amerine M.A., Berg H.W. and Cruess W.V. (1972). The Technology of Wine Making, 3rd Edition, Avi Publications, USA, pp. 357-644.

Battcock, M. and Azam-Ali, S. (2001). Fermented Fruits and Vegetables: A Global Perspective. FAO Agric. Services Bull. 134, Rome, Italy, pp. 96.

Doyle, M.P., Beuchat, L.R. and Montville, T.J. (eds.) (2004). Food Microbiology: Fundamentals and Frontiers, 2nd edn., ASM Press, Washington DC, USA.

Flickinger, M.C. and Drew, S.W. (eds) (1999). Encyclopedia of Bioprocess Technology: Fermentation, Biocatalysis, and Bioseparation, Volumes 1-5, John Wiley & Sons, USA.

Gastineau C.F., Darby W.J. and Turner T.B. (1979). Fermented Food Beverages in Nutrition, Academic Press, New York, USA, pp. 133-186.

Gutcho M.H. (1976). Alcoholic Beverage Processes. Noyes Data Corporation, New Jersey and London, UK, pp. 10-106.

Hammond J.R.M. and Bamforth C.W. (1993). Progress in the development of new barley hop, and yeast variants of malting and brewing. Biotechnol. Gen. Eng. Rev. 11: 147-169.

Hanum, H. and Blumberg S. (1976). Brandies and Liqeuers of the World, Dooubleday Inc, New York, USA.

Hough, J.S. (1985). The Biotechnology of Malting and Brewing. Cambridge University Press, Cambridge, UK, pp. 15-188.

Okafor, N. (1972). Palm wine yeasts of Nigeria. J. Sci. Food Agric. 23: 1399-1407.

Okafor, N. (1978). Microbiology and biochemistry of oil palm wine. Adv. App. Microbiol. 24: 193-200.

Okafor, N. and Aniche, G.N. (1980). Lage beer from Nigerian sorghum. Brew. Distill. Int. 10: 32-33, 35.

Okafor, N (1987). Industrial Microbiology. University of Ife Press, Ile Ife, Nigeria, pp. 201-210.

Okafor, N. (1990). Traditional alcoholic beverages of tropical Africa - Strategies for scale-up. Process Biochem. 25: 213-220.

Packowski, G.W. (ed.) (1978). Distilled Alcoholic Beverages, Encyclopadeia of Chemical Technology, 3rd edition, John Wiley & Sons, New York, USA, pp. 824-863.

Singleton, V.L. and Butzke, C.E. (1998). Wine. Kirk-Othmer Encyclopedia of Chemical Technology. John Wiley & Sons, USA, Article Online Posting Date: December 4, 2000.

Soares, C. (2002). Process Engineering Equipment Handbook. McGraw-Hill, New York, USA.

Taylor, J.J.N. and Dewar, J. (2001). Developments in sorghum food technologies. Adv. Food Nutr. Res. 43: 217-264.

Vogel, H.C. and Tadaro, C.L. (1997). Fermentation of Biochemical Engineering Handbook - Principles, Process Design, and Equipment (2nd edn.) Noyes, USA.

Zeng, A. (1999). Continuous culture. In: Industrial Microbiology and Biotechnology, 2nd edition (eds.) A.L. Demain and J.E. Davies, ASM Press, Washington DC, USA, pp. 151-164.

REFERENCES

[illegible]

14

Aquaculture Biotechnology for Enhanced Fish Production for Human Consumption

*A. Exadactylos and Ioannis S. Arvanitoyannis**

INTRODUCTION

General Aspects

Intensifying of pisciculture over the last few years created products of most excellent quality, their availability can cover the continuously increasing market demand. The application of genetic engineering techniques plays an important role in this growth, which can ensure production and suitable quantity of the most excellent product quality, all year long, covering the rising consumers' demand.

The application field of these techniques, with regard to Fisheries and Aquaculture are: (i) species genetic improvement, (ii) DNA vaccine production, (iii) production of chemical substances and enzymes from marine organisms, and (iv) use of molecular markers for the recording of fisheries reserves and the segregation of natural fish populations.

With regard to pisciculture, the application of genetic engineering techniques pinpoints the need of wise broodstock management aiming at the production of sufficient egg number and good quality larvae, that will contribute to the genetic improvement of generation characteristics, the checking and preservation of fish production characteristics, the most effective control and management of fish feed and stockfarming in general, and finally the production of the most excellent product quality in

Department of Agriculture, Animal Production and Aquatic Environment, School of Agricultural Sciences, Univ. of Thessaly, Volos 38446, Hellas, Greece

**Corresponding author: e-mail: parmenion@uth.gr*

sufficient quantities that will respond to the consumers' demand (Benzie, 2002).

The Role of Biotechnology in Broodstock Management

Biotechnology is a relatively new term, which is used in order to describe all aspects of biological systems to benefit mankind. With regard to broodstock improvement, biotechnology is used to describe the application of new genetic techniques that help in expeditious improvement schemes, or improvement with ways that could not be possible using traditional techniques of selective breeding. These techniques were the basis for the dramatic improvement which has already been accomplished in certain species that are bred in captivity (Refstie, 1990) and will most certainly continue to focus on genetic improvement in the future.

An important challenge today is the combination of newer biotechnological techniques with the already existing ones of selective breeding, for genetic improvement. This will enable the following: (1) estimate the results of possible modifications, (2) accelerate the improvement rhythms, and (3) allow alterations that cannot succeed only with selective breeding. Even if Osteichthyes, molluscs and aquatic plants share a lot of common biotechnological techniques and a lot of useful ideas can be acquired by studying these different systems, we will limit ourselves to pisciculture facilities derived for food consumption. The subjects covered are:

(i) Fish traits that allow application of biotechnological techniques.
(ii) Techniques that are available and how they can be applied.
(iii) The main objectives for pisciculture, and selective examples of biotechnological applications achieving these objectives.
(iv) Probabilities for combination of these techniques for maximization of results.

The progress in broodstock management improvement becomes more effective with the combination of all available biotechnological techniques and applying them in the best way. For example, cloned strains created by handling the set of chromosomes could give excellent material with which important genes for certain vital characteristics could be identified. Similarly, the use of regular strains or hybrids between strains could facilitate the estimation of transgenic fish. The manipulation to create triploids could collide with gene transport in order to recognize species differences in disease resistance. It is vital to emphasize that the success of all these techniques will depend on the farmers' ability to successfully combine traditional material. What is suggested is a more careful, objective, statistical behavioural estimate of selected and control strains.

FISH TRAITS THAT ALLOW APPLICATION OF BIOTECHNOLOGICAL TECHNIQUES

Fishes have a number of characteristics that allow biotechnologists to apply techniques for their genetic improvement (Lutz, 2003). Osteichthyes are the most favourable team within Chordata for experimental application of biotechnological techniques. Fishes have an important reproductive advantage. They produce typically a large number of eggs and certain species can reproduce more than once a year. Their gonads can be managed with ease and sperm can be cryo-preserved for a considerably long time before it can be used for fertilization (Maisse et al., 1998). Sometimes eggs can be equally preserved from already stripped females, something that constitutes an important advantage for a lot of biotechnological applications. The low cost of maintenance of each individual (in relation with small mammals and poultry) is an important advantage for biotechnological applications. Many of these applications are still under experimental protocols and further research is required before they could be used on a commercial basis. The bandwidth of applications that can be evaluated is obviously limited due to the cost of such applications, and in many domestic species the maintenance cost of these animals per individual can be quite an important obstacle for any progress. In any case, smaller fish species constitute an excellent model within Chordata for studying cancer due to the low cost of maintenance per individual (Law et al., 1998; Thorgaard et al., 2002). Fish can bear a breadth of changes in ploidy that can be potentially lethal or subvital for other species within Chordata; such changes lead into important conclusions. Fishes which have suffered changes include viable triploid, tetraploid and inbred ones with both sets of chromosomes derived either from females or males. The variety of genetic diversity that is available for research is another important advantage for the application of biotechnological techniques in fish. All fish species that are bred in captivity today are available from natural populations; they can provide a source of genetic material for future improvement. By using biotechnological approaches, it can be possible to choose a set of desired characteristics from natural populations, without excessive alteration of the improvements already made in cultured populations. Similarly, all closely related species for each cultured species can be used as a genetic reserve for future improvements (Lutz, 2003).

Control and Induction of Maturation and Spawning

While fish mature, they go through certain important changes in their production characteristics, such as growth rate, age or maturation period and colouration; all these may differ considerably between the two sexes. Particularly in rainbow trout and salmon, changes that occur during maturation process concern flesh (fatling) texture with an observed reduction in proteins and fat, malpigmentation and disease sensitivity. Male salmonids develop a protuberant upper jaw, a characteristic that deforms the picture of the product which should have concrete size, form, colour and texture familiar to the consumer. Also, in salmonids male individuals mature a year earlier and are presented in smaller sizes than females. In rainbow trout, during the process of maturation, energy that is received from food is channelled to gonad growth rather than body growth. This parameter is very important for pisciculture, if it is taken into consideration that those gonads in the stage of genital maturity correspond to about 20-30% of the individual body weight. What is the outcome? Only the most productive or attractive sex to be bred and promoted in the market. Even if the output is not influenced by sex, it can be more advantageous if monophyletic individuals are bred, so that uncontrolled genital maturation can be avoided.

IMPORTANT OBJECTIVES IN PISCICULTURE

The most important objectives for pisciculturists are: wide environmental tolerance for cultured species, disease resistance and significant improvement of growth rate and/or meat quality.

Environmental Tolerance

By extending the tolerance level in fish to the prevalent environmental conditions e.g. temperature, salinity, pH, etc. it could well broaden the number of species for stockfarming. The production of triploid hybrids can also help in this direction. By hybridizing a farmed species with another, which exhibits broader tolerance limits, new environments can be exploited. An example is the application of gene transport of an antifreeze protein (it is found in certain marine fish that live in waters with temperature below zero) in Atlantic salmon (Shears et al., 1991). Atlantic salmon that lacks this particular protein represents winter mortality in certain cases when water temperature in the Atlantic coast near Canada falls below zero. Until today, results indicate that the gene was transported successfully in Atlantic salmon but the expression levels are not enough for their protection.

Disease Resistance

Disease resistance is a strong issue in Aquaculture (Chevassus and Dorson, 1990) and a lot of biotechnological techniques are applied to approach this matter. Triploid hybrids could possibly help in this direction. Triploid hybrids of rainbow trout with three other species present increased resistance in viruses related with common trout (Dorson et al., 1991); all three hybrids present resistance in haemorrhagic septicemia. Fish stockfarming for disease resistance provides an ideal modelling plan for application of marker-assisted selection since broodstock resistance in diseases can be created that have never been exposed before in pathogens.

Growth Rate and Meat Quality

Selection for improved growth rates and meat quality is another important objective (Cook et al., 2000). Triploid's sterility is the means to enhance growth rate and quality of mature reproductive fish in salmonids (Thorgaard, 1986) even if this does not appear to be the case in commercial experiments for catfish (Wolters et al., 1991, Lilyestrom et al., 1999). Triploid's sterility also provides a means to enhance new types of hybrids quality. Research studies on gene transport of growth hormones indicate that this is likely to be achieved in a commercial rate (Chen and Powers, 1990; Powers and Chen, 1995; Rocha et al., 2004). Whether the "supporters" of transgenic fish manage to influence the currently applicable international or local regulations and health issues raised vastly by consumers is difficult to anticipate.

Marker assisted selection for growth rate is likely to become a suffix. An important gene that affects faster growth rate and influences considerably rainbow trout's size has been recognized (Allendorf et al., 1986) and similar regions are likely to be recognized in other species in the future (Noguchi et al., 2004).

APPLICATIONS OF GENETIC ENGINEERING IN PISCICULTURE

Techniques That Are Available and How They Can Be Applied

Genetic engineering techniques that intervene in fish genetic material are applied in order to avoid, or better to preclude the undesirable results of genital maturation, to produce a final product of most excellent flesh (fatling) texture, with better stockfarming management (food quantity and convertibility, production timing, disease tolerance).

Techniques applied for broodstock improvement include:

A. Sex determination and control (**A_1.** Production of all-female individuals, **A_2.** Production of all-male individuals).

B. Chromosomal engineering (**B_1.** Polyploidy, **B_2.** Gynogenesis – Androgenesis).

C. Biochemical and molecular markers (**C_1.** Marker assisted selection).

D. Transgenic fish (**D_1.** Gene expression, isolation and cloning, **D_2.** Gene transfer and technology).

These techniques have been used in certain cases for quite a few years, but only recently became widely applicable.

Sex Determination and Control

Production of monophyletic populations or sterile individuals can improve towards success of certain species stockfarming (Donaldson and Hunter, 1982; Phillips et al., 2004). Females frequently mature sexually in older ages than males or produce eggs with high commercial value, something that makes them more attractive for handling. However, in certain species, e.g. catfish, male individuals develop faster than female ones, thus are preferable (Davis et al., 1990). Sterile fish could well be the desirable answer when overpopulations constitute a problem, or when values of growth and quality delay considerably to reach the commercial ones, or even when sexual maturity appear in earlier ages than the marketable size.

Production of all-female salmonid species is a successful example of biotechnological application in a commercial rate (Bye and Lincoln, 1986; Devlin and Nagahama, 2002). This has been achieved with gynogenesis that led to production of male XX individuals that give female progeny. Production of all-male populations descended from egg fertilized with sperm from YY male individuals (Fostier and Chevassus, 1991; Grunina et al., 1995a, b) became feasible when crossing individuals XY after hormonal (Chevassus et al., 1988) or spontaneous (Blanc et al., 1993; Komaru et al., 1998) sex inversion (androgenesis). Gene transport can be applied for fish sterilization, but the need for such an application is not as direct as other ones of transgenic technology, since this can be achieved with the use of hormones and triploidy.

Production of Monophyletic Populations

From all the above, it appears that the unanticipated genital maturation is detrimental for pisciculture enterprises. It can be avoided to some extent by sex controlling. The methods for the production of monosex populations are: (i) screening and selection of specific individuals, and (ii) issuing steroid hormones at early stages of breeding. The first method is disadvantageous economically wise and it presupposes that both sex are distinguishable based on morphological characteristics. The second method is the main method of sex control in fish and is done by issuing steroid hormones (androgens-estrogens) at early stages of breeding (first feeding fry). More specifically oestradiol-17β is granted for production of ova from male individuals (XY) and 17 α-mecyl-testosterone is granted for sperm production from female individuals (XX). Suitable genotypes are selected and with appropriate crossings monophyletic populations are produced (Figs. 14.1 and 14.2). Certain species attribute better when they

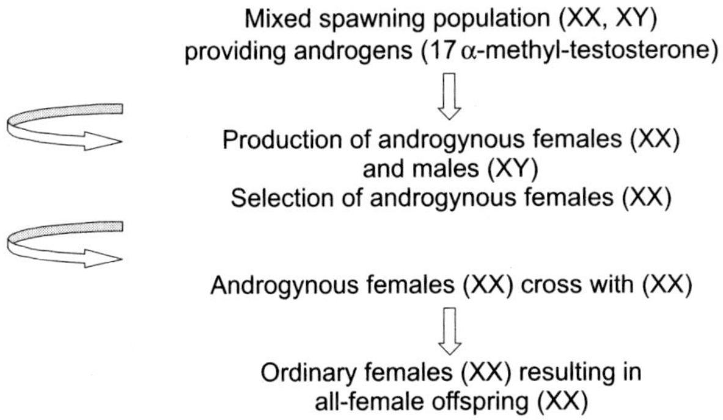
Mixed spawning population (XX, XY)
providing androgens (17 α-methyl-testosterone)
Production of androgynous females (XX)
and males (XY)
Selection of androgynous females (XX)
Androgynous females (XX) cross with (XX)
Ordinary females (XX) resulting in
all-female offspring (XX)

Fig. 14.1. Process of formation of all-female populations

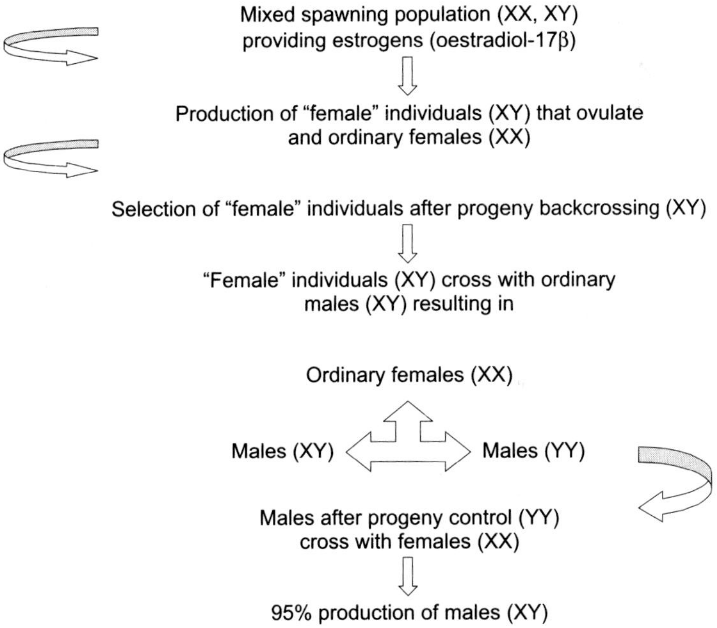
Mixed spawning population (XX, XY)
providing estrogens (oestradiol-17β)
Production of "female" individuals (XY) that ovulate
and ordinary females (XX)
Selection of "female" individuals after progeny backcrossing (XY)
"Female" individuals (XY) cross with ordinary
males (XY) resulting in
Ordinary females (XX)
Males (XY)
Males (YY)
Males after progeny control (YY)
cross with females (XX)
95% production of males (XY)

Fig. 14.2. Process of formation all-male populations

are all male (masculination) such as tilapia, and others attribute better when they are all female (feminization) such as salmonids and seabass. Monophyletic populations that are produced by issuing hormones might be rejected by the consumers mainly because there is a lack of information and inadequate promotion of such a product. Issuing of hormones is done by: (i) injection, (ii) implantation, and (iii) as means of food. The third option is one of the most common and most effective methods of issuing hormones for sex control. Success rate is determined by prescribed dosage in combination with issuing duration and appropriateness of season, which in certain cases can reach 100%. The preparation and issuing process of hormonal dose is as follows: (i) determination of desirable dosage, (ii) dissolution in ethyl alcohol, (iii) mixture with food, (iv) placement in oven for the solvent to evaporate, and (v) issuing to fish fry.

All-male individuals of *Ictalurus punctatus* were produced by issuing synthetic anabolic tranbalone acetate in non-differentiated larvae (Galvez et al., 1995). The heterogeneous size distribution between male individuals of fresh water crayfish *Macrobrachium rosenbergii* caused problems in harvesting and consequently in promoting them market wise (New and Sigholka, 1982). The production of all-male individuals of this species became feasible with the surgical implantation of various tissues from male individuals in female crayfish (Nagamine et al., 1980). However, in the light of the existing information it is appropriate to limit as far the exposure to androgens and estrogens and only authorize those treatments for which no viable effective alternatives exist. In general, there are alternative treatments or strategies available to replace most of the uses of androgens and estrogens for therapeutic or zootechnical purposes. Nonetheless, studies appear to show that at present no viable effective alternatives exist in all the E.U. member states for certain treatments which are currently authorized. In order to allow for the necessary adjustments and in particular for the authorization or the mutual recognition of the pharmaceutical products needed, it is appropritate to maintain the possibility of authorizing, under strict and verifiable conditions so as to prevent any possible misuse and any unacceptable risk for public health, their use for the treatment of certain conditions, which have serious consequences for animal health and welfare. It is necessary to review this possibility within a given time.

Chromosomal Engineering

Handling chromosome set

Chromosome handlings that can become bearable in fish include polyploidy, gynogenesis (only female heritability) and androgenesis (only male heritability) (Ihssen et al., 1990; Gomelsky, 2003). Polyploidy in fish, contrary to mammals, has important applications in genetic improvement. Triploids are the target, because female triploids have shown sterility and they do not develop eggs or secondary sexual characteristics. Male triploids develop secondary sexual characteristics that are connected with sexual maturity, which is not generally desirable (Lincoln and Scott, 1984). However, the use of altered males (XX) can allow production of populations consisting of only female triploid individuals. Another advantage of triploidy is the improved survival rate that is often observed in interspecies triploid hybrids related accordingly with diploid hybrids (Chevassus et al., 1983; Scheerer and Thorgaard, 1983; Scheerer et al., 1991). A few triploid hybrids such as the hybrid of rainbow trout (*Oncorynchus mykiss*) x brook trout (*Salvelinous fontinalis*) can survive when the diploid hybrid is not viable. Triploid hybrids can be important stand alones, or as research material for the characterization of genetic differentiation between species and for the possibility of transporting useful characteristics from one species to another. Triploidy can be caused directly or indirectly. Directly, it is achieved by retaining the second polar particle. Such fishes have two maternal sets of chromosomes and generally have good viability. Vast applications lie in this method, which is used at a commercial level for a lot of species (Bye and Lincoln, 1986). Indirectly, it is achieved with the use of tetraploid progenitors. It is a very attractive method because such methods can be avoided which can damage fertilized eggs; however, until today it is applied successfully only in rainbow trout (Myers and Hershberger, 1991). The initial obstacles as an alternative solution tetraploids x diploids are that tetraploids exhibit low viability, at least the first generation after their production, and it appears that it is considerably more difficult to produce tetraploids than triploids. An additional problem of using tetraploids for the creation of triploids is that male tetraploids exhibit low fertility when crossed with female diploids, obviously because their sperm cannot penetrate diploid fish egg micropyle (Chourrout et al., 1986a, b, c).

Gynogenesis and androgenesis are two techniques that include the suppression of cell division for the creation of viable descendants. In any case, in these techniques, gametes are irradiated before fertilization; only one parent participates in the next generation. In gynogenesis, only the female participates in the next generation, while in androgenesis only the male. In gynogenesis, sperm is irradiated usually with ultraviolet light before fertilization (Piferrer et al., 2004). Thereafter, by applying the same methods of triploidy and tertaploidy, inbred or homozygous fish can be created. An application of these is that in female homogametic species the progeny after gynogenesis will be all female. The second application of gynogenesis lies in the possibility of gene mapping with regard to centromeres in fish after withholding the second polar particle (Allendorf et al., 1986). The third application is the creation of homozygous strains by applying a second gynogenesis circle in those homozygous fish that were produced initially with gynogenesis, which included the suppression of the first cell division (Streisinger et al., 1981; Komen et al., 1991). The fourth application of gynogenesis is fish with two maternal sets of chromosomes, together with chromosome sections from the male parent, if the sperm was previously irradiated with γ rays, where sperm chromosomes are not completely deactivated (Thorgaard et al., 1985). Androgenesis has not been studied as thoroughly as gynogenesis (McKone and Halpern, 2003; Pandian and Kirankumar, 2003). It can be used for the production of cloned strains (Scheerer et al., 1991). The advantage of androgenesis is the possibility of safekeeping and revival of fish strains from their cryopreserved sperm.

Triploidy

The fundamental aim for triploid individuals is sterility. Triploid fish are organisms with regular biological functions, apart from the formation of gametes which is a consequence of chromosomal imbalance. During a regular biological function, when a spermatozoan (1N) enters in a mature ovum (2N), it causes rejection of the 2nd polar particle (1N) from the ovum, during the 2nd meiotic division. The result is the creation of two haploid gametes (ovum, spermatozoan), which are linked for the creation of the zygote. As a consequence, the ovum remains diploid (2N), provided that the 2nd polar particle is not rejected and thus contributes with two lines of chromosomes (2N) in the creation of the zygote. Spermatozoa contribute regularly with one line of chromosomes (1N). The result is the creation of triploid individuals with three lines of chromosomes (3N).

The most usual methods for production of triploid individuals are: (i) thermal methods (heat shock), (ii) mechanical methods (mechanical shock), and (iii) crossing tetraploid with diploid individuals.

According to the thermal method, fertilized eggs are transported in isothermic containers or water baths with a controlled temperature of 26-32°C. Heat shock causes losses even under ideal conditions; a lot of difficulties are usually encountered with this method. The mechanical method is based on the constant practice of high hydrostatic pressure on fertilized eggs, which is friendlier. In both methods there are quite a few parameters that should be taken into consideration: (i) intensity of shock, (ii) timing, (iii) duration, (iv) conditions of incubation, (v) egg quality, and (vi) general hatchery conditions. The applications of triploid individuals exhibit a lot of positively marked arguments. According to them, gonad growth is interrupted; as a result granted food is used for body growth by the organism. Consequently, this method represents great commercial interest, since there are superior feeding management and exploitation, undesirable genital maturity can be precluded, specifically in trout and salmon, and last but not the least flesh quality is properly preserved. The method is applied in quite a few species such as rainbow trout, salmonids, tilapias, sturgeons, European catfish and carp. Five minutes-old fertilized eggs of *Silurus asotus* were shocked at 4°C to induce triploidy by retaining the second polar body. Triploid incidence increased from 1.3 to 98.7%, when the treatment duration was extended from 15 to 60 min (Kim et al., 2001). Hussain et al., (1991) aimed at the identification of treatment optima for triploidy induction in recently fertilized *Oreochromis niloticus* L. eggs by altering the intensity, duration and timing of application of pressure, heat and cold shocks. With regard to coho salmon *Onchorichus kisutch*, Withler et al., (1995) pinpointed that diploid individuals exhibited higher survival rates than triploids from fertilization until incubation (94% opposed to 43%), from incubation till the transport in tanks (92% opposed to 75%) and during breeding in seacages (81% opposed to 60%). Morphological differences between diploid and triploid individuals do not exist. However, triploid individuals can be identified by: (i) blood sample and red blood cell analysis; nuclei of triploid fish are much bigger in size, and (ii) protein electrophoresis, where triploid individuals represent different electrophoretic zymograms.

Tetraploidy

This method has not been applied completely successfully yet (Fig. 14.3). Tetraploid fish exhibit quite low survival rates; they reach genital maturity and are used mainly for the production of triploid individuals when crossed with regular diploids.

Polylpoidy

One of the main reasons for handling polyploidy is for producing fish with faster growth rates, without being influenced by genital maturation or

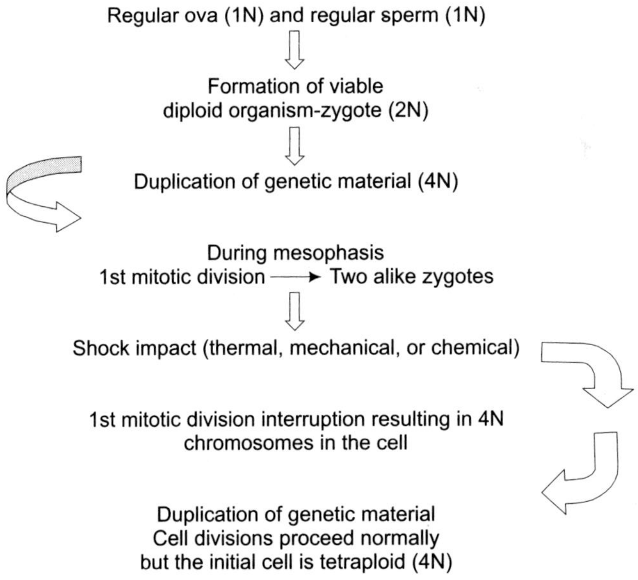

Regular ova (1N) and regular sperm (1N)
Formation of viable diploid organism-zygote (2N)
Duplication of genetic material (4N)
During mesophasis
1st mitotic division → Two alike zygotes
Shock impact (thermal, mechanical, or chemical)
1st mitotic division interruption resulting in 4N chromosomes in the cell
Duplication of genetic material
Cell divisions proceed normally
but the initial cell is tetraploid (4N)

Fig. 14.3. Process of formation of tetraploid populations

reproduction. The benefits of polyploidy include better food convertibility, higher survival rates (because of decreased fish aggressiveness) and higher output with regard to production systems. In fish and ostracea (shellfish), gametes can produce triploid, viable offspring with suitable handling (Gong et al., 2004; Blanc et al., 2005). Processes that are used commonly can induce sterility in the next generation. Polyploidy in fish lies on the fact of individuals with more than two chromosome sequences (Leggatt and Iwama, 2003). In Aquaculture, research has been focused in the development of techniques for producing triploid populations mainly.

Gynogenesis – Androgenesis

Gynogenesis and androgenesis are based on the limited heritability from female or male parents. The key in these two methods is the reduction of genetic contribution of one or the other parent, by interrupting the physiological sequence of events in a new fertilized egg in such a way that 2N diploid gynogenetic or androgenetic larvae are produced. Gynogenesis and androgenesis have small impact on the production of commercial products, but they can be used in mechanisms of sex determination, in the production of monophyletic populations, in the production of clones and populations with higher homozygosity levels than normal in a natural fish population under or near genetic equilibrium.

Gynogenesis

Gynogenesis involves offspring which are derived only from female parents. It includes the following stages:

(i) Sperm exposure in X, γ, or UV radiation during gynogenetic reproduction. The result is loss of the genetic information but preservation of its mobility.
(ii) Fertilization of the genetically inactive but kinetically active spermatozoa with regular eggs. The zygote bears the genetic information of the female parent.
(iii) If there are no exterior interventions, larvae develop normally as haploid and do not live for long.
(iv) If the 2nd polar particle is withheld by applying some sort of shock at the correct timing (2nd meiotic division), then larvae are produced with two chromosome sequences only from the female parent (diploid gynogenetic zygotes).

This technique has been applied successfully in the European catfish, goldfish, *Abramis abramis* and quite a few species of cyprinids. Wiegertjes et al., (1995), used gynogenesis in order to produce homozygote offspring from female carp *Cyprinus carpio* L. Cherfas et al., (1995) revealed diploid maternal chromosomes in *Cyprinus carpio* eggs. The success of this method can be verified by morphological differences, since haploid individuals have distinct malformations in the head, body and fin tails.

Androgenesis

Androgenesis is quite the opposite process of gynogenesis. It includes the following stages:

(i) Egg exposure in X, γ, or UV radiation. The result is loss of genetic information.

(ii) Fertilization of the genetically inactive eggs with regular sperm. The zygote bears haploid genetic information from the male parent.

(iii) Application of some sort of shock at the correct timing and interruption of the 1st mitotic division.

(iv) Production of diploid individuals (2N) that bear genetic information only from male parents (diploid androgenetic zygotes).

The technique of diploid androgenesis has been applied in quite a few cyprinid species, in rainbow trout and in sturgeons with relative success. Marengoni and Onoue (1998) described egg utilization of tilapia *Oreochromis niloticus* fertilized with *Oreochromis aureus* sperm, for the production of androgenetic tilapias. Bercsenyi et al., (1998) fertilized eggs of common carp *Cyprinus carpio* with sperm from three different varieties of goldfish *Carassius auratus*, after having them irradiated. By using heat shock mitotic division was interrupted, thus 1,500 androgenic individuals of common carp were produced without genetic marking. Bongers et al., (1999) reported production of viable YY androgenic common carp, *Cyprinus carpio*. When they crossed them with females with different genetic background, all-male offspring population was produced. Additionally, sperm of these males was used for the production of androgenic generation, of which clones were all males. The main problem in such processes is the effect of radiation in eggs, which results in low egg survival rates. That is the reason why this particular technique of androgenesis is still not correctly applicable in pisciculture, vis-à-vis the other techniques of chromosomal engineering.

Biochemical and Molecular Markers

The production of transgenic fish with any method requires the following confirmations (Chourrout, 1991; Houdebine and Chourrout, 1991; Dunham, 2004). It is required to be sure that the manufactured DNA is present in the genetically modified organism and that it has been integrated in its genome, consequently it is expected to express its action to the right degree and to the right tissues for the commercial use for which it was drawn. In order to locate the presence of transgenes in early ova, a simple method is that of "dot blots". DNA is exported from transgenic ova or larvae and is placed in nylon membranes in a simple provision. With the use of heat and radio-marked DNA, which is an additional of some section of the transgene, one can print in automatic radiographic film and observe where the positive results are. Alternatively, one can use the PCR (Polymerase Chain Reaction) method in order to observe where the transgene can be reproduced, since DNA was formerly extracted from the organism under review. PCR and "dot blots" help us to identify in which particular individuals transgenes are present, but they do not reveal whether these transgenes have been integrated in the chromosome of the organism in question. In reality, because millions of copies of the transgene enter in each effort in eggs, but very little is integrated in the genome, it is certain that they do not lie integrated in the cytoplasm in big numbers. For this reason more specialized techniques have been developed such as the use of "restriction enzyme incubation", "Southern blotting", the use of fluorescent substances and "in situ hybridization" for the verification of transgene embedment in the chromosomes, to which degree it is integrated, but there are still inaccuracies, faults and ambiguities. The most certain method of verification of gene import in the chromosomes of an organism is by controlling the existence of the transgene in the offspring according to Mendelian laws. By simply supposing that the transgene has been integrated in only one chromosome of a genetically modified fish, then only half of the gametes that will be produced by F_0 generation will bear the transgene. Consequently, in F_1 generation it is expected at most half of the offspring to bear the transgene.

How can it be checked if a transgene is present and expressed? It can be simply checked for mRNA from the transgene or still better, the levels of protein that are translated. This can be practiced with immunogenic methodology, where antibodies of a specific protein are produced.

Even if important successes are reported following the traditional methods of genetic improvement, the prospects are much higher with the help of biotechnology. Higher growth rates and fittest genotypes could be achieved with the combination of methods of genetic selection and genetic engineering. Initial experiments already highlight the enormous prospects of this combination. In Israel, Hinits and Moav (1999) accomplished to improve common carp by using growth hormone transgene combined with genetic selection. Similarly, encouraging results are those from Dunham and Liu (2002), who accomplished to increase growth rate in channel catfish and common carp by combining genetic selection, genetic engineering and hybridism, in a degree higher than what was achieved before.

Dolly, the famous first cloned sheep, was produced by transplanting the nucleus from a maternal cell of an adult sheep in a progamete of another from which the nucleus had been removed. Naturally, Dolly is not precisely a real genetic clone but a "chimerical" organism with nDNA from one organism and mtDNA from another. In fish such nuclear transplantations have been achieved in some species but with quite low success rates. What is more intriguing is that the reasons for these attempts are not accurately defined. In fact, real fish clones with identical nuclear and mtDNA are already produced after two generations of gynogenesis (Fingerman and Nagabhushanam, 2000; Arai, 2001).

Marker Assisted Selection

The basic approach is that by recognizing signalled genes or the markers that are perhaps connected with a specific phenotype, it is possible to achieve better selection (Beckmann and Soller, 1988; Beckmann and Soller, 1990; Haley, 1991). A variety of methods are available that are based on DNA for the development of genetic markers (Hallerman and Beckmann, 1988; Takagi, 2001; de Koning and Haley, 2005). The method that is more widely used today is the one that detects polymorphisms from a gene or a DNA region by cutting the total DNA genome with a restriction enzyme, separating DNA in a gel, or blotting DNA on membrane, by carefully examining the gene or the sequence of interest, and finally observing polymorphisms as zymograms in an agarose or polyacrylamide gel. PCR is still under development and scientists have encountered important applications (Michelmore et al., 1991; Liang et al., 1996; Ji et al., 2004; Joerg et al., 2004; Liu et al., 2005; Rodriguez-Lazaro et al., 2005). RAPD (random amplified polymorphic DNA) involves the use of short sequences combined with PCR in order to reveal regions that present differences within and between species (Exadactylos et al., 2003). Differential expression of specific genes can be more thoroughly examined by conventional techniques, such as Northern blotting or quantitative reverse transcriptase PCR (RT-PCR). All methods that are based on PCR have important advantages because only a small quantity of DNA is required, big molecular weights are not needed and finally few stages are involved during the analysis.

Hybridization-based cDNA microarray technology allows simultaneous quantification of gene expression levels for quite a lot of genes among samples of different genotypes, developmental stages, tissues, physiological states, stresses and challenges and various environmental conditions (Karsi et al., 2001; Karsi et al., 2002b). QTL (quantitative trait loci) mapping is becoming a reality. All types of biochemical and molecular markers are potentially useful for QTL mapping. Neutral DNA markers exhibit advantages of being abundant and polymorphic for QTL mapping. VNTR (variable number tandem repeats) microsatellites are particularly good candidates as they are sometimes found within genes, and are highly polymorphic and inherited codominantly. Microsatellites have been identified and can be potentially useful (Karsi et al., 2002a; Kocabas et al., 2002; Bertotto et al., 2005; Webster and Reichart, 2005). New statistical developments combined with the use of highly polymorphic microsatellite DNA markers enable the determination of the population of origin of single fish, resulting in numerous modern research directions and applications in practical management of fish populations (Hansen et al., 2001; Hansen, 2002; Hansen and Jensen, 2005).

Two basic approaches in marker assisted selection are: (i) the targeted gene, i.e. the gene that is suspected for genetic divergence and related with the characteristic is thoroughly researched for the polymorphism to be revealed (with protein electrophoresis, or all different ways of DNA analysis), and (ii) the random marker, at which populations that differentiate for the characteristic in question are thoroughly researched for the random markers to be identified and related to the specific characteristic. If such markers are recognized, they can be used as direct tools for selection. Marker assisted selection can potentially have important use in characteristics such as disease resistance to pathogens; to which broodstock should never be exposed to, from whom progeny are to be used for breeding. By using methods based in molecular markers, it could be possible to determine individuals that would be potentially resistant without the requirement of exposure in pathogens.

TRANSGENIC FISH

Given the enormous diversity of the genome in any of the known species, why genes cannot be transported from one species to another? Time is the answer. Why should emphasis be laid in the process of genetic selection to acquire an organism bearing the desirable gene–a process that requires a lot of labour and time, while the gene of interest can be taken from another species and plant it in the desired organism's genome? The classic methods of genetic selection are still favourable when the characteristic is not a lot different from its natural characteristics; however, genetic engineering is the only easy and fastest way to fill in the blanks that exist between an organism's natural characteristics and the ones required from scientists and the aquaculture sector. The most commonly used gene in transgenic fish is that of growth hormones (GH). Growth hormone transgenes from humans, mice, cattle, salmonids, trout, tilapias and gilthead sea breams (Lu et al., 2002) have been used in trials for rapid growth rates and consequently in reducing the required time for a species to reach marketable size. Another transgene that is broadly used is that responsible for an anti-freeze protein, which is found in fish that live in polar climates, such as the ocean pout and winter flounders surviving in temperatures even below zero. Thus, by planting such a gene in another species its geographical distribution can be broaden and consequently where it could be potentially cultured. For example, the bigger part of Canada's coastline is too cold for culturing salmonids. With the aid of genetic engineering a transgenic salmonid species can be created bearing such genes from winter flounders, which might be successfully cultured in these areas where it was difficult or even impossible until now.

Transgenesis is also developed to provide immunity against various diseases in fish. This can be achieved by intramuscular vaccination of an organism, for protection from pathogens. The question is why should one pursue such a time-consuming process while an organism can be formed that bears in its genome such information exhibiting de novo immunity? It would also be feasible in the future to monitor natural fish populations with the use of genetic labelling; meaning that embedment of DNA segments in their genome will not affect their biology, ecology or behaviour, they will be inactive in other words, but will be easily detected with biotechnological methods. A section of the aquaculture industry concerns ornamental fish that are used for other reasons rather than human consumption. In this production line transgenic fish have already been produced with genes that make them phosphoresce in darkness. An interesting perspective of genetic engineering is the formation of transgenic fish for the production of marketable pharmaceutical substances replacing the ones used today derived from mammals. The most optimistic perspective that the fish geneticists can foresee, at least partly, is the replacement of mice, which are used today in laboratory experiments, with fish.

When arguments are raised about transgenic fish, it should be borne in mind that everything is about genetic transport from one organism to another for the formation (transgenesis) of genetically modified organisms (GMOs). DNA is implanted in a nucleus so that it can take part in the chromosomal replication and is integrated as a hereditable material of the cell. In order to materialize this technique, complicated technology is required. The recovery rates of the new sequence in offspring vary from 0-10%. The promoter has a fundamental role in experimental success that will "run" the gene. Promoter is a DNA fraction that "fixes" how tight RNA polymerase will be connected to DNA, the rate at which RNA synthesis will start (transcription), as well as where a structural gene will be expressed. The likely advantages of the utilization of transgenic fish are faster growth rates, one of the most important characteristics of breeding, tolerance in extreme environmental conditions and diseases. The likely disadvantages generated by these methods are the high investment cost that is required, intensive work rate and technology usage, as well as the likely rejection of genetically engineered products by the consumer. Other major obstacles in handling and culturing these individuals in the aquaculture industry, are the regulation prohibitions with regard to breeding and conservation of genetically modified organisms. Due to the lack of output data for such individuals, it is extremely difficult to report the likely repercussions of genetically modified organisms in natural environments. The technique of transgenesis has been successfully applied in species such as rainbow trout, Atlantic salmon and goldfish. Cook et al., (2000) reported that growth rate in Atlantic salmon was increased to 262-285% in F_2 generation. Rahman and McLean (1999) produced three strains of transgenic individuals of a tilapia species, using growth hormone from chinook salmon *Oncorhychus tshawytscha* and normal sequence from *Macrozoarces americanus*.

Records of Transgenic Fish

With regard to increased growth rates positive results have been obtained, but not in all cases. To date, it has been revealed that increases of growth rates are feasible from 10% to the hyperbolic 3,000% in certain cases.

Even if most trials in transgenic fish have been focused on the increase of growth rates, quite a few focus on cold resistance (Fletcher and Davies, 1991; Shears et al., 1991). The initial experiments were aimed at the formation of salmonids that could be cultured under arctic conditions, but the gene expression levels were not sufficient (Fletcher et al., 1988; Jiang, 1993; Moav et al., 1993). For obtaining encouraging results in this direction thorough experimentation is needed, since the consequences are still poor. Research results are similar, perhaps a little better, with regard to strengthening the fish immunogenic system (Anderson et al., 1996a). Anderson et al., (1996b) reported antivirus activity, however, the most important diseases in cultured species originate from pathogenic bacteria, consequently research should be also focused in this direction.

Results of such experiments indicate some of the problems caused by transgenesis, even if strong steps have been taken for reducing abnormalities that are observed in transgenic fish. In any case, a lot of research is needed before cultured transgenic organisms become acceptable by the consumers and exhibit commercial success (FAO, 2001a; BBC News, 2004). Unfortunately, the likely commercial value of genetic engineering led to research of commercial objectives; research is conducted from private companies that engage their own human resources, who do not need to publish their work results in the broader scientific community. Thus, questions arise with regard to the integrity and the techniques' usage of these companies.

Gene Expression, Isolation and Cloning

The basic stages in transgenesis are: (i) gene selection, (ii) gene isolation, (iii) gene cloning, (iv) gene induction in fertilized egg, and (v) expression of inducted gene.

(i) Growth hormones (GH) are polypeptides composed in vertebrates' hypophysis, and participate in body formation and growth by regulating a lot of metabolic reactions. Metallothioneins are proteins that contain heavy metals in their cells, thus performing a lot of functions in the cell. Their synthesis increases considerably in the presence of heavy metals; thus, whenever fish are exposed to heavy metals, they are capable of increasing their protein synthesis. Their main task is to provide zinc to enzymes that activate only in the presence of metals; their secondary task is detoxification by constraining heavy metals in the cell and excretion of the produced solvate from the liver. According to experiments, the selected genes of rats' mammalian growth hormone recombine with the metallothionein mice promoter, before their entry in fish eggs. Due to the promoter, the growth hormone, being produced in small quantities in hypophysis, initiates synthesis in considerably larger quantities in the liver. Such methods were applied in rainbow trout, where induction of a growth hormone gene recombined with a metallothionein promoter. Its expression and heritability of this sequence have been proved in its offspring. Antifreeze proteins are polypeptides rich in alanine, which run in the blood of many fish species that live in territories of low temperatures (Arctic, Antarctic), acting as a natural antifreeze, decreasing the freezing blood and tissue temperature. The responsible sequences for the synthesis of these proteins have been isolated. The genes which code antifreeze proteins constitute a gene family of 30-40 members, the majority of which are found accidentally repetitive in the genome. In Atlantic salmon, the induction of a gene responsible for the synthesis of an antifreeze protein has been tested (Fletcher et al., 2004). Salmon does not produce antifreeze proteins; fish die in cages when temperature falls below 0.7°C. Disease resistance in certain cases depends on genetic information of all species individuals. Research on isolation of DNA sequences that provide tolerance in their carriers is still progressing (Muller and Brem, 1998; Zbikowska, 2003). Thermo-inducted genes determine proteins that protect cells from extremely high temperatures. Their isolation and induction in the fish genome could increase their tolerance in high temperatures.

(ii) The first step for starting the experimental process is the isolation of the suitable gene sequence along with the adjacent regulatory sequences from a genomic library. The regulatory sequence is supposed to include the promoter's sequence.

(iii) Gene cloning process: (a) gene sequence along with regulatory sequences are cloned in a plasmid or a virus, (b) recombined DNA molecules are induced in suitable bacterial stems, for many copies to multiply, and (c) millions of copies of DNA fragments are isolated, from bacterial cells.

The plasmid that was used for the induction of rat growth hormone gene in fish included also the promoter of mice metallothionein gene. Although most of the experimental work, to date, in fish used cloned DNA designed for other organisms rather than fish, fish genomic libraries increase rapidly.

(iv) It is essential to induce a lot of copies of the new gene in the nucleus at a suitable timing for the cloned DNA to take part in DNA synthesis, in order to achieve an effective induction. Consequently, DNA induction in larvae happens during the early stages and close to DNA replication locus. When this happens before the first mitotic division of the fertilized egg, then the induced material is inherited to all cells. If one or more mitotic divisions take place before the induction, then a mosaic will be shaped, where only specific cells carry the new gene. Fish eggs are ideal for this type of research since they are: (a) available in vast numbers, (b) can be controlled externally to fertilization, (c) easy in macroscopic handlings, and (d) easily incubated. However, they exhibit a hard exterior membrane, which represents an important mechanical obstacle for the appliance of microinjections. This obstacle can be averted either with chemical excerption using proteolytic enzymes, or by piercing with a solid needle, before microinjecting. The techniques for induction of foreign DNA in fish eggs are: (1) microinjection, (2) electroporation, (3) biolistics, (4) germ cell injection, (5) embryonic cell transfer, (6) direct gene transfer in muscular cells, (7) sperm mediated transfer, (8) viral vectors, and (9) lipofection.

(1) Is difficult, time-consuming, highly risky, requires excellent laboratory practice and equipment. DNA that is intended for infusion is suspended in a suitable regulatory buffer solution and is injected either in the nucleus or in the cytoplasm of the fertilized egg. In each egg hundred thousands copies of cloned DNA molecules are injected, since a large number will be deteriorated by various enzymes.

(2) Fertilized fish eggs are subjected to electric impulse of high intensity resulting in destabilization of their cellular structure and cell membranes; cells become temporarily more porous and larger molecules than usual can pass through. Thus, DNA construct molecules can easily pass through the egg. This method has been successfully applied in common carp (Powers et al., 1992), and in catfish (Muller et al., 1993).

(3) This method involves bombardment of cells with microspherical particles, usually gold coated, with DNA construct; it has been applied in trout with 70% success.

(4) Germ cells of a species are implanted in another species' blastula. Chimeric fish were produced by exchanging the upper halves of blastoderms at the blastula stage between goldfish-common carp hybrids with pigmented scales, and goldfish with clear scales, or between goldfish with clear scales and triploid crucian carp with pigmented scales (Yamaha et al., 1997).

(5) This technique was initially applied in *Brachydanio rerio* (Collodi et al., 1992) and includes: embryonic cell culture, induction of foreign DNA in embryonic cells, re-induction of transformed cells in the developing larvae.

(6) Involves microinjection of foreign DNA in fish muscle tissue; it has been successfully applied in seabass, trout (Anderson et al., 1996a) and *Brachydanio rerio* (Tan and Chan, 1997). This technique is also applied for vaccines development in pisciculture.

(7) Includes incubation of spermatozoa with foreign gene before fertilization. It turns out that DNA binds readily to the outer coat of spermatozoa, so at the first sight this looks like a perfect method of getting novel DNA into the egg. This was reported in *Brachydanio rerio* (Khoo et al., 1992).

(8) An alternative gene transfer method is to use a virus as a vector that has been rendered defective. However, there are obvious dangers involved in engineering any virus in a wide range of species.

(9) Synthetic lipids containing foreign DNA can be absorbed from animal cells and this was also tried with fish eggs. However, as with the use of modified sperm foreigner DNA is not highly expressed, since protection mechanisms of the target cell deteriorate it.

(v) The success of gene transfer in larvae will be revealed at the period of intensive DNA replication, after the quick stages of embryonic cell divisions. The simultaneous replication of the inducted DNA together with cell-recipient DNA can be detected by isolating DNA with: (a) the use of probes as marked molecules that bind in controlled sections of the foreign DNA and can be detected with autoradiography, and (b) the digestion of isolated DNA with restriction enzymes followed by electrophoretic segregation, so that the sequences of the foreign DNA molecule are revealed. If foreign DNA has been successfully inducted in the fish genome, restrictive fragments will be greater in length and will exhibit higher molecular weights.

Gene Transfer and Technology

Gene transfer has become a very stirring research subject over the past years (Table 14.1) (Chen and Powers, 1990; McLean and Penman, 1990; Houdebine and Chourrout, 1991; Aerni, 2004; Logar and Pollock, 2005; Petit, 2005). This obviously happened because of the important interest in

Table 14.1. Genetically modified fish species with transgenes and promoters (Beaumont and Hoare, 2003)

Species	Gene	Promoter
Salmon	*hGH, bGH, rGH*	mMT-1
	csGH	opAFP
	sbGH	wfAFP
	coIGF	csMT-1
Trout	*hGH*	SV40
	rGH	mMT-1
	sbGH	cβ-actin
	INHV-G protein	CMV-tk
	cα-globin	cα-globin
Tilapia	*hGH, rGH*	mMT-1
	rGH	cβ-actin
	tiGH	RSV
	tiGH	CMV
	wfAFP	RSV
Catfish	*hGH*	mMT-1
	rtGH, csGH	RSV-LRT
Carp	*hGH*	mMT-1
	rtGH	RSV-LRT

Genes: GH = growth hormone (h = human, b = bovine, r = rat, cs = chinook salmon, sb = seabream, ti = tilapia, rt = rainbow trout), coIGF = coho salmon insulin-like growth factor, IHNV-G protein = infectious hematopoietic necrosis G-protein, cα-globin = carp α-globin, wfAFP = winter flounder antifreeze protein.

Promoters: mMT-1 = mouse metallothionein, opAFP = ocean pout antifreeze protein, wfAFP = winter flounder antifreeze protein, csMT-1 = chinook salmon metallothionein, SV40 = simian virus 40, cβ-actin = carp β-actin, CMV-tk = cytomegalovirus with herpes simplex virus thymidine kinase, cα-globin = carp α-globin, RSV = rous sarcoma virus, CMV = cytomegalovirus, RSV-LTR = rous sarcoma virus with long terminal repeat.

transgenic technology and the identification of fish as the ideal Chordata for such trials. The basic approach that is used for gene transfer is microinjection. Cloned DNA sequences are inducted in eggs after fertilization. Gene transfer is monitored by the presence of foreign DNA in offspring or by their expression. Alternative methods, apart from microinjection, are under trials. Egg DNA electrophoresis after fertilization is one of them, and sperm DNA transfer is another (Tsamis et al., 2001; Hosono et al., 2005). These techniques could well lead to increased production of transgenic animals (Wall, 2002; Smith and Spadafora, 2005).

A number of restrictions followed transgenesis. Even if it is considerably easy to be applied in fish rather than mammals, provisional stages of gene isolation and structuring of gene carriers should be developed. Such studies are very expensive and specialized personnel are required. Another important problem is that of the dangers involved from transgenic animals released accidentally in nature. Natural populations would mix with the cultured ones, which in most of the cases exhibit significant genetic variations leading to genetic pollution (Hindar et al., 1991a, b; Exadactylos et al., 1999; Tufto and Hindar, 2003; Hindar et al., 2004; Naylor et al., 2005). In any case, consumers' viewpoints are of importance and could potentially constitute serious restriction points in breeding transgenic fish in the near future.

Genetic abnormalities have not been reported so far in transgenic individuals, perhaps because the genome of eukaryotic organisms is so big that losses up to 10 kb do not imply serious genetic disorders. However, low survival rates can be observed either in eggs or larvae hatched from them. A few examples are:

- Gene transfer in various fish species began in 1985, where a genomic sequence of human growth hormone was imported with plasmids' help along with a mouse metallothionein promoter in goldfish (*Carassius auratus*) eggs (Niu et al., 1986).
- In 1986 chicken crystalline, one of the simplest ocular proteins, was transferred and expressed in *Oryzias latipes* (Ozato et al., 1986).
- In 1987 a complex gene of metallothionein-human growth hormone was transferred in catfish (*Ictalurus punctatus*), rainbow trout and tilapia (*Oreochromis niloticus*) (Brem et al., 1988).
- In 1988 a gene that regulates the synthesis of an antifreeze protein was carried out in salmon, aiming at tolerance in temperatures below 0°C (Fletcher et al., 1988).
- In 1990 a method was developed for transferring of bacterial gene that exhibits tolerance in neomycin antibiotic in goldfish (Yoon et al., 1990).

- In 1993 a gene of β-galactosidase from a xiphoid species was inducted in seabream, by microinjection in fertilized eggs (Cavari et al., 1993).
- In 1996 a metallothionein trout promoter and a bacterium acetyltransferase gene were inducted in fertilized eggs of *Oryzias latipes* (Kinoshita et al., 1996).
- McKenzie (1996) cloned the human insulin gene in fish, a fact that led the interest of a lot of researchers in the formulation of firm cell forms of transgenic fish that bear the insulin gene.
- Rahman et al., (2000) generated three lines of transgenic tilapia with a construct containing a *lac*Z reporter gene spliced to the regulatory region of a carp *beta actin* gene.
- Chen et al., (2000) generated homogeneous strains of transgenic fish by nuclear transplantation with transgenic early-embryonic cells.
- Aoki et al., (2002) established a transgenic zebrafish line harbouring *rps*L gene for detecting mutagens in the aquatic environment. Using these embryos, mutations induced by chemicals, such as ethyl nitrosourea and benzo[α]pyrene, were detected.
- Wright et al., (2004) have produced transgenic tilapia that expressed a "humanized" tilapia insulin gene. Future improvements on these transgenic fish may allow tilapia to play an important role in clinical islet xenotransplantation.
- Recently zebrafish became an attractive vertebrate model for genetic studies of development, suppressing apoptosis, and cancer (Langenau et al., 2005). Moreover, Alimuddin et al., (2005) demonstrated that fatty acid metabolic pathway in fish can be modified by transgenic techniques, and perhaps this could be applied to tailor farmed fish as even better sources of valuable human food.

The advancement in the applications of selection, hybridization, chromosome engineering, sex control, gene transfer and molecular technologies for enhanced aquaculture productivity has played a pivotal role in increasing the world food production through revolutions in plant and animal sciences. Though attention on fisheries has been inadequate (Lakra, 2001), production and crossings of transgenic fish are limited, up to date, under laboratory conditions. Consequently, scientific evidence on behaviour of transgenic organisms in the natural environment, or the likely implications from their crossing with natural populations, is lacking (Sin, 1997). Although it is still a speculation, quite a few believe that populations of transgenic fish will dominate the natural ones, during an unrestrained release scenario in the natural environment, because of their disease resistance, their tolerance to extreme environmental conditions, etc. as such, deliberate or accidental release of transgenic fish in the natural environment is prohibited, since the implications in ecosystems' balance are unpredictable.

What could be possibly done to prevent such practices?

- Culturing of transgenic fish should be run quite far from natural habitats.
- Clear-cut safety measures should be taken in cages or reservoirs, in order to avoid potential thefts and vandalisms.
- Water flow exits in cages should be examined before drainage in the main system flow, so that we are certain that no transgenic fish has escaped.
- Sterilization of transgenic fish is of major importance, in order to prevent crossing over with natural populations in the worst case scenario.

GM FISH FEED – RISK ASSESSMENT

Presence of GM in Fish Feed

Although intensive fish farming has been considered over a long time as the best solution for helping in depleting wild fish stocks, the scientists involved in the Endangered Seas Program have recently severely refuted this. In fact, the new approach is that only through better fisheries management the pressure on current fisheries can be relieved. Today's farming of carnivorous species such as salmon and trout is not sustainable, as it consumes four times as much wild-caught fish as it produces farmed fish (http://www.mindfully.org/GE/Over-Fishing-Solution.htm).

By using large amounts of wild-caught fish to feed farmed fish, the European aquaculture industry is actually increasing the pressure on ocean fish populations. The demand for wild-caught fish for fish feed puts pressure on wild fish stocks like the blue whiting, and fears have been expressed that the growing fish farming industry will contribute to a further pressure on already heavily exploited fish species. In an attempt to reach a sustainable and "hygienic" aquaculture, the following measures can be recommended:

(i) A better fisheries management including reduction in fleet size, ending harmful subsidies and adoption of an ecosystem based approach to management.

(ii) The aquaculture industry must, as it is the largest consumer of fish oil and a major consumer of fish meal, make sure it only buys oil and meal from healthy, sustainable and well managed fish stocks.

The industry should make an effort to find more sustainable alternatives, preferably fish offal and fish waste or fish feed from certified fisheries.

(iii) Farming of carnivorous fish species should not be encouraged. However, species from lower down the food web like fish with herbivorous diets or filter feeders such as oysters can potentially be sustainable.

(iv) Use of fish waste and fish offal from fisheries and fish processing to be used for fish meal and fish oil, while at the same time controlling levels of contaminants.

(v) Ensure that when research and testing on future alternative feed resources is conducted, the precautionary principle be implemented.

Alternatives to fish meal and fish oil, e.g. soybean protein, are being and have been investigated for some time as ingredients of fish feed, as raw materials traditionally processed for fish feed are becoming increasingly limited in terms of their availability. Vegetable protein sources like soybeans and maize gluten are good substitutes for these traditional materials and are currently being used in fish feeds. However, the world leading producers of soybeans and maize (Argentina, China, Canada and USA) now grow a considerable fraction of their soybeans and maize as transgenic varieties. The consequences of changing fish diets on final product safety and quality need to be investigated. In part, this responsibility has forced Norwegian fish-feed producers to avoid incorporation of genetically modified plant products (GMPPs) into their feed (http://www.biotech.dnsalias.net/en/2005/02/3916.shtml)

However, feed companies, active particularly in EU, struggle to get guaranteed GM-free ingredients. The feed business operator is required by legislation, to label compound feed containing GM feed ingredients. At present, demands on labelling are made in Norway, when the GM part of a feed ingredient exceeds 2%. However, according to new EU legislations (1829/2003 and 1830/2003), a threshold labelling at 0.9% for the adventitious presence of approved GM material is demanded.

Effects to be taken into account for feeding GM plant products to fish should be considered in terms of two aspects: (a) fish safety; physiognomic, i.e. does the change in feed composition lead to a change, most particularly, in fish growth rate, weight, size, etc., and (b) food safety; can transgenic sequences become transferred to and traced within the fish itself (Sanden et al., 2004).

Although the potential effect of feeding GM feed to poultry and cattle has been studied quite extensively (Einspanier et al., 2001; Hohlweg and

Doerfler, 2001), there are only two available publications (Padgette et al., 1995; Hammond et al., 1996) in the case of fish feed. In both publications the effect of using GM ingredients in catfish feed, in terms of final fish weight and other physiognomic parameters, was investigated. Their conclusions were similar since the feeding values of GM soybeans and conventional soybeans were not found to be different. A more recent publication (Sanden et al., 2004) was focussed both on: (1) the fate of selected GM soy DNA fragments from feed to fish and on their survival through the fish gastrointestinal (GI) tract, and (2) whether the DNA could be traced in a variety of fish tissues. Fish were fed in three experimental diets for six weeks, which were formulated from defined components and represented either GM or non-GM materials (17.2% of the fish meal was replaced with either GM or non-GM soy). A control diet composed of fish meal as the only protein source was used for comparison purposes. The transgenic sequences (120 and 195 bp) and the lectin gene (180 bp) could be detected in the GM soy feed. In the fish GI tract, however, only the smaller DNA fragment (120 bp) could be amplified from the content of the stomach, pyloric region, mid intestine and distal intestine. No transgenic or conventional soy DNA fragments could be detected in liver, muscle or brain tissues dissected from sacrificed fish. The sensitivity limit of the method was evaluated to be 20 copies. Their data indicated that though GM soy transgenic sequences may survive passage through the GI tract, they could not be traced in fish tissues.

The risk assessment of present GM food or feed has been recently assessed with the aid of FMEA (Failure Mode and Effect Analysis). Scipioni et al., (2002, 2005) in two publications intended to develop a system capable of assuring the production of food not containing or produced from GMOs with reasonable certainty and relative simplicity in a food firm and, in the mean time, propose methodological suggestions for its possible implementation also in other food firms. To reach this target, the methodological approach suggested by the HACCP (Hazard Analysis Critical Control Point) and FMEA methods proved particularly useful. The concept of risk assessment based on the calculation of Risk Priority Number (RPN) [RPN=Severity (S) x Occurrence (O) x Detection (D), where S, O and D are in a 1-10 scale] was extended to the whole of the food chain upstream of the interested firm considering the food chain as a single macro-process made up of various processes (the single firms) communicating among them. As the authors clearly stress, such result was possible only through sensitization of the participants in the food chain, thus leading to a real sharing of the "NON GMO" policy adopted by the top management.

Transgenic Fish and Regulation

Worldwide, demand for wild fish continues to increase at a rate which is higher than its natural population growth. By the late 1990s, worldwide consumption of fish per capita had almost tripled since 1960 (FAO, 2001b). During the same period, the annual global fish catch reached a plateau at approximately 95 million metric tonnes (FAO, 2001b); the Food and Agriculture Organization expects this stagnation to continue into the future (FAO, 2002). To meet the demand and ease the pressure on natural fisheries, commercial aquaculture – fish farming – grew rapidly as a cost effective alternative to traditional fisheries (Naylor et al., 2000), with aquaculture's contribution to global supplies of fish, molluscs and crustaceans increasing from 3.9% of total production by weight in 1970 to 27.3% in 2000 (FAO, 2002). Beyond traditional aquaculture, transgenic fish provide a possible means for meeting seafood demand more efficiently. For example, AquaBounty estimates that its AquAdvantage salmon, a modified Atlantic salmon, reach commercial size in one-third of the time required for non-transgenic salmon (Fletcher et al., 2001).

In 1999, AquaBounty Inc., submitted a proposal to the US Food and Drug Administration (FDA) for approval of the first transgenic fish for human consumption: the AquAdvantage salmon. This application triggered a new regulatory approval process by the FDA (Logar and Pollock, 2005). America's decision regarding the regulation of transgenic fish may influence how other countries or international bodies such as the FAO or World Health Organization approach the issue. Although the US is rather a small player in the world aquaculture market, if other countries follow the US lead in approving transgenic fish for commercial consumption, it is likely that these other countries, such as China (the world leader in aquaculture), will feel the lion's share of the impacts from farming transgenic fish. Given this possibility, the implications of the decision on how to regulate transgenic fish in the US could be felt worldwide. Although according to Logar and Pollock (2005) the AquaBounty proposal is not likely to be favourably voted in the US, especially since the Americans' support for transgenic organisms has recently (in 2002) gone down to 61% from 78% since 1997, as the controversy over genetic engineering has intensified (IFIC, 2002). Despite the fact that consumer and environmental groups in the US and elsewhere have targeted transgenic products in their campaigns and are communicating their concerns to their supporters and the general public, they are confronted with great and occasionally unsurmountable difficulties. This is due to the current US laws, according to which citizens are denied access to official information regarding the FDA's evaluation of these GM organisms.

APPROACH OF FAO, FDA AND EU TOWARDS FISH SAFETY AND FISH FEED SAFETY

During its 8th session, the COFI Sub-Committee on Fish Trade1 (COFI:FT) recommended further strengthening of FAO's work on fish safety and quality, in order to meet the increasing need for capacity-building, especially in developing countries, and to promote a more rapid harmonization of fish safety and quality standards and systems in accordance with the rules of the SPS and TBT agreements. FAO was requested to strengthen its support to improve safety and quality management in aquaculture, particularly through the implementation of HACCP principles in the production chain. Equivalency of safety management systems was recognized as an area where progress was relatively slow and which required specific attention, including capacity-building. The work of Codex was highlighted and the importance of active participation in this work was emphasized. The 25th Session of COFI2 requested FAO to further pursue its work in this field, with particular mention of HACCP, dioxins, residues and fish meal, capacity building and institutional strengthening in the field of WTO multilateral trade negotiations relating to fish trade. The Sub-committee on Fish Trade was requested to avoid duplication and coordinate with the Sub-committee on Aquaculture, especially with regard to safety and trade of aquaculture products.

During both the 8th Session of COFI:FT and the 25th Session of COFI, many Members expressed serious concern regarding the maintaining of restrictions on trade and use of fish meal for animal feed, on the grounds of alleged link to the Bovine Spongiform Encephalopathy (BSE). However, the FAO report endorsed by the 8th Session of COFI:FT was rather reassuring since no epidemiological data linking fish meal to BSE were reported.

In fisheries, there are five broadly defined axes along which a strategy in support of a food chain approach to food safety should be based:

- Fish safety and quality from a food chain perspective should incorporate the three fundamental components of **risk analysis**– *assessment, management and communication* – and, within this analysis process, there should be an **institutional separation** of science-based risk assessment from risk management – which is the regulation and control of risk.
- **Tracing techniques** *(traceability)* must be improved from the primary producer (including animal feeds and therapeutics used in aquaculture), through post-harvest treatment, processing and distribution to the consumer.

- **Harmonization of fish quality and safety standards**, implying increased development and wider use of internationally agreed scientifically based standards.
- **Equivalence in food safety systems** – achieving similar levels of protection against fish-borne hazards and quality defects, whatever means of control are used must be further developed as no such agreements have as yet been put in place.
- Increased emphasis on **risk avoidance or prevention at source** within the whole food chain *–from farm or sea to plate–*, including development and dissemination of good aquaculture practices (GAP), good manufacturing practices (GMP) and safety and quality assurance systems (HACCP, ISO 22000, ISO 9001:2000) (FAO, 2004).

US – FDA

In the US, implementation of the Federally Mandated Seafood Rule 6, along with the Good Manufacturing Practices GMP (21 CFR part 110) and Sanitation Control Procedures (21 CFR part 123) were continued. Likewise, application of the updated *Fish and Fishery Products Hazards and Controls Guide* issued by Food and Drug Administration, FDA to assist the fish industry was broadened. The Seafood HACCP Alliance, a national education programme designed to complement the Guide, was strengthened. This programme involved academic and regulatory expertise in every state plus numerous international training efforts. Risk assessment work for specific pathogens of concern for seafood was carried out. Of particular interest is the 2003 FDA Interim Final Regulation (21 CFR Parts 1 and 20) promulgated under the *Public Health Security and Bio-terrorism Preparedness and Response Act*. This regulation requires that domestic and foreign facilities that manufacture/process, pack or hold food for human or animal consumption in the US register with FDA and submit electronically prior notice to FDA before the shipment is due to arrive into the US. It is feared by several fish exporting countries, that the implementation of these requirements, at least in the beginning, may disrupt fish trade flows from exporting countries into the US (FAO, 2004).

FDA has issued control guidance for fish and fishery product hazards (FDA, 2001). This guidance consists of the following contents in an attempt to cover, at a generic level, most of the hazards (physical, chemical and microbiological) occurring either in aquaculture or in the fish and seafood industry;

(a) Steps in developing HACCP plan
(b) Potential species–related and process-related hazards

(c) Pathogens from the harvest area
(d) Parasites, natural toxins, scombrotoxin (histamine) formation
(e) Other decomposition – related hazards
(f) Environmental chemical contaminants (methyl mercury, aquaculture drugs)
(g) Pathogen growth and toxin formation as a result of time/temperature abuse and inadequate drying
(h) *Clostridium botulinum* toxin formation
(i) *Staphylococcus aureus* toxin formation in hydrated batter mixes
(j) Pathogen survival through cooking or pasteurization
(k) Introduction of pathogens after pasteurisation and specialized cooking processes
(l) Allergens/Food intolerance substances and prohibited food and colour additives
(m) Physical hazards (metal and glass inclusion)

The main reason for this document is the extensive pollution (mostly chemical and microbiological) reported in several publications and reports. In particular, every year millions of tons of agrichemical pesticides are applied to soils in and around aquatic ecosystems that eventually pollute rivers, streams, lakes, wetlands, and municipal water supplies. In addition, agricultural soils are routinely sterilized with methyl bromide, a chemical that destroys atmospheric ozone and increases the amount of ultraviolet light radiation which results in the degradation of aquatic systems (http://www.wfrc.usgs.gov/research/contaminants/STRodriguez4.htm). Another issue is the high mercury content in the US and to be more specific, the status and dynamics of mercury contamination associated with past hydraulic mining operations. By comparison, little work is being conducted on mercury contamination in historically dredged environments in urbanized areas (e.g. the cities of Folsom, Gold River, and Rancho Cordova) within the American River watershed and neighbouring lakes. Preliminary results from 22 fish samples collected from Lake Natoma during August 2000 indicated that the mercury content may approach or exceed guidelines for human consumption. Thus, the USBR has requested that USGS conduct a survey of mercury contamination in edible portions (muscle tissue or fillets) of selected sport fishes from Lake Natoma (US) (http://www.wfrc.usgs.gov/research/contaminants/STSaiki5.htm).

Aflatoxicosis is a disease that can affect many species of fish, and results when feed contaminated with aflatoxins is eaten by the fish. Aflatoxins are chemicals produced by some species of naturally occurring fungi (*Aspergillus flavus* and *Aspergillus parasiticus*) commonly known as moulds.

Aflatoxins are common contaminants of oilseed crops such as cottonseed, peanut meal, and corn. Wheat, sunflower, soybean, fish meal, and complete feeds can also be contaminated with aflatoxins (http://www.aquafeed.com). Four major aflatoxins (AFB1, AFB2, AFG1 and AFG2) are direct contaminants of grains and finished feeds. Factors that increase the production of aflatoxins in feeds include environmental temperatures above 27°C, humidity levels greater than 62%, and moisture levels in the feed above 14%. The extent of contamination varies with geographic location, feed storage practices and processing methods. Improper storage is one of the most important factors favouring the growth of aflatoxin-producing moulds, and is a major element that can be controlled by the fish producer. Aflatoxin B1 (AFB1) is one of the most potent, naturally occurring, cancer-causing agents in animals. The first documented incidences of aflatoxicosis affecting fish health occurred in the 1960s in trout hatcheries. Domesticated rainbow trout (*Oncorhynchus mykiss*) that were fed a pelleted feed prepared with cottonseed meal contaminated with aflatoxins, developed liver tumours. As many as 85% of the fish died in these hatcheries. Although cottonseed meal is no longer used as a major ingredient in feed formulations, poor storage of other feed ingredients and complete feeds can lead to contamination with aflatoxins. Although Aflatoxicosis is now rare in the rainbow trout industry due to strict regulations enforced by the US Food and Drug Administration (FDA) for aflatoxin screening in oilseeds, corn and other feed ingredients, the interest in the toxic effects on cultured warm-water fishes (such as in *tilapia* and channel catfish) has increased as diets for these species are now being formulated to contain more plant and less animal sources. This increases the potential for development of aflatoxicosis in these species because, as noted earlier, plant ingredients have a higher potential than animal ingredients for contamination with aflatoxins (http://www.edis.ifas.ufl.edu/FA095).

EU Legislation

EU laws for undesirable substances and dioxins

EU legislation appears in the following forms: Regulations (mandatory and legally binding, no change of national law is required), Directives (mandatory and legally binding but change of national law is required), Decisions (specific rules addressed to Member States binding in their entirety) and Recommendations and Opinions (not binding on Member States) (Forsythe and Hayes, 1998).

The Council Directive 2001/102/EC of 27 November 2001 amending Council Directive 1999/29/EC on undesirable substances and products in animal nutrition (http://www.food.gov.uk/foodindustry/regulation/ria/ria2001/71541) was one of the relatively new Directives aiming at better control and monitoring of animal health nutrition. The Feeding Stuffs Regulations 2000 mainly implement harmonized EC Directives. Provisions cover controls on the marketing and composition of feeding stuffs including a list of additives that can be used in feeds, maximum permitted levels for certain undesirable substances (contaminants), and rules on the labelling of compound (i.e. mixed feeds) and feed materials (ingredients and feeds fed singly). The Regulations cover England and there are parallel Regulations for Scotland, Wales and Northern Ireland. The Regulations now require amendment to implement or provide enforcement machinery for Council Directive 2001/102/EC introducing maximum permitted levels (MPLs) for dioxins in a range of animal feeding stuffs.

Dioxins are toxic substances produced during various combustion and incineration processes and are also unwanted by-products in the manufacture of certain chemicals. Dioxin-like polychlorinated biphenyls (PCBs) have been used since the early 1930s, mainly in electrical equipment, but their manufacture and general use stopped in the 1970s and are no longer permitted in the UK. However, both dioxins and PCBs are persistent organic pollutants which degrade slowly and are extremely widespread in the environment and accumulate in the food and feed chain. Previously, the Undesirable Substances Directive (1999/29/EC) included only one MPL for dioxins in citrus pulp, introduced in 1998 following the discovery of elevated levels of dioxins in citrus pulp pellets imported from Brazil. In 1999 there was major dioxin contamination of animal products in Belgium due to the mixing of transformer oil with animal fats for use in animal feeds. In the wake of this incident, the Council agreed to amend the annexes to Directive 1999/29 to introduce dioxin MPLs for a range of feeding stuffs. [Due to the susceptibility of some feed binders (e.g. certain clays) to dioxin contamination, the Commission has already introduced MPLs for these in the Feed Additives Directive (70/524/EEC)]. The intention was to exclude grossly contaminated material from the food chain and to reduce dioxin exposure by setting MPLs for dioxins in various types of feed materials (feed ingredients and feeds that can be fed singly) and compound feeds (manufactured feeds). This will be of benefit to the health of both animals and the ultimate consumers of animal products.

During negotiation of the measure, however, the UK unsuccessfully pressed the Commission a number of times to undertake a risk assessment of the impact of the MPLs. Our view is that the primary objective of any dioxins strategy must be to reduce the release of dioxins into the environment – the main source of dioxins in feed and food which is essential if the MPLs are to be met or further reduced in future. In the absence of a risk assessment, it is a matter of concern that although the MPLs might exclude consignments of feeds with gross contamination, they might also remove entirely from the market some feed supplies which did not pose significant risks to consumers of animal products. Furthermore, no MPLs for PCBs have been set, although data from UK surveys has shown that dioxin-like PCBs contribute at least the same toxic load as dioxins in most foods and should be included in control measures, including those for feed. In recognition of Member-States' concerns, the MPLs were reviewed for the first time in December 2004, with a particular view to the inclusion of dioxin-like PCBs in any new levels by early 2006. The MPLs will be further reviewed no later than 31 December 2006 with the aim of significantly reducing them. The Commission has also agreed to bring forward proposals for a strategy to reduce environmental dioxin emissions.

There appear to be three main possibilities in terms of the EU Directive implementation by state members: (a) non-implementation or partial implementation of the measure, (b) full implementation of the measure, and (c) delayed implementation.

Non-implementation or partial implementation of the Directive would ensure that most of the range of animal feed stuffs remained available. Dioxin levels would instead be controlled by the established requirement for feed compounders to exercise due diligence over their manufacturing processes, and by existing legislative measures requiring that feed stuffs should not be deleterious to public health or animal health. Partial implementation might involve non-implementation of certain MPLs which were considered particularly onerous. However, with both non-implementation and partial implementation, there would be concerns that feed safety measures were being ignored, including the added protection that MPLs provide in excluding any cases of gross contamination. Non-implementation or partial implementation would also result in legal proceedings against any state member in the European Court of Justice, as the terms of the Directive require implementation of all provisions. Costs of non-implementation would include the cost of infraction proceedings to the member-state government. There might also be a loss of EU markets to member-state companies as they could not market non-complying feeds in other Member States.

Full implementation of the Directive would meet most member-states' mandatory obligation to implement EC measures. It might benefit the member-states' manufacturers and suppliers because compliance with the MPLs would provide a greater assurance of the safety status of their products. It is not possible, however, to determine the exact benefit or the costs to the industry of demonstrating compliance. Delaying implementation might defer any implementation costs. However, this would still leave the member states open to infraction proceedings and there would be concerns that feed safety related measures were being delayed. Taking into account the above factors, it is advisable to opt for full implementation of the various measures without any further delay.

EU Legislation for Heavy Metals and Sampling Methods in Foodstuffs

The most recently published European Community (EC) Laws are as follows:

(a) Commission Regulation (EC) No. 78/2005 of 19 January 2005 amending Regulation (EC) 466/2001 with regard to heavy metals.

(b) Commission Directive 2005/4/EC of 19 January 2005 amending Directive 2001/22/EC laying down the sampling methods and the methods of analysis for the official control of the levels of lead, cadmium, mercury and 3-MCPD in foodstuffs.

Furthermore, there is an ongoing Commission Working Group on Dioxins and Dioxin-like PCBs having as objective the total revision of acceptable limits for Dioxins.

(c) Monitoring Recommendation (2004/705/EC) and target levels. A new Recommendation will replace Commission Recommendation 2004/705/EC on the monitoring of background levels of dioxins and dioxin-like PCBs. The new Recommendation, which will be discussed in detail at the next Working Group meeting, will include the new Member States and will have a provision for action levels only for food (feed action levels are included in the animal feeds limit directive).

(d) Commission Regulation (EC) No. 78/2005, amending Regulation 466/2001. The European Commission is currently carrying out a review of Commission Regulation 466/2001. The review included proposals for a number of changes to the fish categories in respect of heavy metals and a proposed revised higher limit for cadmium in swordfish. The proposals were agreed and adopted. The list of fish has been modified to include a list of the main traded species and the maximum limit for cadmium in swordfish is now 0.3 mg kg^{-1} wet weight. Commission Regulation 78/2005 which makes provision for these revisions has recently been published in the Official Journal of the European Union (OJ) – Reference 20.1.2005 L 16/43. The Regulation, which is directly applicable in Member States, will enter into force on the 20th day following the date of its publication.

The Commission Regulation amending Regulation 466/2001 with regard to PAHs and the allied enforcement sampling and analysis Directive have not yet been published. Although the majority of Member States supported the proposal that for a transitional period the current dioxin limits would be maintained in addition to the Total TEQ (Toxic Equivalent) limits, some Member States continued to voice their concerns. The Commission explained that this was a transitional measure to ensure that there was no relaxation of the current position and that in future the legislation would only limit Total TEQ. Other Member States voiced concerns about a lack of laboratory capacity to measure PCBs. Other concerns included the proposed fish limit as applied to eels and whether there was a public health concern in respect to this species, which is not eaten widely. The Commission stated that the inclusion of Toxic Equivalent Factor (TEF) in the new regulation would be helpful, since the limits were based on this particular set, and other Member States expressed support.

The Working Group agreed that the Total TEQ limit for fish oil should be returned to the originally proposed 10 pg g^{-1} fat to reflect the current situation – an earlier draft Working Paper had proposed a limit of 8 pg g^{-1} fat. The Commission agreed to return the marine oil limit (which includes fish oil) to 10 pg g^{-1} fat, in line with the approach to other foods and animal feeds. Sweden and Finland have derogation, until 31 December 2006, on the current dioxins legislation [refer to Council Regulation 2375/2001 Article 1(1a) for details]. The Commission presented two options – either a general derogation for Sweden and Finland to market fish from the Baltic, or a derogation covering certain fish species – notably Baltic salmon, Baltic Sea herring and river lamprey. There was also a provision for the derogation to be extended to Estonia. The three Member States concerned signalled that they preferred the second option, which was accepted by the Commission. The other Baltic States (Latvia, Poland and Lithuania) have also asked for the derogation. The Commission will consider these requests further.

EU legislation for GM

This guidance document (2003) for the risk assessment of genetically modified plants and derived foods and feeds (6-7 March 2003) prepared for the Scientific Steering Committee by *The Joint Working Group on Novel Foods and GMOs* became the new Regulation (EC) No. 1829/2003 of the European Parliament of 22 September 2003 on genetically modified food and feed for the use of risk assessors and notifiers (generic reference to the official body submitting the notification) who intend to apply for the commercial release of genetically modified plants and derived cultivars and/or for the commercial authorization of genetically modified (GM) food or feed, i.e. food or feed containing, consisting of or produced from genetically modified plants. These topics were previously covered by Regulation (EC) 258/97 on Novel foods (http://www.europa.eu.int/eurlex/pri/en/oj/dat/2001/l_106/l_10620010417en00010038.pdf), and Regulation (EC) No. 258/97 of the European Parliament and of the Council of 27 January 1997, concerning novel foods and novel food ingredients (Official Journal of the European Communities L43: 1-7).

The two 2000 regulations (49/2000/EC and 50/2000/EC) related to GM labelling were mainly triggered by the industrialists. The regulation 49/2000/EC, referring to the 1139/98/EC specified that foods had to be labelled if the percentage of transgenic material in the ingredient exceeded 1%. However, these new regulations unfortunately did not help to solve the confusion but rather contributed to it. According to the new Regulation (EC) 1830/2003, the threshold value (1%) was further reduced to 0.9% for varieties approved for placing in the EU and to 0.5% for positively assessed but not yet approved varieties. For non-assessed or non-approved varieties the rejection threshold remained at 0% (Cheftel, 2005)

The main points of Regulation (EC) 1830/2003 could be summarized as follows:

1. Obligation of manufacturers not to place products derived from or containing GM ingredients on the market unless authorized.
2. Products to be labelled if the food is or contains a GMO.
3. The labelling requirement for products derived from or containing a GMO.
4. 0.9% threshold for labelling products containing an EU approved GM material.
5. 0.5% threshold for labelling of EU positively assessed GM material.
6. Rejection of all non-approved GM material (0% tolerance).
7. Labelling requirement of products derived from GM material even if DNA is no longer detectable (e.g. refined soybean oil).

The above-mentioned points were agreed on 10th December 2002, when the EU Environmental Council agreed on the requirement of traceability of individual GMOs contained in food or feed and exemption of traceability only for technically unavoidable adventitious contamination (Popping, 2003).

The application of this new EU Regulation (1830/2003) has resulted in the repealing of Regulation (EC) No. 1139/98, Regulation (EC) No. 49/2000, Regulation (EC) No. 50/2000 and several amendments in specific articles and clauses of Regulation (EC) No. 258/97 and Directive 2001/18/EC.

EU Sets Out Food Traceability Guidelines

The EU has issued new guidelines to facilitate the traceability of food products across all member states [http://www.foodqualitynews.com/news/news-ng.asp?n=57779-eu-sets-out Accessed 2005 June 29]. This follows the passing of the General Food Law, which entered into force on 1 January 2005. The Standing Committee on the Food Chain and Animal Health, consisting of representatives of the Member States, agreed on this common guidance document to make harmonized implementation in all Member States easier.

The specific requirements covered in the guidance document include the traceability of food products, withdrawal of dangerous food products from the market, operator responsibilities and requirements applicable to imports and exports. While being welcomed by the food industry and consumer groups, this drive towards complete supply chain traceability is nonetheless putting incredible pressure on food manufacturers, especially private-label processors (http://www.foodqualitynews.com/news/news-ng.asp?n=57779-eu-sets-out; Accessed on 2005 June 29).

Various definitions have been suggested for traceability, including a EU general food law regulation definition according to which "traceability is the ability to trace and follow a food, feed food-producing animal or substance through all stages of production and distribution" (EU Regulation Food Law 8/05/2001 quoted by Furness and Osman, 2003). The respective ISO (International Organizational for Standardization) definition is as follows "Traceability is the ability to trace the history, application or location of an entity by means of recorded information" (ISO 8402:1994). Implementation of traceability is expected to play an important role in rapid response to food safety incidents, improving effectiveness of food residue surveillance programmes, risk assessment from food exposure, fraud prevention, food wastage reduction and in improving meat and fish hygiene over processing and handling (Furness and Osman, 2003)

The new mandatory traceability requirement applies to all food, animal feed, food-producing animals and all types of food chain operators from the farming sector to processing, transport, storage, distribution and retail to the consumer. The guidance document lays down detailed implementing rules for operators. Information on the name, address of producer, nature of products and date of transaction must be systematically registered within each operator's traceability system. This information must be kept for a period of five years and on request, it must immediately be made available to the competent authorities.

The drive towards complete supply chain traceability in Europe is inexorable. In addition to the legislation, retail giants are beginning to roll out RFID (radio frequency identification) mandates to all their suppliers. As far as these retail giants are concerned, their primary motivation is building customer loyalty and trust. In driving through these traceability measures down the supply chain, they are putting into place guarantees that if there is ever a recall, then any problem could be contained quickly without losing credibility. The best way of ensuring this is to make sure that suppliers have full control. This is why retailers, as well as legislators, are leaning heavily on manufacturers to install technology that will guarantee complete traceability. Indeed, the new EU guideline specifically defines the criteria that would trigger the withdrawal or recall of a dangerous product from the market. Situations where operators are required to inform competent authorities of this withdrawal are also specified. Ultimately, the guideline underlines the fact that food manufacturers are responsible for the safety of the food that they produce and put on the market.

TRACEABILITY SYSTEMS DEVELOPMENT

The main driving force behind traceability is to assure the consumer over food safety, hygiene and authenticity. However, traceability is threatened by various factors among which adulteration has always been considered of primary importance. Adulteration can be classified in the following categories (Uberth, 2003; Arvanitoyannis et al., 2005):

(A) Non-compliance with legal requirements (product standards) such as:
  - (i) Max/min content of water/frozen water in deep frozen fish and seafood.
  - (ii) Geographical origin of fish or seafood (national or international level).
  - (iii) GM fish or seafood.
  - (iv) Farmed fish or wild caught.

(B) Erroneous addition of certain ingredients in processed fish and seafood:
  - (i) Addition of GM fish or GM plant product.
  - (ii) Non-authorized preservatives or at higher than prescribed concentration.
  - (iii) Mixing fish or seafood of different species.
  - (iv) Mixing farmed fish with wild caught.

(C) Non-compliance in terms of employment of certain technological processes for processed fish and seafood such as:

(i) Specific heat treatment.
(ii) Inappropriate packaging methods/materials.

Traceability Monitoring

In a supply chain where numerous new food items or ingredients enter the food chain, it is of great importance to establish primary identification aiming at continuously monitoring the authenticity of the food entities concerned. Nowadays, a plethora of identification techniques are available depending on the nature of the product/ingredient and property under study. Some of the most effective and accurate are the following: immunoassays (DNA or protein based, metabolic fingerprinting), spectroscopy (NMR, FT-IR, NIR), and chromatography (HPLC, GC, GC-MS). Since ideally, any raw material/product should be identified individually or by category, there arises the need for an additional code or identifier so that the information is accessible remotely by electronic means.

Nowadays, several coding systems and item numbers are available, among which the most important ones are: EAN.UCC (International Numbering Association-Universal Code Council), GTIN (Global Trade Item Number), Serial Shipping Container Code (SSCC), Global Location Number (GLN), Global Returnable Asset Identifier (GRAI), Global Individual Asset Identifier (GIAI), Global Service Relation Number (GSRN), and Data carrier technologies [linear and two-dimensional bar codes, composite codes, radio frequency identification (RFID)]. These coding systems are summarized with regard to their use and applicability in Table 14.2.

Traceability in Fish Processing

EU has already sponsored two EU concerted actions entitled "Fish Quality labelling and Monitoring, FAIR PL98-4174" and "TraceFish QLK 1-2000-00164" aiming at improving the traceability system's effectiveness and reliability. One of the problems still occurring in EU and requiring urgent solution is the difference in attitudes among the EU countries as to traceability management. The Nordic countries firmly believe that a government based system will best ensure traceability and origin of the fish and seafood, whereas the UK and the southern part of Europe think that industry and certification bodies will prove to be the most credible traceability managers (Larsen, 2003). Frederiksen et al., (1997; 2002) reported on two traceability systems; a Danish and a British. The former operates with Internet and is known as "Info-fish" (a demonstration model for fresh fish chain), whereas the latter is built around the weight and

Table 14.2. Presentation of currently employed coding systems; EAN.UCC and Data Carrier Technologies and their applications (Furness and Osman, 2003; http: //www.maff.go.jp/trace/guide_en.pdf; http://www.namm.com/techcenter/GLN%20Implementation%20Guide.pdf; http://www.eanmalaysia.com)

Coding System	Digits	Interpretation	Applications
EAN.UCC			Identification of items, locations and services
a) GTIN	14		Items and packaged items
EAN/UCC-13	13	(a) Company prefix nr (b) Item number [a + b = 12 digits] (c) Check digit	Item identification at point of sale
UCC-12	12	Similar to EAN/UCC 13	In use only in the US and Canada
EAN/UCC-8	8	(a) Truncated prefix nr (b) Item number [a + b = 12 digits] (c) Check digit	In store item identification Unique identification when processed in a 14 data field with zeros for non-available values
b) SSCC	18	(a) Extension digit [a = 0-9] (b) Company prefix nr (c) Serial number (d) Check digit	More digits to accommodate larger item reference numbers and supplementary info
c) GLN	13	(a) Company prefix (b) Location reference (c) Check digit	Unique numbering system for locations (address, contact person, fax, tel, bank account)
d) GRAI	14-16	Mainly a EAN/UCC-13 with optional extension to 16	For reusable entities such as containers and totes (food transportation and storage)
e) GIAI	Up to 30	(a) Company prefix (b) Individual asset reference	Any article belonging to a company's inventory

contd...

Table 14.2 contd...

f) GSRN	18		Relationship or action identification
Data carrier technologies			
a) Linear (1D) bar code	—	Series of narrow and wide black bars and light spaces	Most widely applied due to its simplicity
b) 2D bar code		Multi row or matrix symbol	Due to low cost attractive for shipping details (large data payload, more than 1,500 characters)
c) EAN/UPC bar code data carriers	Bar code and digits	UPC-A, UPC-E, EAN-13, EAN-8	Omni-directional scanning at point of sale or retail (standard use)
d) Interleaved two of five (ITF)	Five bars and five spaces. Interleaved. The symbology is surrounded by bearer bar.	ITF is normally well adapted to materials and printing conditions normally used for non-retail items, particularly fibreboard cartons.	ITF symbology is applied when an item is not scanned at the retail checkout. Since it is larger it does not demand high print quality as EAN-13 for machine readability.
e) UCC/EAN-128	Variant of code 128 (exclusively licenced to EAN.UCC)	Variable length, alpha numeric symbology, high flexibility	Not to be scanned at point of retail but at other areas of supply chain
f) RFID (radio frequency identification)		Tags (microchips)	Advantages: great amount of info data and high MW penetration. Disadvantages: high cost, only 5 frequencies, not yet fully ISO standardized.

grading system. A common characteristic of both systems is the use of barcodes. The Danish system appears to be more informative towards the consumers since the information is conveyed both in terms of barcode and full text label (catching date, ground, fish species, weight, etc.). The retailer is expected to play a dominant role in traceability since the fish processing industry hears his 'voice'. One, however, should also bear in mind the practical problems the fishermen are confronted with depending on the kind of fisheries they are involved in; only large fish can be individually marked, whereas small fish can only be marked in a recipient/package (Larsen, 2003).

CONSUMER BEHAVIOUR TO GM FISH AND SEAFOOD PRODUCTS

Non-Governmental Organizations (NGOs)

The NGOs active in the GM debate are almost universally opposed to implementation of the technology, and often radically so. The head of Greenpeace in Europe, Benedikt Haerlin, has been reported as saying that it would be preferable to let people die from starvation than to grow GM crops: "In remarks to reporters afterward, Mr. Haerlin dismissed the importance of saving 10 African or Asian lives at the risk of spreading a new science that he considered untested. The Greenpeace position, he said, was that research could continue as long as no seeds or animals were ever released" (1999). While there are many issues that underlie NGO opposition to biotechnology, the emotional roots of the argument appear to derive from distrust of government, opposition to corporate "globalization," a mistrust of science, and a belief that saving the planet can be achieved only by reversing the tide of technology and returning to a simpler, "organic" lifestyle. Perhaps the only unifying thread among these groups is an undercurrent of fear regarding the modern world. Despite diffuse ideals and the absence of concrete goals, most of these groups are certainly clear about the fact that opposition to GM technology offers greater hope of political influence and organizational power. Fear is an easy product to sell, and in this campaign NGOs have willing and supportive sales partners in the mass media (http://www.idea.iastate.edu/aqua/05pdf).

The influence of NGOs varies considerably by country and region. In North America they have less persuasive power, and tend to be more moderate; in Europe, they have proved to be both radical and politically powerful. Although some of these groups may be amenable to modifying their positions (e.g. the Centre for Science in the Public Interest has come out in favour of GM technology), gaining acceptance of GM products by these groups may prove to be impossible under any circumstances.

The question for producers of GM products, therefore, is not how to convince these groups, but how to marginalize their influence when it derives from unsubstantiated and technically unsound claims. Accomplishing this will require a convincing demonstration of the safety of GM technology, which can only occur over a period of time. Institutions and groups trusted by the public must vocally support the technology based on facts, the media must be convinced to balance its approach, and the developing safety record of agbiotech foods must be exemplary. Opinion leaders must also be convinced, especially in Europe, to heed opinion rather than ponder over it. Some of this will occur only glacially, so the effort must be continuous and geared for the long run.

Consumers' Behaviour towards GM Fish

Given the public debate and conflict among institutions and opinion leaders, it is perhaps surprising that gene modification has been as widely accepted by the general public as it has been in recent years. One often tends to forget that major changes require overcoming innate conservatism that fear of the unknown is not any less powerful today than it was in previous eras. In the present circumstances, when central cultural values such as food and nature are at stake, these fears are understandably high. Simple proofs of safety are not sufficient to overcome these fears: the proofs have to be repeated many times over, the cultural institutions that are trusted by the majority of people have to offer continuous support, and consumers must recognize a benefit that clearly outweighs any sense of risk. The questions consumers ask are rational and require answers:

(1) Whom do I believe in assessing the safety of the product?

(2) Why should I buy this if there is any risk at all? Why change my present buying patterns?

(3) Why don't the big companies label these products?

(4) Gene modification is scary: have we ever done this before?

Answering these questions will require patience and a willingness on the part of those developing this technology to engage in long-term efforts to increase the public understanding. Educational efforts and a willingness to share product information beyond what may be legally required, or even considered prudent under the conditions of normal competition, may be necessary. Products that confer obvious benefits must have the highest priority for development. And finally, commercial companies must maintain a willingness to view the long-term benefits of this technology and not allow obstacles in the immediate path to dissuade them from continuing.

In the literature, one can identify six categories of factors related to consumers' attitudes and beliefs, which influence the overall consumer behaviour *vis-à-vis* GM foods (the classification is made by the authors and does not constitute a ranking-by-importance of these factors): (1) awareness, pre-existing beliefs towards food technology and its risks and other general attitudes, (2) perceived GM consumption benefit, (3) sociodemographic profile, (4) social and moral consciousness, (5) perceived food quality and trust, and (6) other secondary influential factors. This classification assumes a conflict between the beliefs of 'risks' and 'benefits'. Arvanitoyannis and Krystallis (2005) have recently published an article on the behavior of Greek consumers towards GM foods. This research, though preliminary in nature, evolves around the internationally identified conflict between 'perceived safety' and 'perceived benefits' of GM products. Low real awareness, negative overall attitude, high perception of risk and limited ability of the benefits of GM foods to overcome these negative perceptions are among the major findings of this study. On the contrary, it was also revealed that GE adoption by the food industry is now perceived as an 'unavoidable reality' or even a 'necessity'. A high social status segment of 'early adopters' were rather positive towards the concept of GM foods. Overall, Greek consumers appear to make decisions based on their preconceived attitudes, a fact that explains the consensus of attitudes of consumers with different social status. In the international context, the numerous surveys implemented so far also differ in their objectives and methodology (Zimmerman et al., 1994; Bredahl, 1999, 2001; Kerr, 1999; Baker and Burnhum, 2002; Lahteenmaki et al., 2002). This research showed, beyond any doubt, that prior to the implementation of new surveys in Greece and also internationally, the development of a solid theoretical framework is required in order to assure the accurate analysis of consumer behaviour towards GM food. This framework should focus on the relation between perceived hazards and benefits, with the maximum degree of certainty possible. The way in which perceptions about GM foods are related to other, wider concepts, such as consumers' personal values, remains obscure, as does the manner with which consumers evaluate specific GM food products. The evaluation of individual GM food products can further increase the ability of marketing surveys to accurately predict the purchasing and selection decision of consumers buying GM foods (Kerr, 1999) and constitute the basis of more efficient marketing strategies. Georgakopoulos and Thompson (2005) reported on interviews with salmon farmers exploring their decision as to whether to adopt organic production methods as an antidote to GM salmon. Organic salmon farming has the potential to considerably reduce the social, environmental and economic risks associated with salmon farming. Salmon farming is an industry subjected to intense scrutiny and is highly controversial. The combination of these two factors was expected to reveal the use of environmental accounting in evaluating this potentially difficult, expensive strategic decision, responding to the barrage of public criticism, driven by changing environmental regulations and a potential value shift by key actors. However, interviews revealed that opting for organic farming was regarded as a normal agricultural decision, largely based on price forecasts. The shift to organic was relatively easy, unproblematic and not too expensive. The environmental pressure groups campaigns had very little impact on this decision and it was not subjected to systematic accounting evaluation. The interviews described a sector exhibiting many of the characteristics of Beck's Risk Society thesis (Beck, 1992, 1995, 1996). Decision makers' risk perception is identified as important for considering what factors are thought to be legitimate/illegitimate and powerful/weak in the decision making process. Unless the decision makers consider risks 'real', then the associated costs/benefits of doing or not doing something are not going to figure in the decision heuristics, regardless of the nature of their calculation. Environmental accounting could play a part, not necessarily at the individual farm level, but as part of a reflexive process in reconstructing the underlying knowledge of the social, environmental and economic risks of salmon farming as a whole.

In essence, Finucane and Holup (2005) suggested that the two main qualitative features apparently responsible for risk perceptions are: unknown risk (known vs. unknown), and dread risk (not dreaded vs. dreaded). The unknown risk factor reflects the extent to which a hazard is unknown, unobservable, unfamiliar, and has delayed consequences. The dread risk factor reflects the extent to which a hazardous activity or technology is seen as dreaded, uncontrollable, fatal, not equitable, high risk to future generations, not easily reduced, involuntary, and potentially catastrophic. In the domain of food risks, this two-dimensional structure has been revealed through psychometric work (Fife-Schaw and Rowe, 1996) showing that public perceptions are related primarily to awareness and severity. The importance of the unknown and dread risk characteristics in perception of food risks can explain, in part, public opposition to GM foods. First, GM foods present unknown risk because they are based on a relatively new science (thus, scientists do not know enough to estimate the risks accurately); second, the inadvertent introduction of harmful changes in DNA structure is not immediately obvious (the effects are delayed); and third, consumers do not necessarily know when they are exposed because GE is not obvious to the casual observer and they are not well informed about which products contain GM ingredients (Marris, 2000). A good example is GM soy, the first transgenic product launched in the European market, which is present in many foods and thus difficult to avoid. GM soy is perceived by some to be a concealed ingredient whose use in processed food is often not understood by the public (Moses, 1999). A very clear and beyond any doubt evidence of Europeans' perception of their own lack of knowledge or understanding about GM products was pretty much apparent in the 1996 Eurobarometer, in which a large percentage answered 'Don't Know' to questions about the applications of biotechnologies. A staggering 51% said they had never talked to anyone about biotechnology before, whereas around 80% of respondents to the 1999 Eurobarometer said they were 'insufficiently informed' about biotechnology (Gaskell et al., 2000).

According to Myhr and Dalmo (2005), the most prominent challenge in implementing Genetic Engineering in a sustainability context is how to ensure environmental protection and at the same time achieve economic benefits. This challenge involves active investigation of the proposed benefits and the potential adverse effects in a research and policy agenda that encourages broad and long-term thinking. Sustainability also influences socio-economic and ethical issues; these aspects highlight the need to involve stakeholders in risk governance of GE in aquaculture. Such extended peer-review may help to identify: (a) at the concept stage, what sorts of GE applications may be acceptable, (b) during the development stage, what sort of risk-associated research needs to be initiated, and (c) at the commercialization stage, socio-economic, ethical and policy implications that are important to the stakeholders.

Overall, a systematic assessment of reliable cross-cultural differences in perceptions, values, attitudes, and behaviours regarding GM food will help to fill in current knowledge gaps and to respond to conflicts encountered in international negotiations over GM foods. Decision making and communication about GM food risks will only be successful if they are based on a thorough understanding of the psychological and socio-cultural determinants of risk

CONCLUSION

The development of aquaculture industry over the years generated excellent quality products, in low prices and in large quantities to meet consumers' demand. The quality uniformity of these products contributed positively to their promotion on an increased demand. Genetic engineering methods described above constitute the recent scientific experimental efforts and their development over the last decades. Their application answered a lot of questions raised by early genital maturation for broodstock management, aiming at the production of sufficient number of eggs and good quality larvae. These will, hence, contribute to the genetic improvement of generations' characteristics, to the control check and preservation of fittest fish production characteristics and to the control and management of fish feeding and breeding in general.

While fish mature, they experience certain important changes in their productive characteristics. These characteristics such as the growth rate, age or maturation time, pigmentation and flesh texture, differ considerably between the two sexes and are influenced considerably during this period. Energy uptake by food during genital maturation period is channelled to gonads growth and not for body growth. Moreover, individuals develop sensitivity in diseases. All these parameters influence product's figure, which should have concrete size, formation, colour and texture. What is also very important for the viability of a fish farm enterprise is the production cost, since gonads during stages of genital maturity correspond to 20-30% of the individual's body weight.

For the above reasons, it is substantial to readily comprehend fish reproductive, biological cycles, so as to decide on the correct genetic engineering method to be applied. The economic projections of such applications are clear-cut. Fish farmers are interested to culture fish with high growth rates, superior food convertibility, tolerance to extreme environmental conditions and disease resistance, a better overall breeding management, low production cost, prolepsis of genital maturation, superior flesh quality and texture, efficient maintenance of production characteristics, high survival rates and high production systems attribution. Consumers on the other hand are interested in products of high nutritional value, of standard quality, at satisfactory low prices and correct service timing. It is up to the scientists along with politicians to produce and select the suitable practice of genetic engineering, and determine the limits that should not be exceeded.

The applications of genetic engineering constitute a new dynamic and in a lot of cases effective tool for improvement of, mainly commercial species, aquaculture. Results from laboratory genetic research are to be transferred in applied aquaculture know-how, after being quality checked. Enactment of legislation is a prerequisite, regarding the allowance limit of hormones, antibiotics, radiation practice, aiming at consumers' protection. Only when products are quality certified, one could assure absolutely no hazard for public health. Thus, collaboration of all involved parts is essential, including the scientific community; their main concern should be to minimize potential problems that could evolve while practicing genetic engineering in the aquaculture industry sector.

REFERENCES

Aerni, P. (2004). Risk, regulation and innovation: The case of aquaculture and transgenic fish. Aquatic Sci. 66 (3): 327-341.

Alimuddin, J.L., Yoshizaki, G., Kiron, V., Satoh, S. and Takeuchi, T. (2005). Enhancement of EPA and DHA biosynthesis by over-expression of masu salmon Delta 6-desaturase-like gene in zebrafish. Transgenic Res. 14(2): 159-165.

Allendorf, F.W., Seeb, J.E., Knudsen, K.L., Thorgaard, G.H. and Leary, R.F. (1986). Gene centromere mapping of 25 loci in rainbow trout. J. Hered. 77 (5): 307-312.

Anderson, E.D., Mourich, D.V. and Leong J.C. (1996a). Gene expression in rainbow trout (*Oncorhynchus mykiss*) following intramuscular injection of DNA. Mol. Mar. Biol. 5: 105-113.

Anderson, E.D., Mourich, D.V. and Leong, J.C. (1996b). Genetic immunization of rainbow trout (*Oncorhynchus mykiss*) against infectious hemotopoietic necrosis virus. Mol. Mar. Biol. Biotechnol. 5: 114-122.

Aoki, Y., Sato, H. and Amanuma, K. (2002). Detection of environmental mutagens using transgenic animals. Bunseki Kagaku 51(6): 373-380.

Arai, K. (2001). Genetic improvement of aquaculture finfish species by chromosome manipulation techniques in Japan. Aquaculture 197(1-4): 205-228.

Arvanitoyannis, I.S., Tsitsika, E. and Panagiotaki, P. (2005). Multivariate analysis in conjunction with quality control methods towards fish authenticity. Int. J. Food Sci. Technol. 40(3): 237-263.

Arvanitoyannis, I.S. and Krystallis, A. (2005). Consumers' beliefs, attitudes and intentions towards genetically modified foods, based on the 'perceived safety vs. benefits' perspective. Int. J. Food Sci. Technol. 40: 343–360.

BBC News, UK Edition (2004). Available at: http://www.news.bbc.co.uk/1/hi/sci/tech/3660289.stm.

Baker, G.A. and Burnhum, T.A. (2002). The market for genetically modified foods: Consumer characteristics and policy implications. International Food and Agribusiness Management Review 4: 351–360.

Beaumont, A.R. and Hoare, K. (2003). Biotechnology and Genetics in Fisheries and Aquaculture. Blackwell Publishing, Oxford, UK.

Beck, U. (1992). Risk society: Towards a new modernity. Sage Publications, London, UK.

Beck, U. (1995). Ecological politics in the age of risk. Polity Press, Cambridge, UK.

Beck, U. (1996). World risk society as cosmopolitan society? Ecological questions in a framework of manufactured uncertainties. Theory, Culture and Society 13(4): 1–32.

Beckmann, J.S. and Soller, M. (1988). Detection of linkage between marker loci and loci affecting quantitative traits in crosses between segregating populations. Theoret. App. Genet. 76(2): 228-236.

Beckmann, J.S. and Soller, M. (1990). Toward a unified approach to genetic mapping of eukaryotes based on sequence tagged microsatellite sites. Biotechnology 8 (10): 930-932.

Benzie, J.A.H. (2002). Genetics in Aquaculture VII. Elsevier, Amsterdam, The Netherlands.

Bercsenyi, M., Magyary, I., Urbanyi, B., Orban, L. and Horvath. L. (1998). Hatching out goldfish from common carp eggs: Interspecific androgenesis between two cyprinid species. Genome 41(4): 573-579.

Bertotto, D., Cepollaro, F., Libertini, A., Barbaro, A., Francescon, A., Belvedere, P., Barbaro, J. and Colombo, L. (2005). Production of clonal founders in the European sea bass, *Dicentrarchus labrax* L., by mitotic gynogenesis. Aquaculture 246(1-4): 115-124.

Blanc, J.M., Poisson, H., Escaffre, A.M., Aguirre, P. and Vallee, F. (1993). Inheritance of fertilizing ability in male tetraploid rainbow trout (*Oncorhynchus mykiss*). Aquaculture 110(1): 61-70.

Blanc, J.M., Maunas, P. and Vallee, F. (2005). Effect of triploidy on paternal and maternal variance components in brown trout, *Salmo trutta* L. Aquaculture Res. 36(10): 1026-1033.

Bongers, A.B.J., Zandieh-Doulabi, B., Richter, C.J.J. and Komen, J. (1999). Viable androgenesis YY genotypes of common carp (*Cyprinus carpio* L.). J. Hered. 90(1): 195-198.

Bredahl, L. (1999). Consumers' cognitions with regard to genetically modified foods. Results of a qualitative study in four European countries. Appetite 33: 343–360.

Bredahl, L. (2001). Determinants of consumer attitudes and purchase intention with regard to genetically modified foods – results of a cross-national survey. J. Consumer Policy 24: 23-61.

Brem, G., Brenig, B., Horstgen-Schwark G. and Winnacker, E.L. (1988). Gene transfer in Tilapia (*Oreochromis niloticus*). Aquaculture 68: 209-219.

Bye, V.J. and Lincoln, R.F. (1986). Commercial methods for the control of sexual maturation in rainbow trout (*Salmo giardneri* R). Aquaculture 57 (1-4): 299-309.

Cavari, B., Funkenstein, B., Chen, T.T., Gonzalezvillasenor, L.I. and Schartl, M. (1993). Effect of growth hormone on the growth rate of the gilthead seabream (*Sparus aurata*), and use of different constructs for the production of transgenic fish. Aquaculture 111(1-4): 189-197.

Cheftel, J.C. (2005). Food and nutrition labelling in the European Union. Food Chemistry 93(3): 531-550.

Chen, T.T. and Powers, D.A. (1990). Transgenic fish. Trends Biotechnol. 8(8): 209-215.

Chen, S.P., Zhao, H.B., Sun, Y.H., Zhang, F.Y. and Zhu, Z.Y. (2000). Nuclear transplantation with early-embryonic cells of transgenic fish. Prog. in Nat. Sci. 10(12): 925-930.

Cherfas, N., Gomelsky, B., Ben-Dom, N. and Hulata, G. (1995). Evidence for the heritable nature of spontaneous diploidization in common carp (*Cyprinus carpio* L.) eggs. Aquaculture Res. 26(4): 289-292.

Chevassus, B., Guyomard, R., Chourrout, D. and Quillet, E. (1983). Production of viable hybrids in salmonids by triploidization. Genetics Selection Evolution 15 (4): 519-532.

Chevassus, B., Devaux, A., Chourrout, D. and Jalabert, B. (1988). Production of YY rainbow trout males by self fertilization of induced hermaphrodites. J. Hered. 79(2): 89-92.

Chevassus, B. and Dorson, M. (1990). Genetics of resistance to disease in fishes. Aquaculture 85(1-4): 83-107.

Chourrout, D., Chevassus, B. and Guyomard, R. (1986a). Genetic improvement of fish. Recherche 17(180): 1028-1038.

Chourrout, D., Chevassus, B., Krieg, F., Happe, A., Burger, G. and Renard, P. (1986b). Production of 2nd generation triploid and tetraploid rainbow trout by mating tetraploid males and diploid females – Potential of tetraploid fish. Theoret. App. Genet. 72(2): 193-206.

Chourrout, D., Guyomard, R. and Houdebine, L.M. (1986c). High efficiency gene transfer in rainbow trout (*Salmo gairdneri* R) by microinjection into egg cytoplasm. Aquaculture 51(2): 143-150.

Chourrout, D. (1991). Gene transfer in fish. Bulletin de la Societe Zoologique de France-Evolution et Zoologie 116(2): 151-157.

Collodi, P., Kamei, Y., Ernst, T., Miranda, C., Buhler, D.R. and Barnes, D.W. (1992). Culture of cells from zebrafish (*Brachydanio rerio*) embryo and adult tissues. Cell Biol. Toxicol. 8(1): 43-61 Jan-Mar 1992.

Cook, J.T., Mc Niven, M.A., Richardson, G.F. and Sutterlin, A.M. (2000). Growth rates body composition and feed digestibility/conversion of growth-enhanced transgenic Atlantic salmon *Salmon salar*. Aquaculture 188(1-2): 15-32.

Davis, K.B., Simco, B.A. and Goudie, C.A. (1990). Plasticity of sex determination in channel catfish. American Zoologist 30(4): A41-A41.

de Koning, D.J. and Haley, C.S. (2005). Genetical genomics in humans and model organisms. Trends Genet. 21(7): 377-381.

Devlin, R.H. and Nagahama, Y. (2002). Sex determination and sex differentiation in fish: an overview of genetic, physiological, and environmental influences. Aquaculture 208(3-4): 191-364.

Donaldson, E.M. and Hunter, G.A. (1982). Sex control in fish with particular reference to salmonids. Can. J. Fisheries Aquatic Sci. 39(1): 99-110.

Dorson, M., Chevassus, B. and Torhy, C. (1991). Comparative susceptibility of three species of char and of rainbow trout x char triploid hybrids to several pathogenic salmonids viruses. Diseases of Aquatic Organisms 11(3): 217-224.

Dunham, R. and Liu, Z.J. (2002). Gene mapping, isolation and genetic improvement in catfish. In: Aquatic Genomics: Steps Toward a Great Future, (eds.) N.Shimizu, T. Aoki, I. Hirono and F. Takashima, Springer-Verlag, New York, USA, pp. 45-60.

Dunham, R.A. (2004). Aquaculture and Fisheries Biotechnology. Genetic Approaches. CABI Publishing, Cambridge, UK.

Einspanier, R., Klotz, A., Kraft, J., Aulrich, K., Poser, R., Schwaegle, F., Jahreis, G. and Flachowsky, G. (2001). The fate of forage plant DNA in farm animals, a collaborative case-study investigating cattle and chicken fed recombinant plant material. Eur. Food Res. Technol. 212: 129-134.

Exadactylos, A., Geffen, A.J. and Thorpe, J.P. (1999). Growth and genetic variation in hatchery-reared larval and juvenile Dover sole, *Solea solea* (L.). Aquaculture 176(3-4): 209-226.

Exadactylos, A., Geffen, A.J., Panagiotaki, P. and Thorpe, J.P. (2003). Population structure of Dover sole *Solea solea*: RAPD and allozyme data indicate divergence in European stocks. Mar. Ecol. Progr. Ser. 246: 253-264.

EU 1829/2003 Regulation (EC) No. 1829/2003 of the European Parliament and of the Council on genetically modified food and feed: http://europa.eu.int/eur-lex/pri/en/oj/dat/2003/l_268/l_26820031018en00010023.pdf

EU 1830/2003 Regulation (EC) No 1830/2003 of the European Parliament and of the Council concerning the traceability and labelling of genetically modified organisms and the traceability of food and feed products produced from genetically modified organisms and amending Directive 2001/18/EC: http://www.europa.eu.int/eur_lex/pri/en/oj/dat/2003/l_268/l_26820031018en00240028.pdf.

FAO. (2001a). Genetically modified organisms, consumers, food safety, and the environment. http://www.fao.org/DOCREP/003/X9602E/X9602E00.htm.

FAO. (2001b). The State of World Fisheries and Aquaculture 2000, Rome, Italy.

FAO. (2002). The State of World Fisheries and Aquaculture 2002, Rome, Italy.

FAO. (2004). COFI:FT/IX/2004/4 Committee on fisheries, safety and quality, with particular emphasis on fish meal and BSE. Bremen, Germany, 10-14 February 2004, pp.1-8.

FDA (US Food and Drug Administration Center for Food Safety and Applied Nutrition). Fish and Fisheries Products Hazards and Controls Guidance: Third Edition June 2001 In http://www.cfsan.fda.gov/~comm/haccp4k.html (Accessed on 15th July 2005).

Fife-Schaw, C. and Rowe, G. (1996). Public perceptions of everyday food hazards: a psychometric study. Risk Analysis. 16: 487–500.

Fingerman, M. and Nagabhushanam, R. (2000). Recent Advances in Marine Biotechnology Volume 4: Aquaculture Part B, Fishes. Science Publishers Inc., Enfield, NH, USA.

Finucane, M.L. and Holup, J.L. (2005). Psychosocial and cultural factors affecting the perceived risk of genetically modified food: an overview of the literature. Social Sci. and Med. 60: 1603–1612.

Fletcher, G.L., Shears, M.A., King, M.J., Davies, P.L. and Hew, C.L. (1988). Evidence of anti freeze protein gene transfer in Atlantic salmon (*Salmo salar*). Can. J. Fisheries Aquatic Sci. 45: 352-357.

Fletcher, G.L. and Davies, P.L. (1991). Transgenic fish for aquaculture. Genetic Eng. 13: 331-369.

Fletcher, G.L., Goddard, S.V., Shears, M.A., Sutterlin, A. and Hew, C.L., (2001). Transgenic salmon: potentials and hurdles. In: Molecular Farming, Proceedings of the OECD Workshop, (eds.) J.P. Toutant and E. Balazs, La Grande Motte (France), 3-6 September 2000, INRA editions, Paris, France, pp. 57-65.

Fletcher, G.L., Shears, M.A., Yaskowiak, E.S., King, M.J. and Goddard, S.V. (2004). Gene transfer: Potential to enhance the genome of Atlantic salmon for aquaculture. Australian J. Exp. Agric. 44(11): 1095-1100.

Forsythe, S.J. and Hayes, P.R. (1998). Food Hygiene, Microbiology and HACCP, 3rd edition, Aspen Publication, Gaithersburg, Maryland, USA, pp. 394-421.

Fostier, A. and Chevassus, B. (1991). Genetics and reproduction of salmonids. Biofutur (106): 31.

Frederiksen, M., Popescu, V. and Olsen, K. (1997). Integrated quality assurance of chilled food fish at sea. In: Seafood From Producer to Consumer, Integrated Approach to Quality (eds.) J.B. Luten, T. Borresen and J. Oehlenschlaeger. Elsevier, Amsterdam, The Netherlands, pp. 87-96.

Frederiksen, M., Oesterberg, C., Silberg, S., Larsen, E. and Bremner, H.A. (2002). Info-fish: Development and validation of an Internet based traceability system in a Danish domestic fresh fish chain. J. Aquatic Food Product Technol. 11(2): 13-34.

Furness, A. and Osman, K.A. (2003). Developing traceability systems across the supply chain In: Food Authenticity and Traceability (ed.) M. Lees, Woodhead Publishing Ltd., Cambridge, UK, pp. 473-495.

Galvez, J.I., Mazik, P., Phelps, R.P. and Mulvaney, D.R. (1995). Masculinization of channel catfish *Ichtalus punctatus* by oral administration of trenbalone acetate. J.World Aquacul. Soc. 26 (4): 378-383.

Gaskell, G., Allum, N., Bauer, M.W., Durant, J., Allansdottir, A., Bonfadelli, H., Boy, D., de Cheveigne, S., Fjaestad, B., Gutteling, J., Hampel, J., Jelsoe, E., Jesuino, J., Kohring, M., Kronberger, N., Midden, C., Nielsen, T.H., Przestalski, A., Rusanen, T., Sakellaris, G., Torgersen, H., Twardowski, T. and Wagner, W. (2000). Biotechnology and the European public. Nat. Biotechnol. 18(9): 935-938.

Georgakopoulos, G. and Thompson, I. (2005). Organic salmon farming: Risk perceptions, decision heuristics and the absence of environmental accounting. Accounting Forum 29: 49–75.

Gomelsky, B. (2003). Chromosome set manipulation and sex control in common carp: a review. Aquatic Living Resources 16 (5): 408-415.

Gong, N., Yang H., Zhang, G., Landau, B.J. and Guo, X. (2004). Chromosome inheritance in triploid Pacific oyster *Crassostrea gigas* Thunberg. Heredity 93(5): 408-415.

Grunina, A.S., Neyfakh, A.A. and Gomelsky, B.I. (1995a). Investigations on diploid androgenesis in fish. Aquaculture 129 (1-4): 218-219.

Grunina, A.S., Recoubratsky, A.V., Emelyanova, O.V. and Neyfakh, A.A. (1995b). Induced androgenesis in fish: Production of viable androgenetic diploid hybrids. Aquaculture 137 (1-4): 149.

Hallerman, E.M. and Beckmann, J.S. (1988). DNA level polymorphism as a tool in fisheries science. Can. J. Fisheries Aquatic Sci. 45(6): 1075-1087.

Haley, C.S. (1991). Use of DNA fingerprints for the detection of major genes for quantitative traits in domestic species. Animal Genetics 22(3): 259-277.

Hammond, B.G., Vicini, J.L., Hartnell, G.F., Naylor, M.W., Knight, C.D., Robinson, E.H., Fuchs, R.L. and Padgette, S.R. (1996). The feeding value of soybeans fed to rats chicken, catfish and dairy cattle is not altered by genetic incorporation of glyphosate tolerance. J. Nutr. 126: 717-27.

Hansen, M.M., Kenchington, E. and Nielsen, E.E. (2001). Assigning individual fish to populations using microsatellite DNA markers. Fish and Fisheries 2: 93-112.

Hansen, M.M. (2002). Estimating the long-term effects of stocking domesticated trout into wild brown trout (*Salmo trutta*) populations: an approach using microsatellite DNA analysis of historical and contemporary samples. Mol. Ecol. 11(6): 1003-1015.

Hansen, M.M. and Jensen, L.F. (2005). Sibship within samples of brown trout (*Salmo trutta*) and implications for supportive breeding. Conservation Genetics 6(2): 297-305.

Hindar, K., Ryman, N. and Utter, F. (1991a). Genetic effects of cultured fish on natural fish populations. Can. J. Fisheries Aquatic Sci. 48(5): 945-957.

Hindar, K., Ryman, N. and Utter, F. (1991b). Genetic effects of Aquaculture on natural fish populations. Aquaculture 98(1-3): 259-261.

Hindar, K., Tufto, J., Saettem, L.M. and Balstad, T. (2004). Conservation of genetic variation in harvested salmon populations. ICES J. Mar. Sci. 61 (8): 1389-1397.

Hinits, Y. and Moav, B. (1999). Growth performance studies in transgenic *Cyprinus carpio*. Aquaculture 173: 285-296.

Hohlweg, U. and Doerfler, W. (2001). On the fate of plant or other foreign genes upon the uptake in food or after intramuscular injection in mice. Mol. Genet. Genomics 265: 225–233.

Hosono, M., Sugawara, S., Ogawa, Y., Takayanagi, M. and Nitta, K. (2005). DNA aggregating protein from carp (*Cyprinus capio*) eggs. Biol. Pharmac. Bull. 28(1): 13-18.

Houdebine, L.M. and Chourrout, D. (1991). Transgenesis in fish. Experientia 47(9): 891-897.

Hussain, M.G., Chatterji, A., McAndrew, B.J. and Johnstone, R. (1991). Triploidy induction in Nile tilapia, *Oreochromis niloticus* L using pressure, heat and cold shocks. Theoret. Appl. Genet. 81(1): 6-12.

Ihssen, P.E., McKay, L.R., McMillan, I. and Phillips, R.B. (1990). Ploidy manipulation and gynogenesis in fishes – Cytogenetic and fisheries applications. Transact. Am. Fish. Soc. 119(4): 698-717.

International Food Information Council (IFIC). (2002). Food Biotechnology Survey Questionnaire (24 September 2002): US Consumer Attitudes Towards Food Biotechnology. IFIC, Washington, DC, USA.

Ji, N.N., Peng, B., Wang, G.Z., Wang, S.Y. and Peng, X.X. (2004). Universal primer PCR with DGGE for rapid detection of bacterial pathogens. J. Microbiol. Methods 57(3): 409-413.

Jiang, Y. (1993). Transgenic fish – Gene transfer to increase disease and cold resistance. Aquaculture 111(1-4): 31-40.

Joerg, H., Asai, M., Graphodatskaya, D., Janett, F. and Stranzinger. G. (2004). Validating bovine sexed semen samples using quantitative PCR. J. Animal Breed. Genet. 121(3): 209-215.

Karsi, A., Moav, B., Hackett, P. and Liu, Z.J. (2001). Effects of insert size on transposition efficiency of the Sleeping Beauty transposon in mouse cells. Marine Biotechnol. 3(3): 241-245.

Karsi, A., Cao, D., Li, P., Patterson, A., Kocabas, A., Feng, J., Ju, Z., Mickett, K.D. and Liu, Z. (2002a). Transcriptome analysis of channel catfish (*Ictalurus punctatus*): Initial analysis of gene expression and micro satellite-containing cDNAs in the skin. Gene 285(1-2): 157-168.

Karsi, A., Patterson, A., Feng, J.N. and Liu, Z.J. (2002b). Translational machinery of channel catfish: I. A transcriptomic approach to the analysis of 32 40S ribosomal protein genes and their expression. Gene 291(1-2): 177-186.

Kerr, W.A. (1999). Genetically modified organisms, consumer scepticism and trade low: Implications for the Organization of International Supply Chains. Supply Chain Management 4: 67–74.

Khoo, H.W., Ang, L.H., Lim, H.B. and Wong, K.Y. (19920. Sperm cells as vectors for introducing foreign DNA into zebrafish. Aquaculture 107: 1-19.

Kim, D.S., Cho, H.J., Park, I.S., Choi, G.C. and Nam, Y.K. (2001). Cytogenetic traits and gonad development of induced triploidy in far eastern catfish, *Silurus asotus*. Korean J. Genet. 23(1): 55-62 MAR.

Kinoshita, M., Toyohara, H., Sakaguchi, M., Inoue, K., Yamashita, S., Satake, M., Wakamatsu, Y. and Ozato, K. (1996). A stable line of transgenic medaka (*Oryzias latipes*) carrying the CAT gene. Aquaculture 143(3-4): 267-276.

Kocabas, A.M., Kucuktas, H., Dunham, R.A. and Liu, Z.J. (2002). Molecular characterization and differential expression of the myostatin gene in channel catfish (*Ictalurus punctatus*). Biochimica et Biophysica – Gene Structure and Expression 1575(1-3): 99-107.

Komaru, A., Kawagishi, T. and Konishi, K. (1998). Cytological evidence of spontaneous androgenesis in the freshwater clam *Corbicula leana prime*. Development Genes and Evolution 208(1): 46-50.

Komen, J., Bongers, A.B.J., Richter, C.J.J., Vanmuiswinkel, W.B. and Huisman, E.A. (1991). Gynogenesis in common carp (*Cyprinus carpio* L). 2. The production of homozygous gynogenetic clones and F_1 hybrids. Aquaculture 92(2-3): 127-142.

Lahteenmaki, L., Grunert, K., Oydis, U., Astrom, A., Arvola, A. and Bech-Larsen, T. (2002). Acceptability of genetically modified cheese presented as real product alternative. Food Qual. Pref. 13: 523–533.

Lakra, W.S. (2001). Genetics and molecular biology in aquaculture – Review. Asian-Australasian J. Animal Sci. 14(6): 894-898.

Langenau, D.M., Jette, C., Berghmans, S., Palomero, T., Kanki, J.P., Kutok, J.L. and Look, A.T. (2005). Suppression of apoptosis by bcl-2 overexpression in lymphoid cells of transgenic zebrafish. Blood 105(8): 3278-3285.

Larsen, E. (2003). Traceability in fish processing. In: Food Authenticity and Traceability. (ed.) M. Lees, Woodhead Publishing Ltd., Cambridge, UK, pp. 507-517.

Law, J.M., Bull, M., Nakamura, J. and Swenberg, J.A. (1998). Molecular dosimetry of DNA adducts in the medaka small fish model. Carcinogenesis 19(3): 515-518.

Leggatt, R.A. and Iwama, G.K. (2003). Occurrence of polyploidy in the fishes. Rev. Fish Biol. Fisheries 13(3): 237-246.

Liang, Y.H., Cheng, C.H.K. and Chan, K.M. (1996). Insulin-like growth factor Iea2 is the predominantly expressed form of IGF in common carp (*Cyprinus caprio*). Mol. Marine Biol. Bioetchnol. 5: 145-152.

Lilyestrom, C.G., Wolters, W.R., Bury, D., Rezk, M. and Dunham, R.A. (1999). Growth, carcass traits, and oxygen tolerance of diploid and triploid catfish hybrids. North American J. Aquacul. 61(4): 293-303.

Lincoln, R.F. and Scott, A.P. (1984). Sexual maturation in triploid rainbow trout, *Salmo gairdneri* R. J. Fish Biol. 25(4): 385-392.

Liu, S.F., Liu, B.W., He, S., Zhaol, Y, and Wang, Z. (2005). Cloning and characterization of zebra fish SPATA4 gene and analysis of its gonad specific expression. Biochemistry-Moscow 70(6): 638-644.

Logar, N. and Pollock, L.K. (2005). Transgenic fish: Is a new policy framework necessary for a new technology? Environ. Sci. and Policy 8(1): 17-27.

Lu, J.K., Fu, B.H., Wu, J.L. and Chen, T.T. (2002). Production of transgenic silver sea bream (*Sparus sarba*) by different gene transfer methods. Marine Biotechnol. 4(3): 328-337.

Lutz, C.G. (2003). Practical Genetics for Aquaculture. Blackwell Publishing Co. Oxford, UK.

Maisse, G., Labbe, C., de Baulny, B.O., Calvi, S.L. and Haffray, P. (1998). Fish sperm and embryos cryopreservation. Production Animales 11 (1): 57-65.

Marengoni, N.G. and Onoue, Y. (1998). Ultraviolet induced androgenesis in Nile Tilapia *Oreochromis niloticus* (L.) and hybrid Nile x Blue Tilapia *O. aureus* (Steindachner). Aquaculture Res. 29(5): 359-366.

Marris, C. (2000). Swings and roundabouts: French public policy on agricultural GMOs 1996-1999 (2000-2002). Guyancourt, Politeia, France.

McKenzie, D. (1996). Doctors farm fish for insulin. New Scientist 2056: 20.

McKone, M.J. and Halpern, S.L. (2003). The evolution of androgenesis. American Naturalist 161(4): 641-656.

McLean, N. and Penman, D. (1990). The application of gene manipulation to Aquaculture. Aquaculture 85(1-4): 1-20.

Michelmore, R.W., Paran, I. and Kesseli, R.V. (1991). Identification of markers linked to disease resistance genes by bulked segregant analysis – A rapid method to detect markers in specific genomic regions by using segregating populations. Proc. Natl. Acad. Sci. U.S.A. 88(21): 9828-9832.

Moav, B., Liu, Z.J., Caldovic, L.D., Gross, M.L., Faras, A.J. and Hackett, P.B. (1993). Regulation of expression of transgenes in developing fish. Transgenic Res. 2(3): 153-161.

Moses, V. (1999). Biotechnology products and European consumers. Biotechnol. Adv. 17(8): 647–678.

Muller, F., Lele, Z., Varadi, L., Menczel, L. and Orban, L. (1993). Efficient transient expression system based on square pulse electroporation and in vivo luciferase assay of fertilized fish eggs. FEBS 234: 27-32.

Muller, M. and Brem, G. (1998). Transgenic approaches to the increase of disease resistance in farm animals. Revue Scientifique et Technique de l' Office International des Epizooties 17(1): 365-378 APR 1998.

Myers, J.M. and Hershberger, W.K. (1991). Early growth and survival of heat shocked and tetraploid derived triploid rainbow trout (*Oncorhynchus mykiss*). Aquaculture 96(2): 97-107.

Myhr, A.I. and Dalmo, R.A. (2005). Introduction of genetic engineering in aquaculture: Ecological and ethical implications for science and governance. Aquaculture: 250(3-4): 542-554.

Nagamine, C., Knight, A.W., Maggenti, A. and Paxman, G. (1980). Masculinization of female *Macrobrachium rosenbergii* (de Man) (Decapoda, Palaemonidae) by androgenic gland implantation. Gen. Comparat. Endocrinol. 41(4): 442-457.

Naylor, R.L., Goldburg, R.J., Primavera, J.H., Kautsky, N., Beveridge, M.C.M., Clay, J., Folke, C., Lubchenco, J., Mooney, H. and Troell, M. (2000). Effect of aquaculture on world fish supplies. Nature 405: 1017-1023.

Naylor, R., Hindar, K., Fleming, I.A., Goldburg, R., Williams, S., Volpe, J., Whoriskey, F., Eagle, J., Kelso, D. and Mangel, M. (2005). Fugitive salmon: Assessing the risks of escaped fish from net-pen aquaculture. Bioscience 55(5): 427-437.

New, M.B. and Singholka, S. (1982). Freshwater Prawn Farming. A Manual for the Culture of *Macrobrachium rosenbergii.* FAO Fisheries Technical Paper No. 225, Food and Agriculture Organization of the United Nations, Rome, Italy.

Niu, L.C., Xue, G.X., Yang, J. and Niu, M.C. (1986). Messenger RNA mediated transfer of genetic information in goldfish eggs. Federation Proceedings 45(6): 1703

Noguchi, A., Takekawa, N., Einarsdottir, T., Koura, M., Noguchi, Y., Takano, K., Yamamoto, Y., Matsuda, J. and Suzuki, O. (2004). Chromosomal mapping and zygosity check of transgenes based on flanking genome sequences determined by genomic walking. Experimental Animals 53(2): 103-111.

Ozato, K., Kondoh, H., Inohara, H., Iwamatsu, T., Wakamatsu, Y. and Okada, T.S. (1986). Production of transgenic fish – Introduction and expression of chicken delta crystallin gene in medaka embryos. Cell Differentiation 19(4): 237-244.

Padgette, S.R., Kolacz, K.H., Delannay, X., Re, D.B., LaVallee, B.J., Tinius, C.N., Rhodes, W.K., Otero, Y.I., Barry, G.F., Eichholtz, D., Peschke, V.M., Nida, D.L., Taylor, N.B. and Kishore, G.M. (1995). Development, identification and characterisation of a glyphosate-tolerant soybean line. Crop Sci. 35: 1451–1461.

Pandian, T.J. and Kirankumar, S. (2003). Androgenesis and conservation of fishes. Curr. Sci. 85 (7): 917-931.

Petit, H. (2005). A fluo transgenic fish that hasn't finished speaking about himself. Biofutur (254): 38-40.

Phillips, R.B., Park, L.K., Devlin, R.H. and Thorgaard, G.T. (2004). Sex chromosomes and sex linkage groups in Pacific salmon and trout. Biol. Reprod. 137 Sp. Iss.

Piferrer, F., Cal, R.M., Gomez, C., Alvarez-Blazquez, B., Castro, J. and Martinez, P. (2004). Induction of gynogenesis in the turbot (*Scophthalmus maximus*): Effects of UV irradiation on sperm motility, the Hertwig effect and viability during the first 6 months of age. Aquaculture 238(1-4): 403-419.

Popping, B. (2003). Identifying genetically modified organisms (GMOs). In: Food Authenticity and Traceability (ed.) M. Lees, Woodhead Publishing Ltd., Cambridge, UK, pp. 415-425.

Powers, D.A., Herefird, L., Creech, K., Chen, T.T., Lin, C.M., Knight, K. and Dunham, R. (1992). Electroporation a method for transferring genes into the gametes of zebrafish (*Brachydanio rerio*), channel catfish (*Ichtalurus punctatus*) and common carp (*Cyprinus carpio*). Mol. Mar. Biol. Biot. 1: 301-308.

Powers, D.A. and Chen, T.T. (1995). Genetic engineering to improve the growth characteristics of finfish and shellfish for aquaculture in Pacific Rim countries: Cloning, expression, and gene transfer of growth hormone and insulin-like growth factors. Aquaculture 135(1-3): 240.

Rahman, M.A. and McLean, N. (1999). Growth performance of transgenic tilapia containing an exogenous piscine growth hormone gene. Aquaculture 173(1-4): 333-346.

Rahman, M.A., Hwang, G.L., Razak, S.A., Sohm, F. and Maclean, N. (2000). Copy number related transgene expression and mosaic somatic expression in hemizygous and homozygous transgenic tilapia (*Oreochromis niloticus*). Transgenic Res. 9(6): 417-427.

Refstie, T. (1990). Application of breeding schemes. Aquaculture 85(1-4): 163-169.

Rocha, A., Ruiz, S., Estepa, A. and Joll, J.M. (2004). Application of inducible and targeted gene strategies to produce transgenic fish: A review. Marine Biotechnol. 6(2): 118-127.

Rodriguez-Lazaro, D., Jofre, A., Aymerich, T., Garriga, M. and Pla, M. (2005). Rapid quantitative detection of *Listeria monocytogenes* in salmon products: Evaluation of pre-real-time PCR strategies. J. Food Protect. 68(7): 1467-1471.

Sanden, M., Bruceb, I.J., Azizur Rahmanb, M. and Gro-Ingunn Hemrea, M. (2004). The fate of transgenic sequences present in genetically modified plant products in fish feed, investigating the survival of GM soybean DNA fragments during feeding trials in Atlantic salmon, *Salmo salar* L. Aquaculture 237: 391–405.

Scheerer, P.D. and Thorgaard, G.H. (1983). Increased survival in salmonid hybrids by induced triploidy. Can. J. Fisheries Aquatic Sci. 40(11): 2040-2044.

Scheerer, P.D., Thorgaard, G.H. and Allendorf, F.W. (1991). Genetic analysis of androgenetic rainbow trout. J. Exp. Zool. 260(3): 382-390.

Scipioni, A., Saccarola, G., Centazzo, A. and Arena, F. (2002). FMEA methodology design, implementation and integration with HACCP system in a food company. Food Control 13: 495–501.

Scipioni, A., Saccarola, G., Centazzo, A. and Arena, F. (2005). Strategies to assure the absence of GMO in food products application process in a confectionery firm. Food Control 16: 569–578.

Shears, M.A., Fletcher, G.L., Hew, C.L., Gauthier, S. and Davies, P.L. (1991). Transfer, expression, and stable inheritance of antifreeze protein genes in Atlantic salmon (*Salmo salar*). Mol. Marine Biol. Biotechnol. 1: 58-63.

Sin, F.Y.T. (1997). Transgenic fish. Reviews in Fish Biology and Fisheries 7(4): 417-441.

Smith, K. and Spadafora, C. (2005). Sperm-mediated gene transfer: Applications and implications. Bioessays 27(5): 551-562.

Streisinger, G., Walker, C., Dower, N., Knauber, D. and Singer F. (1981). Production of clones of homozygous diploid zebra fish (*Brachydanio rerio*). Nature 291 (5813): 293-296.

Takagi, M. (2001). Genetic and breeding studies on the hypervariable DNA markers in fish. Nippon Suisan Gakkaishi 67(4): 610-613.

Tan, J.H. and Chan, W.K. (1997). Efficient gene transfer in zebrafish skeletal muscle by intramuscular injection of plasmic DNA. Mol. Mar. Biol. Biot. 6: 98-109.

Thorgaard, G.H., Scheerer, P.D. and Parsons, J.E. (1985). Residual paternal inheritance in gynogenetic rainbow trout – Implications for gene transfer. Theoret. Appl. Genet. 71(1): 119-121.

Thorgaard, G.H. (1986). Ploidy manipulation and performance. Aquaculture 57(1-4): 57-64.

Thorgaard, G.H., Bailey, G.S., Williams, D., Buhler, D.R., Kaattari, S.L., Ristow, S.S., Hansen, J.D., Winton, J.R., Bartholomew, J.L., Nagler, J.J., Walsh, P.J., Vijayan, M.M., Devlin, R.H., Hardy, R.W., Overturf, K.E., Young, W.P., Robison, B.D., Rexroad, C. and Palti, Y. (2002). Status and opportunities for genomics research with rainbow trout. Comp. Biochem. Physiol. B 133(4): 609-646.

Tsamis, V., Mamuris, Z., Panagiotaki, P. and Kouretas, D. (2001). Proteins from fish eggs that protect DNA from acid precipitation and inhibit DNA synthesis. Comp. Biochem. Physiol. C 129(4): 369-376.

Tufto, J. and Hindar, K. (2003). Effective size in management and conservation of subdivided populations. J. Theoret. Biol. 222(3): 273-281.

Uberth, F. (2003). Milk and Dairy products. In: Food Authenticity and Traceability (ed.) M. Lees, Woodhead Publishing Ltd., Cambridge, UK, pp. 357-377.

Wall, R.J. (2002). New gene transfer methods. Theriogenology. 57(1): 189-201.

Webster, M.S. and Reichart, L. (2005). Use of microsatellites for parentage and kinship analyses in animals. Molecular Evolution: Producing the Biochemical Data, Part B Methods in Enzymology 395: 222-238.

Wiegertjes, G.F., Stet, R.J.M. and van Muiswinkel, W.B. (1995). Divergent selection for antibody production to produce standard cap *Cyprinus carpio* L. lines for the study of disease resistance in fish. Aquaculture 137(1-4): 257-262.

Withler, R.E., Beacham, T.D., Solar, I.I. and Donaldson, E.M. (1995). Freshwater growth, smolting, and marine survival and growth of diploid and triploid coho salmon *Oncorhychus kisutch*. Aquaculture 136(1-2): 91-107.

Wolters, W.R., Lilyestrom, C.G. and Craig, R.J. (1991). Growth, yield, and dress out percentage of diploid and triploid channel catfish in earthen ponds. Progressive Fish Culturist 53(1): 33-36.

Wright, J.R., Pohajdak, B., Xu, B.Y. and Leventhal, J.R. (2004). Piscine islet xenotransplantation. Ilar J. 45(3): 314-323.

Yamaha, E., Mizuno, T., Hasebe, Y. and Yamazaki, F. (1997). Chimeric fish produced by exchanging upper parts of blastoderms in goldfish blastulae. Fisheries Sci. 63(4): 514-519.

Yoon, S.J., Hallerman, E.M., Gross, M.L., Liu, Z., Schneider, J.F., Faras, A.J., Hackett, P.B., Kapuscinski, A.R. and Guise, K.S. (1990). Transfer of the gene for neomycin resistance into goldfish, *Carassius auratus*. Aquaculture 85(1-4): 21-33.

Zbikowska, H.M. (2003). Fish can be first – advances in fish transgenesis for commercial applications. Transgenic Res. 12(4): 379-389.

Zimmerman, L., Kendall, P., Stone, M. and Hoban, T. (1994). Consumer knowledge and concern about biotechnology and food safety. Food Technology, 48: 71-77.

http://www.europa.eu.int/eurlex/pri/en/oj/dat/2001/l_106/l_10620010417en00010038.pdf.

http://www.food.gov.uk/foodindustry/regulation/ria/ria2001/71541.

http://www.biotech.dnsalias.net/en/2005/02/3916.shtml Accessed on 2005 July 3.

http://www.edis.ifas.ufl.edu/FA095.

http://www.wfrc.usgs.gov/research/contaminants/STRodriguez4.htm Accessed on 19 June 2005.

http://www.wfrc.usgs.gov/research/contaminants/STSaiki5.htm.

http://www.eanmalaysia.com.my/p_ne_it.asp?NewsID Accessed on 3 July 2005.

http://www.foodqualitynews.com/news/news-ng.asp?n=57779-eu-sets-out Accessed on 2005 June 29.

http://www.idea.iastate.edu/aqua/05pdf Accessed on 29 June 2005.

http://www.maff.go.jp/trace/guide_en.pdf "Guidelines for Introduction of Food Traceability Systems" Accessed on 3 July 2005.

http://www.mindfully.org/GE/Over-Fishing-Solution.htm Accessed on 3 July 2005.

http://www.namm.com/techcenter/GLN%20Implementation%20Guide.pdf Accessed on 2 July 2005.

http://www.aquafeed.com.

http://www.agriculture.gov.ie/legislation/SI2004/s827-04.pdf.

http://www.opsi.gov.uk/si/si2006/20060755.htm

http://www.foodlaw.rdg.ac.uk/pdf/com2003_0489.pdf.

15

Microbial Processing of Agricultural Residues for Production of Food, Feed and Food-Additives

Ramesh C. Ray[*,1], *Anup K. Sahoo*[1], *Kozo Asano*[2] *and Fusao Tomita*[2]

INTRODUCTION

Agriculture is the world's largest economic sector. On worldwide basis, more people are involved in agriculture than in all other occupations combined (Estabrook, 2000). However, after the processing of agricultural crops, millions of tonnes of low-value by-products are generated annually, which in general are called agricultural residues (Hayes, 2003). The volume and diversity of these residues represent an enormous and underutilized renewable resource, which can create adverse environmental and cost impacts for disposal. Bioprocessing of these agricultural residues by using microorganisms for industrial purposes is a much more environmental friendly practice than many residue disposal methods currently in use. Until recently, many farmers disposed of agricultural wastes by burning or land filling them (Boopathy, 2005). Therefore, using agricultural residues as an industrial feedstock is a critical channel for drastically reducing the environmental problem. The major advantages for using agricultural residues are three-fold: economic, environmental, and technological.

Economic: For the farmer, agricultural residues can be a cash crop, too. Traditionally, farmers have harvested grains and burnt or otherwise

[1]*Regional Centre of Central Tuber Crops Research Institute, Bhubaneswar 751 019, India*
[2]*Laboratory of Applied Microbiology, Hokkaido University, Sapporo, Japan*
**Corresponding author: rc_ray @rediffmail.com*

disposed of straw and other residues (stalks, stover, etc.), but the heightened interest in making use of agricultural waste in harnessing profit such as cultivating mushrooms means that farmers can reap a "second harvest" from grain plantings (Mandeel et al., 2005).

***Environmental*:** Farmers will also amass environmental dividends; studies have shown that the burning of agricultural wastes causes air pollution, soil erosion, and decrease in soil biological activity, which eventually lead to soil crusting and may lower yields (Boopathy, 2005). An increase in farm earnings will diminish the need for farm subsidies, which will eventually save money for taxpayers and reduce farmers' reliance on the government for support (Morris, 1997).

***Technological*:** Current research and testing indicate that use of agricultural residues is best achieved at small and local industries. The local industries using agricultural residues will offer a number of jobs to the region where it is difficult to start big industries (Al Wong, 1996).

AGRICULTURAL RESIDUES AND THEIR TYPES (BERTON, 2005)

Field and seed crops: Field and seed crop residues are the materials remaining above the ground after harvesting, including straw or stubble from barley, beans, oats, rice, rye, and wheat, stalks, or stovers from corn, cassava, cotton, sorghum, soybeans and alfalfa. The moisture content (wet basis) of these residues ranges from 8-80%.

Fruit and nut crops: Fruit and nut crop residues include orchard pruning and brushes. The types of fruit and nut crops include almonds, apples, apricots, avocados, cherries, dates, figs, grapes, lemons, limes, olives, oranges, peaches, pears, plums, prunes and walnuts. The moisture content (wet basis) of these residues ranges from 35-55%.

Vegetable crops: Vegetable crop residues consist mostly of vines and leaves that remain on the ground after harvesting. The types of vegetable crops include such plants as artichokes, asparagus, cucumbers, lettuce, melon, potatoes, squash, and tomatoes. The moisture content (wet basis) of these residues is usually greater than 90%.

Nursery crops: Nursery crop residues include the pruning and trimming taken from the plants during their growth and in the preparation for market. There are more than 30 different species of nursery crops (e.g. flowers and indoor plants, etc.) that are grown. The moisture content of these residues is usually dependent on the type of crop that is being grown.

DEGRADATION OF LIGNOCELLULOSE

All plant residues are made of lignocelluloses which consist of lignin, hemicellulose and cellulose. Table 15.1 compiled from various authors shows the typical compositions of the three components in various lignocellulosic materials. Due to the difficulties in dissolving lignin without destroying it, its exact chemical structure is difficult to ascertain (Howard et al., 2003). In general, lignin contains three aromatic alcohols (coniferyl, sinapyl and p-coumaryl). In addition, grass and dicot lignin also contains large amounts of phenolic acids such as p-coumaric and ferulic acid which are esterified to alcohol groups of each other and to other alcohols such as sinapyl and p-coumaryl.

The microbiology of lignocellulose degradation has been discussed by Durrant (Chapter 13) and Ward and Singh (Chapter 15 in Volume I of this series–Microbial Biotechnology in Agriculture and Aquaculture) and hence not elaborated further in this article.

Table 15.1. Lignocelluloses containing of common agricultural residues and waste matter

Lignocellulosic materials	Cellulose (%)	Hemicellulose (%)	Lignin (%)
Hardwood stems	40-55	24-40	18-25
Softwood stems	45-50	25-35	25-35
Nut shells	25-30	25-30	30-40
Corn cobs	45	35	15
Paper	85-99	0	0-15
Wheat straw	30	50	15
Rice straw	32.1	24	18
Sorted refuse	60	2	20
Leaves	15-20	80-85	0
Cotton seed hairs	80-95	5-20	0
Newspaper	40-55	25-40	18-30
Waste paper from chemical pulp	60-70	10-20	5-10
Primary wastewater solid	8-15	NA	24-29
Fresh bagasse	33.4	30	18.9
Swine waste	6	28	NA
Solid cattle manure	1.6-7.1	1.4-3.3	2.7-5.7
Coastal Bermuda grass	25	35.7	6.4
Sweet potato bagasse	45	31.4	12.0
S32 ray grass (early leaf)	21.3	15.8	2.7
S32 ray grass (seed setting)	26.7	25.7	7.3
Orchard grass (medium maturity)	32	40	4.7
Cassava fibrous residue	45-55	25-40	15-20
Sugarcane bagasse	40-42	16-20	12-18

NA: Data not available
Betts et al., 1991; Sun and Cheng, 2002; Howard et al., 2003 and Durrant, 2005

POTENTIAL BIOPRODUCTS FROM AGRICULTURAL RESIDUES

Biomass can be considered as the mass of organic material from any biological material, and by extension, any large mass of biological matter. A wide variety of plant biomass resources are available (Table 15.1) on our planet for conversion into bioproducts. These may include whole plants, plant parts (e.g. seeds and stalks), plant constituents (e.g. starch, lipids, protein and fibre), processing by-products (distiller's grains, corn solubles), phytoplanktons of aquatic and marine origin, and fruit and vegetable processing wastes (Smith et al., 1987; Ward et al., 2006). These resources can be used to create biomaterials (Sonnino, 1994; Piccinini, 1998; Das and Singh, 2004) and these will require an intimate understanding of the composition of the raw material whether it is whole plant or constituents, so that the desired functional elements can be obtained for bioproduct production, e.g. hemicellulose fraction which yields xylitol (sweetener). There is excellent and comprehensive literature (Wong and Saddler, 1992a, 1992b; Kuhad and Singh, 1993; Bhat, 2000; Beauchemin et al., 2001, 2003; Beg et al., 2001; Subramaniyan and Prema, 2002; Sun and Cheng, 2002; Durrant, 2005) available on the different potential bioproducts and their many applications.

Bio-fuels

The demand for ethanol has the most significant market where it is used as a chemical feedstock or as an octane enhancer or petrol additive. Global crude oil production is predicted to decline from 25 billion barrels to approximately 5 billion barrels in 2050 (Campbell and Laherrere, 1998). The production of ethanol from sugars and starch impacts negatively on the economics of the process, thus making ethanol more expensive compared with fossil fuels (Ward and Singh, 2005; Ward et al., 2006). Hence, the technology development focus on the production of ethanol has shifted towards the utilization of residual lignocellulosic materials to lower production costs. This aspect has been covered by Ward and Singh (2005) in Chapter 15 in Volume I of this series.

Organic Chemicals

Bioconversion of agricultural residues makes a significant contribution to the production of organic chemicals such as acetone-butanol, ethylene, propylene, benzene, etc. This aspect has also been dealt with by Durrant (2005) in the Chapter 13 in Volume I of this series.

Other High Value Bioproducts

A number of bioproducts such as enzymes, organic acids, amino acids, vitamins and bacterial and fungal polysaccharides (e.g. dextran, xanthan, pullulan, etc.) are produced by fermentation using glucose as the base substrate, but theoretically these very products could be manufactured from "lignocelluloses waste". Based on the predicted catabolic pathway and the known metabolism of *Phanerochaete chrysosporium* lignin, Ribbons (1987) presented a detailed discussion of the potential value added products which could be derived from lignin. For example, vanillin and gallic acid are two most frequently discussed monomeric potential products which have attracted interest. Vanillin extraction from vanilla pods cost between US$ 1,200 to 4,000 kg^{-1}, whereas synthetic vanillin costs less than US$ 15 kg^{-1} (Walton et al., 2003). Vanillin is used for various purposes including being an intermediate in the chemical and pharmaceutical industries for the production of herbicides and antifoaming agents or drugs, such as papaverine, L-dopa and the anti-microbial agent, trimethoprim. It is also used in household products such as air-fresheners and floor polishers (Walton et al., 2003). The high price and limited supply of natural vanillin have necessitated a shift towards the production from other sources (Priefert et al., 2001).

Hemicelluloses are of particular industrial interest since these are readily available bulk source of xylose from which xylitol and furfural can be derived (Nigam and Singh, 1995). Xylitol used instead of sucrose in food as a sweetener, has odontological application such as teeth hardening, remineralization, and as an antimicrobial agent, it is used in chewing gum and toothpaste formulations (Parajo et al., 1998; Roberto et al., 2003). The yield of xylans as xylitol by chemical means is only about 50-60% making xylitol production expensive. Various bioconversion methods, therefore, have been explored for the production of xylitol from hemicellulose using microorganisms or their enzymes (Nigam and Singh, 1995; Winkelhausen and Kuzmanova, 1998). Furfural is used in the manufacture of furfural-phenol plastic, varnishes and pesticides (Montané et al., 2002). Over 200,000 tonnes of furfural with a market price about US$ 1,700 per tonne (Montané et al., 2002) is annually produced (Zeitch, 2000).

BIOPROCESSING OF LIGNOCELLULOSIC MATERIALS

Bioconversion of lignocellulosic material to useful, higher value products normally require multi-step processes which include: (i) pre-treatment (mechanical, chemical or biological) (Grethlein and Converse, 1991; Kuhad and Singh, 1993), (ii) hydrolysis of polymers to produce readily metabolizable molecules (e.g. hexose or pentose sugars),

(iii) bio-utilization of these molecules to support microbial growth or to produce chemical products and (iv) the separation and purification (Smith et al., 1987). To date, the production of cellulase and other lignocellulosic enzymes have been widely studied in submerged culture processes in the laboratory, ranging from shake flask to 15,000-l fermentor (Haltrich et al., 1996; Kim et al., 1997, 1998; Xia and Len, 1999). A stirred–tank reactor widely used for the production of cellulase and other lignocellulosic enzyme is known to have shear problems which rupture mycelial cells and may deactivate the enzymes (Wase et al., 1985; Durrand, 2003). Alternative bioreactors such as the air-lift and bubble-column, which have a lower shear stress, seem to produce better results. For example, studies of cellulase and xylanase production by *Aspergillus niger* in various bioreactors showed that in general, better yield and productivity were shown in a bubble-column and an air-lift reactor than in the stirred tank reactor (Kim et al., 1997).

Fermentation Process

Many of these traditional fermented food and industrial enzymes are based on the *koji* process that belongs to a major technical fermentation category known as [solid substrate (state) fermentation (SSF)] differing from either submerged (SF) or immobilization fermentations in the amount of free water that they contain.

Solid State Fermentation (SSF)

SSF is a process used for the cultivation of edible mushrooms and production of fermented food, animal feed, fuel, enzymes and pharmaceuticals, which involves the growth of microorganisms (mainly fungi and thermophilic bacteria) on moist solid substrates in the absence of free-flowing water. Growth patterns of fungi in SSF have been detailed in many cases and these can be summarized in two phases: (a) germination, germ tube elongation and mycelial branching to loosely cover most of the substrate; and (b) increase in mycelial density with aerial and penetrative hyphae development (Mitchell, 1992).

SSF processes exhibit several advantages over submerged fermentation (SF) including lower cost, improved product characteristics, higher product yields and productivities, easier product recovery and reduced energy requirements. Since SSF processes have been used for centuries, a great number of references are available and many excellent reviews have also been published (Doelle et al., 1992; Soccol and Vandenberghe, 2003; Suryanarayan, 2003; Krishna, 2005).

Two main types of processes are used for the solid support in SSF. (1) SSF processes that use natural solid substrates like starch, or (ligno) cellulose residues, or agro-industrial sources such as grains and grain by-products, cassava, potato, rice, beans and sugar beet pulp/residues. In these cases the substrate is also used as the source of carbon and nutrients for microbial growth (Nigam and Singh, 1995; Tengerdy and Szakacs, 2003). (2) SSF processes that use inert natural (sugarcane bagasse, cassava fibrous residues, potato waste, etc.) or artificial (perlite, amberlite, polyurethane foam and others) solid supports. In these latter processes, support is used only as an attachment structure for the microorganisms (Ooijkaas et al., 2000).

It has been reported that the SSF is an attractive alternative process to produce fungal microbial enzymes using lignocellulosic material from agricultural waste matter due to its lower capital investment and lower operating cost (Chahal et al., 1996; Haltrich et al., 1996; Jecu, 2000). SSF process, for the reasons stated will be ideal for developing countries. Solid-state fermentations are characterized by the complete or almost complete absence of free liquid. Water, which is essential for microbial activities, is present in an absorbed or in complexed form with the solid matrix and the substrate (Cannel and Moo-Young, 1980). These cultivation conditions are especially suitable for the growth of fungi, known to grow at relatively low water activities. As the microorganisms in SSF grow under conditions closer to their natural habitats, they are more capable of producing enzymes and metabolites which will not be produced or will be produced only in low yield in submerged conditions (Jecu, 2000). SSFs are practical for complex substrates including agricultural, forestry and food-processing residues and waste matters which are used as carbon sources for the production of lignocellulolytic enzymes (Haltrich et al., 1996). Compared with the two-stage hydrolysis-fermentation process during ethanol production from lignocellulosic, Sun and Cheng (2002) reported that SSF has the following advantages: (1) increase in hydrolysis rate by conversion of sugars that inhibit enzyme (cellulase) activity; (2) lower enzyme requirements; (3) higher product yield; (4) lower requirement for sterile conditions since glucose is removed immediately and ethanol is produced; (5) shorter process time; and (6) less reactor volume. In a recent review, Malherbe and Cloete (2003) reported that the primary objective of lignocellulosic treatment by the various industries is to access the potential of the cellulose encrusted by lignin within the lignocellulose matrix. They express the opinion that a combination of SSF technology with the ability of an appropriate fungus to electively degrade lignin will make possible industrial-scale implementation of lignocellulose-based biotechnology.

Like all technology, SSF has its disadvantages and these have been duly recognized (Doelle et al., 1992). Problems commonly associated with SSF are heat build up, bacterial contamination, scale-up, biomass growth estimation and control of substrate content. However, the process has been used for the production of many microbial products, and the engineering aspects and scale-up will depend on the bioreactor design and operation (Lonsane et al., 1992). A recent technical report in 2002 (http://www.igu.und.edu/project/outline.cfm?trackID=1934) on "The science and engineering for a bio-based industries and economy" has adequately discussed some of the strategies in lignocellulose bioconversion process. Other lignocellulose bioprocessing strategies include anaerobic treatment, composting, production of single cell protein for ruminant animal feeding and mushroom cultivation.

Submerged Fermentation (SF)

In contrast to SSF, SF is the process of choice for industrial operations due to the very well-known engineering aspects, such as fermentation modelling, bioreactor design and process control (Gutierrez-Correa and Villena, 2003). Food additives and enzyme production by SSF is not of wide use by commercial manufacturers mainly due to difficulties in bioreactor design, although the advantages of this form of microbial cultivation as compared to SF are well known.

In bioconversion of agricultural and food wastes to valuable chemicals, separation and purification of the products from the bulk liquid represented the highest percentages of the manufacturing cost. Therefore, the economic feasibility of reusing waste will strongly depend on the down steaming processing efficiency. In recent years, considerable advances have been achieved in biomolecular purification technologies, one of the key goals is to achieve more selective, more efficient, and safer separation routes (Angenent et al., 2004). For example, a single-step direct lactic acid separation from fermentation broths has been successfully implemented using anionic fluidized-bed columns (Sosa, 2001). Superficial fluid technology (i.e. the use of fluid at a temperature and pressure that are grater than its critical levels) has also proved to be a useful tool for achieving separation of compounds directly from cultures, and is an attractive option because separation is usually performed at ambient temperatures using non-toxic, non-flammable solvents while achieving product crystallization (Angenent et al., 2004).

In this chapter, the following four aspects of lignocellulose bioprocessing are discussed:

- Mushroom cultivation
- Single cell protein (SCP)
- Food and feed enzymes, and
- Production of food additives

MICROBIAL TECHNOLOGY FOR MUSHROOM PRODUCTION

Mushrooms are one of the most important fungi used directly as food source grown on agricultural residues (Table 15.2). Conversions of plant residues and by-products by edible fungi have several advantages (Zadrazil and der Wiesche, 1999; Quimio, 2006):

Table 15.2. Cultivated edible mushrooms in the world (Thomas et al., 1998; Quimio, 2006)

Mushroom Species	Common Name	Fruiting Temp. (°C)	Substrate
1. *Agaricus brunnescens*	Button mushroom, champignon	15-18	Straw, corn leaves, sugar cane leaves
2. *Agrocybe aegerita*	Yanagi-matsutake	13-18	Sawdust
3. *Auricularia polytricha*	Kukurage, ear fungus	21-30	Sawdust, logs
4. *Coprinus comatus*	Shaggy mane	18-24	Straw, cased
5. *Flammulina velutipes*	Enokitake, golden mushroom	10-16	Sawdust
6. *Ganoderma lucidum*	Reishi, lingchi	21-27	Sawdust
7. *Grifolia frondosa*	Maitake, hen of the woods	13-18	Sawdust
8. *Hericium erinaceus*	Monkey head, lion's mane	18-24	Sawdust
9. *Hypholoma capnoides*	Brown-gilled Hypholoma	10-16	Sawdust
10. *Hypholoma sublateritum*	Kuritake	10-16	Sawdust
11. *Hypsizygus tessulatus*	Buna-shimeji	13-18	Sawdust
12. *Hypsizygus ulmarius*	Shirotamogitake	13-18	Sawdust
13. *Lentinula edodes*	Shiitake	16-18	Sawdust
14. *Morchella angusticep*	Black morel	4-16	Burnt compost
15. *Pholiota nameko*	Nameko	13-18	Sawdust
16. *Pleurotus citrinopileatus*	Yellow pleurotus	21-29	Sawdust
17. *P. cystidiosus*	Abalone	21-27	Sawdust
18. *P. djamor*	Pink hiratake	20-30	Sawdust
19. *P. eryngi*	King oyster	15-21	Sawdust
20. *P. euosmus*	Tarragon mushroom	21-27	Sawdust
21. *P. ostreatus*	Oyster mushroom, hiratake	10-21	Sawdust, corn cobs
22. *P. pulmonarius*	Phoenix mushroom	18-24	Sawdust, straw
23. *P. tuber regnum*	Sclerotial mushroom	24-32	Sawdust
24. *Polyporus umbellatus*	Chorei-maitake	10-16	Sawdust
25. *Psilocybe cyanescens*	Fantasi-take	10-18	Sawdust
26. *Stropharia rugoso-annulata*	King Stropharia	16-21	Sawdust
27. *Volvariella volvacea*	Straw mushroom	27-32	Straw, banana leaves

- Agricultural plant residues (e.g. straw, bagasse, saw dust, etc.) can efficiently be reintegrated into the ecosystem through natural process.
- Solid and liquid waste matters (e.g. sugarcane and cassava bagasse) can be converted into fungal substrate.
- Lignocellulosics, which are a source of carbon of unsuitable low digestibility, can be transformed into consumable biomass (i.e. fruit bodies).
- Fruit bodies (pure microbial biomass) can be harvested from the surface of the substrate without expensive separation and purification, as in other fermentation processes.
- Edible fungi represent a well-characterized microbial biomass, generally accepted by the consumers/humans.

There are at least 12,000 species of fungi that can be considered as mushrooms with at least 2,000 species showing various degrees of edibility (Chang, 1999a). Furthermore, over 200 species of mushroom have been collected from the wild and utilized for food and various traditional medical purposes mostly in the Far East. To date, about 27 edible mushroom species have been cultivated commercially and among these about 20 species are cultivated on an industrial scale (Quimio, 2006). The majority of cultivable species belong to: *Pleurotus, Agaricus, Auricularia, Volvariella, Lentinus edodes* and *Flammulina velutipes*. Besides these types, few are medicinal mushrooms (i.e. possessing some medicinal properties). Two of the major medicinal mushrooms, i.e. *Ganoderma lucidum* and *Trametes (Coriolus)* are distinctly inedible (Chang, 1999b, c).

Cultivation Technology

Mushrooms can be cultivated through a variety of methods, but some methods are extremely simple and demand little or no technical expertise. On the other hand, cultivation, which requires aspects of sterile handling technology, is much more technically demanding (Stamets, 1993; Royse, 1996). Detailed descriptions of the many growing techniques can be found in the following references (Royse, 1996; Kozak and Krowczy, 1999; Quimio, 2002, 2006) and a brief description of the cultivation procedure is given below.

Strain Selection and Maintenance

The first stage in any mushroom cultivation process is to obtain a pure mycelial culture of the specific mushroom strain and such cultures should be purchased from mushroom specialists or mushroom institutes (Chang, 1996).

***Substrate Preparation*:** Agricultural residues or organic by-products from other primary industries are used in the preparation of a selective nutrient-rich medium in which mushrooms are grown. Raw materials used in mushroom substrate preparation include rice or wheat straw, poultry manure, stable waste matter, cottonseed hull and cottonseed meal (Miles and Chang, 1997). After these materials are mixed, the substrate is then composted and pasteurized.

***Spawning*:** Spore (the seed of the mushroom) is used to produce grain spawn under sophisticated sterile conditions. Constant checks ensure that temperature, air composition and humidity are kept at the right levels while the spawn grows through the substrate to produce a thick network of fine cotton-like threads known as mycelium.

The production of spawn is the most important step in the process of mushroom cultivation. The successive steps of spawn production include the following stages (Zadrazil and der Wiesche, 1999):

(a) Spawn carrier: Generally different cereal grains such as wheat, maize, sorghum, etc. and in some countries sawdust/bran mixtures, are generally used as carriers for fungal mycelium. The use of cereal grains takes into account the following requirements: they allow fast mycelial development, easy handling and steeliness during sterilization.

(b) Steps in spawn production:

Cleaning/calibration of grains – The quality of the cereal grains used plays an important role in guaranteeing a consistent production process and product quality. The stored grains are mechanically cleaned and mixed thoroughly to achieve a homogeneous substrate for the start of spawn production.

Cooling/steaming of grains – One of the most important factors influencing the physical and chemical properties of the grain substrate, the fungal growth and the storage life of the final product (spawn), is the water content. Water content between 40-50% is ideal for mycelium development and storage life. For moistening, the grain is boiled in a pressure proof container using steam for heating. At the end of the boiling, the grain must be soft but not structurally collapsed (Zadrazil and der Wiesche, 1999).

Mixing with additives – Additives such as gypsum and calcium carbonate are mixed into the cooled seed in order to adjust the pH to the desired value (between 6.5-7.5) and to prevent the grain particle from clotting.

Sterilization – After filling the grain into glass or plastic bottles which are closed with cotton or special filter plugs, the sterilization is carried out in autoclave at a pressure of 15 lb and temperature of 120-130°C for two to three hours.

Preparation of inoculum – Usually, starter cultures consisting of wooden sticks carrying the mycelium are aseptically placed into the spawn substrate. Grain starter or liquid 'gelations' starter cultures are also utilized.

Inoculation – The inoculation of the sterilized spawn substrate with the starter culture takes place in special rooms with laminar airflow hoods or at clean benches under axenic condition.

Incubation – Incubation follows inoculation, which is executed in rooms with controlled humidity, and temperature – key factors for product quality. Occasional shaking (by hand or by mechanical device) ensures homogeneous growth of the mycelium throughout the substrate.

Quality control – Due to the high sensitivity of the spawn during the production process (inoculation and incubation) to competitive microorganisms, sterility and the quality of the spawn have to be ensured.

Storage – Finished containers of some mushroom (*Agaricus bisporus, Pleurotusi* spp., *Flammulina velutips* and *Lentinus edodes*) spawns are stored in the refrigerator at a constant temperature of 2-5°C. Some fungi (i.e. *Volvariella volvacea*) are sensitive to low temperature near 0°C.

Transport – Spawn should be transported in airconditioned vehicles because every fluctuation in temperature retards the product quality.

Casing: After the mycelium has spread throughout the substrate, a surface layer of peat is added known as casing. The casing layer stimulates the mycelium to fruit and produce mushrooms. The first pinhead-sized mushrooms appear after about 12 days from casing (Royse, 1996). At every stage of the process, the environment is carefully monitored and hygiene is strictly controlled.

Growing: Commercial production of mushrooms requires strictly sophisticated environmentally controlled conditions (Yamanaka, 1997). After they appear on the beds, the pinhead-sized mushrooms double in size every 24 hours and will be ready for harvest about three weeks after spawning.

Harvesting: Harvesting is generally done by hand with teams of trained pickers. Each tray or bag of substrate produces three commercially harvestable crops (called "flushes") over a period of about six weeks.

Spent Substrate Use: Once the crop has been harvested, the used substrate provides an organic and nutritionally balanced product ideal for addition to high quality potting mixes or as garden mulch. Thus, nothing is wasted and the mushroom industry can claim to be the ultimate recycler of agricultural waste matter.

Oudemansiella tanzanica nom. prov. mushroom

The edible mushroom *Oudemansiella tanzanica* nom. prov., which is new to science, has been studied as a potential crop to reduce agricultural solid wastes and increase domestic mushroom production. The substrates sawdust, sisal waste and paddy straw supplemented with chicken manure, resulted in the highest biological efficiencies of any mushroom cultivated in Tanzania so far. In addition, this mushroom has one of the shortest cultivation cycles at 24 days. Despite the fact that the mushroom extracts substantial amounts of nutrients (Maziero et al., 1995; Sede and Lopez, 1999; Magingo et al., 2004), the spent substrate can be used as fodder, as a soil conditioner and fertilizer and in bioremediation (Magingo et al., 2004).

BIOPROCESSING OF AGRICULTURAL RESIDUES INTO ANIMAL FEED

Agricultural waste recycling has been advanced as an effective method for providing animal feed [Single Cell Protein (SCP)]. The term SCP refers to dead, dry cells of microorganism such as yeast, bacteria, fungi and algae that grow on different carbon sources including agricultural wastes. The growth of microorganisms, more rapid than that of higher plants, makes them very attractive as high-protein crops while one or two grain crops can be grown per year, a crop of yeast or moulds may be harvested weekly and bacteria may be harvested daily (Anupama and Ravindra, 2000). The use of microorganisms as a source of protein for animal feed as silage has a high content of microorganism to which their nutritional properties are partly due. However, there is very little data available on animal feeding trials using SCP produced from agricultural residues, particularly on the amino acid composition and protein quality (Ebine, 2003).

Properties of SCP (Israelidis, 2003)

One of the main advantages of SCP compared to other types of protein is the small doubling time of cell as shown in Table 15.3. Due to this property, the productivity of protein production from microorganisms is greater than that of traditional protein (Table 15.4). It is assumed that the growth occurs without any restriction. Other advantages of SCP over conventional

Table 15.3. Mass doubling time

Organism	Mass Doubling
Bacteria and yeast	10-20 minutes
Mould and algae	2-6 hours
Grass and some plants	1-2 weeks
Chicken	2-4 weeks
Pigs	4-6 weeks
Cattle	1-2 months

Table 15.4. Efficiency of protein production of several protein sources in 24 hours

Organism (1,000 kg)	Amount of protein
Beef cattle	1.0 kg
Soya beans	10.0 kg
Yeast	100.0 tonnes
Bacteria	100 × 10,000,000 tonnes

protein source are: (i) it is independent of land and climate; (ii) it works on a continuous basis; (iii) it can be genetically controlled; and (iv) it causes less pollution.

There are five factors that impair the usefulness of SCP: (i) non-digestible cell wall (mainly algae); (ii) high nucleic acid content; (iii) unacceptable colouration (mainly with algae); (iv) disagreeable flavour (part in algae and yeasts); and (v) cells should be killed before consumption. The SCP is treated with several methods in order to kill the cells, improve the digestibility and reduce the nucleic acid content (Bellamy, 1974).

Nutritional Value of SCP

For the assessment of the nutritional value of SCP, factors such as nutrient composition, amino acid profile, vitamin and nucleic acid content as well as palatability, allergies and gastrointestinal effect should be taken into consideration. Also long term feeding trials should be undertaken for toxicological effect and carcinogenesis. Table 15.5 shows the average cell composition of the major group of microorganisms.

Bacterial protein is similar to fish protein, yeast's protein resembles Soya and the fungal protein is somewhat lower than that of yeast's. Ofcourse microbiological proteins are deficient in the sulphur amino acid i.e. cysteine and methionine and require supplementation, while they exhibit better levels of lysine of conventional protein sources. Acceptability and palatability results on nutritional trials were not always encouraging.

Table 15.5. Average composition of main groups of microorganisms (% dry weight)

	Fungus	Algae	Yeast	Bacteria
Protein	30-45	40-60	45-55	50-60
Fat	2-8	7-20	2-6	1.5-3.0
Ash	9-14	8-10	5-9.5	3-7
Nucleic acid	7-10	3-8	6-12	8-12

Problems of Nucleic Acid

About 70-80% of total cell nitrogen is represented by amino acids while the rest in nucleic acids. This concentration of nucleic acids is higher than other conventional protein and is characteristic of all fast growing organisms. The problem which occurs from the consumption of protein with high concentration of nucleic acids (78-25 g 100 g^{-1} protein dry weight) is the high level of uric acid in the blood, sometimes resulting in the disease gout. Uric acid is a product of purine metabolism. Most mammals, reptiles and molluscs possess the enzyme uricase, and the end product of purine metabolism is allantoin. While the uric acid limits the daily intake of SCP for mono-gastric animals such as pigs and chickens, ruminants such as cattle, sheep and goats can tolerate higher levels. There are several methods available for removal of uric acids: (i) chemical treatment; (ii) treatment of cell with 10% NaCl; and (iii) thermal shock. These methods aim to reduce the RNA content from about 7% to 1% which is considered within acceptable levels.

The vitamins of microorganism are primarily of the B type, B12 occurs mostly in bacteria, while vitamin A is usually found in algae. Other nutritional properties which evaluate the quality of a given SCP are: the biological value (BV), the protein efficiency ratio (PER) and the net protein utilization (NPU) (Israelidis, 2003).

SCP from Starchy Crops

Starchy materials, more specifically cassava (*Manihot esculanta* Crantz), sweet potato (*Ipomoea batatas* L.) and banana refuse in tropical regions, or potatoes in the temperate climate, are of obvious interest because of high productivity per hectare and excellent rate of conversion to biomass by microorganisms (Table 15.6). Cassava roots with 25-30% starch and 0.17% protein content is a good source for production of SCP (Adeyemi and Sipe, 2004). Further, in the manufacture of starch from cassava, four types of wastes are produced as outer skin (peel), inner rinds, fibrous residues and wastewater (Sriroth et al., 2000; Ray, 2004). These waste matters constituting about 20-30% by weight of the cassava roots processed, are good substrates for protein enrichment by culturing microorganism such as *Saccharomyces cerevisiae* and *Candida tropicalis* through solid-state fermentation (Anita and Mbongo, 1994; Soccol, 1996; Adeyemi and Sipe, 2004). Wastewater obtain from cassava processing industry was used to produce SCP. A number of yeast strains i.e. *C. tropicalis, Schwaaniomyces occidentalis, Torulopsis wickerham, Endomycopsis fibuligera, Candida utilis* and *Saccharomyces* spp. when applied individually or in co-culture resulted in protein enrichment. *C. tropicalis* gave the highest protein enrichment (Jamuna and Ramakrishna, 1989).

Table 15.6. Protein enrichment of various starchy crops and residues (Senez, 2003; Ray and Ravi, 2005)

	Initial composition		Final product	
	Protein %	Carbohydrate %	Protein %	Carbohydrate %
Cassava	2.5	90	18	30
Banana	6.4	50	20	25
Banana waste	6.5	72	17	33
Potato	5.0	90	20	35
Potato waste	5.0	65	18	28
Sweet potato waste	2.5	80	18	35

SCP from Lignocelluloses

The lignocellulosic wastes, mainly from agriculture, constitute the most abundant substance for SCP which is also renewable. The world production of straw, for example, reaches 600 million tonnes every year. For the utilization of lignocelluloses, a pre-treatment is usually necessary. Many pre-treatment methods have been reported which vary from alkali or acid treatment, steam exploitation or even x-ray radiation. Among the substrates suitable in cost and supply, special emphasis is usually laid on cellulosic materials, but at the moment, the many attempts made in this direction have had little success, the main difficulty being the lack of cellulolytic organisms with an adequate growth rate, although *Trichoderma* spp. are vastly employed for production of SCP (Lo Curto et al., 1992; Sriram and Ray, 2005).

Economic Problems

In the long term, any process must generate enough profit above the operating cost to attract investment capital. Production of SCP from agricultural waste matters is no exception to this general rule. SCP must be produced at a price that is competitive with the major conventional protein source of plant and animal origin. If there is a guaranteed payment for waste removal, the production cost will be reduced accordingly. But no credit can be taken for waste removal in the absence of a firm and long-term commitment. Some factors that will contribute to reducing cost of SCP production from agricultural waste are the following:

- Payment for waste disposed (negative cost of raw materials).
- More effect and less expensive pretreatment.
- Superior microorganisms with a more rapid growth rate.
- Superior microorganisms that can rapidly utilize natural lignocelluloses without pretreatment.

- Methods of SCP production that do not require stirred tank fermentor.
- Method that will decreases the capital investment.

It is obvious that the value placed on waste elimination is variable and will depend upon the laws and customs in each region. Local laws providing credit for waste disposal are an effective method for stimulating SCP production. There is very little work in progress on a low-cost technology, labour intensive, low-capital SCP system. SCP production is an area with great potential for the developing countries, but of lesser interest to the industrial countries or to the private sector of industrial countries because it does not appear to have a large potential market. It is in this area of controlled fermentation under non-sterile conditions with minimum equipment that methods may be developed for using agricultural wastes at the small farm level. Instead of being grown under sterile conditions in expensive stirred fermentors, the moulds should be grown in a plastic container under conditions where the desired organisms predominate.

PRODUCTION OF MICROBIAL ENZYMES

Industrial enzymes represent the heart of biotechnology process. The field of industrial enzymes now is experiencing major initiatives, resulting in the development of a number of new products and in improvement in the process and performance of existing products. The global market for industrial enzymes is estimated as US$ 1,050 million in 2004 and is expected to rise at an average annual growth rate (AAGR) of 3.3% to US$ 2.0 billion in 2009 (Fig. 15.1) (http://www.mindbranch.com/products/R2-853.html). The three major application sectors of industrial enzymes are – technical, food and animal feed enzymes. While technical enzymes for detergent and pulp and paper manufacturing, among others, are the largest segment with a 52% share, food enzymes cover around 40% share. Growth of the animal feed enzyme sector is somewhat higher, at nearly a 4% of AAGR, held by increased use of phytase enzyme to high phosphate pollution.

Food Enzymes

For many thousand years, man has used naturally occurring microorganisms–bacteria, mould and yeasts – and the enzymes they produce are used for making bread, cheese, beer and wine. Today, enzymes are used for an increasing range of applications: bakery, cheese making, starch processing and production of fruit juices and other drinks (Table 15.7). Here they can improve texture, appearance and natural value,

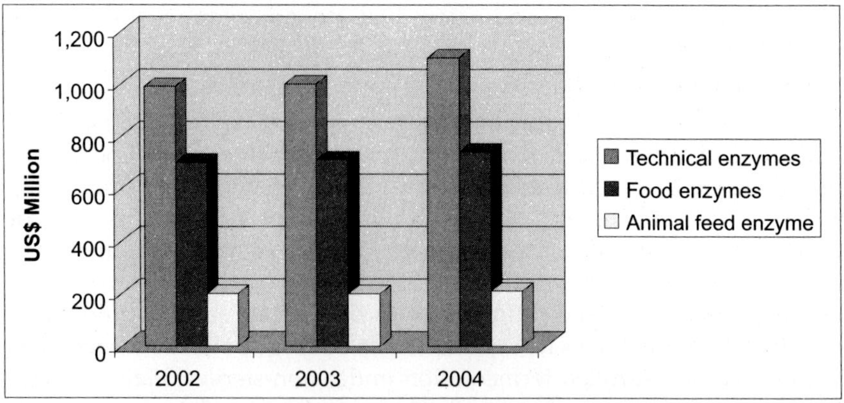

Fig. 15.1. Global enzyme market based on the application sector (US$ million)(http://www.mindbranch.com/products/R2-853.html. Accessed on August 2004)

Table 15.7. Some use of the microbial enzymes in food production

Market	Enzyme	Purpose/Function
Dairy	Protease	Coagulating in cheese production
	Lactase	Hydrolysis of lactose to give lactose free milk products
	Protease	Hydrolysis of whey proteins
	Catalases	Removal of hydrogen peroxide
Brewing	Cellulases, β-glucanases, α-amylases, proteases, maltogenic amylases	For liquefaction, clarification and to supplement malt enzymes
Alcohol production	Amyloglucosidases	Conversion of starch to sugar
Bakery	α-amylases	Breakdown of starch, maltose production
	Amyloglucosidases	Saccharification
	Maltogen amylase (Novamyl)	Delays process by which bread become stale
	Protease	Breakdown of proteins
	Pentosanase	Breakdown of pentosan leading to reduced gluten production
	Glucose oxidase	Stability of dough
Wine and fruit juice	Pectinase	Increase of yield and juice clarification
	Glucose oxidase	Oxygen removal
	β-glucanases	Clarification of fruit juice
Meat	Proteases	Meat tenderizing
Protein	Proteases, trypsin, amino peptidases	Breakdown of various components of protein
Starch	α-amylase, glucoamylases, hemicellulases, maltogenic amylase, glucose isomerases, dextranases, β-glucanases	Modification and breakdown of starch and conversion into dextrose or high fructose syrups
Inulin	Inulinases	Production of fructose syrups

and may generate desirable flavour and aroma. Currently used food enzymes sometimes originate in animals and plants (e.g. amylase, a starch-digesting enzyme, can be obtained from germinating barely seeds) but usually they originate from a range of beneficial microorganisms (the main types include species of *Bacillus, Aspergillus, Streptomyces* and *Kluyveromyces*).

In food production, enzymes have a number of advantages:

- They are welcomed as alternatives to traditional chemical based technology, and can replace synthetic chemicals in many processes. This can allow real advantages in the environment performance of production process, through lower energy consumption and biodegradability.
- They are specific in their action compared to synthetic chemicals. Processes which use enzymes, therefore, have fewer side reactions and waste by-products, giving higher quality products and reducing the likelihood of pollution.
- They allow some processes to be carried out which would other wise be impossible. An example is the production of clear apple juice concentrate, which relies on the use of enzyme, pectinase.

Cellulase and hemicellulases have numerous application and biotechnological potentials for various industries including food, brewery and wine and animal feed (Bhat, 2000; Sun and Cheng, 2002; Beauchemin et al., 2001, 2003; Sriram and Ray, 2005). In the bakery industry, xylanases are used for improving desirable texture, loaf volume and shelf life of bread. A xylanase Novozyme 867 has shown excellent performance in the wheat separation process (Christopherson et al., 1997).

Feed Enzymes

There is a huge potential market for fibre-degrading enzymes for the animal feed industry and over the years a number of commercial preparations have been produced (Beauchemin et al., 2001, 2003). The use of fibre-degrading enzymes for ruminants such as cattle and sheep for improving feed utilization, milk yield and body weight have evinced considerable interest. Steers fed with an enzyme mixture containing xylanase and cellulase shared an increased live weight gain of approximately 30-36% (Beauchemin et al., 1995). In dairy cows the milk yield increased in the range of 4-10% on various commercial fibrolytic enzyme treated forages (Beauchemin et al., 2001). Similarly, many alkaline proteases were used in feed technology (Ganeshkumar and Takagi, 1999) for the production of amino acids or peptides (Kida et al., 1995)

Enzymes through Gene Technology

Companies producing enzymes since the early 1980s have been using genetic engineering techniques to improve production efficiency and quality and to develop new products. Some of the enzymes produced through genetic engineering are show in Table 15.8.

When enzymes are produced from microorganisms, these are grown by fermentation either by SSF or SF. As in other parts of the food chain, strict rules of hygiene are followed. The crude enzyme produced through fermentation is purified by passing it through a series of filters to remove impurities and extract the enzymes. Table 15.9 shows the production of some enzymes using agricultural residues as substrates and carbon sources.

MICROBIAL TECHNOLOGY FOR PRODUCTION OF FOOD ADDITIVES

Foods, microorganisms, and humans have had a long and interesting association that developed long before the beginning of recorded history. Foods are not only of nutritional value to those who consume them, but often are an ideal culture medium for microbial growth (Doyle et al., 1997) Therefore, due to its perishable nature, processing and preservation have

Table 15.8. Enzymes produced and marketed using genetic engineering technology

Principal Enzyme Activity	Application
α-acetolactate decarboxylase	Brewing
α-amylase	Baking, brewing, distilling, starch, production of maltodextrin, corn syrup
Catalase	Mayonnaise
Chymosin	Cheese
β-glucanase	Brewing
α-glucanotransferase	Starch
Glucose isomerase	Starch
Hemicellulase	Baking
Glucose oxidase	Baking egg mayonnaise
Lipase	Fats, oils
Maltogenic amylase	Baking, starch
Microbial rennet	Dairy
Phytase	Starch
Proteases	Baking, brewing, dairy, distilling, fish, meat, starch, vegetable
Pullulanase	Brewing, starch, high fructose corn syrup
Xylanase	Braking, starch

Table 15.9. Some examples of microbial enzymes produced using agricultural residues as substrates in solid-state fermentation

Enzyme	Organism	Substrate	Reference
α-amylase	*Bacillus coagulans*	Mustard oil cake	Babu and Satyanarayana, 1995
α-amylase	*Bacillus subtilis*	Banana peel and waste	Kaukab et al., 2003; Krishna and Chandrasekaran, 1996
α-amylase	*Bacillus circulans*	Wheat bran	Palit and Banerjee, 2001
Amylase	*B. subtilis*	Wheat bran	Baysal et al., 2003
Amylase	*Aspergillus niger*	Rice bran	Akpan et al., 1999
Amylase	*Rhizopus niger*	Cassava bagasse	Ray, 2004
Glucoamylase	*A. niger*	Rice husk	Radhakrishnan and Pandey, 1993
Amyloglucosidase	*A. niger*	Wheat bran	Ashraf et al., 2002
Multi enzyme (pectinase, cellulase and Xylanase)	*A. niger*	Orange peels	Abdel-Mohsen and Ismail, 1996
Cellulase/ hemicellulase	*Aspergillus terreus*	Sugarcane bagasse, wheat straw	Abdel-Fattaeh et al., 1987
β-glycosidase	*A. niger* and *Trichoderma reesei*	Waste paper	Juhasez et al., 2003
Cellulase (endoglucanase I and cellobiohydrolase II)	*Trichoderma reesei*	Cotton cellulase	Klemanlayer et al., 1996
Cellulase	*A. niger*. *T. viride* and *Penicillum funiculosum*	Waste paper	Vanwyk et al., 2001
Cellulase	*T. reesei*	Wheat bran	Xia and Len, 1999
Cellulase	*T. reesei*	Cassava fibrous residue	Singhania et al., 2005
Cellulase	*Aspergillus flavus*	Sawdust, bagasse and corn cob	Ojumu et al., 2003
Pectinase	*Moniliella* SB9, *Penicillium* sp. EGC5	Wheat bran, Orange bagasse, Sugarcane bagasse	Martin et al., 2004
Xylanase	*B. circulans*	Wheat straw	Dhillon and Khanna, 2000

contd...

Table 15.9 contd...

Phytase	*A. niger*	Wheat bran	Papaginni et al., 2001
Lignolytic enzyme	*Nematoloma flowardii*	Wheat straw	Vyas et al., 1994
Lignolytic enzyme	*Phlebia radiata*	Wheat straw	Vares et al., 1995
Chitonase	*A. niger, Penicillium citrinum, Fusarum oxysporum* and *Rhizopus oryzae*	Rice stalk	Khalaf, 2004
Exoglucanase	*B. subtilis*	Banana peel	Shafique et al., 2004
Protease and amylase	*Streptomyces rimosus*	Sweet potato residue, Peanut meal	Yang and Wang, 1999
Lipase	*Rhizopus oligosporous*	Almond meal	Awan et al., 2003
Lipase	*Candida rugosa*	Coconut cake	Benjamin and Pandey, 1997

to be applied by certain chemical substances, which are known as food additives. In its broadest sense, legally the term also refers to "*any substances added intentionally to foodstuffs to perform certain technological functions*" (Wood, 1998). These are meant to enhance the overall acceptability of food in appearance, colour, flavour, texture, mouth feel, and preservation, and in some cases the nutritive value.

General Aspects of Food Additives

Generally additives are used in foods for four main reasons:

- To maintain product consistency
- To improve or maintain nutritional value
- To maintain palatability and wholesomeness
- To provide leavening or control acidity/alkalinity

Food Additives of Fermentation Origin

Acidulents

These are a group of acidic substances used to provide a desired sour taste, and serve to intensify or enhance the overall flavour of the product. Usually, they help in the reduction of the pH to prevent or retard the growth of microorganisms and the germination of spores, and to increase the lethality of the process (Kuntz, 1995). These occur naturally in many foods, e.g. malic acid in apples, tartaric acid in grapes, tamarind and citric acid in lemons, limes and oranges. Some examples of acidulents are mentioned below with their food engineering characteristics.

Acetic acid (Vinegar): Generally, vinegar contains about 4% aqueous solution of acetic acid. It has widespread applications as an acidifier, flavour enhancer, flavouring agent, pH-controlling agent, pickling agent in kitchen and most of the food industries (Ebner, 1996). Due to its antimicrobial properties in foods, it is frequently used in salad dressing, ketchups, etc. It is produced from sugar rich material such as apples, grapes, pineapples, coconuts and their processing wastes by successive anaerobic (alcoholic) and aerobic (acetic acid) fermentations (Battcock and Azam-Ali, 2001). The first step involves conversion of sugar to ethyl alcohol and CO_2 through yeast (*Saccharomyces cerevisiae*) enzymes. The second step involves conversion of ethanol to acetaldehyde and then into acetic acid by *Acetobacter*.

Citric acid: Traditionally it was manufactured from lemon juice and pineapple waste, but now commercially it is produced exclusively by fermentation, either by surface process, solid state or *koji* process, immobilized process or by the submerged process (Goldberg et al., 1991). Its production by fungi is one of the first fungal fermentation processes of industrial importance discovered a century ago (Wehmer, 1893). Different species of *Aspergillus, Candida, Trichoderma* and various strains of yeasts accumulate citric acid in the medium along with iso-citric acid from various carbon sources.

Lactic acid: It is manufactured commercially either by fermentation or by chemical synthesis. Basically homofermentative bacteria viz. *Lactobacillus delbrueckii, Lb. bulgaricus,* and *Lb. leichmanii* produce lactic acid exclusively or predominantly from carbohydrates and are used commercially. Generally, different starchy substrates are used for production of lactic acid by fermentation process. Most lactobacilli cannot use starch. Some species, i.e. *Lb. amylophilus, Lb. manihotivorans* and *Lb. amylovorus* are able to ferment starch to lactic acid (Ray and Ward, 2006). Table 15.10 shows the production of some organic acids from various agricultural and food processing waste matter.

Gluconic acid: It is prepared by fermentation using *A. niger* and less commonly *Acetobacter (Glucanobacter) suboxidans* and same *Penicillium* spp. (Pai, 2003). *Aspergillas niger* produces gluconic acid at high levels (97-99%) with negligible by-products. Glucose is converted to glucano δ-lactone by glucose oxidase enzyme. The lactone dehydrolases to gluconic acid.

Other acids: Apart from the above-mentioned acids, other acids, i.e. fumaric acid, malic acid (72%), succinic acid (67%), etc. yield from agricultural waste material (Moresi et al., 1992). These acids also have a lot of importance in most of the food industries.

Table 15.10. Organic acid production using various agricultural waste matters as substrate

Agricultural waste matter	Microorganism used	Reference
Acetic acid		
Pine apple, sweet orange, coconut waste	*Acetobacter*	Battcock and Azam-Ali, 2001
Lactic acid		
Cheese whey	*Lactobacillus helveticus*	Ghaly et al., 2003
Starch waste (derived from the production of potato chips)	*Aspergillus niger, Lactococcus lactis, Lactobacillus delbrueckii*	Chau, 2002
Potato processing waste	Amylolytic LAB	Wang and Yuksel, 2004
Alfalfa fibre, soya fibre, corn cob and wheat straw	*Lactobacillus plantarum, Lactobacillus delbrueckii*	Sreenath et al., 2001
Wheat bran	*Lactobacillus amylophilus*	Naveena et al., 2004
Cassava bagasse	LAB	Rajan et al., 2005; Ray et al., 2005
Grape vine shoots	*Lactobacillus pentosus, Lb. rhamnosus*	Bustos et al., 2005
Citric acid		
Sweet potato residue	*Aspergillus niger*	Lu et al., 1995, 1997; Bindumole et al., 2005
Cassava bagasse	*Aspergillus niger*	Vandenberghe et al., 2004

LAB: Lactic acid bacteria

Antimicrobial Agents or Preservatives

Food preservation has become somewhat of a lost art in the past few decades (Dalton, 2002). The selection of a preservative will depend on the process conditions, in particular the "pH" (acidity), the "water activity" (water is essential for microbial growth) and the type of organisms that may be present in the food (Kuntz, 1995).

Bacteriocins

Bacteriocins are defined as *"any proteinaceous substances (usually peptides) produced by bacteria that has bactericidal action against closely related or limited range of organisms and help the host survive and establish an ecological niche"*. Most bacteriocins kill sensitive cells due to the dissipation of the proton motive force, ATP depletion and leakage of small ions and molecules from the cell (Chikindas, 1995). Although bacteriocins are highly active antimicrobial agents, they are not antibiotics. However, many bacteriocins are active against food borne pathogens and food spoilage microorganisms (Tan et al., 2000), and stable over a wide range of environmental conditions (http://www://anka.livestock.ith.se2080/list-cin.htm-5k). Therefore, in future these molecules should be considered as food preservatives, especially when used as a part of hurdle technology (Abee et al., 1995).

Lactobacter, Leuconostoc, Pediococcus spp. are lactic acid bacteria (LAB) that are commonly associated with foods and are used as starter bacteria in some dairy fermentations (Stiles, 1993). Lactic acid bacteria are inhibitory to other bacteria because of pH, organic acids, hydrogen peroxide, and other chemicals produced during their growth, including bacteriocins. Some examples are given below:

Lantibiotics: These chemicals have a broad spectrum of antimicrobial activity, and are active against a number of Gram-positive bacteria (Daeschel, 1989). They derive their name from the inclusion of sulphur containing amino acids, lanthionine and β-methyllanthionin, in their structures. The examples of lantibiotics are nisin, subtilin, epidermin, gallidermin, etc. Currently, nisin is the only LAB bacteriocin that is commercially used as a food preservative for decades in more than 50 countries.

Agricultural and animal waste matter such as mussel processing waste is used for production of lantibiotics (Guerra and Pastrana, 2002).

Pediocins: Generally, the bacteriocins produced from *Pediococcus* are called pediocins and are of importance since they have the ability to inhibit the growth of a wide range of pathogens and Gram-negative organisms, and maintain adequate microbiological safety (Bamby-Smith, 1992). Some of them are also very sensitive to proteolytic enzymes, resistance to heat (up to 121°C) (Stiles and Hastings, 1991) and organic solvents, and stable in the pH range of 2.5-9.0 (AcH from *P. acidilactici* H).

Microbial Antioxidants

Antioxidants are a type of preservative. Most commonly used antioxidants are butylated hydroxytoluene, butylated hydroxyanisole, *tert*-butylhydroquinone, and propyl gallate (Foulke, 2003) which stop the chemical breakdown of food that occurs in the presence of oxygen. These are amongst the safest chemicals known; a recent survey includes 'vitamin C' and 'vitamin E' which inhibit the formation of carcinogenic nitrosamines, stimulates the immune system, protects against chromosome breakage, and regenerates as part of the antioxidant defence system (Kromhout, 1999).

Enzymatic Removal of Oxygen

The only commercial use of enzymes as antioxidants is currently limited to removal of oxygen using glucose-oxidase in conjugation with catalase. In the net reaction, one mole of oxygen is removed for two moles of glucose that are oxidized. This enzyme combination is extracted from *A. niger*, and enjoys GRAS (Generally Regarded As Safe) status in several countries and has been shown to be effective in preventing off-flavour development in various food products such as fruit juice, salad dressing and mayonnaise (Sagi and Mannheim, 1990; Jafar et al., 1994). Other oxygen scavenging enzymes such as alcohol oxidase, thiol oxidase, galactose oxidase, pyranose oxidase and hexose oxidase have been envisaged as potentially useable food antioxidants (Fuglsang et al., 1995; Freedomyou, 2005).

TBHQ and Lycopene

These two antioxidants are the recent technological development of one Chinese company (Guangzhou Youbao Industry Co.), under the brand 'U-BEST' and are titled with the names of 'The Excellent Products in China' and 'Excellent New Products', respectively. They are widely used in the food industry. U-BEST brand TBHQ is a very effective and heat-resistant antioxidant, preventing the oxidation of edible oil, fat and oil-containing food, and thus extending their shelf life (Shong, 2005). An extremely small dosage brings a very good effect when it is used in food processing industries.

Tocopherols and Ascorbic acid

The above two chemicals are major fat-soluble antioxidants. Of these, tocopherols generally protect edible fats, vegetable oils and salad dressings from turning rancid. They come under the group of 'Vitamin E', whereas ascorbic acid is generally derived from corn, and helps in the colour preservation of freshly cut fruits and vegetables (The Big Carrot, 2004).

Microbial Stabilizers, Thickeners and Gelling Agents

Recently, a great deal of interest has been shown in the ability of microorganisms to produce exopolysaccharides (EPS). Pseudoplastic rheology and high stability at high temperature within a broad pH range and at high ion concentrations make microbial EPS of great potential applicability as food additive in form of either stabilizer, thickener or gelling agent (Sutherland, 1996a; Wang and McNeil, 1996). Most of the microbial polysaccharides (Table 15.11) are produced by using low-value and low-cost raw materials such as agricultural residues by either SF or SSF.

The use of microbial polysaccharides in food industry probably represents just less than 10% of the usage of all types of polysaccharides. Their role may be as thickening agents, for gel formation, or because of their colloidal properties (Table 15.12).

Table 15.11. Some examples of microbial EPS produced using agricultural residues as substrate and carbon source

Microbial Polysaccharides	Agricultural Residue	Microorganisms	Reference
Dextran	Wheat bran	*Leuconostoc mesenteroides*	Shamala and Prasad, 1995
Alternan	Corn fibre, corn condensed distillers soluble (CDS)	*L. mesenteroides*	Leathers, 1998
Scleroglucan	Cassava and sweet potato flour	*Sclerotium glucanicum*	Selbmann et al., 2002
Pullulan	CDS, Jerusalem artichoke tuber, carob pods	*Aureobasidium pullulans*	Leathers and Gupta, 1994; Roukas and Biliaderis, 1995
Xanthan	Citrus waste, sugar beet molasses, cassava fibrous waste	*Xanthomonas campestris*	Bilanovic et al., 1994; Roukas, 1998; Woiciechowski et al., 2004

Table 15.12. Use of microbial exopolysaccharides in food

1. Thickening agent: In sauces and syrups
2. Gelating agents: In milk-based desserts, confectionery and jellies, pie and pastry filling
3. Colloids: Stabilizer in ice creams, salad dressing and fruit drinks
4. Synthetic gel formation: In synthetic meat gels, etc.
5. Ice or sugar crystal retardation: Ice cream, ice-lollies, etc.
6. Stability at low pH: Salad dressings

In addition to their use in foods for human consumption, microbial polysaccharides are incorporated into pet foods, particularly semi-moist preparations where they act as thickening agents and binders (FMC Biopolymers, 2005). When incorporated into foodstuffs, microbial polysaccharides should be compatible with the salts, pH and other components found naturally in the foods. In this respect, xanthan (discussed below) is highly satisfactory (Sutherland, 1996b). An additional property of xanthan which is utilized on its food additive role, is the synthetic viscosity increase found with galactomannans. Some examples of microbial EPSs are given below.

Fungal Polymers

Scleroglucan

The EPSs from the fungal genus *Scerotium* have been termed scleroglucan. The polymer is basically a linear β-(1,3) linear D-glucan to which single D-glucopyranosyl groups are linked (Selbmann et al., 2002).

Pullulan

The EPS from *Aureobasidium pullulans* is an α-D-glucan in which maltotriose or maltotetraose units are coupled through α-1, 6 bonds to form a polymer. It has numerous patented uses in various industries such as manufacturing, pharmaceuticals, electronics and food industries (Leathers and Gupta, 1994; Ray and Moorthy, 2006).

Bacterial Polymers

Curdlan

Several microbial polysaccharides have the capacity to form gels. One of these is curdlan, the linear (α-1, 3)-D-glucan from *Alcaligenes faecalis* var. *myxogenes* and related bacteria.

Dextran

Dextrans are α-D-glucan produced by *Leuconostoc mesenteroides*. They are widely used as a blood plasma extender, in food industry, photo film manufacture, as a chromatographic medium, soil conditioner, etc.

Alternan

Alternans are also α-D-glucan distinguishable from dextran by backbone structure of regular, alternating α-(1,6), α-(1,3) linkages (Leathers et al., 1997). Although many strains of *L. mesenteroides* produce dextrans, only few naturally occurring strains produce alternan. The unique structure of alternan is thought to be responsible for the distinctive physical properties of high solubility and low viscosity. Alternans and its derivatives have functional properties that resemble commercially valuable glucans such as gum aerobic, malto dextrins and polydextrose (Leathers, 1998).

Bacterial Alginates

Most microbial polysaccharides, although possessing physical properties similar to those of the plant and algal gums, differ considerably from them in their chemical structures. An exception is bacterial alginate. These polysaccharides are co-polymers of D-mannuronic acid and L-guluronic acid. The microbial products do differ from the alginate produced from marine algae in possessing O-acetyl groups; they are thought to be associated only with D-mannuronic acid.

Xanthan

This is the only EPS currently produced on a large scale by several suppliers. The polysaccharide is derived from *Xanthomonas campestris* which is marketed under a variety of trade names: Keltrol and Kenkel (Kelco Inc., San Diego, USA) and Rhodogel (Rhone Poulenc S.A., Melle,

France). This EPS, the structure of which has been elucidated as substituted cellulose, produces high viscosity aqueous solutions which are highly pseudoplastic. Xanthan has been proposed for a very wide range of applications. They are to be found in food, pharmaceutical and other industries (Table 15.13).

Flavours

Flavours and fragrances are extremely important for the food, feed, cosmetic and pharmaceutical industries (Saltmarsh, 2000). The chemical industries produce a wide range of flavours synthetically. However, a significant drawback is that chemical flavours do not often closely match the natural food flavour, being described as 'synthetic' or 'chemical' by flavour panels (Grohol, 2005). Hence, there is a premium value on 'natural' flavouring and the so-called 'nature identical' flavouring compounds. The 'natural' flavours produced by microorganisms have been reviewed by Janssens et al., (1992).

Esters

Flavour production using immobilized lipase from yeast *Candida cylindraceae* in a non-aqueous system has been studied for production of a

Table 15.13. Some of the patented uses of xanthan (Sutherland, 1996b)

Application	Patent Holder	US Patent
Oil drying muds	Exxon	3251768(1966)
Stabilizer	Tennecon Chemicals	3438915(1969)
Stabilizer – water-based paints	Kenkel	3481889(1969)
Suspending agent for laundry starch	Henkel	3692552(1972)
Carrier for agro chemicals	Kelcon	3717452(1973)
Agricultural and herbicidal sprays	Kelcon	3659026(1972)
Metal picking baths	Diamond Shamrock	3594181(1971)
Gelled detergents	Chemed	3655579(1972)
Gelled explosives	Kelcon	3326733(1967)
Waterproof dynamite	Ashed Oil and Refining	3383307(1968)
Clay coating for paper finish	Kelcon	3279934(1966)
Flocculant for water clarification	Ashed Oil and Refining	3342732(1967)
Alkylene glycol ester for foods, cosmetics	Kelcon	3256271(1966)
Carboxy alkali ethers for paper, textiles	Kelcon	3236831(1966)
Dialkaliminoalkoxy ether for textile sizing	Kelcon	3244695(1966)
Hydroxy alkyl ether for cosmetics	Kelcon	3349077(1967)
Sulphate for glue thickening	Kelcon	3446796(1969)

broad range of esters including ethyl butyrate, iso-amyl acetate and iso-butyl acetate (Deosen Corporation Ltd., 2005). Ethyl butyrate, which has a pineapple-banana like flavour, has a large market demand. Currently, studies are being carried out on the kinetics of AATase (alcohol acetyl transferase–the enzyme that leads to ester production by yeast, and iso-amyl acetate hydrolase, and the enzyme that breaks esters down). Ultimately, these studies intend to combine genetic manipulation of such enzyme activities with the selection of mutant microbes with an increased ability to process amino acids directly to their equivalent fusel alcohol. In this way, microbial strains will be developed that produce flavour esters at levels that are commercially viable (Murphy, 2000).

Terpenes

Terpenes are often the most important components responsible for the characteristic odours of essential oils. They are composed of isoprene units and can be cyclic, open chain, saturated, unsaturated, oxidized, etc. Most of the terpene-producing microorganisms are fungi. An example is the genus *Ceratosystis* which has been the object of intensive research.

Cassava bagasse, apple pomace, amaranth and soyabean have been found to be adequate substrates for aroma (terpenes) production by *Ceratocystis fimbriata* in solid-state fermentation (Table 15.14) (Bramorski et al., 1998).

Pyrazines

This chemical provides a roasted or nutty flavour on heating. When culturing *Bacillus subtilis* in the media containing sucrose, asparagines and metal ions are known to produce tetramethyl pyrazine. *Corynebacterium glutanicum* also has the capability to produce pyrazine, therefore, this organism offers an opportunity of investigating the possibility of co-producing glutamate and tetramethylpyrazine in a fermentation process.

Table 15.14. Aroma and volatile compound production by *Ceratocystis fimbriata* on different solid media comprising agricultural residues (Bramorski et al., 1998)

Substrate	Aroma intensity	Total volatiles (μmole ethanol litre^{-1})
Cassava bagasse + soyabean	Fruity++++	1243
Cassava bagasse + soyabean + soyabean oil	Fruity++++	1037
Apple pomace + cassava bagasse + soyabean	Fruity++++	1120
Amaranthus grains	Pineapple+++	966

Lactones

Basically this chemical compound imparts fruity, coconut like, buttery sweet and nut like flavour to various foods. Microbial production offers an advantage over the chemical method, in the sense that it produces optically pure form, and the microbes used for this production are *Candida* and *Saccharomyces, Penicillium notatum, Cladosporium butyricum, C. suaveolens* and *Sarcina lutea. Trichoderma viride,* a soil fungus, generates a strong coconut aroma on a simple growth medium. The character impact compound is 6-pentyl-2-pyrone, which is produced in a concentration of 170 mg $litre^{-1}$ (Janssens et al., 1992).

Flavour Potentiators

Monosodium glutamate (MSG)

Various strains of glutamic acid producing bacteria are widely distributed in nature and are listed in Table 15.15. These organisms can use various carbohydrate sources including acetic acid, glucose, fructose, sucrose, maltose and xylose, which may be derived from various substrates such as starch, cane molasses, beet sugar or molasses (Freedomyou, 2005). Sugar molasses would be an economic feedstock, except that it needs processing to depress excess biotin. The continuous production of MSG using a fluidized bed and immobilized bacteria has also been reported (Henkel et al., 1990). Advances in bioreactor technology as well as R-DNA technology and cell fusion would result in more efficient production of MSG via fermentation technology.

Natural Food Colours

Due to surrounding controversy of synthetic colours and about their toxicological implications, day-by-day the list is shrinking and the demand for natural colours is progressively increasing. Some of the natural colours used as food additives are discussed below.

Table 15.15. Microbial strains producing L-glutamic acid

Genus	Species
Corynebacterium	*C. glutamicum, C. lilium, C. callunae, C. herculis*
Brevibacterium	*B. divaricatum, B. aminogenus, B. flavum, B. roseum, B. saccharolyticum, B. lactofermentum, B. ammoniagenes, B. thiogenitalis*
Micobacterium	*M. flavum* var. *glutamicum, M. salicinovolum, M. ammoniaphilum*
Arthrobacter	*A. glubiformis, A. aminofaciens*

Ref: Kawakita et al., 1990

Carotenoids

It is the foremost natural colour, especially β-carotene (of plant origin). Besides being a natural orange pigment (carrots, mango, papaya, winter squash, etc.) it is converted in the body to vitamin A and has antioxidant powers. It is believed to have a beneficial effect in reducing the risk of some cancers and perhaps heart disease (Mohapatra et al., 2006).

Anthocyanin

Anthocyanins are of plant origin, having antioxidant properties (Panda et al., 2006). Some sources of anthocyanins, besides red grapes, are elderberries, red cabbage, blood orange, the less familiar black chokeberry, and the sweet potato.

Microbial Colour

Astaxanthin (3, 3′-dihydroxy-β-carotene-4, 4′-dione) is a yellowish red pigment found in animal kingdom, but rarely found in microorganisms. A yeast species, *Phaffia rhodozyma*, is known to produce astaxanthin, the optimization of which has been reported using CDS (corn soluble distillate) as the substrate (Hayman et al., 1995). Other low-cost substrates used were alfalfa residual juice (Okagbue and Lewis, 1984), grape pomace (Meyer and Du Prez, 1994) and molasses (Haard, 1988).

Microbial Sweeteners

There are many low-caloric sweeteners of microbial and plant origin, i.e., sugar alcohols (xylitol, erythritol, mannitol, lactitol, sorbitol, etc.), aspartame (a derivative of amino acid, phenyl analine), sucralose, high fructose syrup, etc. and out of these, xylitol, a five-carbon sugar alcohol, is produced from hemicellulose fraction of agricultural residues (Mussatto and Roberto, 2005). Primary interest in xylitol is due to its properties and potential uses as: (1) an anti-carcinogenic sweetener with low-caloric value, (2) as a sugar substitute for diabetics, (3) a food edulocorant with clinical, food additive, (4) considerable pharmaceutical applications, and (5) for the treatment of glucose-6-phosphate dehydrogenase deficiency (Mussatto and Roberto, 2002; 2005). It is currently produced by the chemical reduction of pure xylose. However, microbial production especially from hemicellulose hydrolysate would not require pure xylose and would be cheaper. Several yeast strains including *Candida* spp., *Pachysolen tannophilus, Debaryomyces hansenii, Kluyveromyces fragilis* and various bacterial and mycelial fungi can produce xylitol. Among these, the best yields of xylitol are from *Candida guilliermondii, C. tropicalis* and a transformed strain of *Saccharomyces cerevisiae* (Sijtsma, 2005).

Xylitol productions have been reported from various agricultural residues, i.e. rice straw (Silva and Roberto, 1999, 2001; Mussatto and Roberto, 2005), sugarcane bagasse (Pessoa et al., 1996), aspen wood hydrolysates (Preziosi-Belloy et al., 2000), etc. These agricultural residues contain a large amount of hemicellulose fraction that can be easily hydrolyzed to its constituent carbohydrates, a mixture of sugars containing xylose (Roberto et al., 2003). Xylose solution obtained can be used as a fermentation medium to produce xylitol (Silva and Roberto, 2001; Mussatto and Roberto, 2003).

CONCLUSION AND PERSPECTIVES

Microorganisms use both constitutive and inducible enzymes to degrade and synthesize a wide array of compounds not only for their survival and multiplication, but also for their secondary metabolism. Such "bioconversion" processes are successfully used in many areas like conversion of agricultural residues to animal feed and food, bioenergy, manufacturing of enzymes and various food additives for human use. The recent impact of biotechnology research, especially in the fields of molecular biology, genetic engineering and protoplast fusion, is very apparent in modifying microorganisms to enhance their metabolic potential. Successful processing technology should give emphasis on the bioconversion technology, its control strategies, process engineering, scale-up, cost analysis, toxicity elimination, socio-economic and environmental factors.

REFERENCES

Abdel-Fattah, A.F., Abdel-Naby, M.A. and Ismail, A.S. (1987). Purification and properties of xylan degrading enzyme from *Aspergillus terreus* 603. Biol. Wastes 20: 143-151.

Abdel-Mohsen and Ismail, A.S. (1996). Utilization of orange peels for the production of multienzyme complexes by some fungal strains. Process Biochem. 31: 645-650.

Abee, T., Krockel, L. and Hill, C. (1995). Bacteriocins: Modes of action and potentials in food preservation and control of food poisoning. Int. J. Food Microbiol. 28(2): 169-185.

Adeyemi, O.A. and Sipe, B.O. (2004). In vitro improvement in the nutritional composition of whole cassava root meal by fermentation. Indian J. Animal Sci. 74: 321-323.

Akpan, I., Bankole, M.O. and Adesemowo, A.M. (1999). Production of amylase by *Aspergillus niger* in a cheap solid medium using rice bran and agricultural materials. Trop. Sci. 39: 77-79.

Al Wong. (1996). The Agri-pulp Newsprint Alternative. Presented at the Newspaper Association of America Super Conference, Miami, Fl., USA, March 6.

Angenent, L.T., Karim, K., Al-Dahhan, M.H., Wrenn, B.A. and Dimiguez-Espinosa, R. (2004). Production of bioenergy and biochemicals from industrial and agricultural waste water. Trends Biotechnol. 22: 477-485.

Anita, S.P. and Mbongo, P.M. (1994). Utilization of cassava peels as substrate for crude protein formation. Plant Food Human Nutr. 46(4): 345-351.

Anupama and Ravindra, P. (2000). Value-added food: Single cell protein. Biotechnol Adv. 18(6): 459-479.

Ashraf, H.I., Omar, S. and Qadeer, M.A. (2002). Biosynthesis of amyloglucosidase by *Aspergillus niger* using wheat bran as substrate. Pakistan J. Biol. Sic. 5(9): 962-964.

Awan, F.U., Mirza, S., Shafiq, K., Ali, S., Rehman, U.A. and Haq, U.I. (2003). Mineral constituents of culture medium for lipase production by *Rhizopus oligosporous* fermentation. Asian J. Plant Sci. 2(12): 913-915.

Babu, K.R. and Satyanarayana, T. (1995). Alpha-amylase production by thermophilic *Bacillus coagulans* in solid state fermentation. Process Biochem. 30(4): 305-309.

Bamby-Smith, F.M. (1992). Bacteriocins: Application in food preservation. Trends Food Sci. Technol. 3(6): 133-143.

Battcock, M. and Azam-Ali, S. (2001). Fermented Fruits and Vegetables: a Global Perspective. FAO Agric. Services Bull. 134, Rome, Italy, pp. 96.

Baysal, Z., Uyarb, F.A. and Aytekina, C. (2003). Solid state fermentation for production of α-amylase by a thermotolerant *Bacillus subtilis* from hot-spring water. Process Biochem. 38: 1665-1668.

Beauchemin, K.A., Rode, L.M. and Sewalt, V.J.H. (1995). Fibrolytic enzymes increase fibre digestibility and growth rate of steers fed dry forages. Can. J. Anim. Sci. 75:641-644.

Beauchemin, K.A., Morgavi, D.P., Mcallister, T.A., Yang, W.Z. and Rode, L.M. (2001). The use of enzymes in ruminant diets. In: Recent Advances in Animal Nutrition (eds.) J. Wiseman and P.C. Garnsworthy, Nottingham University Press, Nottingham, UK, pp 296-322.

Beauchemin, K.A., Colombatto, D., Morgavi, D.P. and Yang, WZ. (2003). Use of exogenous fibrolytic enzymes to improve animal feed utilization by ruminants. J. Anim. Sci. 81(E. Suppl.2): E37-E47.

Beg, Q.K., Kapoor, M., Mahajan, L. and Hoondal, G.S. (2001). Microbial xylanases and their industrial applications: A review. Appl. Microbiol. Biotechnol. 56: 326-338.

Bellamy, W.D. (1974). Single-cell proteins from cellulosic wastes. Biotechnol. Bioeng. 16: 869-880.

Benjamin, S. and Pandey, A. (1997). Production of lipase by *Candida rugosa* by using coconut cake as solid substrate. Acta. Biotechnol.1760: 241-251.

Berton, F. (2005). Agricultural and Forest Waste Feasibility Study: Agricultural Residues. Accessed from, ULC: http://www.ciwmb.ca.gov/Organics/Conversion/AgForestRpt/agriculture/default .htm. Accesed on 25 February 2005.

Betts, W.B., Dart, R.K., Ball, A.S. and Pedlar, S.L. (1991). Biosynthesis and Structure of lignocellulose. In: Biodegradation: Natural and Synthetic Materials (eds.), W.B. Betts, Springer-Verlag, Berlin, Germany, pp. 139-155.

Bhat, M.K. (2000). Cellulases and related enzymes in biotechnology. Biotechnol. Adv. 18: 355-383.

Bilanovic, D., Shelef, G. and Green, M. (1994). Xanthan fermentation of citrus waste. Biores. Technol. 48: 169-172.

Bindumole, V.R., Sasikiran, K. and Balagopalan, C. (2000). Production of citric acid by the fermentation of sweet potato using *Aspergillus niger*. J. Root Crops, India 26(1): 38-42.

Boopathy, R. (2005). Use of microbes in managing post-harvest crop residues: Sugarcane residue as a case study. In: Microbial Biotechnology in Agriculture and Aquaculture, Volume I (ed.) R.C. Ray, Science Publishers, Enfield, New Hampshire, USA, pp. 433-448.

Bramorski, A., Soccol, C.R., Christen, P. and Revah, S. (1998). Fruity aroma production by *Ceratocystis fimbriata* in solid-state cultures from agro-industrial wastes. Rev. Microbiol. 29: 1-10.

Bustos, G., Moldes, A.B., Cruz, J.M. and Dominguez, J.M. (2005). Production of lactic acid from vine-trimming wastes and viticulture lees using a simultaneous saccharification and fermentation method. J. Sci. Food Agric. 85: 466- 472.

Campbell, C.J. and Laherrere, J.H (1998). The end of cheap oil. Sci. Am. 3: 78-83.

Cannel, E. and Moo-Young, M. (1980). Solid-state fermentation systems. Process Biochem. 15: 24-28.

Chahal, P.S., Chahal, D.S. and Le, G.B.B. (1996). Production of cellulose in solid – state fermentation with *Trichoderma reesei* MCG 80 on wheat straw. Appl. Biochem. Biotechnol. 57/58: 433-442.

Chang, S.T. (1996). Mushroom Research and Development – Equality and Mutual Development. In: Mushroom Biology and Mushroom Products (ed.) D.J. Royse, Pennsylvania State University, Pennsylvania, USA, pp. 1-10.

Chang, S.T. (1999a). Global impact of edible and medicinal mushrooms on human welfare in the 21st Century: Non-green revolution. Int. J. Medicinal Mushrooms 1: 1-7.

Chang, S.T. (1999b). World production of cultivated edible and medicinal mushrooms in 1997 with emphasis on *Lentinus edodes* (Berk.) Sing. China. Int. J. Medicinal Mushrooms 1: 291-300.

Chang, S.T. (1999c). World production of edible and medicinal mushrooms in 1997. http:// www.food science. Psu. Edu/research_ iSMS_ 03.pdf.

Chau, K. (2002). Biochemistry of lactic acid fermentation from starchy wastes. http://www. Cheques.uq.edu.au/ugrad/these/2000/pdf.

Chikindas, M.L. (1995). Bacteriocins: Natural Antimicrobials as Effective Food Preservatives. Accessed from, ULC: http://www.aesop.rutgers.edu/Foodadditives/ Bacteriocins N.htm.

Christopherson, C., Anderson, E., Jokobsen, T.S. and Wagner, P. (1997). Xylanases in wheat separation. Starch 49: 5-12.

Daeschel, M.A. (1989). Antibacterial substances from lactic acid bacteria for use as food preservatives. Food Technol. 43: 164-167.

Dalton, L. (2002). What's that stuff? Food preservatives. Chemical & Engineering News. 80(45): 40.

Das, H. and Singh, S. (2004). Useful byproducts from cellulose wastes of agriculture and food industry- a critical appraisal. Crit. Rev. Food Sci. Nutr. 44(2): 37-89.

Deosen Corporation Ltd. (2005). ULC: http://www.foodprocessing-technology.com/ contractors/ingredients/deosen/office.html. Accessed on 25 February 2005.

Dhillon, A. and Khanna, S. (2000). Production of a thermostable alkali-tolerant xylanase from *Bacillus circulans* A316 grown on wheat straw. World J. Microbiol. Biotechnol. 27: 325-327.

Doelle, H.W., Mitchell, D.A. and Rolz, C.E. (eds.) (1992). Solid Substrate Cultivation. Elsevier Science Publication, New York, USA.

Doyle, M.P., Beuchat, L.R. and Montville, T.J. (1997). Food Microbiology: Fundamentals and Frontiers, ASM Press, Washington, DC, USA.

Durrand, A. (2003). Bioreactor design for solid-state fermentation. Biochem. Eng. J. 13: 113-125.

Durrant, L.R. (2005). Biotechnology applied for the utilization of lingo cellulose biomass. In: Microbial Biotechnology in Agriculture and Aquaculture, Volume I (ed.) R.C. Ray, Science Publishers, Enfield, New Hampshire, USA, pp. 405-432.

Ebine, H. (2003). Integrated research on agricultural waste reclamation. ULC: http:// www.unu.edu/unupress/unupbooks/80434e/80434E0s.htm, Accessed on 12 August 2005.

Ebner, H. (1996). Vinegar. In: Ullmann's Encyclopedia of Industrial Chemistry. Vol. A 27, B. Elvers and S. Hawkins (eds.) VCH Verlagsgesellschaft GmbH, Weinheim, Germany, p. 403.

Estabrook, R. (2000). Agriculture and food production. Food Insight Media Guide on Food Safety and Nutrition. International Food Information Council (IFIC) Foundation, Washington D.C., USA.

FMC BioPolymer, (2005). ULC: http://www.fmcbiopolymer.com/Biopolymer/V2/Product/0,1424,LOB=2&LOBName=Food+Ingredients&TYPE=4&Key=388,00.html Accessed on 12 August 2005.

Foulke, J.E. (2003). Fresh look at food preservatives. ULC: http://www. openseason .com/annex/library/cic/fresh look at food preservatives.htm Accessed on 22 August 2005.

Freedomyou. (2005). ULC: http://www.freedomyou.com/nutrition_book/enriched_fortified_synthetic_food.htm Accessed on 22 August 2005.

Fuglsang, C.C., Johansen, C., Christgau, S. and Alder-Nissen, J. (1995). Antimicrobial enzymes: Applications and future potential in the food industry. Trends Food Sci. Technol. 6 (12): 390- 395.

Ganeshkumar, C. and Takagi, H. (1999). Microbial alakaline proteases: From a bioindustrial viewpoint. Biotechnol. Adv. 17: 561-594.

Ghaly, A.E., Tango, M.S.A. and Adams, M.A. (2003). Enhanced lactic acid production from cheese whey with nutrient supplement addition. http;// dspace.library.cornell.edu/bitstream/1813/122/51/FP+02+009a+Ghaly.pdf Accessed on 22 August 2005.

Goldberg, I., Peleg, Y. and Roken, J.S. (1991). Citric, fumaric and malic acids. In: Biotechnology of Food Ingredients (eds.) I. Goldberg and R.A.Williams, Van Nostrand Reinhold, New York, USA, p. 349-378.

Grethlein, H.E. and Converse, A.O. (1991). Common aspects of acid prehydrolysis and steam explosion for pretreating wood. Biores. Technol. 36: 77-82.

Grohol, J. (2005). Food additives. ULC: http://psychcentral.com/ psypsych/List_of_food_additives/anticaking agent.htm Accessed on 22 August 2005.

Guerra, N.P. and Pastrana, I. (2002). Nisin and pediocins production on mussel-processing waste supplemented with glucose and five nitrogen sources. Lett. Appl. Microbiol. 34: 114-118.

Gutierrez-Correa, M. and Villena, G.K. (2003). Surface adhesion fermentation: A new fermentation category. Rev. Peru Biol. 10(2): 113-124.

Haard, N.F. (1988). Astaxanthin formation by the yeast *Phaffia rhodozyma* on molasses. Biotechnol. Lett. 10: 609-614.

Haltrich, D., Nidetzky, B. and Kulbe, K.D. (1996). Production of fungal xylanases. Biores. Technol. 58: 137-161.

Hayes, M. (2003). Agricultural residues: A promising alternative to virgin wood fiber. ULC: http://www.woodconsumption.org/index.html - Accessed on 22 August 2005.

Hayman, G.T., Mannarelli, B.M. and Leathers, T.D. (1995). Production of carotenoids by *Phaffia rhodozyma* grown on media composed of corn wet-milling co-products. J. Ind. Microbiol. 14: 389-395.

Henkel, H.J., Johl, H.J., Trosch, W. and Chmiel, H. (1990). Continuous production of glutamic acid in a three phase fluidized bed with immobilized *Corynebacterium glutamicum*. Food Biotechnol. 4: 149-152.

Howard R.L., Abotsi, E., Jansen van Rensburg, E.L. and Howard, S. (2003) Lignocellulose biotechnology: Issues of bioconversion and enzyme production African J. Biotechnol. 2(12): 602- 619.

Israelidis, C.J. (2003). http:SCP/C_J_Israelides-NUTRITION-SINGLE CELL PROTEIN.htm Accessed on 5 September 2005.

Jafar, S.S., Hultin, H.O., Bimbo, A.P., Crowther, J.B. and Barlow, S.M. (1994). Stabilization by antioxidants of mayonnaise made from fish oil. J. Food Lipids 1: 295-299.

Jamuna, R. and Ramakrishna, S.V. (1989). SCP production and removal of organic load from cassava starch industry waste by yeasts. J. Ferment. Bioeng. 67: 126-131.

Janssens, L., de Pooter, H.L., Schamp, N.M. and Vandamme, E.J. (1992). Production of flavours by microorganisms. Process Biochem. 27: 195-215.

Jecu, L. (2000). Solid-state fermentation of agricultural wastes for endoglucanase production. Industrial Crops and Products. 11: 1-5.

Juhasez, I., Kozma, K., Szengyel, Z. and Reczey, K. (2003). Production of β-glucoamylase in mixed culture of *Aspergillus niger* BKMF1305 and *Trichoderma reesei* RUTC30. Food Technol. Biotechnol. 41(1): 49-53.

Kaukab, S., Asghar, M., Rehman, K., Asad, M.J. and Adedayo, O. (2003). Bioprocessing of banana peel for α-amylase production by *Bacillus subtilis*. Int. J. Agri. Biol. 5: 36-39.

Kawakita, T., Sano, C., Shioya, S., Takehar, M. and Yamaguchi, S. (1990). Monosodium glutamate. In: Ullman's Encyclopedia of Industrial Chemistry. Vol. A16 (eds.) B. S. Hawkins and G. Schulzs, VCH Verlagsgesellschaft GmbH, Weinheim, Germany, p. 711.

Khalaf, A.S. (2004). Production and characterization of fungal chitosan under solid state fermentation conditions. Int. J. Agri. Biol. 6(6): 1033-1036.

Kida, K., Morimura, S., Noda, J., Nishida, Y., Imai, T. and Otagiri, M. (1995). Enzymatic hydrolysis of the horn and hoof of cow and buffalo. J. Ferment. Bioeng. 80: 478- 484.

Kim, H., Goto, M. and Jeong, H.J. (1998). Functional analysis of a hybrid endoglucanase of bacterial origin having a cellulose binding domain from a fungal exoglucanase. Appl. Biochem. Biotechnol. 75: 1938-2040.

Kim, S.W., Kang, S.W. and Lee, J.S. (1997). Cellulase and xylanase production by *Aspergillus niger* KKS in various bioreactors. Biores. Technol. 59: 63-67.

Klemanlayer, M.K., Siika-Aho, M., Teeri, T.T. and Kent Kirk, T. (1996). The cellulase endoglucanase I and cellobiohydrolase II of *Trichoderma reesei* act synergistically to solubilize native cotton cellulase but not to decrease its molecule size. Appl. Environ. Microbiol. 62(8): 2283-2887.

Kozak, M. and Krawczyk, J. (1999). Growing Shiitake Mushrooms in a Continental Climate. Field and Forest Production, Peshtiga, Wisconsin, USA.

Krishna, C. and Chandrasekaran, M. (1996). Banana waste as substrate for α-amylase production by *Bacillus subtilis* BTK (106): Solid-state fermentation. Appl. Microbiol. Biotechnol. 46: 106-111.

Krishna, C. (2005). Solid state fermentation systems- an overview. Crit. Rev. Biotechnol. 25: 1-30.

Kromhout, D. (1999). Are atioxidants effective in primary prevention of coronary heart disease? In: *Natural Antioxidants and Anticarcinogens in Nutrition, Health, and Disease* (eds.) J. T. Kumpulainen and J. T. Salonen, Royal Society of Chemistry, UK, Special Publication No. 240, pp. 20-26.

Kuhad, R.C. and Singh, A. (1993). Lignocellulose biotechnology: Current and future prospects. Crit. Rev. Biotechnol. 13: 151-172.

Kuntz, L.A. (1995). Acidulants in hot-pack products. ULC: http://www.foodproductdesign.com/archive/1995/0795AP.html#top Accessed on 31 August 2005.

Leathers, T.D. and Gupta, S.C. (1994). Production of pullulans from fuel ethanol byproducts by *Aureobasidium* sp. strain NRRL Y-12, 974. Biotechnol. Lett. 16: 1163-1166.

Leathers, T.D., Ahlgren, J.A. and Cote, G.L. (1997). Alternansucrase mutants of *Leuconostoc mesenteroides* strain NRRL B-21138. J. Ind. Microbiol. 18: 278-283.

Leathers, T.D. (1998). Utilization of fuel ethanol residues in production of the biopolymer alternan. Process Biochem. 33: 15-19.

Lo Curto, R., Tripodo, M.M., Leuzzi, U., Giuffre, D. and Vaccarino, C. (1992). Flavonoid recovery and SCP production from orange peel. Biores. Technol. 42: 83-87.

Lonsane, B.K., Saucedo–Castaneda, G. and Raimbault M. (1992). Scale-up strategies for solid fermentation system. Process Biochem. 27: 259-273.

Lu, M., Maddox, I.S. and Brooks, J.D. (1995). Citric acid production of *Aspergillus niger* in solid substrate fermentation. Biores. Technol. 54: 235-239.

Lu, M., Brooks, J.D. and Maddox, I.S. (1997). Citric acid production by solid state fermentation in a packed bed reactor using *Aspergillus niger*. Enz. Microb. Technol. 21: 392-397.

Magingo, F.S., Oriyo, N.M., Kivaisi, A.K. and Danell, E. (2004). Cultivation of *Oudemansiella tanzanica* nom. prov. on agricultural solid wastes in Tanzania. Mycologia 96(2): 197–204.

Malherbe, S. and Cloete, T.E. (2003). Lignocellulose biodegradation: Fundamentals and applications: a review. Environ. Sci. Biotechnol. 1: 105-114.

Mandeel, Q.A., Al-Laith, AA. and Mohamed, S.A. (2005). Cultivation of oyster mushrooms (*Pleurotus* spp.) on various lignocellulosic wastes. World J. Microbiol. Biotechnol. 21: 601-607.

Martin, N.de, Souza, R.S., da Silva, R. and Gomes, E. (2004). Pectinase production by fungal strains in solid state fermentation using agro-industrial bioproduct. Brazilian Arch. Biol. Technol. 47(5): 813-819.

Maziero, R., Bononi, V.L., Adami, A. and Cavazzoni, V. (1995). Exopolysaccharide and biomass production in submerged culture by edible mushrooms. In: Science and Cultivation of Edible Fungi. Mushroom Science XV. Vol. 2 (ed.), T.J. Elliot, Balkema, Rotterdam, Netherlands, pp. 887–892.

Meyer, P.S. and Du Prez, J.C. (1994). Astraxanthin production by a *Phaffia rhodozyma* mutant on grape juice. World J. Microbiol. Biotechnol. 10: 178-183.

Miles, P.G. and Chang, S.T. (1997). Mushroom Biology – Concise Basics and Current Developments. World Scientific, Singapore, pp. 194.

Mitchell, D.A. (1992). Growth patterns, growth kinetics and the modelling of growth in solid-state cultivation. In: Solid Substrate Cultivation (eds.) H.W. Doelle, D.A. Mitchell and C.E. Rolz, Elsevier Science Publication, New York, USA, pp. 87-114.

Mohapatra, S., Panda, S.H., Sahoo, S.K., Sivakumar, P.S. and Ray, R.C. (2006) β-Carotene rich sweet potato curd: Production, nutritional and proximate composition. Int. J. Food Sci. Technol. 41: in press.

Montané, D., Salvadó, J., Torrasm C. and Farriol, X. (2002). High-temperature dilute-acid hydrolysis of olive stones for furfural production. Biomass Bioenergy 22: 295-304.

Moresi, M., Parente, E., Petruccioli, M. and Federici, F. (1992). Fumaric acid production from hydrolysates of starch-based substrates. J. Chem. Technol. Biotechnol. 54: 283-290.

Morris, D. (1997). "The Coming Fiber Revolution" speech given at the Future Fibers Conference, Monterey, CA. June 2.

Murphy, G. (2000). Flavour of the mouth. ULC: http://www.irishscientist.ie/2000/flavour_of_the_mouth.htm. Accessed on 12 September 2005.

Mussatto, S.I. and Roberto, I.C. (2002). Xylitol: a sweetener with benefits for human health. Braz. J. Pharm. Sci. 38: 401-413.

Mussatto, S.I. and Roberto, I.C. (2003). Xylitol production from high xylose concentration: Evaluation of the fermentation in bioreactor under different stirring rates. J. Appl. Microbiol. 95: 331-337.

Mussatto, S.I. and Roberto, I.C. (2005). Evaluation of nutrient supplementation to charcoal-treated and untreated rice straw hydrolysate for xylitol production by *Candia guilliermondii*. Braz. Arch. Technol. 48(3): 497-502.

Naveena, B.J., Altaf, M., Bhadrayya, K. and Reddy, G. (2004). Production of L(+) lactic acid by *Lactobacillus amylophilus* GV6 in semi-solid state fermentation using wheat bran. Food Technol. Biotechnol. 42: 147-152.

Nigam, P. and Singh, D. (1995). Processes for fermentative production of xylitol – a sugar substitute: A review: Process Biochem. 30(2): 117-124.

Ojumu, T.V., Solomon, B.O., Betiku, E., Loyoku, S.K. and Amigun, B. (2003). Cellulase production by *Aspergillus flavus* L. isolate NSPR 101 fermented in sawdust, bagasse and corn cob. African J. Biotechnol. 2: 145-152.

Okagbue, R.N. and Lewis, M.J. (1984). Use of alfalfa residual juice as a substrate for propagation of the red yeast *Phaffia rhodozyma*. Appl. Microbiol. Biotechnol. 20: 33-39.

Ooijkaas, L.P., Weber, F.J., Buitelaar, R.M., Tramper J. and Rinzema, A. (2000). Defined media and inert supports: Their potential as solid-state fermentation production systems. Trends Biotechnol. 18: 356-360.

Pai, J.S. (2003). Applications of microorganisms in food biotechnology. Indian J. Biotechnol. 2: 382-386.

Palit, S. and Banerjee, R. (2001). Optimization of extraction parameters for recovery of α-amylase from the fermented bran of *Bacillus circulans* GRS 313. Braz. Arch. Biol. Technol. 44(1): 1-8.

Panda, S.H., Naskar, S.K. and Ray, R.C. (2006). Production, proximate and nutritional evaluations of anthocyanin- rich sweetpotato curd. J. Food Sci. Agric. Environ. 4(1): 124-127.

Papaginni, M., Nokes, E.S. and Filer, K. (2001). Submerged and solid state phytase fermentation. Food Technol. Biotechnol. 39(4): 319-326.

Parajó, J.C., Domínquez, H.D. and Domínquez, J.M. (1998). Biotechnological production of xylitol. Part 1: Interest of xylitol and fundamentals of its biosynthesis. Biores. Technol. 65: 191-201.

Pessoa, A. Jr., Mancilha, I.M. and Sato, S. (1996). Cultivation of *Candida tropicalis* in sugarcane hemicellulosic hydrolysate for microbial protein production. J. Biotechnol. 51: 83-88.

Piccinini, A. (1998). Gli Agricoltori Europei Tra QuoteE. Mercato, Milano, Italy: Piccinini Franco Angeli s.r.l.

Preziosi-Belloy, L., Nolleau, V. and Navarro, J.M. (2000). Xylitol production from aspen wood hemicellulose hydrolysate by *Candida guilliermondii*. Biotechnol. Lett. 22: 239-243.

Priefert, H., Rabenhorst, J. and Steinbüchel, A. (2001). Biotechnological production of vanillin. Appl. Microbiol. Biotechnol. 56: 296-314.

Quimio, T.H. (2002). Tropical Mushroom Cultivation. National Book Store, Manila, Philippines.

Quimio, T.H. (2006). Mushrooms: production, disease management and processing. In: Microbial Biotechnology in Horticulture, Volume 1 (eds.) R.C. Ray and O.P. Ward, Science Publishers, Enfield, New Hampshire, USA, pp. 473-516.

Radhakrishnan, S. and Pandey, A. (1993). The production of glucoamylase by *Aspergillus niger* NCIM 1245. Process Biochem. 28: 305-309.

Rajan, P.J., Nampoothiri, K.M. and Pandey, A. (2005). Cassava bagasse–an inexpensive carbon source for L(+) lactic acid production using mixed culture of lactobacilli. In: Proceedings of National Seminar on Achievements and Opportunities in Post-harvest management and Value Addition in Root and Tuber Crops, 19-20 July, 2005, Central Tuber Crops Research Institute, Thiruvanathapuram, India, Abstract, p. 88.

Ray, R.C. (2004). Extracellular amylase(s) production by fungi *Botryodiplodia theobromae* and *Rhizopus oryzae* grown on cassava starch residue. J. Environ. Biol. 25(4): 489-495 .

Ray, R.C. and Ravi, V. (2005). Post harvest spoilage of sweet potato and control measures. Crit. Rev. Food Sci. Nutr. 45(8): 623-644.

Ray, R.C., Sharma, P., Panda, S.H. and Sahoo, A.K. (2005). Lactic acid production in submerged fermentation by *Lactobacillus plantarum* using cassava fibrous residue as the carbohydrate source. In: Proceedings of National Seminar on Achievements and Opportunities in Post-harvest management and Value Addition in Root and Tuber Crops, 19-20 July 2005, Central Tuber Crops Research Institute, Thiruvanathapuram, India, Abstract, p. 123.

Ray, R.C. and Moorthy, S.N. (2006). Microbial exopolysaccharide production by *Aureobasidium pullulans* from cassava fibrous residue. Eng. Life Sci. 3: in press.

Ray, R.C. and Ward, O.P. (2006). Post harvest microbial biotechnology of root and tuber crops. In: Microbial Biotechnology in Horticulture Volume 1 (eds.) R.C. Ray and O.P. Ward, Science Publishers, Enfield, New Hampshire, USA, pp. 345-396.

Ribbons, R.W. (1987). Chemicals from lignin. Phil. Trans. R. Soc. Lond. Ser. A. 321: 485-494.

Roberto, I.C., Mussatto, S.I. and Rodrigues, R.C.L.B. (2003). Dilute-acid hydrolysis for optimization of xylose recovery from rice straw in a semi-pilot reactor. Indust. Crops Prod. 17: 171-176.

Roukas, T. and Biliaderis, C.G. (1995). Evaluation of carob pod as a substrate for pullulan production by *Aureobasidium pullulans.* Appl. Biochem. Biotechnol. 55: 27-44.

Roukas, T. (1998). Pretreatment of beet molasses to increase pullulan production. Process Biochem. 33(8): 805-810.

Royse, D.T. (1996). Specialtry mushrooms. In: Progress in New Crops (ed.) J. Janick, ASHS Press, Arlington, USA, pp. 464-475.

Sagi, I. and Mannheim, C.H. (1990). The effect of enzymatic oxygen removal on quality of unpasteurized and pasteurized orange juice. J. Food Process. Preserv. 14: 253-256.

Saltmarsh, M. (ed.) (2000). Essential Guide to Food Additives. Leatherhead Food RA Publishing, USA, pp. 1-322.

Sede, S.M. and Lopez, S.E. (1999). Cultural studies of *Agrocybe cylindrica, Gymnopilus pampeanus* and *Oudemansiella canarii* (Agaricales) isolated from urban trees. Mycotaxon 70: 377-386.

Selbmann, L., Crognale, S. and Petruccioli, M. (2002). Exopolysaccharide production from *Sclerotium glucanicum* NRRL 3006 and *Botryosphaeria rhodina* DABAC-P 82 on raw and hydrolyzed starchy materials. Lett. Appl. Microbiol. 34: 51-55.

Senez, J.C. (2003). Solid state fermentation of starchy crops. http://www:/Bioconversion of organic residues for rural communities.com Accessed on 5 September 2005.

Shafique, S., Asgher, M., Sheikh, M.A. and Asad, M.J. (2004). Solid state fermentation of banana stalk for exoglucanase production. Internal. J. Agric Biol. 6(3): 488-491.

Shamala, T.R. and Prasad, M.S. (1995). Preliminary studies on the production of high and low viscosity dextran by *Leuconostoc* spp. Process Biochem. 30: 237-241.

Shong, S. (2005). YOUBAO–Antioxidants TBHQ and lycopene, food additives. ULC: http://www.foodprocessing-echnology.com/contractors/ingredients/ youbao/office.html. Accessed on 12 September 2005.

Sijtsma, L. (2005). Bioconversion of the hemicellulose fraction of lignocellulose to xylitol. www.//Agricultural and fisheries- Contract Ni 19/0009.htm. Accessed on 12 September 2005.

Silva, C.J.M.S. and Roberto, I.C. (1999). Statistical screening method for selection of important variables on xylitol biosynthesis from rice straw hydrolysate by *Candida guilliermondii* FTI 200037. Biotechnol. Tech. 13: 743-747.

Silva, C.J.M.S. and Roberto, I.C. (2001). Improvement of xylitol production by *Candida guilliermondii* FTI 200037 previously adapted to rice straw hemicellulosic hydrolysate. Lett. Appl. Microbiol. 32: 248-252.

Singhania, R.R., Sukumaran, R.K. and Pandey, A. (2005). Cellulase production using cassava bagasse as a substrate. In: Proceedings of National Seminar on Achievements and Opportunities in Post-harvest Management and Value Addition in Root and Tuber Crops, 19-20 July 2005, Central Tuber Crops Research Institute, Thiruvanathapuram, India, Abstract, p. 89.

Smith, J.E., Anderson, J.G. and Senior, E.K. (1987). Bioprocessing of lignocelluloses. Phil. Trans. R. Soc. Lond. Ser. A. 321: 507-521.

Soccol, C. R. (1996). Biotechnology products from cassava root by solid-state fermentation. Bull Grain Technol. 55: 358-364.

Soccol, C.R. and Vandenberghe, L.P.S. (2003). Overview of applied solid-state fermentation in Brazil. Biochem. Eng. J. 13: 205-218.

Sonnino, A. (1994). Agricultural biomass is an energy option for the future. Renewable Energy 5: 857-865.

Sosa, A.V. (2001). Fluidized bed design parameters affecting novel lactic acid downstream processing. Biotechnol. Prog. 17: 1079-1083.

Sreenath, H.K., Koegel, R. and Straub, R. (2001). Lactic acid production from agricultural residues. Biotechnol. Lett. 23: 179-184.

Sriram, S. and Ray, R.C. (2005). *Trichoderma*: Systematics, molecular taxonomy and agricultural and industrial applications. In: Microbial Biotechnology in Agriculture and Aquaculture, Volume I (ed.) R.C. Ray, Science Publishers, Enfield, New Hampshire, USA, pp. 335-376.

Sriroth, K., Chollakup, R., Choineeranat, S., Piyachomkwan, K. and Oates, C.G. (2000). Processing of cassava wastes for improved biomass utilization. Biores. Technol. 71(1): 63-70.

Stamets, P. (1993). Growing Gourmet and Medicinal Mushrooms. Ten Speed Press, PO Box 7123, Berkeley, California, USA.

Stiles, M.E. and Hastings, J.W. (1991). Bacteriocin production by lactic acid bacteria: potential for use in meat preservation. Trends Food Sci. Technol. 2 (10): 247-249.

Stiles, M.E. (1993). Bacteriocins Produced by *Leuconostoc* Species. J. Dairy Sci. 77 (9): 24-27.

Subramaniyan, S. and Prema, P. (2002). Biotechnology of microbial xylanases: enzymology, molecular biology, and application. Crit. Rev. Biotechnol. 22: 33-64.

Sun, Y. and Cheng, J. (2002). Hydrolysis of lignocellulosic material from ethanol production: A review. Biores. Technol. 83: 1-11.

Suryanarayan, S. (2003). Current industrial practice in solid state fermentations for secondary metabolite production: the Biocon India experience. Biochem. Eng. J. 13: 189-195.

Sutherland, I.W. (1996a). Extracellular polysaccharides. In: Biotechnology, Vol. 6 (eds.) H.J. Rehm and G. Reed, VCH Publishers, Weinheim, Germany, pp. 615-657.

Sutherland, I.W. (1996b). Microbial biopolymers from agricultural products: Production and potential. Int. Biodet. Biodeg. 38: 249- 261.

Tan, J.D., Galvez, F.C.H. and Tomita, F. (2000). Isolation and characterization of bacteriocins producing microorganisms from *agos-os*. J. Food Safety 20(3): 177-192.

Tengerdy, R.P. and Szakacs, G. 2003. Bioconversion of lignocellulose in solid substrate fermentation. Biochem. Eng. J. 13: 169-179.

The Big Carrot, (2004). Common additives. ULC: http://www. thebigcarrot.ca/agriculture/additives.htm. Accessed on 15 March 2005

Thomas, G.V., Prabhu, S.R., Reeny, M.Z. and Boopaih, B.M. (1998). Evaluation of lignocellulose biomass from coconut palm as substrate for cultivation of *Pleurotus sajor-caju* Singer. World J. Microbiol. Biotechnol. 14: 879-882.

Vandenberghe, L.P.S., Soccol, C.R., Prado, F.C. and Pandey, A. (2004). Comparison of citric acid production by solid-state fermentation: Flask, column, tray and drum bioreactors. Appl. Biochem. Biotechnol. 118: 293-304.

Vanwyk, J., Mogale, A. and Seseng, T. (2001). Bioconversion of waste papers to sugar by cellulase from *Aspergollus, niger*, *Trichoderma viride* and *Penicillum funiculosum*. J. Solid Waste Technol Manage 27: 82-86.

Vares, T., Kalsi, M. and Hatakka, A. (1995). Lignin peroxidase, manganese peroxidase and other lignolytic enzyme produced by *Phlebia radiata* during SSF of wheat straw. Appl. Environ. Microbiol. 61: 3515-3520.

Vyas, B.R.M., Vole, J. and Sasek, V. (1994). Ligninolytic enzyme of selected white rot fungi cultivated on wheat straw. Folia Microbiol. 39: 235-240.

Walton, N.J., Mayer, M.J. and Narbad, A. (2003). Molecules of interest: Vanillin. Photochemistry 63: 505-515.

Wang, T. and Yuksel, G.U. (2004). Production of lactic acid from off-grade and small potatoes processing waste using amylolytic acid bacteria. http://www.Ift.confex.com/ift/2004/techprogram/session_3150.htm_12k. Accessed on 18 September 2005.

Wang, Y. and McNeil, B. (1996). Scleroglucan. Crit. Rev. Biotechnol. 16: 185- 215.

Ward, O.P. and Singh, A. (2005). Microbial technology for bioethanol production from agricultural and forestry wastes. In: Microbial Biotechnology in Agriculture and Aquaculture, Volume I (ed.) R.C. Ray, Science Publishers, Enfield, New Hampshire, USA, pp. 449- 479.

Ward, O.P., Singh, A. and Ray, R.C. (2006). Production of renewable energy from agricultural and horticultural substrates and wastes. In: Microbial Biotechnology in Horticulture, Volume 1 (eds.) R.C. Ray and O.P. Ward, Science Publishers, Enfield, New Hampshire, USA, pp. 517-557.

Wase, D.A.J., McManamey, W.J., Raymahasay, S. and Vaid, A.K. (1985). Comparison between cellulose production by *Aspergillus fumigatus* in agitated vessels and in an air-lift fermentor. Biotechnol. Bioeng. 27: 1166-1172.

Wehmer, C. (1893). Notes sur la fermentation citrique. Bull Soc. Chim., Fr. 9: 728.

Winkelhausen, E. and Kuzmanova, S. (1998). Microbial conversion of D-xylose to xylitol. Ferment. Bioeng. 86: 1-14.

Woiciechowski, A.L., Soccol, C.R., Rocha, S.N. and Pandey, A. (2004). Xanthan gum production from cassava bagasse hydrolysate with *Xanthomonas campestris* using alternative sources of nitrogen. Appl. Biochem. Biotechnol. 118: 305-312.

Wong, K.K.Y. and Saddler, J.N. (1992a). Applications of hemicellulases in the food, feed and pulp and paper industries. In: Hemicellulose and Hemicellulases (eds.), P.P. Coughlan and G.P. Hazlewood, Portland Press, London, UK, pp. 127-143.

Wong, K.K.Y. and Saddler, J.N. (1992b). *Trichoderma* xylanases: their properties and applications. In: Xylans and their Xylanases (eds.) A. Visser, Elsevier, Amsterdam, The Netherlands, pp. 171-186.

Wood, J.B. (ed.) (1998). Microbiology of Fermented Foods, Blackie Academic and Professional, London, UK.

Xia, L. and Len, P. (1999). Cellulose production by solid-state fermentation on lignocellulosic waste from the xylose industry. Process Biochem. 34: 909-912.

Yamanaka, K. (1997). Production of cultivated mushrooms. Food Review Int. (13): 327-333.

Yang, S.S. and Wang, J.Y. (1999). Protease and amylase production of *Streptomyces rimosus* in submersed and solid sate cultivations. Bot. Bull. Acad. Sin. 40: 259-265.

Zadrazil, F. and der Wiesche, C. (1999). Production of edible mushrooms. In: Biotechnology: Food Fermentation, Volume I (eds.) V.K. Joshi and A. Pandey, Educational Publishers, New Delhi, India, pp. 865- 894.

Zeitch, K.J. (ed.) (2000). The Chemistry and Technology of Furfural and Its Many By-Products. Elsevier, Amsterdam, The Netherlands, pp. 358.

INDEX